Polymer Science and Technology
for Scientists and Engineers

Polymer Science and Technology
for Scientists and Engineers

Professor Richard A. Pethrick

Emeritus Burmah Professor of Physical Chemistry at University of Strathclyde
DSc CChem FRSC FRSE FIMMM

Whittles Publishing

Published by
Whittles Publishing,
Dunbeath,
Caithness KW6 6EY,
Scotland, UK

www.whittlespublishing.com

ISBN 978-1904445-40-1

Typeset by
iPLUS Knowledge Solutions Private Limited, Chennai-32, India.

Printed by
Bell & Bain Ltd., Glasgow.

Contents

Preface xv

1 What are plastics? 1

1.0 Introduction 1
1.1 A brief history of the development of plastics 1
1.2 What are plastics, polymers and macromolecules? 3
1.3 A simple analogy for a polymer chain 3
1.4 What factors influence the physical properties of thermoplastic materials? 4
1.5 How are polymers made? 4
1.5.1 Polyethylene: the simplest linear polymer chain 6
1.5.2 Step growth polymerisation 7
1.5.3 Ring opening polymerisation 8
1.6 Effect of chemical structure on physical properties of polymers 9
1.6.1 Ability of polymer molecules to pack together 9
1.7 Copolymers 14
1.8 Polymer design for application 17
1.9 Polymer classification 17
1.10 Molar mass and molar mass distribution 17
1.10.1 Molar mass averages 18
1.10.2 How does the method of synthesis influence the molar mass distribution? 19
Brief summary of chapter 20
References and additional reading 20

2 Mechanical properties of polymeric materials 21

2.0 Introduction 21
2.1 Assessment of mechanical properties 21
2.2 Stress–strain measurements 21
2.2.1 Volume change and shear 22
2.2.2 The process of simple shear 22
2.2.3 Principal elastic moduli 23
2.2.4 Energy 24
2.2.5 Stretching of a wire 24
2.2.6 Bending a thin beam 25
2.2.7 Cantilever 26
2.2.8 Beam loaded in the middle 27
2.2.9 Poisson's ratio 27
2.3 Dynamic modulus 27
2.4 Methods of measurement of mechanical properties 30
2.4.1 Tensile testing 30
2.4.2 Clamping of samples 31
2.4.3 Problems with sample clamping 31
2.4.4 Extensometers and measurement of strain 33
2.4.5 Properties measured 33

2.4.6 Strain ... 34
2.4.7 Ultimate tensile strength ... 34
2.4.8 Flexural modulus ... 34
2.4.9 Shear modulus ... 34
2.4.10 Pendulum impact tests ... 35
2.4.11 Drop tests ... 36
2.4.12 Instrumented methods ... 36
2.4.13 Ductile vs brittle fracture ... 37
2.4.14 Creep properties ... 37
2.4.15 How to determine stress relaxation? ... 38
2.4.16 Fatigue testing ... 38
2.4.17 Compression testing ... 38
2.4.18 Modes of deformation in compression testing ... 39
2.4.19 Other mechanical property measurements in polymer systems ... 40
2.4.20 Dynamic mechanical thermal analysis ... 40
2.5 Thermal expansion coefficient measurements ... 42
2.5.1 Molar mass dependence of glass transition point ... 43
2.5.2 Influence of chemical structure on the glass transition point ... 43
2.5.3 Plasticisation ... 47
2.5.4 Examples of T_g calculations ... 48
2.5.5 Other molecular mass effects ... 48
2.5.6 What is happening at the critical molecular mass? ... 49
2.5.7 Effects of chain entanglement on the mechanical properties of thermoplastics polymers ... 50
2.6 Viscoelastic behaviour ... 50
2.6.1 Maxwell model ... 50
2.6.2 Kelvin–Voigt model ... 51
2.6.3 More complex models ... 53
2.6.4 Standard linear solid: Zener solid ... 53
2.6.5 Use of the time-dependent modulus approach ... 55
2.7 What does the experimental data look like for a real polymer system? ... 57
2.8 Other mechanical properties of polymer systems ... 60
2.9 Effects of water ... 61
2.10 Environmental stress crazing ... 62
Brief summary of chapter ... 63
References and further reading ... 63

3 Crystallinity and polymer morphology ... 65

3.1 Introduction ... 65
3.2 Crystallography and crystallisation ... 65
3.3 Single crystal growth ... 68
3.3.1 Habit of polymer crystals ... 70
3.4 Crystal lamella and other morphological features ... 71
3.5 Melt crystallised lamellae ... 71
3.6 Polymer spherulites ... 71
3.7 Differential scanning calorimetry ... 76
3.8 Polytetrafluororethylene ... 77
3.9 Other types of morphology in semicrystalline polymer systems ... 77

3.10 Copolymers and phase separation ... 78
3.10.1 What are the implications of phase separation? ... 80
3.10.2 Alternating, random copolymers and blends of two polymers ... 80
3.10.3 Block copolymers of incompatible monomers ... 80
3.10.4 Varying the styrene–butadiene ratio in triblock copolymers ... 82
3.11 Why do we need to be able to change the modulus of polymeric materials? ... 83
3.12 Polyurethanes ... 84
3.13 High-temperature polymers ... 86
Brief summary of chapter ... 88
Additional reading ... 88

4 Chemistry of polymer processing ... 89

4.1 Introduction ... 89
4.2 Processing thermoplastic materials ... 89
4.3 Thermosets: elastomers ... 89
4.3.1 Rubbers and vulcanisation ... 89
4.3.2 Siloxanes ... 91
4.3.3 Rubber elasticity ... 92
4.4 Thermoset polymers: rigid materials ... 93
4.5 Cure of thermoset resins and time temperature transformation diagrams ... 95
4.5.1 How do we effectively cure resins? ... 98
4.5.2 Thermoset cure resins ... 98
4.5.3 Mechanical properties of the cured resin system ... 98
4.6 Commercial thermoplastic polymers ... 99
4.6.1 Antioxidants ... 99
4.7 Fillers ... 99
4.7.1 Carbon black ... 100
4.7.2 Quartz, silica and clay fillers ... 101
4.7.3 What is the structure of a clay? ... 101
4.8 Plasticisers ... 102
Brief summary of chapter ... 102
Additional reading ... 102

5 Polymer processing: thermoplasticsand thermosets ... 103

5.1 Introduction ... 103
5.2 Processing thermoplastics ... 103
5.3 Rotational moulding ... 103
5.3.1 Moulding process ... 104
5.3.2 Theory of the rotational moulding process ... 105
5.3.3 Powder deposition ... 105
5.3.4 Powder melting in contact with the heated surface ... 105
5.3.5 Bubble removal ... 107
5.3.6 Behaviour of the polymer melt ... 108
5.3.7 Degradation effects on the melt ... 110
5.3.8 Solidification ... 110
5.3.9 Moulding cycle ... 110

5.4 Injection moulding 110
5.4.1 Extruder 111
5.4.2 Feed or mixing zone 112
5.4.3 Compression zone 112
5.4.4 Metering zone 113
5.4.5 Analysis of flow in extruder 113
5.4.6 Drag flow 113
5.4.7 Pressure flow 114
5.4.8 Leakage flow 115
5.4.9 Free-flow condition 117
5.4.10 Flow into a mould or die 117
5.4.11 Extruder volume efficiency 118
5.4.12 Power requirements 118
5.4.13 Location of melt front 118
5.4.14 Twin-screw extruders 119
5.4.15 Use of vented barrel 120
5.4.16 Simplest use of extruder 120
5.4.17 Fabrication of simple, continuous profile materials 122
5.4.18 Polymer-coated products 123
5.4.19 Blow moulding 123
5.4.20 Moulding of bottles 125
5.4.21 Tensar process 128
5.5 Compression moulding 128
5.5.1 Vacuum and compression forming 129
5.5.2 Pressure forming process 129
5.6 Injection moulding 130
5.6.1 Plunger-type injection moulding machine 130
5.6.2 Extruders used for injection moulding 130
5.6.3 Selection of plastics for extruder applications 131
5.7 Plastisol processes 133
5.8 Thermoset processing 134
5.8.1 Hand lay-up process 134
5.8.2 Spray lay-up method 135
5.8.3 Vacuum bagging 135
5.8.4 Resin transfer moulding 135
5.8.5 Resin infusion processes 136
5.9 Composite fabrication 136
5.9.1 Autoclave prepreg moulding 137
5.9.2 Filament winding 138
5.9.3 Pultrusion 139
5.10 Cure monitoring 141
5.11 Repair of composite parts 142
5.11.1 Basic repair process 142
5.11.2 Types of repair 143
5.11.3 Damage assessment 144
5.12 General physical characteristics of composites 145
Brief summary of chapter 146
References and additional reading 146

6 Composites 147

6.1 Introduction 147
6.2 Classification of composites 147
6.2.1 Why do we need composite materials? 147
6.3 Particle-reinforced composites 149
6.3.1 Fibre-reinforced fillers 149
6.3.2 Structures 150
6.3.3 Sandwich structures 151
6.4 Prediction of characteristics of filled composite materials 152
6.4.1 Volume fractions 152
6.5 Fibre-reinforced composites 153
6.5.1 Fibre performance 154
6.5.2 Influence of fibre length 154
6.5.3 Influence of fibre orientation and concentration 155
6.5.4 Fibre phase 157
6.6 Fabrication 158
6.7 Failure 158
6.8 Factors influencing the performance of composites 159
6.8.1 Adhesive properties 159
6.8.2 Mechanical properties 159
6.8.3 Microcracking 160
6.8.4 Fatigue resistance 160
6.8.5 Water ingress 161
6.8.6 Osmosis 161
6.8.7 Adhesive properties of the resin system 161
6.9 Uses of plastic composites 162
6.10 Elastic behaviour of composite materials 162
6.10.1 Different types of anisotropic materials 163
6.10.2 Monoclinic materials 163
6.10.3 Orthotropic material 163
6.10.4 Unidirectional material 164
6.10.5 Isotropic materials 165
6.10.6 Moduli of elasticity 166
6.10.7 Uniaxial tension or compression 166
6.10.8 Shear modulus 166
6.10.9 Spherical compression or tension 167
6.11 Elastic behaviour of composite materials 167
6.11.1 Transverse Young's modulus 169
6.11.2 Longitudinal Poisson ratio 170
6.11.3 Longitudinal shear modulus 170
6.11.4 Halpin–Tsai equations 171
6.12 Orthotropic composites 171
6.12.1 Elasticity relations for an off-axis orientation 172
6.12.2 Off-axis tensile testing 174
6.13 Fracture mechanisms induced in composite materials 176
6.13.1 Unidirectional composite subjected to longitudinal tension 177
6.13.2 Fracture mechanisms induced in composite materials 178
6.13.3 Practical composite structures 179

6.13.4 Fracture in laminate structures 179
6.13.5 Failure criteria 181
6.13.6 Maximum stress criterion 181
6.13.7 Off-axis failure criterion 182
6.13.8 Interactive criteria 183
6.13.9 Hill's criterion 183
Brief summary of chapter 184
References and additional reading 184
Appendix 185

7 Case studies 187

7.1 Introduction 187
7.2 Environmental stress cracking: some case studies 187
7.2.1 Failure in 'sight glasses' 187
7.2.2 High density polyethylene blow moulded containers 188
7.2.3 Cracks in communication wiring 189
7.2.4 Failure of polycarbonate electronic housing 189
7.2.5 Environmental stress cracking in polyethylene 190
7.2.6 Model for environmental stress cracking failure 190
7.2.7 How is resistance to environmental stress cracking assessed? 191
7.3 Energy absorption and vibration damping 194
7.3.1 Which materials are useful for energy damping? 195
7.3.2 Rubber balls and tyres 195
7.3.3 Tyre technology 195
7.3.4 Effect of cross-linking on rubber characteristics 197
7.4 Adhesion and adhesives 198
7.4.1 Polymers as adhesives 198
7.4.2 Adhesion mechanisms 199
7.4.3 Examples of specific interactions which can occur in surfaces 202
7.4.4 How does surface tension help to achieve a good bond? 204
7.4.5 Issues which arise during the bonding process 208
7.4.6 Adhesive bond design 209
7.4.7 Stresses in adhesive joints 211
7.4.8 Axially loaded butt joints: tensile measurements 211
7.4.9 Single lap joints 212
7.4.10 Fracture mechanics 212
7.4.11 Service life of an adhesive joint 218
7.4.12 Toughing of adhesives 220
7.5 Polymers in corrosion protection 220
7.6 Gas diffusion through polymer matrices 222
7.7 Selection of polymeric materials for particular applications 222
Brief summary of chapter 224
References and additional reading 224

8 Polymer chemistry and synthesis 225

8.1 Introduction 225
8.2 Condensation polymerisation 225
8.2.1 Degree of polymerisation and molar mass 227

8.2.2 Molar mass distribution 227
8.2.3 Molar mass control 229
8.3 Vinyl polymerisation 230
8.3.1 Initiation 230
8.3.2 Kinetics of vinyl or addition polymerisation 230
8.3.3 Kinetics of free radical polymerisation 233
8.3.4 Experimental measurement of polymerisation kinetics 235
8.3.5 Molecular weight and $\overline{DP}_n$ 236
8.3.6 Inhibition and retardation 237
8.3.7 Determination of absolute rate constants 237
8.4 Free radical copolymerisation 242
8.4.1 Kinetic of copolymerisation 243
8.4.2 Mean sequence length 245
8.5 Methods of polymerisation 246
8.6 Specialist chemical reactions 247
8.7 Heterogeneous catalysis 249
8.8 Homogeneous catalysis 251
8.8.1 Homogeneous metallocene catalysts 251
8.8.2 Atom transfer radical polymerisation 253
8.8.3 Group transfer polymerisation 253
8.8.4 Cobalt-catalysed polymerisation 254
8.9 Polymer degradation 255
8.9.1 Analysis of polymer degradation: thermogravimetric analysis 256
8.9.2 Kinetics of polymer degradation: the random scission model 256
8.9.3 Degradation of polyethylene 257
8.9.4 General mechanism of radical depolymerisation 258
8.9.5 Depolymerisation versus transfer 259
8.9.6 Degradation of polyvinylchloride 260
8.9.7 Polyvinyl acetate 263
8.9.8 Polymethylmethacrylate 263
8.9.9 Degradation routes for alkylmethacrylate polymers 265
8.9.10 Degradation of polyethyleneterephthalate 265
8.9.11 Polystyrene 266
8.9.12 Hydrolysis 267
8.9.13 Importance of β hydrogen in degradation 267
8.10 Polymers and fire 267
8.10.1 Cone calorimeter 268
8.10.2 Experimentally measurable parameters 269
8.10.3 Improved fire retardancy of polymers 270
8.10.4 Flame chemistry 271
8.10.5 Effect of various fillers on limiting oxygen index 272
8.10.6 Stabilisers 272
8.10.7 Final comments on fire issues 272
8.10.8 Use of nanofillers to form nanocomposites 272
8.11 Polymer identification 273
8.11.1 Tests to identify an unknown polymer 273
8.11.2 Burning tests 276
Brief summary of chapter 276
References and additional reading 276

9 Polymer physics: models of polymer behaviour 277

9.1 Introduction 277
9.2 Simple statistical models of isolated polymer molecules in solution 277
9.3 Freely jointed random coil model 277
9.4 Valence constrained random coil model 280
9.5 Rotational isomeric states model 282
9.6 Long-range interactions: excluded volume 283
9.7 Comparison of the theoretical models 284
9.7.1 Dynamic response of polymer solutions 285
9.7.2 Theories of polymer dynamics 286
9.7.3 Rouse model 286
9.7.4 Zimm model 288
9.7.5 Dynamics of polymer molecules in oscillatory shear 289
9.7.6 Rouse model: mode theory 290
9.7.7 Zimm model: theory 291
9.8 Dynamic rheological behaviour of polymers with molar mass above M_c 292
9.8.1 Relaxation times 294
9.8.2 Stress relaxation and viscosity 294
9.9 Rubber elasticity 296
9.9.1 Separation of energetic and entropic terms 298
9.9.2 Unentangled rubber elasticity: affine network model 299
9.10 Polymer crystal growth 301
9.10.1 Thermodynamic of polymer molecule in the melt 302
9.10.2 Nucleation 302
9.10.3 Minimum energy conditions and simple theory of growth 303
9.10.4 Nature of chain folding 305
9.10.5 Crystals grown from the melt and lamellae stacks 306
9.10.6 Location of chain ends 307
9.10.7 Crystallisation kinetics 308
9.10.8 Equilibrium melting temperature 310
9.10.9 General Avrami equation 311
9.10.10 Comparison of experiment with theory 314
9.11 Determination of molar mass and size 314
9.11.1 Absolute method of determination 314
9.11.2 Number average molar mass 314
9.11.3 Scattering methods for molar mass determination 315
9.11.4 Light scattering by small particles (size compared to the wavelength of light) 315
9.11.5 Light scattering from molecules larger than $\lambda/20$ 316
9.11.6 Viscosity measurements 317
9.11.7 Relative methods of molar mass determination: gel permeation chromatography and size exclusion chromatography 318
Brief summary of chapter 321
Additional reading 322

10 Polymers for the electronics industry 323

10.1 Introduction 323
10.2 Lithographic materials 325

10.2.1 Semiconductor processing....325
10.2.2 Front-end processing....326
10.2.3 Metal layers....326
10.2.4 Photolithography....327
10.2.5 Wafer processing....329
10.2.6 Chemistry of photoresists....330
10.2.7 Applications of lithography....338
10.3 Intrinsically conducting polymers....339
10.4 Organic light-emitting polymers....340
Brief summary of chapter....342
Additional reading....342

11 Medical applications of polymers....343

11.1 Applications in medical devices....343
11.1.1 Polymers used in devices and therapy....343
11.1.2 Silicone breast implants....343
11.1.3 Hip joints....344
11.1.4 Heart valve replacement....345
11.1.5 Contact lenses....347
11.1.6 Polymers used in devices and therapy....347
Brief summary of chapter....347
Additional reading....348

12 Recycling of plastics and environmental issues....349

12.1 Introduction....349
12.2 Recycling plastics....349
12.3 Issues of plastic identification....350
12.4 Why do we need to recycle plastics?....350
12.5 Methods for recycling plastics....351
12.6 Degradation and bioplastics....352
12.7 Bioplastics....352
12.8 Polyhydroxyalkonates....352
12.9 Polylactides and polyglycolides....353
12.10 Issues with recycling....353
12.11 Feedstock recycling....354
12.12 Conclusions....354
Brief summary of chapter....354
Additional reading....354

Index....355

Preface

Plastics as commodity materials have been around for over 50 years. In the 1950s, plastics were considered to be specialist materials only finding applications in niche applications or being used by artists to create new concepts. With the greater availability of plastics in the 1960s, a number of engineers started to explore their application in a variety of areas. Ford Motors explored the application of plastic moulded handles for winding windows up and down. The initial impact was good but after several months of in-field use the problem of the components ageing in a warm environment became apparent. The designers had assumed that plastics would behave similarly to metal alloy components and had not allowed for the shrinkage which can occur in plastic mouldings after a period of time. Shrinkage of the plastic around the metal bar which joined the handle to the door locks led to the handles breaking! There were several other bad experiences with plastics which caused many engineers to be sceptical about their usefulness. However, with a greater understanding of the advantages and disadvantages of plastics, engineers overcame the initial problems and satisfactorily used the materials in a range of applications. Bumpers for cars were one of the initial areas of extensive application. The initial injection moulded structures were very brittle and even a minor bump would cause them to fracture and require replacement. Subsequent improvements in the materials being used have allowed the creation of structures which are capable of taking a significant impact before being significantly damaged.

For engineers to be able to successfully use plastics they need to recognise that, unlike metals, alloys and ceramics, the physical properties of polymers can vary significantly with temperature. The temperature dependence exhibited by plastics can be both an advantage and a disadvantage. With the temperature dependence comes the viscoelastic nature of the materials, which can be usefully exploited to provide vibration damping and a wide range of useful applications. However, the down side of the temperature dependence is that the material may loose its rigidity on heating. A plastic therefore has a range of temperature over which it can be considered to be useful for a specific application and the application defines this 'work temperature range'.

This textbook aims to introduce to engineers the molecular–materials science which governs the physical–mechanical properties of polymers. The text is based on a series of lectures on polymers which were delivered at the University of Strathclyde to engineering students. The text also includes elements of courses on semiconductor manufacture which were given to chemistry and physics undergraduates and to postgraduates studying for an MSc in optoelectronics. The aim of the text is to provide the engineer with a sufficient understanding of the properties of polymers so that they can effectively design components which will be fit for purpose. This is an introductory text and students are encouraged to consult the texts listed at the end of each chapter for a more in-depth presentation of specialist topics.

The text can be read either as a developing story or as individual chapters. The first six chapters represent the core of the subject and are written to be understood by students who may have a limited background in chemistry. Chapter 3, which deals with morphology, is somewhat more specialist than the others, but attempts to provide a more in-depth understanding of the variety of structures which plastics naturally create. The chapters concerned with the chemistry and physics of polymers presume a slightly greater understanding of the core subjects and provide students with a greater core ability in these subjects, an insight into the synthesis of polymers and the ability to theoretically model their molecular behaviour. Chapter 10, on plastics for the electronics industry, illustrates how knowledge of both the chemistry and physics of polymers has allowed the creation of materials without which the fabrication of semiconductor devices would be impossible. Chapter 11 briefly introduces the use of polymers in medical engineering, and Chapter 12 discusses the important topic of recycling.

It would be impossible to cover all aspects of polymers in great detail and this text should only be considered as an introduction for some topics. For instance, finite element analysis plays a pivotal role in engineering design. Finite element analysis is a topic in its own right and it would be impossible to discuss its application to polymers in any detail. However, the algorithms used to describe the viscoelastic properties of polymeric materials are based on the models discussed in Chapter 2. It is hoped that the reader will have gained sufficient understanding of the properties of polymers to be able to constructively question the validity of any calculation they may perform and in particular understand the relevance of the variation in temperature for the predicted properties. Finite element predictions are only as good as the physical data on which they are based and if that has limitations then this must be recognised when viewing the predictions from a specific calculation.

It is my pleasure to acknowledge the contribution which various colleagues have made to this text. They include Professor W.M. Banks, who some 20 years ago invited me to teach the polymer part of a course on polymers and composites to his engineering students. About 20 years ago I inherited a course from Professor A.M. North which formed the basis of the elements of this text which focus on the physical chemistry of polymers. The section on polymer degradation was developed from notes provided by my colleague Dr J.J. Liggat. Over the years this course material has been added to and modified as a result of discussions with many other polymer scientists with whom I have had the pleasure of working. They include: Professors R.W. Richards, D.C. Sherrington, J. Stanford, N.B. Graham, Dr S. Affrossman and Dr G. Eastmond. The course material was originally produced using the books referenced in the additional reading at the end of each chapter and has been added to and revised over the years. The material presented is core polymer science and specific reference to individual papers has been omitted. The author wishes to acknowledge a debt to the many polymer scientists with whom he has had contact over the years and who have, through discussion, contributed to his understanding of the subject.

If the use of this book helps one engineer to avoid designing a product which is not fit for purpose then the exercise will have been worthwhile. Plastics are marvellous materials when correctly used but can lead to disastrous consequences when wrongly applied. The requirement to create strong, lightweight structures will always require that plastics are considered as potential materials and it is hoped that with their sensible use many new innovative solutions to engineering problems will be found. Plastics used to be considered to be cheap disposable materials. In the future they are likely to be viewed as high-performance materials for specialist applications.

R.A. Pethrick
Strathclyde

Author's note

Throughout the book the use of a dash associated with a chemical formula indicates that this is a repeat unit, e.g. polyethylene will be written as $(-CH_2CH_2-)_n$, indicating that the group $-CH_2CH_2-$ has been repeated n times. In certain structures it is more convenient to use square brackets. The dashes are the connecting chemical bonds between the groups.

1

What are plastics?

1.0 Introduction

Early human history is a catalogue of our ability to use materials. The creation of stone implements heralded the *Stone Age* and assisted our ancestors to hunt, fish and cultivate the land. The development of early metallurgical processes, fashioning bronze and iron into tools, marked, respectively, the *Bronze Age* and *Iron Age*. In the 20th century, polymers and plastics have made a major impact on our lives and mark the *Age of Plastics*.

1.1 A brief history of the development of plastics

It is useful to briefly summarise the incredible growth in the availability and use of plastics over the last century or so. In 1839, Charles Goodyear discovered how to vulcanise natural rubber and in 1855 Alexander Parkes mixed pyroxylin, a partially nitrated form of cellulose, with alcohol and camphor and produced a hard, flexible, and transparent material, which he named Parkesine. In 1909 Leo Baekeland combined phenol and formaldehyde and made Bakelite.

Synthetic polymers or *plastics* were first created towards the end of the 19th century, by the modification of naturally occurring materials such as cotton and straw, and were based on cellulose (see Section 10.1). Notable examples included cellulose nitrate, which was used extensively in the production of early movie films, and cellulose acetate, which was preferred for clear glass-like objects (Morawetz, 1985). Modified cellulose dissolved in a solvent was cast to make combs, letter openers, trinket boxes, bowls, dishes and other objects. Cellulose nitrate mixes were able to simulate turtle shell and dyed cellulose acetate allowed artists to create smooth, light and aesthetically pleasing objects.

At the beginning of the 20th century, chemists developed an understanding of the atomic molecular nature of matter. Initially, plastics were thought to be large clusters of small molecules. However, careful investigations carried out by Staudinger in Freiburg in the 1930s, proved that plastics were made up of large single molecules (Feast, 1999). This pioneering research on the size of polymer molecules ultimately created the new discipline of *polymer science*. Plastics, or more correctly, *macromolecules* are all around us (proteins, DNA, RNA, plant fibre, collagen and rubber are just a few examples of naturally occurring macromolecules).

Polystyrene and polyvinylchloride (PVC) were developed after the First World War by IG Farben, in Germany. In the late 1930s Wallace Carruthers, at Du Pont, created nylon, a *synthetic* plastic, by the chemical reaction of smaller molecules and was awarded a patent for it in 1938. Since 1940, the production of plastics has grown year-on-year reflecting the broad diversity of materials which can be produced and their increasing use in everyday life (see Figure 1.1).

Synthetic nylon fibre, which is very strong and flexible, rapidly replaced natural bristles in toothbrushes and allowed the creation of synthetic silk stockings. By 1944, a total of 50 factories were manufacturing synthetic rubber and produced a volume of the material twice that of the world's natural rubber production before the beginning of the Second World War.

By 1936, polymethylmethacrylate (PMMA), was produced as a tough substitute for glass. Plexiglas was used during the Second World War to build aircraft canopies. Polyethylene was

Figure 1.1 Examples of a variety of objects produced from plastics.

discovered in 1933 by Gibson and Fawcett at Imperial Chemical Industries (ICI) in the UK. Polypropylene was discovered in the early 1950s by Zeigler and Natta. Polytetrafluoroethylene was developed as a fluorinated analogue of polyethylene and had superior chemical resistance. Polyurethanes were invented in 1937 by Bayer and Farben. Epoxy chemistry led to the creation of the new class of polymers, the thermosets, and ultimately to the first composite materials.

Polyethyleneterephthalate (PET) was developed at Manchester by Whinfield and Dickson in 1941. PET is less gas-permeable than other low-cost plastics and is a popular material for making bottles for carbonated soft drinks. Polycarbonate was developed by General Electric in the 1970s and at about the same time Du Pont produced Kevlar, a very tough plastic material which could stand high temperatures.

The precise materials used to create each of the objects in Figure 1.1 are slightly different. Some of the objects have to be very rigid, whereas others have to be flexible, some require high impact strength and others do not. Some of the objects, such as cutlery, have to be sterilised or capable of being reused whereas others will be disposable. Plastics may be exposed to different types of fluids: water, hydrocarbons or polar solvents. One plastic may be suitable for one application but unsuitable for another. By understanding the nature of the environment in which an object is to be used, it is possible to select the material which is fit for purpose. In order to be able to make that selection it is necessary to understand the factors which influence the physical properties of polymeric materials. In the course of studying this textbook it is hoped that the reader will gain an insight into the factors which control whether or not a particular plastic may be fit for purpose and gain confidence in designing functional and structural components in these materials.

The growth of the use of plastics has been encouraged by engineers seeking solutions to practical problems. The earliest biplanes had wings and fuselage constructed from tensioned cotton fabric. The rigidity of the fabric skin was increased by coating it with a solution of cellulose. When dried, the cellulose matrix bridges the cotton fibres and produces a skin with improved aerodynamic characteristics and mechanical properties like those of a drum skin. The rigid skin absorbed less moisture, produced less drag and was an early polymer composite. During the Second World War, natural rubber was in short supply and synthetic rubber was created to fill that gap. Vulcanised rubber had been used since the 1900s for the construction of tyres and fuel pipes in motor cars and aircraft. The synthetic substitute was more consistent than the natural materials and soon replaced its natural counterpart.

The requirement to incorporate windows in the fuselage of passenger carrying aircraft created the need for a light, durable, clear material as a substitute for glass. Acrylic polymers were the obvious choice. Windscreens in automobiles are usually created as a sandwich of glass and a clear plastic, this laminated material will craze rather than shatter on impact. Plastics are used for household plumbing, window frames and internal fittings for automobiles. Carbon and glass fibre composites are used in transportation, wind turbines, aircraft, bridges and numerous civil engineering applications. Packaging for food relies heavily on the use of plastic film to achieve the desired storage life. Computer chips are constructed using polymer resists and encapsulated in a potting resin to protect them from the environment. Sports activities such as skiing, sailing and golfing are reliant on the use of composite materials to produce ski boards, boats and golf clubs. Polymers are used extensively in medical applications and it would now be very difficult to imagine a modern world which did not have plastics.

Plastics can either be rigid or flexible. Unlike metals and ceramics, the physical properties of polymers can change dramatically with temperature and pressure. With so many polymeric materials available, it is important to be able to classify plastics and understand how to select a material which is fit for purpose. In the 1960s, plastic was used to replace aluminium in the window handles in automobiles. The design was a direct copy of the aluminium handles and did not allow for the creep characteristics of plastics. After a short period of time the handles started to split as the thin section suffered fatigue. It took ten years for car manufacturers to regain confidence in the use of plastics. Subsequent designs eliminated the high stress points and allowed for possible creep of the plastic with time. With an understanding of how chemical structure and molar mass influence physical properties it is possible to design articles in plastics with confidence.

1.2 What are plastics, polymers and macromolecules?

The terms *plastic, polymer* or *macromolecule* are used to describe a range of materials which have one feature in common: they are all large molecules created by chemically linking smaller entities (IUPAC Recommendations, 1980). The term *polymer* is derived from two terms: *poly* meaning many and *mer*, which is an abbreviation of the word *monomer*. A monomer is the primary building block from which the polymer is created. The alternative name for a polymer is a *macromolecule*, indicating that it is a high molar mass species. The term *macromolecule* does not necessarily imply that all the elements along the backbone of the molecule are the same. Biomacromolecules are often created by linking many different nucleic acids together and their specific characteristics are a reflection of the diversity of the monomers used. The term *plastic* has come to be used to describe a wide range of *synthetic* macromolecules. Polymers can be either rigid or flexible and may either be brittle or very elastic. Unlike metals and ceramics, change of temperature can convert a brittle rigid plastic into a soft and extensible *elastomer*. Whilst most polymers are based on linking molecules that have a carbon–carbon bonded structure, there are very important materials based on inorganic bonded structures, these include: silicon–oxygen $(R_2Si0_2)_n$, phosphorus–nitrogen $(P\text{–}N)_n$, boron–nitrogen $(B\text{–}N)_n$ and sulphur $(S)_n$ and other chemistries. These materials have specialist applications and are beyond the scope of this text.

1.3 A simple analogy for a polymer chain

An easy way to understand the properties of polymers is to consider the behaviour of a poppet bead necklace (see Figure 1.2). The poppet bead, which was popular in the 1970s, is a hollow sphere of plastic attached to a tail terminating in a small solid sphere. The smaller sphere can be pushed through a hole in the larger hollow sphere forming a link between two hollow beads rather like a chemical bond. The link between the beads gives the necklace a high degree of flexibility and mimics properties found in polymer chains.

A polymer chain may typically have between 1,000 and 5,000 monomer units (individual beads) linked together. A polymer chain created with a single link between the beads is like a single-stranded necklace and resembles a linear chain.

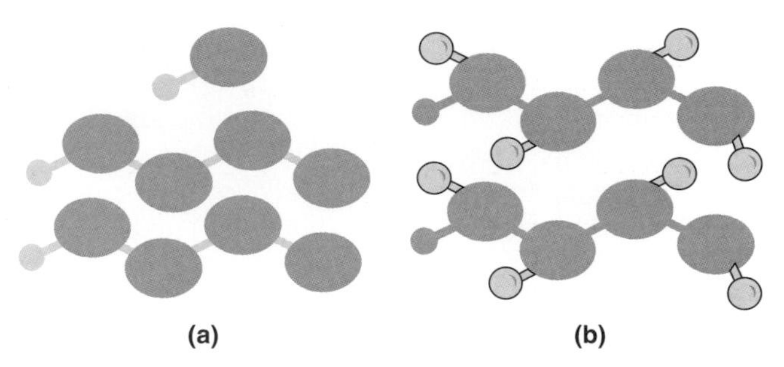

Figure 1.2 A schematic of a 'poppet' bead, (a) top; an element of a necklace, (a) middle; a close packing of neighbouring chains, (a) middle and bottom; and the effects of nodules - pendant groups, keeping the chains apart, (b).

The necklace can be unravelled and twisted to take on a new form. Such a single-stranded chain has many of the properties of a *thermoplastic* polymer. The name *thermoplastic* implies that heating the material to a high temperature will produce a free flowing liquid which can be cast into any desired shape. However, if each bead is capable of bonding to more than one other bead, a network can be created. The three-dimensional network formed by the creation of more than one bond per bead generates a network which cannot be reshaped without breaking links (the chemical bonds) and is known as a *thermoset*. The name *thermoset* implies that the monomers react together to form a fused three-dimensional structure.

Careful arrangement of strings of beads can create a close-packed structure with a high degree of order (see Figure 1.2(a)), which is typical of a *crystalline* polymer. If the poppet beads have little nodules attached to their surface, known as *pendant groups* (see Figure 1.2(b)), close packing is inhibited, the chains are disordered and the material is regarded as being *amorphous*. Pendant groups can stiffen the chain, influence its ability to pack together and produce materials with different physical properties.

A polymer chemist can design and synthesise a polymer material which can be used for a particular purpose provided the engineer can define the critical properties that are required (see Section 7.7).

1.4 What factors influence the physical properties of thermoplastic materials?

A single polymer chain will have a limited load bearing capability determined by the strength of the individual chemical bonds. However, distribution of the load across a large number of chains reduces the stress on any particular link and increases the load bearing capability. Close packing strings together results in a bundle which is stiffer. However, nodules inhibiting packing retain flexibility but reduce the load bearing characteristics. Although most synthetic polymers have a carbon–carbon backbone, their physical properties can vary significantly as a consequence of the way in which the neighbouring chains interact. Since the strength of the chain–chain interactions will vary with the distance between the chains, it is not surprising to find that plastics have physical properties that vary with temperature. Increasing the temperature will increase the chain–chain separation, flexibility and reduce the polymer's load bearing capability.

1.5 How are polymers made?

Why are polymers often available in different grades? To answer this question we must consider how polymers are produced. Any molecule that contains a carbon–carbon double bond, known as a *vinyl* bond, is capable of being converted into a polymer. Examples of typical vinyl monomers with their corresponding polymeric names are given in Table 1.1. The first column contains the common and IUPAC name of the monomer. In some cases, such as polyvinylchloride,

Table 1.1 Structure of monomer and corresponding polymer for some vinyl polymers

Monomer: common – IUPAC name	Structure	Polymer	Structure
Ethylene – ethane	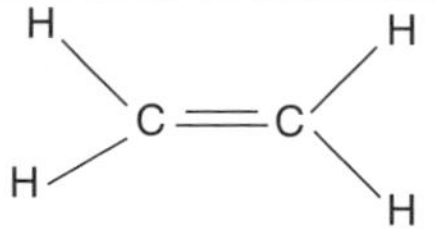	Polyethylene (PE®)	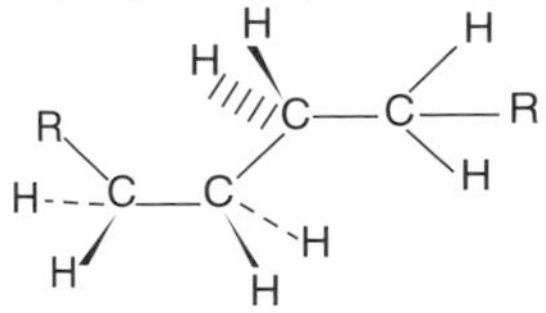
Vinyl chloride – chloroethene	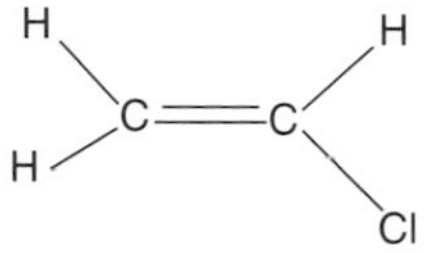	Polyvinylchloride (PVC®)	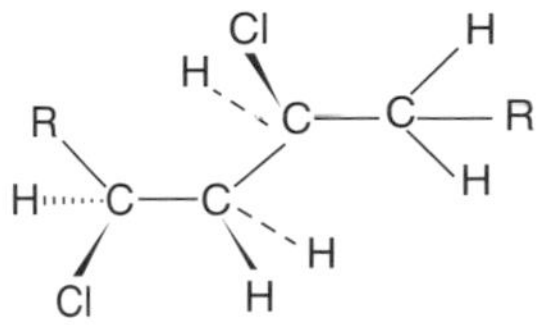
Propylene – propene	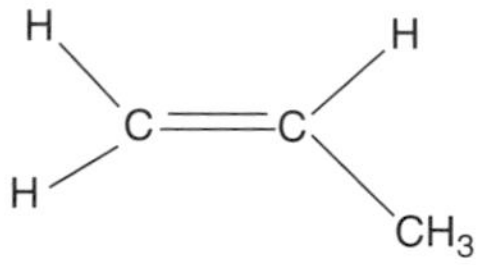	Polypropylene (Propylene®)	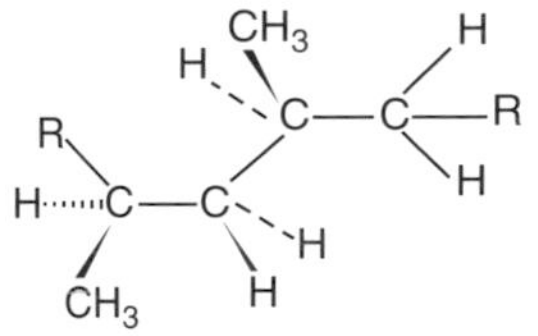
Styrene – phenylethene	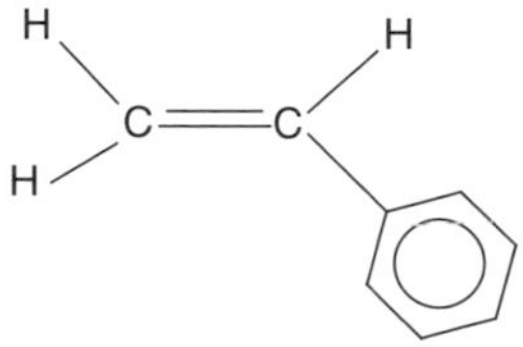	Polystyrene (Styrene®)	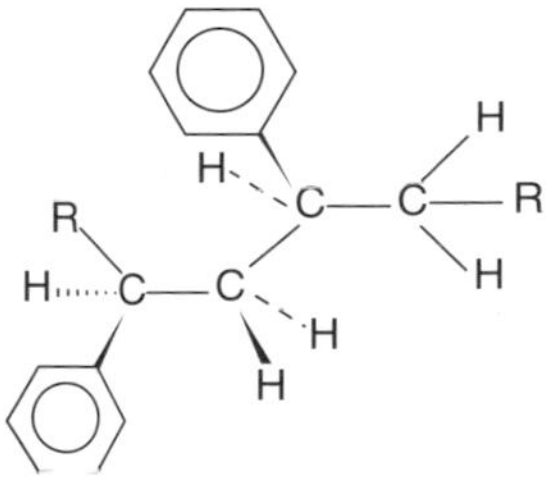
Methacrylate	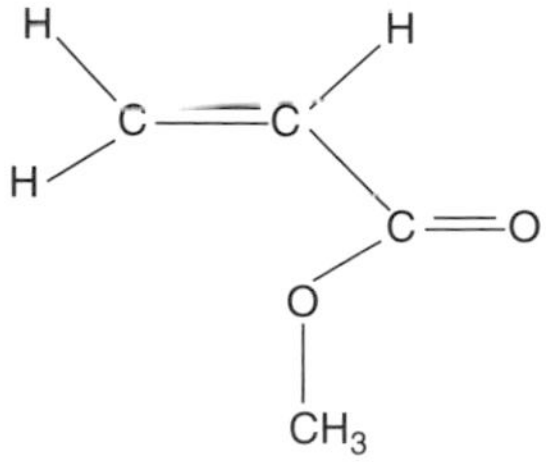	Polymethylacrylate (Acrylic®)	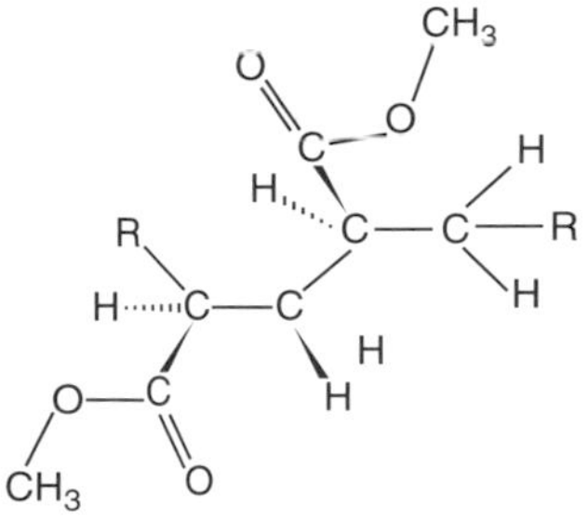
Methylmethacrylate–methyl-2-methylpropenoate	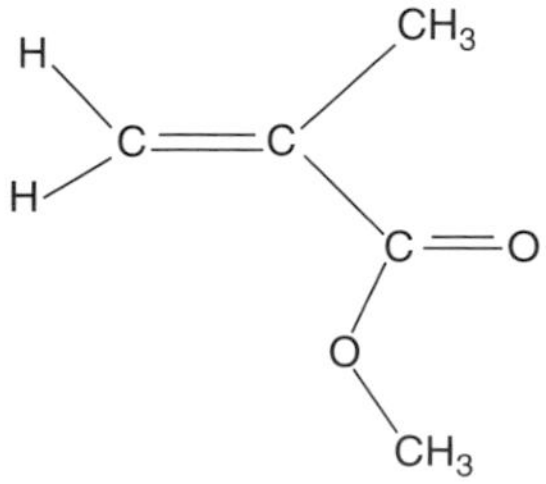	Polymethyl-methacrylate (Perspex®)	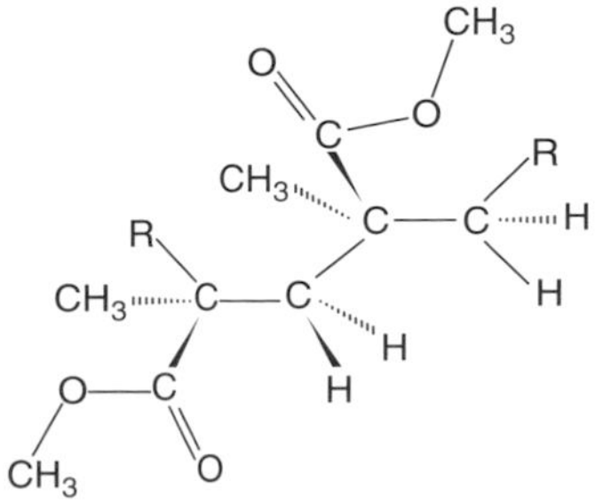

an abbreviation is used (PVC). In other cases, the polymer may be known by its tradename: polymethylmethacrylate has the tradename Perspex®. These polymers created using a single monomer are known as *homopolymers*.

Any catalogue of materials will include a number of grades of polymer and materials created from more than one monomer, known as *copolymers*. Many commercial polymers are copolymers, a second monomer being introduced at a low level to impart some desirable attribute to the original polymer. Commercial acrylic copolymers are created by polymerisation of methylmethacrylate and methacrylate or some other monomer combination. These copolymers will usually have slightly different properties from their parent homopolymers.

The length of a chain of *vinyl polymers* can be varied resulting in different grades of material. It is therefore common practice to refer to a material as having an average chain length and to define a parameter which reflects the spread of the distribution of chain lengths.

1.5.1 Polyethylene: the simplest linear polymer chain

Polyethylene is the result of the polymerisation of the monomer *ethylene* (see Figure 1.3). The ethylene or more correctly *ethene* molecule has a σ- or *single* bond which connects the hydrogen atoms to the carbon atoms and the two carbon atoms together. The C–H σ-bond has two electrons and is formed by sharing one electron from a hydrogen atom and one from a carbon atom. The σ-bond is very strong and it is necessary to heat the molecule to several hundred degrees Centigrade to break the bond. Since ethylene has only two hydrogen atoms and a carbon atom joined to each carbon atom, there is a spare electron on each of the carbon atoms which can participate in bonding. In the ethylene monomer, these spare electrons on the carbon atoms combine to form a π-bond (see Figure 1.3).

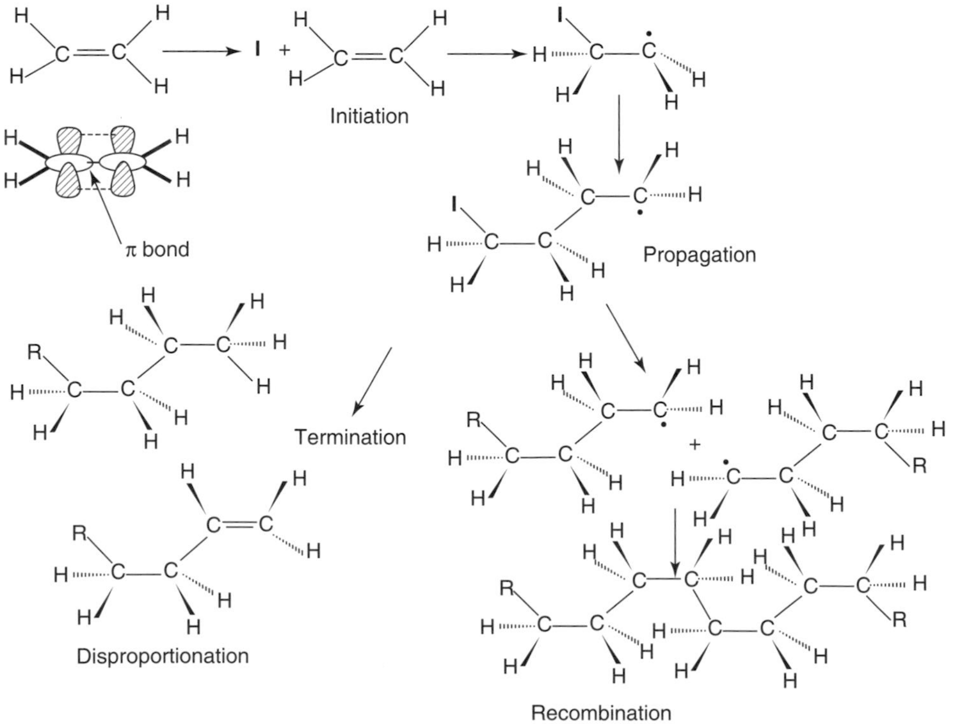

Figure 1.3 Schematic diagram of the polymerisation process for polyethylene.

In the ethylene molecule, the carbon atoms are said to be doubly bonded. The weaker π-bond can be opened by suitable chemistry in which a molecule with a spare electron approaches one of the carbon atoms and forms a stable σ-bond. The process forming a σ-bond with one carbon atom creates a *free* electron on the other carbon atom. The *free* electron is designated by a dot close to the carbon atom (see Figure 1.3) and is called a *radical*. The free electron can approach another ethylene monomer to form a bond and propagate the polymerisation reaction (see Figure 1.3). The polymerisation process creates long chains of linked ethylene units, hence polyethylene. This is called a *free radical vinyl polymerisation reaction*.

To start the *polymerisation* process, a species with an excess of electron density has to be added to the reaction mixture and is called the *initiator* (designated I in Figure 1.3). Once the reaction has been initiated, polymerisation will continue until the free electron is removed by reaction. The polymerisation process is *terminated* by either *recombination* or *disproportionation*. If two of the growing chains can combine to form a stable molecule (see Figure 1.3), this process is called *recombination*, indicating two free electrons have combined to form a stable σ-bond. This process usually requires a third molecule to remove excess energy and create a stable molecule. If the energy is not removed, the encounter may not lead to the formation of a stable bond. Alternatively, the free electron can abstract a hydrogen atom from a neighbouring chain and form a stable molecule. The chain from which the hydrogen has been abstracted has now two free electrons which will form a double π-bond, this process is termed *disproportionation*. If all the chains were to grow to the same length before terminating, then the two different termination processes would lead to some of the polymer chains being approximately twice the length of others and some of the shorter chains would retain reactivity, because they have terminal double bonds. The mix of shorter and longer chains can vary with the conditions used in the synthetic process and will influence the resulting physical properties of the polymer.

The length of the chain will depend on how quickly it grows relative to the efficiency of the termination process and the number of chains initiated in the polymerisation processes. The process of polymerisation will be discussed again in Sections 8.3 and 8.4. Changes in the way the polymer is produced can significantly affect the physical properties. These differences in the physical properties allow selection of materials with the best properties for a particular application or processing method. A cautionary note, which cannot be emphasised too strongly, is that different grades of material can have very different physical properties: one may be ideal for a particular application and another totally unsuitable. Use of the incorrect materials can cause major problems and give rise to failure of the object in service. Selection of the wrong polyethylene for a gas distribution pipe could lead to premature failure and an explosion. The selection of a material will be considered in the case studies (see Section 7.7).

1.5.2 Step growth polymerisation

Whilst many polymers are created by addition reactions, there are a large number which are formed by a *step growth* polymerisation, the process Carruthers originally used to create nylon. Many biopolymers are formed by step growth processes and unlike the vinyl polymerisation process, the chemistry pauses after the addition of each monomer. The synthesis of polyethylene adipate is an example of this process in which the diol (ethylenediol) is reacted with difunctional adipic acid (see Figure 1.4).

The esterification reaction involves reaction of the alcohol with and an acid, eliminating water to form an ester. This reaction is reversible and polyesters are susceptible to *hydrolysis* when exposed to water for a long time. The susceptibility to hydrolysis differentiates these materials from vinyl polymers where such a process does not exist. If a material is subjected to immersion in water for long periods of time, then hydrolytic susceptibility can become an important issue.

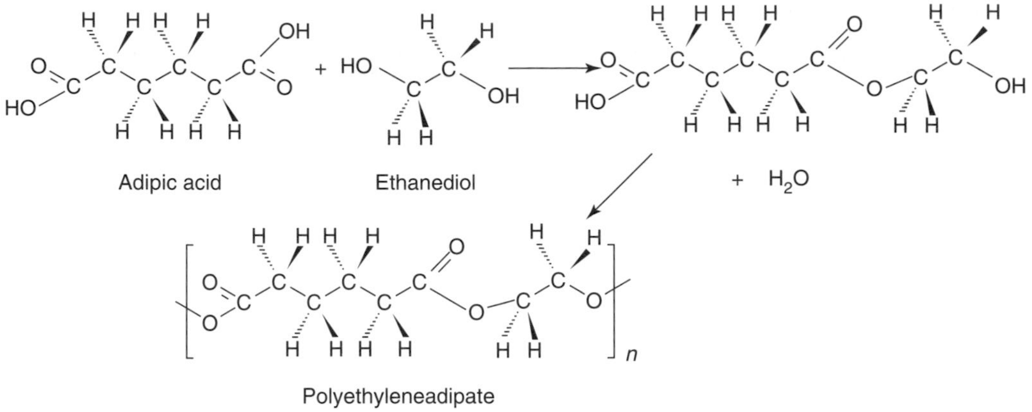

Figure 1.4 Schematic for the synthesis of poly(ethyleneadipate).

In principle, in a *condensation polymerisation process*, a stable molecule is created at each step of the polymerisation process, but the molecule formed retains its ability to undergo further reaction this is known as *step growth polymerisation*. If the reaction mixture contains equal amounts of the two reactants, then 50% reaction of the acid and alcohol groups creates an average chain length of just over two monomers! To make a long chain, it is essential that a very high percentage of the acid and alcohol groups are consumed. This method produces material with a broad distribution of chain lengths, in contrast to the vinyl polymerisation process where a relatively narrow distribution of chain lengths is produced.

The *step growth* process, shown in Figure 1.4, can be achieved using a range of chemical functionalities (see Table 1.2).

If the monomer contains two reactive functions then a linear *thermoplastic* polymer is formed. If, however, the monomer has more than two functionalities then a three-dimensional *thermoset* network structure is created. Condensation polymerisation mimics the way nature produces complex molecules such as DNA and collagen.

1.5.3 Ring opening polymerisation

A number of organic ring structures can be opened to produce polymers. Depending on the method used, the reaction can mimic either chain or step growth polymerisation. Ethylene oxide can be polymerised to form polyethyleneoxide (see Figure 1.5). This polymer is usually dihydroxy terminated and can undergo reaction with other entities such is diisocyanates to form polyurethanes. If the ethyleneoxide ring is attached to a larger molecule such as in the diglycidyl ether of bisphenol A – 'epoxy resin' (see Figure 1.6), then polymerisation can be used to create a thermoset material and is the basis of *epoxy resin* chemistry.

Polymerisation of the epoxy ring can be achieved in two different ways. The epoxy ring can be opened using an amine, the *two-pack epoxy resin process* (see Figure 1.6). The epoxy, usually a white sticky liquid, is cured with the light yellow smelly amine.

On mixing, the components react to produce a permanently cross-linked *thermoset*. This ring opening chemistry is used extensively in engineering applications and is the basis of many composite materials. Alternatively, the ring opening process can be initiated by the use of a Lewis acid or a base-forming ether-linked material. The latter process, once initiated, quickly creates long chains and resembles a chain propagation process. The curing characteristics and physical properties of these materials are very different.

Table 1.2 Summary of common organic reactions that can be used to create polymers

Function	Structure	Function	Structure	Polymer	Structure
Acid	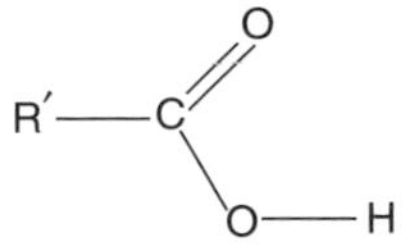 R′—C(=O)—O—H	Alcohol	HO—R″	Polyester	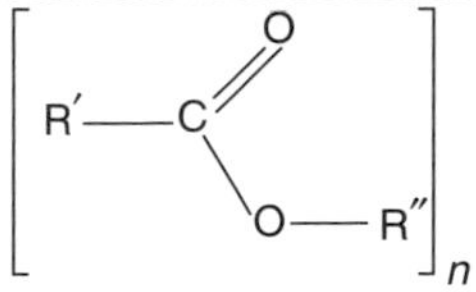 [R′—C(=O)—O—R″]$_n$
Acid	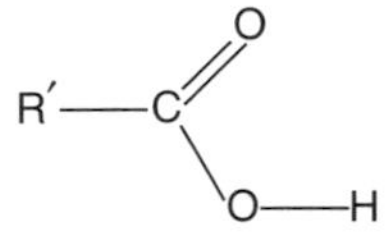 R′—C(=O)—O—H	Amide	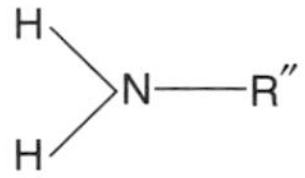 H$_2$N—R″	Polyamide – nylon	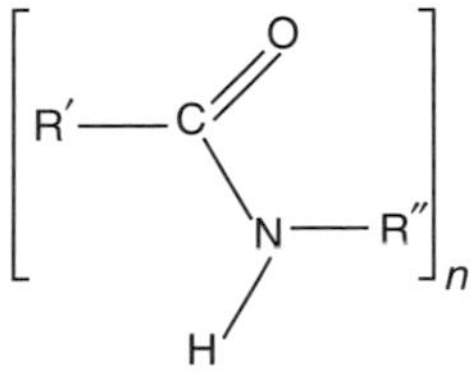 [R′—C(=O)—N(H)—R″]$_n$
Isocyanate	R′—NCO	Alcohol	HO—R″	Polyurethane	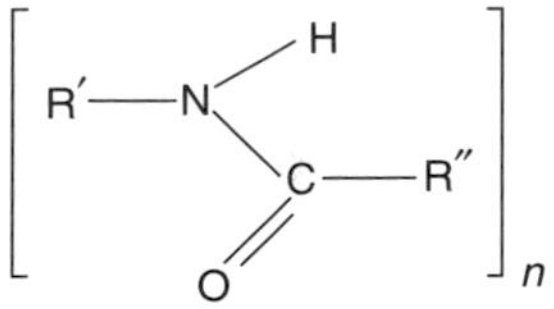 [R′—N(H)—C(=O)—R″]$_n$
Isocyanate	R′—NCO	Amine	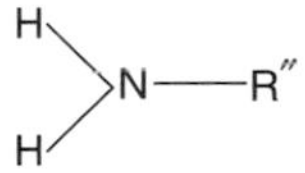 H$_2$N—R″	Polyurea	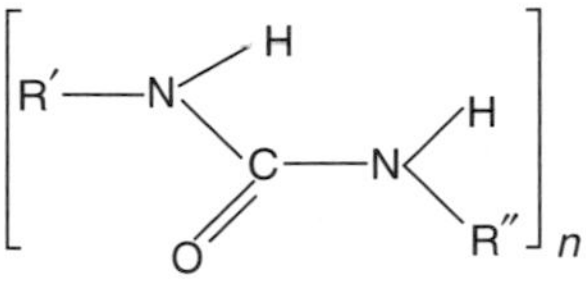 [R′—N(H)—C(=O)—N(H)—R″]$_n$
Anhydride	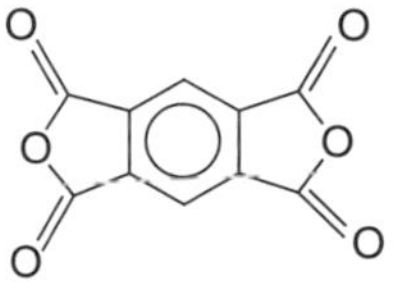 (pyromellitic dianhydride: benzene ring with two O=C—O—C=O anhydride rings)	Amine	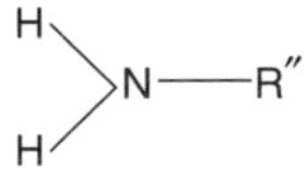 H$_2$N—R″	Polyimide	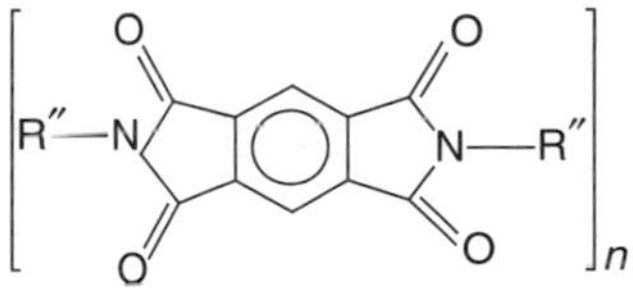 [R″—N(diimide ring, O O O O)—N—R″]$_n$

Ethyleneoxide → Polyethyleneoxide

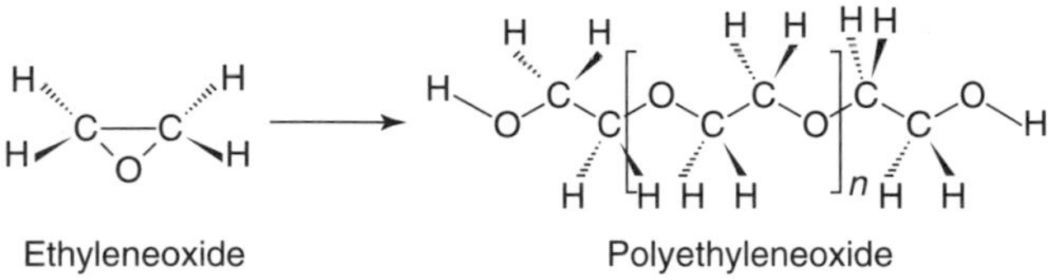

Figure 1.5 The polymerisation of ethyleneoxide to form polyethyleneoxide.

1.6 Effect of chemical structure on physical properties of polymers

1.6.1 Ability of polymer molecules to pack together

Polyethylene has a very simple chemical structure: -(CH_2–CH_2)$_n$-. The backbone of the polymer is made up of carbon atoms connected in a linear fashion and resembles a piece of string. If we allow string to fly about, it will rapidly start to coil into a ball. In terms of the *conformation*, the loops that are formed are *gauche* structures and correspond to an increased entropic contribution to the total free energy (see Figure 1.7). The lowest energy form is the fully stretched form, the all *trans* structure. The higher energy or *gauche* form involves the chain bending back on itself.

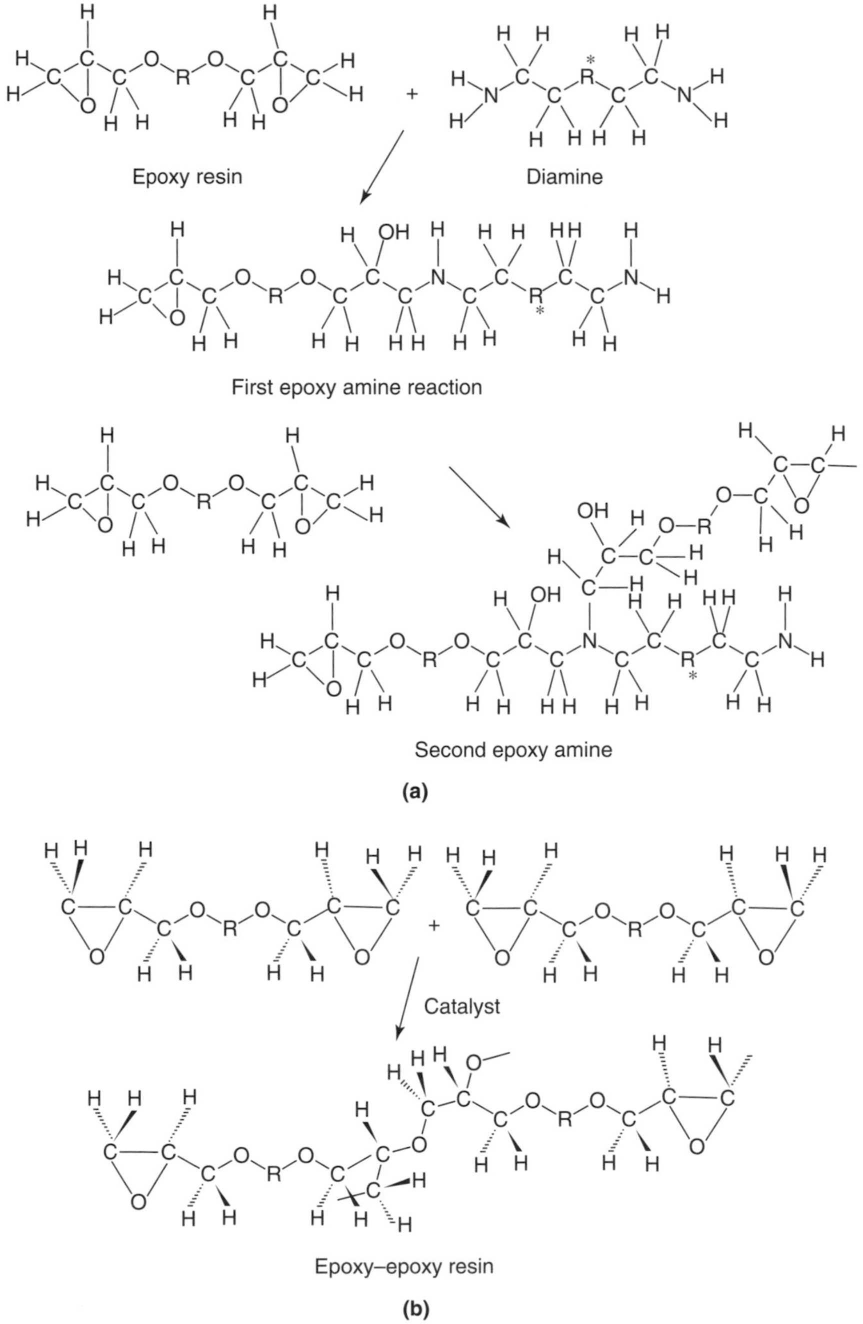

Epoxy resin

Diamine

First epoxy amine reaction

Second epoxy amine

(a)

Catalyst

Epoxy–epoxy resin

(b)

Figure 1.6 The chemistry of epoxy resins to form thermoset materials: (a) reaction of a difunctional epoxy molecule with a four functional amine; (b) reaction of epoxy with itself.

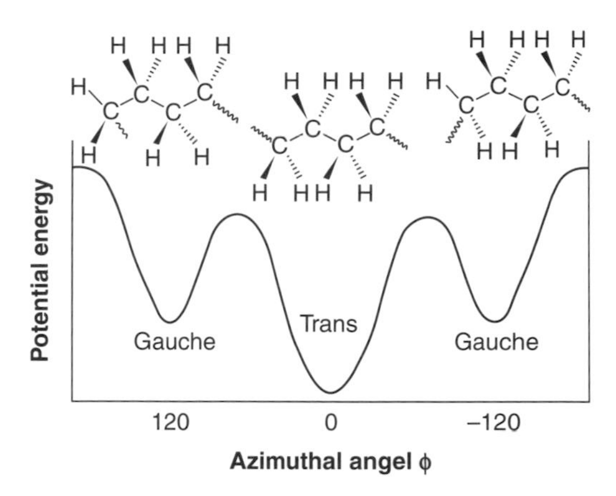

Figure 1.7 Potential energy variation with azimuthal angle for polyethylene.

The *azimuthal* angle is the projection of the C–C bonds forming the backbone on the next bond. The *trans* structure has all the bonds in a straight line and the azimuthal angle is 0°. The *gauche* structure has the bonds at 120° and is a bent *conformation*. The *gauche* conformation can be different depending on whether the angle is 120° positive to the 0° or negative but, in the case of polyethylene, these conformations have the same energy. At any temperature the polymer chain will adopt a distribution of *trans* and *gauche* conformations: the more *gauche* conformations which exist along the chain the more tightly folded will be the structure. The distribution of conformations is dictated by thermodynamics: increasing the temperature will increase the number of higher energy conformations.

The exchange of conformations between the *trans* and *gauche* forms depends on the eclipsing of atoms on neighbouring atoms, thus there is an energy barrier to this rotation process. The rotation about the backbone produces a change in the potential energy (see Figure 1.7). Increasing the temperature will promote more *gauche* conformations and the polymer chain will shrink to a more ball-like structure.

A piece of string subjected to random motion will eventually coil up on itself and form a ball (see Figure 1.8); the polymer exhibits similar behaviour. As the melt is cooled, the proportion of *trans* conformations will increase and the chains will straighten, thus increasing the size of the coil (see Figure 1.8). The straight sections of the chain can interact and will nucleate the formation of crystalline regions.

Polyethylene, because it forms an ordered phase on cooling, is classed as a *crystalline* polymer. The *crystalline* phase is limited in its range and will not extend across the whole of the solid. The organisation of crystalline polymers is discussed in Chapter 3.

Polypropylene is related to polyethylene, one of the hydrogen atoms being replaced by a bulkier methyl group (see Figure 1.9). As a result of this substitution, one of the *gauche* states is higher in energy than the other and the interaction of neighbouring methyl groups shifts the

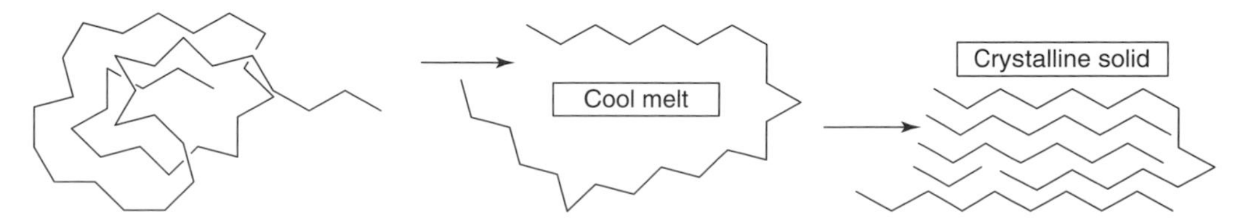

Figure 1.8 The ball of string like polymer in the melt straightens and forms a crystalline solid on cooling.

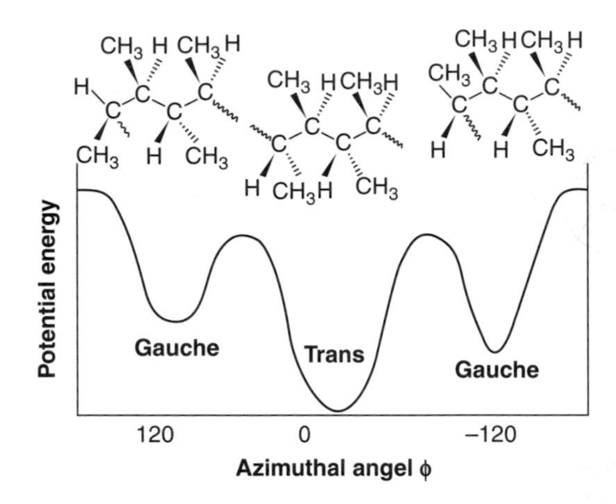

Figure 1.9 Potential energy surface for polypropylene.

energy minimum from 0° and imposes a twist on the polymer backbone. The helices which result from the twisted backbone can pack together to form a rigid, crystalline material. Whether or not the correct twist is created depends on the way in which the methyl groups are distributed relative to the plane of the polymer backbone (see Section 8.6).

The backbone carbon atom that contains the methyl group has the possibility of linking to the next bond in either a right- or left-handed sense and is termed *chiral* (see Figure 1.10). The method of synthesis can produce a backbone in which pendant groups are distributed. They can be all right- or all left-handed or just at random. The polymer produced with a defined *chirality* is termed *tactic*. Polypropylene can be produced in three different forms: *isotactic* with all the methyl groups lying on one side of the plane running down the polymer backbone, *syndiotactic* when the methyl groups alternate about the backbone, and *atactic* when they are randomly distributed down the polymer backbone (see Figure 1.10).

Isotactic polypropylene is a rigid crystalline material used for hot water pipes. The syndiotactic and atactic materials are unable to pack to form ordered structures and are rubbery and can be used as additives to improve the performance of lubricants. Engineers need to be aware that apparently the same polymer can have markedly different physical properties.

Polystyrene has a very bulky phenyl ($-C_6H_5$) group attached to the polymer backbone. The phenyl group is so large that it makes it almost impossible for the backbone to form a regular structure and is very difficult to crystallise. However, in the *syndiotactic* polymer, the interaction between the phenyl groups is minimised, a helical structure is formed and a crystalline material results. Normal polystyrene is usually *atactic* and exhibits a very disordered solid phase structure. Polystyrene forms a *polymer glass*, characterised by a high degree of transparency. Glassy polymers have very different characteristics from those of the crystalline materials and are termed *amorphous* solids.

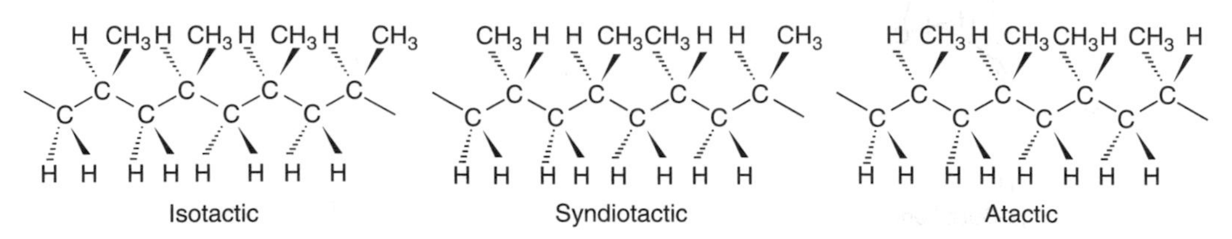

Figure 1.10 The three tactic forms of polypropylene: isotactic, syndiotactic and atactic.

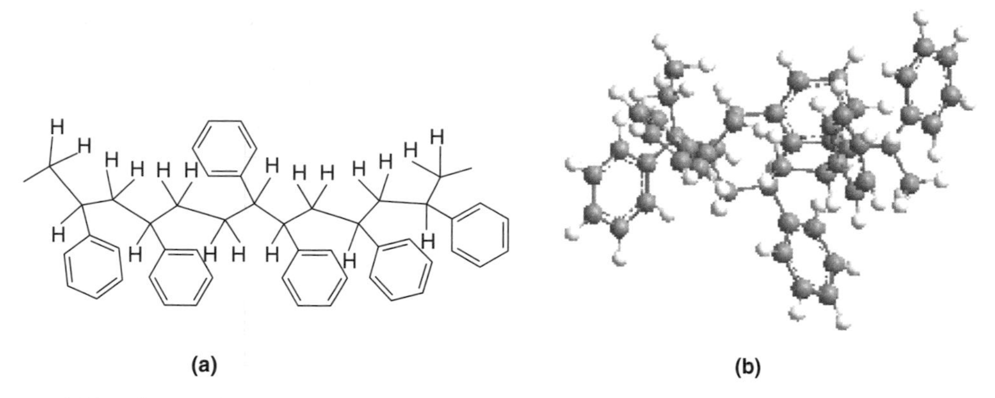

Figure 1.11 The one-dimensional representation of (a) the chemical structure and (b) the minimised three-dimensional structure of a polystyrene chain.

The pendant phenyl group in polystyrene is capable of rotating without requiring the backbone to execute rotational motion. This phenyl side group rotation can start below 0 °C and gives polystyrene improved impact resistance compared with normal glass.

It is difficult to picture the structure of a polymer from a chemical structure drawing (see Figure 1.11(a)). The distribution in space is only truly appreciated in a three-dimensional picture (see Figure 1.11(b)) in which we see the phenyl rings pointing in various directions and the twisted nature of the polymer backbone.

Polybutadiene is synthesised from the monomer butadiene, which is ethylene with another ethylene molecule replacing one of the hydrogen atoms: $-(CH_2{=}CH{-}CH{=}CH_2-)_n$. Polymerisation can create a number of different types of polymer depending on the way in which the addition reaction proceeds (see Figure 1.12).

In the simple reaction scheme, the addition to the double bond occurs across the first vinyl bond, leading to the 1,2 addition products. This polymer has pendant groups that inhibit the rotation about the backbone and is rigid compared with the 1,4 addition product that has a single bond between each unit and is rubbery. The 1,4 addition can create the *trans* and *cis* configurations that have different degrees of crystallinity. Thus, polybutadiene can be obtained with different physical characteristics.

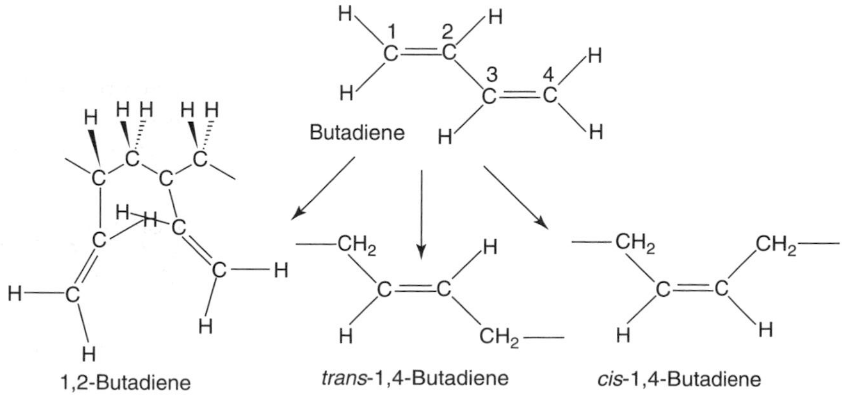

Butadiene

1,2-Butadiene | *trans*-1,4-Butadiene | *cis*-1,4-Butadiene

Figure 1.12 Various configurations that can occur on polymerisation of butadiene.

1.7 Copolymers

So far we have considered *homopolymers*, those polymers which can be synthesised using only one monomer. However, in an attempt to create materials that are fit for purpose, it is common for manufacturers to combine more that one monomer into the polymer chain to create materials with very different physical characteristics. These materials are known as *copolymers*, indicating they are produced from more than one monomer. An example would be the copolymerisation of styrene with butadiene, which is used commercially. Styrene is a glassy, rather brittle solid that has a fairly high modulus and softens to a rubbery solid at about 100°C, whereas butadiene, depending on the configuration, will be a hard or a soft rubber. Combining butadiene and styrene produces a tough rubbery material. Combining the monomers can be carried out in the following ways:

Diblock copolymers

These are created by a *living* polymerisation technique (see Section 8.8). The process involves polymerisation of the first monomer, which is totally consumed, but each polymer retains an *active site* and can be further polymerised. A second monomer is added to create a second block structure and the process is terminated. The polymer is shown in Figure 1.13(a).

The styrene and butadiene blocks are thermodynamically incompatible and will phase separate (see Section 3.10). As a consequence, the solid is made up of regions that are rich in either butadiene or styrene. A number of different arrangements can be created (see Figure 1.14). The type of structure, the *morphology*, depends on the ratio of the styrene to butadiene in the copolymer. If the dominant phase is butadiene then the material will be rubbery and has the ability to creep (to flow under pressure). If the dominant phase is styrene, then the material is glassy but it will have better impact properties than pure styrene (see Section 3.10).

Triblock copolymers

The disadvantageous creep characteristics found in the diblock copolymer can be overcome with triblock copolymers (see Figure 1.13). In the case of styrene–butadeine–styrene we can envisage that chains starting in one domain will end up in another. This anchoring of the chains in

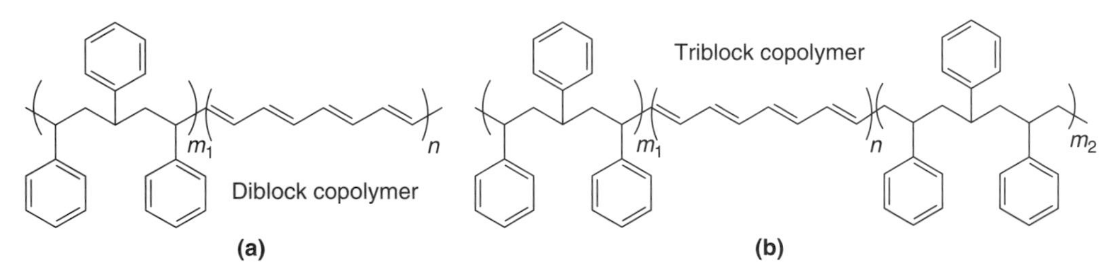

Figure 1.13 Structures of (a) styrene–butadiene diblock and (b) triblock copolymers.

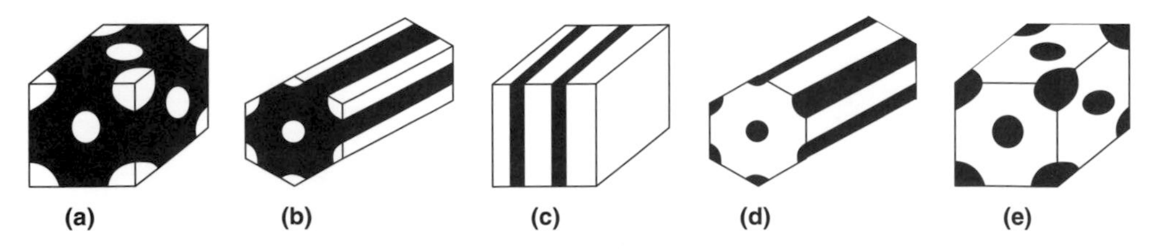

Figure 1.14 Effect of composition on block copolymer morphology: (a) spheres of A in matrix of B; (b) cylinders of A in matrix of B; (c) alternating A and B lamellae; (d) cylinders of B in matrix of A; and (e) spheres of B in matrix of A.

neighbouring domains creates a structure that can both have a high modulus and rubbery characteristics yet is not subject to the creep observed with the diblock copolymers (see Section 3.10).

Alternating copolymers

It is possible to create certain polymers with a structure in which monomers have an alternating sequence. These materials are rare and have very different properties from those of the constituent monomers. Styrene–maleic anhydride copolymer is an alternating copolymer. Maleic anhydride does not form a polymer by itself and will only form with styrene as a copolymer.

Random copolymers

Some polymerisation processes allow the uncontrolled reaction of monomers and create random copolymers. The properties of these materials will tend to be a mixture of those of the monomers and are directly proportional to the amounts of the individual monomers that are incorporated in the polymer. Thus the random copolymer of styrene and butadiene will be more rigid than butadiene but more rubbery than pure styrene (see Section 3.10).

Blends of polymers

Polymers which are thermodynamically compatible can be mixed to give materials with averaged properties. Some materials are compatible at high temperatures but demix when the temperature is lowered. Blends can be divided into two types: compatible blends in which the two different polymers form a homogenous mixture, and incompatible blends where the properties of the material are influenced by the way in which the two materials phase separate. It is possible to use the phase separation of one polymer in another to advantage and create a material which has improved fracture or impact properties (see Sections 3.10 and 7.4.12).

Other architectures

Some structures arise by the addition of low concentrations of multifunctional monomers to the polymerisation mixture (see Figure 1.15), and produce *branched, hyperbranched* or *dendrimer* materials. Because the backbone and side chains are often produced from different polymers, the molecules can bridge between different phases and are used as compatiblisers for polymer blends, surfactants and are a major constituent of nondrip paints.

By careful synthesis using controlled condensation reactions, it is possible to create *dendrimers*. These molecules have structures which resemble biological molecules and can mimic viruses. It is difficult to visualise these molecules in two-dimensional space but it is found that

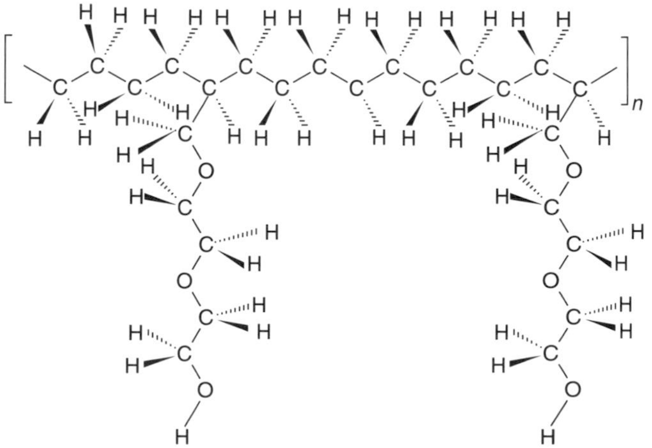

Figure 1.15 Branched chain copolymer of poly(ethylene-co-ethylene oxide).

for a particular system there will be a level of growth at which any further reaction becomes sterically inhibited and a very tight, close packed structure is formed (see Figure 1.16(a)), which can be appreciated from the three-dimensional projection (see Figure 1.16(b)).

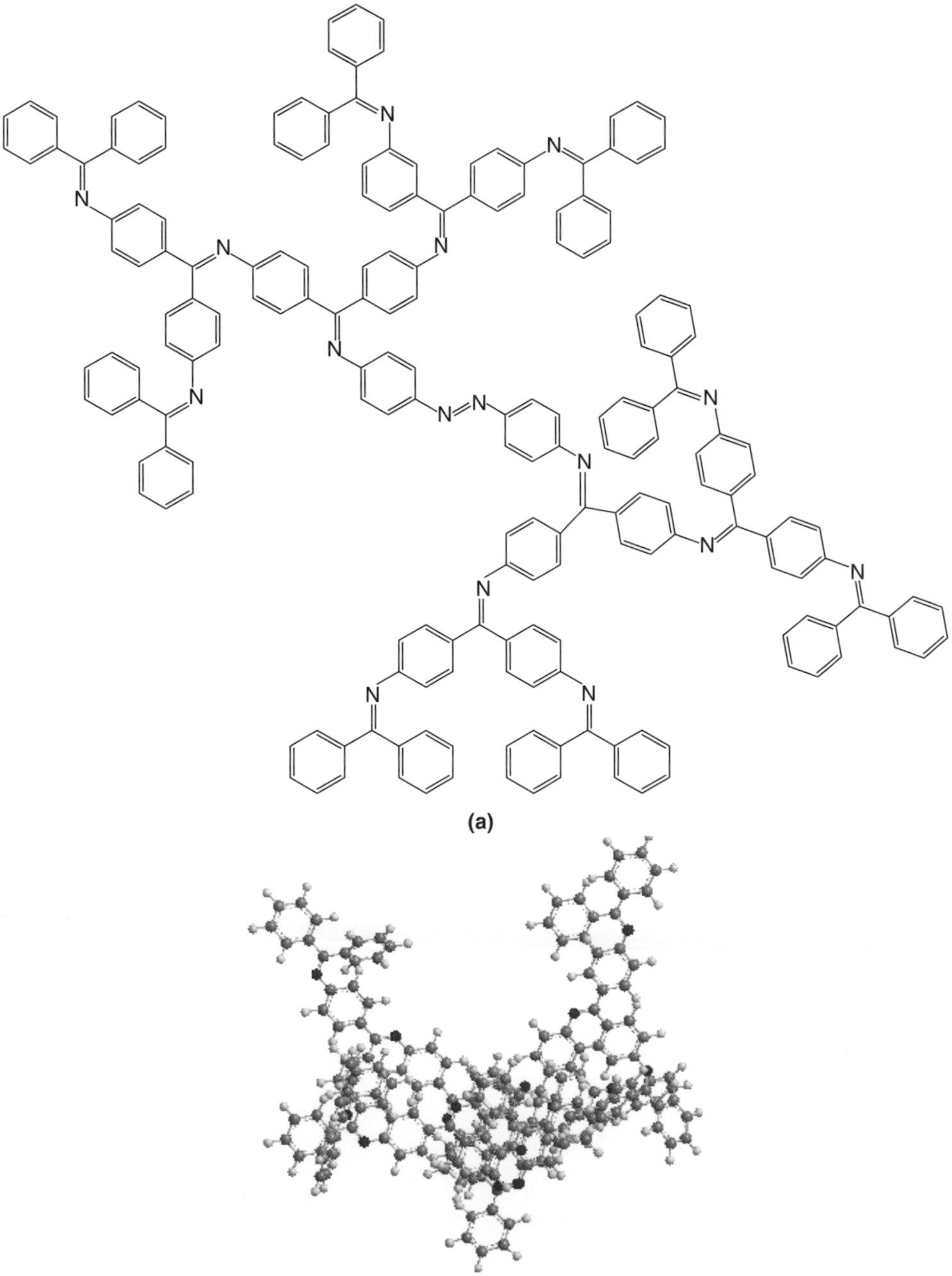

(a)

(b)

Figure 1.16 Two-dimensional representation of (a) a dendrimer and (b) the equivalent three-dimensional picture of the same structure.

As a result of steric interactions between neighbouring arms of the molecule, *dendrimers* form almost spherical particles.

1.8 Polymer design for application

At room temperature many polymers may appear to have similar mechanical properties. However, it is only when we explore how these characteristics vary over a temperature range that the differences emerge. Polymers which apparently have the same chemical structure may have different physical properties as a consequence of differences in chain length (or its equivalent, the molar mass). Knowing how the physical properties change with temperature allows a working temperature range to be defined for that material and it is then possible to determine whether or not the material is fit for purpose. However, it is useful to classify polymers according to some simple characteristics. The selection of a polymer material for a particular application is considered in a case study (see Section 7.7).

1.9 Polymer classification

Polymers can be classified into subgroups according to various chemical and physical characteristics (see Figure 1.17) and simple tests are discussed in Section 11.2.

Simply heating the material will usually indicate whether it is a thermoplastic or a thermoset. Chemical analysis will usually be used to differentiate between vinyl and condensation polymers and if relevant the configuration (*stereochemistry*) adopted by the elements of the polymer chain. Polymers with the same chemical structure will often have different chain lengths and this is reflected in the molar mass.

1.10 Molar mass and molar mass distribution

The degree of polymerisation is defined as the number of monomers incorporated in a particular polymer chain and is designated n. For low values of the degree of polymerisation the materials are called *oligomers* or *telomers*. As the value of n is increased, so the physical properties will change in a systematic manner. The melting point of a homologous series (a set of polymers with the same chemical structure but with different chain lengths) will increase with molecular weight. Paraffins are the same group of materials as polyethylene. Butane, $n = 2$, is a liquid at room temperature, n = 14–16 are soft solids and for n above 20 the solids are waxy and eventually become partially crystalline (see Figure 1.18). Whilst the melting point increases with molar mass for the shorter chain materials, a point is reached where it becomes independent of the molar mass. Short chain polymers are known as *oligomers*. The molar mass effect occurs because the ends of the chain are more flexible and less constrained than the main chain. As the proportion of the extended main chain relative to the more mobile ends is increased, so the melting point increases. The chain length is clearly a very important parameter when defining the properties of a polymer.

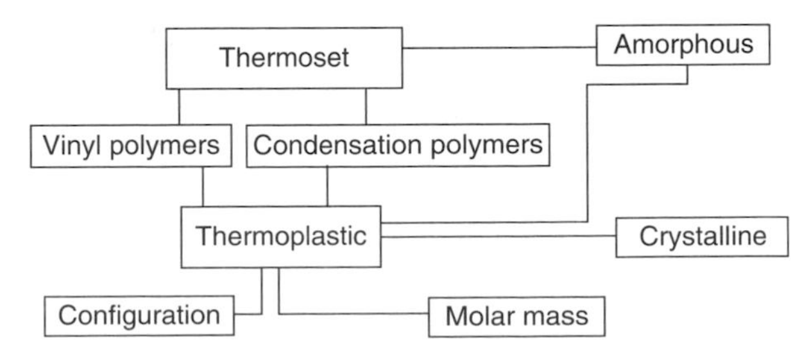

Figure 1.17 Scheme for the classification of polymeric materials.

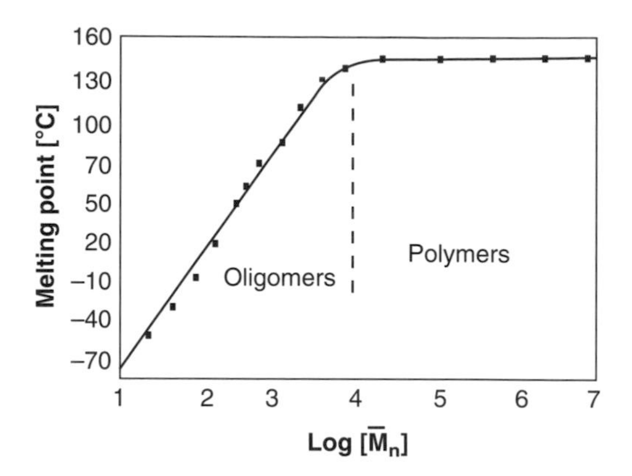

Figure 1.18 Variation of the melting point with molar mass for the hydrocarbon series – polyethylene.

The chain length (or molar mass) can be measured via a number of different techniques (see Section 9.12). A polymer will have a distribution of chain lengths and it is usual to consider the average molar mass. The two most common molar mass averages used are the number and weight averages.

1.10.1 Molar mass averages

To understand the concept of a number average consider taking spaghetti of different lengths and asking: what is the mean length of the spaghetti in a particular sample? The length of a piece of spaghetti is equivalent to a particular molar mass and hence the average will be obtained by taking the sum of the product of the number with a particular length multiplied by that length and then dividing by the total number (see Figure 1.19). In molar mass terms this is expressed as the *number average molar mass* (Griffiths and Thomas, 1983).

Number average molecular weight

Consider the chains of a particular chain length as labelled i and having a molecular weight, M_i. The number average molecular weights will be:

$$\bar{M}_n = \frac{\sum M_i n_i}{\sum n_i} \tag{1.1}$$

The molecular weight of a polymer chain is obtained by multiplying the mass of a monomer unit by the number of monomer units in the chain. This is equivalent to sorting the strands of spaghetti according to their length (see Figure 1.19).

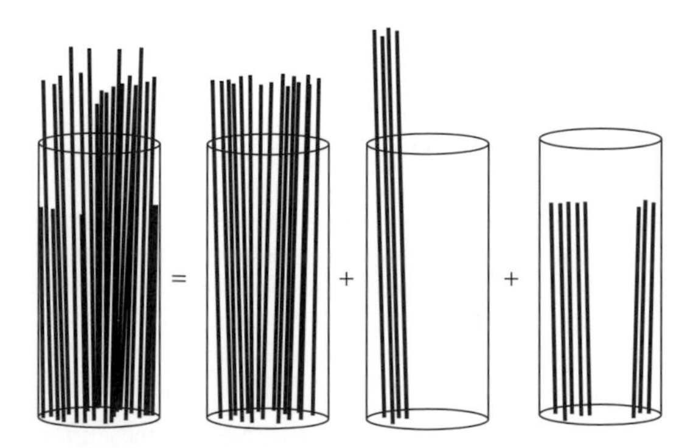

Figure 1.19 Schematic of the number average distribution – lengths of spaghetti.

Weight average molecular weight

Analysis of the polymer molecules by their weight w_i will then be given by:

$$\bar{M}_w = \frac{\sum M_i w_i}{\sum w_i} = \frac{\sum M_i M_i n_i}{\sum M_i n_i} \qquad w_i = M_i n_i \tag{1.2}$$

Note that the weight of a given size of polymer chain is simply described by the product of the molar mass of that chain multiplied by the number of chains of that size that are present. This average will be different from the number average as the longer chains will make a greater contribution to the average than shorter chains. The weight average will always be higher than the number average.

Molar mass distribution

It is usual to gauge the breadth of the distribution of molar masses in terms of the ratio of the weight and number average distributions:

$$\frac{\bar{M}_w}{\bar{M}_n} = \frac{\sum M_i M_i n_i}{\sum M_i n_i} \frac{\sum n_i}{\sum M_i n_i} \qquad \bar{M}_w \approx \bar{M}_n \qquad \frac{\bar{M}_w}{\bar{M}_n} \approx 1 \tag{1.3}$$

Thus, as the values of M_w and M_n approach one another, the chains all have the same length, the ratio will approach one.

Worked example: molar mass calculation

A sample of a polymer was analysed and was found to contain three components of ideal molar mass distribution ($M_w/M_n = 1$). The three components have values of molar mass of 15,000, 25,000 and 60,000 and are present as, respectively, 20%, 45% and 35%. What are the values of $\bar{M}_n$, $\bar{M}_w$ and $\bar{M}_n / \bar{M}_w$?

Number average molar mass:

$$\bar{M}_n = \frac{100}{\frac{20}{15,000} + \frac{45}{25,000} + \frac{35}{60,000}} = \frac{100}{0.001,33 + 0.001,8 + 0.000,583} = \frac{100}{0.003,713} = 26,932$$

Weight average molar mass:

$$\bar{M}_n = \frac{20 \times 15,000 + 45 \times 25,000 + 35 \times 60,000}{100} = \frac{300,000 + 1,125,000 + 2,100,000}{100} = 35,250$$

$M_w/M_n = 35{,}200/26{,}932 = 1.30$

1.10.2 How does the method of synthesis influence the molar mass distribution?

The method of synthesis influences both the average chain length and the distribution of chain lengths. The *chain growth* mechanism associated with *vinyl polymerisation* creates high molar mass chains every time a polymerisation is initiated. Alternatively, the *condensation polymerisation* process involves the successive coupling of small units, the chains grow very slowly and the final polymer material will retain significant traces of the original monomers and oligomers. These differences are illustrated by the distributions in Figure 1.20. For a typical condensation polymerisation, the molar mass distribution may be in the range 3–20 or higher, whereas for a vinyl polymerisation the values will typically be in the range 1.05–3.0. The narrowest molar mass distribution is observed with *cationic* or *anionic* initiated polymerisations (see Section 8.7) and broader distributions are obtained with *radical* initiated systems. Molar mass effects are observed with all polymer systems but they are very important when we consider *amorphous* polymer systems.

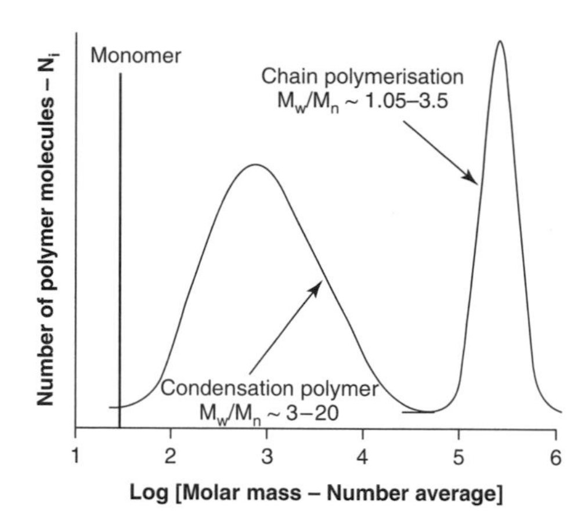

Figure 1.20 Typical molar mass distributions for vinyl and condensation polymers.

Brief summary of chapter

Polymers/plastics are very useful engineering materials and to understand how to use them we must understand how the following factors influence their physical properties:

- Method of synthesis: Vinyl addition, step growth and ring opening polymerisation are the most common routes to polymer formation but create polymers which have distinctly different characteristics.
- Effect of functionality of the monomer: If the monomer is bifunctional then the polymer will have a linear chain structure and will be *thermoplastic*. If the monomer has a higher functionality, it will form a cross-linked polymer network and is a *thermoset*.
- Symmetry and size of the groups attached to the polymer backbone: If the groups are small or if they are symmetrically distributed then the polymer can form a regular structure in the solid and the material will have crystalline characteristics. If the groups are not symmetrically distributed down the polymer chain or are bulky then the chains cannot form an orderly packed structure in the solid phase and the material will be an *amorphous* glass.
- Molar mass effects: Changes in the molar mass, for short chain polymers, will change their physical properties. Short chains will behave like liquids, whereas longer chains can exhibit superior mechanical and physical properties at ambient temperatures. Many of the important physical properties associated with polymers are dependent on the molar mass and hence will depend on the method used for the production of the polymer.
- Polymer blend and copolymers: Blending of polymers and making polymers by incorporation of monomers with different structures allows their physical properties to be engineered.

References and additional reading

Batzer H. and Lohse F. *Introduction to Macromolecular Chemistry*, Wiley, Chichester, UK, 1976.
Ebewelle R.O. *Polymer Science and Technology*, CRC Press, Boca Raton, FL, USA, 2000.
Elias H.G. *Macromolecules, Structures and Properties*, 2nd edn., Plenum, New York, NY, USA, 1984.
Feast J. *The Age of Molecules*, Royal Society of Chemistry Publishing, Cambridge, UK, 1999.
Griffiths P.J.F. and Thomas J.D.R. *Calculations in Advanced Physical Chemistry*, Arnold, London, UK, 1983.
IUPAC Stereochemical Definitions and Notations Relating to Polymers (IUPAC Recommendations, 1980). *Pure and Applied Chemistry* 1981, **53**, 733–752.
Morawetz H. *Polymers: the Origins and Growth of a Science*, Wiley, New York, NY, USA, 1985.
Rodriquez F., Cohen C., Ober K. and Archer L.A. *Principles of Polymer Systems*, Taylor and Francis, London, UK, 2003.

2

Mechanical properties of polymeric materials

2.0 Introduction

In engineering applications, the ability to support or transfer a load is an important characteristic of a material. Because of their molecular nature, polymers exhibit a temperature dependence in their physical properties, which is not found in ceramics or metals. It is therefore important to know at what temperature a certain polymeric material is going to be used, before it is possible to determine whether or not it is fit for purpose. The temperature range over which a polymer can be safely used is known as its *work range*.

2.1 Assessment of mechanical properties

In order to be able to define the mechanical properties of a polymer, it is important to make both static and dynamic measurements. Because of their molecular nature, polymeric solids exhibit many physical properties which are sensitive to the speed or rate at which a measurement is performed. In the case of metals and ceramics, the static and dynamic properties are essentially identical and it is not important to differentiate between these two different types of measurement. However, in the case of polymers significant differences can be observed between static and dynamic measurements.

2.2 Stress–strain measurements

The proportionality constant between the stress and strain is the modulus of a material. For a Hooke's law solid, the strain (extension, ε) is proportional to the stress (load, σ):

$$\sigma(\textit{stress})/\varepsilon(\textit{strain}) = \text{modulus} \tag{2.1}$$

which is the Young's modulus, Y, for a simple solid. For a wire of cross-section a and an applied force f, the stress σ is defined as:

$$\sigma = f/a \tag{2.2}$$

Increasing the force by an amount δf increases the stress by an amount $\delta f/a$, and produces an extension δl. The strain is $\delta l/l$, where l is the initial length. The modulus is:

$$Y = [\partial f/a]/[\partial l/l] = (l/a)(\partial f/\partial l) \tag{2.3}$$

and in the limit as δf and δl approach zero:

$$Y = (l/a)(\partial f/\partial l) \tag{2.4}$$

A plot of stress against strain has a slope which is proportional to the modulus. The dimensions of force being $[MLT^{-2}]$, it follows that the dimensions of E are: $E = [LL^{-2}MLT^{-2}] = [ML^{-1}T^{-2}]$. The dimensions of stress are also $[ML^{-1}T^{-2}]$. Since strain is a ratio of two similar quantities which are always zero, it is usual to express strain as a percentage. In the case of polymers, values can vary from being very small for rigid materials at ~0.01%, to rubbers where values can be as high as ~600%.

2.2.1 Volume change and shear

Following the approach outlined by Starling and Woodall (1950), if a body has similar properties in all directions then the material is said to be *isotropic*. If the properties differ with change in direction, the material is said to be *anisotropic*. To understand how the application of stress influences the material, consider the way in which a simple lattice is distorted by the application of a stress. In Figure 2.1, the strained lattice (a) is expanded to (b). In the process of expansion, the distances between the lattice points have all been increased by an equal amount.

If the increase is in the ratio $(1+\alpha)$:1, assuming that the expansion in all directions is the same, then the volume is increased in the ratio $(1+\alpha)^3$:1. In Figure 2.1(c), the layers are being sheared with respect to each other, the distance between the layers remains constant and the distance between particles at right angles to the plane is unchanged. Thus the side AB in Figure 2.1(a) is changed in the direction to A′B′ (see Figure 2.1(c)), and moves through an angle θ. This distortion is called a simple shear stress and the volume of the body is unchanged. Both these distortions of a simple solid are termed *homogeneous strains*, and are defined as changes in which equal and parallel lines in the unstrained body become equal and parallel lines in the body when strained. The length of the equal parallel lines change in the act of strain by rotation through an angle θ.

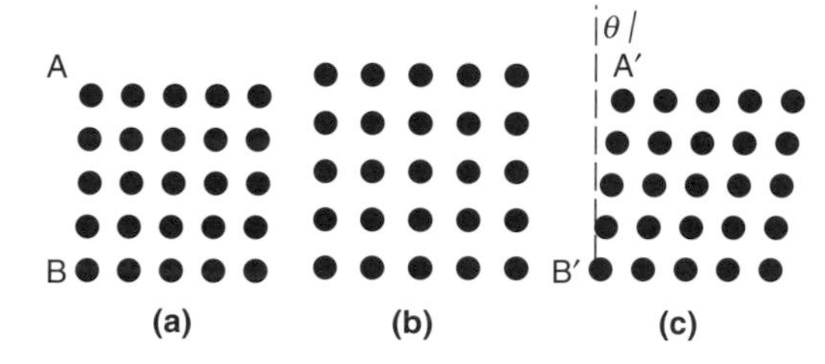

Figure 2.1 Expansion and shearing of a simple lattice.

2.2.2 The process of simple shear

A cube with faces ABCD is subjected to a shear force, so that the edges AD and BC are displaced whilst the edges AB and DC remain parallel and retain their original separation (see Figure 2.2).

Then $AA_1 = BB_1 = l\theta$. The value of θ is a small reflection, a small distortion of the matrix. It follows that the diagonal DB has increased to DB_1 and the increase is found from:

$$DB_1^2 - DB^2 = l^2 + (l + l\theta)^2 - 2l^2 = 2l^2 + 2l^2\theta - 2l^2 = 2l^2\theta \tag{2.5}$$

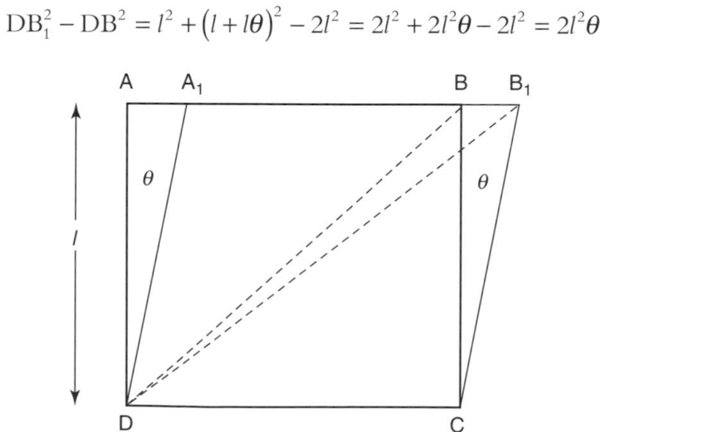

Figure 2.2 Cube subjected to a simple shear force.

since θ is very small and terms in θ^2 are negligible. Now:

$$DB_1^2 - DB^2 = (DB_1 - DB)(DB_1 + DB) \tag{2.6}$$

and if second-order terms are neglected, then:

$$DB_1 + DB = 2\sqrt{2l} \quad DB_1 - DB = 2l^2\theta / 2\sqrt{2l} = l\theta / \sqrt{2} \tag{2.7}$$

If we divide the extension DB_1 by the original dimension DB, we obtain the extension coefficient which is:

$$(DB_1 - DB)/\sqrt{2l} = l\theta/\sqrt{l}\sqrt{l} = \theta/2 \tag{2.8}$$

A similar process shows that the diagonal AC has undergone a contraction $\theta/2$. Thus the shear θ is equivalent to an expansion $\theta/2$ and an equal contraction in the directions at right angles to each other, both being at right angles to the axis about which shearing takes place.

2.2.3 Principal elastic moduli

When pressure is applied equally on all sides of an isotropic cubic body, the strain is observed as a change in volume, without change of shape. If the cube has an original volume v and δv is the change in volume produced by the pressure change δp then the strain is $\delta v/v$ for the stress δp and the appropriate modulus of elasticity is the *bulk modulus*, *K*, defined as:

$$K = \delta p\left(v / \partial v\right) = v\left(\partial p / \partial v\right) \tag{2.9}$$

or in the limit when δp and δv are infinitesimally small, then:

$$K = -v\left(\partial p / \partial v\right) \tag{2.10}$$

the negative sign is used because ∂p and ∂v are always of opposite sign. Simple shear, on the other hand, does not involve a change in volume. The strain is the angle θ in Figure 2.2 and the stress required to produce this strain may be a force applied parallel to AB the face DC being fixed. The implication is that there are really two forces; one in the direction AB and the other CD. If the force per unit area of the top face AB is f, then the modulus of elasticity is f/θ and is called the *rigidity modulus* or *shear modulus* (G). The Young's or *tensile modulus* is a simple pull which produces elongation of the sample, which corresponds to lateral contraction, and the specimen becomes thinner. The ratio of the lateral contraction to the longitudinal extension is known as *Poisson's ratio* and is designated ξ.

Consider a body being subjected to three mutually perpendicular stresses which are parallel to the three axes x, y and z (see Figure 2.3). These forces will be considered to be positive when they cause dilation and negative when they cause compression.

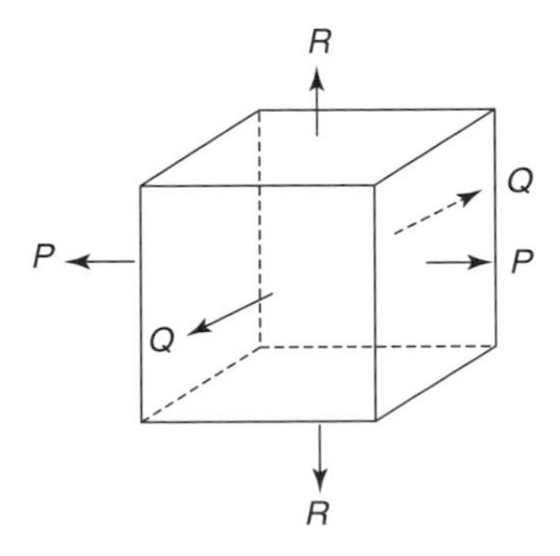

Figure 2.3 Forces acting on an isotropic cube.

The stresses P cause an elongation P/Y per unit length in the direction of x, and a contraction βP in the directions y and z at right angles to P. If the material is isotropic, the stress Q will cause an elongation Q/Y in the direction of y and contraction βQ along x and z. Similarly, R causes extension R/Y and contraction βR. The three moduli of elasticity can be expressed in terms of the stresses and strains. The total extension in the direction of x is $(P/Y)x - \beta Qx - \beta Rx$, y is $(Q/Y)y - \beta Py - \beta Ry$ and z is $(R/Y)z - \beta Pz - \beta Qz$. For the case of compression or dilation then $P = Q = R$, then the extensions are $Px(1/Y - 2\beta)$, $Py(1/Y - 2\beta)$, $Pz(1/Y - 2\beta)$, and the original volume xyz becomes $xyz(1 + P(1/Y - 2\beta))^3$. For small strains, the squares and cubes of the extensions may be neglected and the increase in volume is $3xyz(1 + P(1/Y - 2\beta))^3$. The bulk modulus, K, is therefore: $3xyzP(1/Y - 2\beta)$. But Poisson's ratio, ξ, is the ratio of lateral contraction βP to the longitudinal extension P/Y, that is $\xi = \beta P(1/Y)$ or $\beta = \xi / Y$ and therefore $K = (1/3[(1/Y) - (2\xi/Y)]) = Y/(3(1-2\xi))$. If now $R = 0$ and $P = -Q$, there is an extension $(P/Y) + \beta P$ in the x-direction and a compression $(P/Y) + \beta P$ in the y-direction and these being equal, the strain is a shear for which the angle is $2P((1/Y) + \beta)$ about the z- axis (see Figure 2.2). Therefore $G = [P/(2P(1/Y) + \beta) = 1/(2[1/Y + \xi/Y])] = Y/2(1+\xi)$. There are now two equations connecting G, K, Y and ξ. By eliminating σ we have $(9/Y) = (3/G) + (1/K)$ and $2G + 2\xi G = Y$, therefore $\xi = (Y - 2G)/2G$ There are a number of ways of determining the Young's and shear moduli with considerable accuracy, but owing to the difficulties of measuring the bulk modulus and Poisson's ratio, it is better to find the value of these from the former two parameters. The bulk modulus can be obtained from ultrasonic wave propagation measurements, although equipment to perform this type of measurement is not generally available.

2.2.4 Energy

Whenever strain occurs, the stress produces a displacement and work is done. For a perfectly elastic body, the work is recovered when the stress is removed and the sample recovers to its original dimensions. When a body is strained beyond its elastic limit or, in the case of a polymer, when the chain order is disrupted, then some of the work is released in the form of heat and the process becomes nonreversible and the original dimensions are not recovered.

2.2.5 Stretching of a wire

The force producing the stretching of the wire is $Y(\partial l/l)a = f$. The stretch δl increases from zero to some final value as the stress is applied. For an infinitesimal increase $d(\delta l)$ in the stretch, the work done is:

$$f\partial(\delta l) = (Ya/l)\delta l \partial(\delta l) \tag{2.11}$$

In this equation although δl is small, it represents a finite length. Equation (2.11) may also be written $(1/2)al(Y\partial l/l)(\partial l/l)$. The volume of the wire is al, $\partial l/l$ is the strain and $Y(\partial l/l)$ is the stress. Thus the work done per unit volume is 1/2(stress × strain). For a volume strain, the stress is δp and the strain $\partial v/v$ and $\partial P = K(\partial v/v)$, v being the original volume. The work done for a small change is:

$$\delta p\partial(\delta v) = K(\partial v/v)\partial(\delta v) \tag{2.12}$$

and the work done increases as δv increases or decreases from zero to its final value:

$$\int_0^{\delta v} K(\partial v/v)\partial(\delta v) = 1/2K\left((\partial v)^2/v\right) = 1/2K(\partial v/v)(\partial v/v)v \tag{2.13}$$

and the work per unit volume is 1/2(stress × strain). For shear, consider the force acting upon side AB in Figure 2.2 is $f = aG\theta$. The work done associated with the displacement is $fl\partial\theta = aGl\theta\partial\theta$, and the work done in this displacement is:

$$alG\int_0^{\theta} \theta\partial\theta = 1/2alG\theta^2$$

Since the volume is *al*, work per unit volume is $1/2G\theta.\theta$ which is again equal to 1/2(stress × strain).

2.2.6 Bending a thin beam

In a number of applications polymeric materials are subjected to bending and therefore the physical properties associated with this displacement are often important (see Figure 2.4).

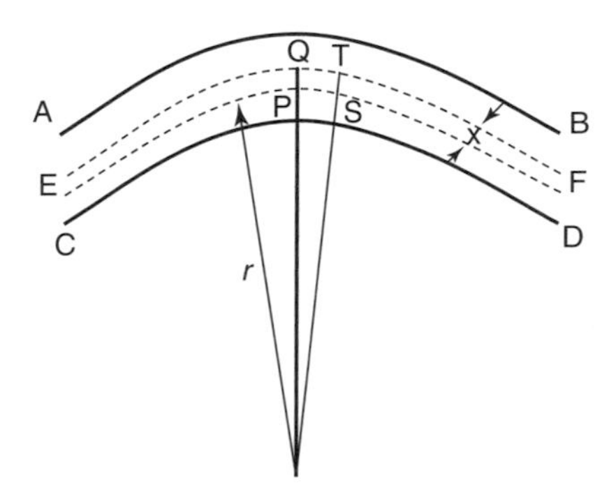

Figure 2.4 Thin beam under deformation.

When a solid strip is bent, the outside or convex surface is stretched and therefore under tension and the inner or concave surface is under compression and the centre contains a zone of material that is unstressed. This zone is called the *neutral axis* and if its radius at the point P is OP = r and the radius QO is $r + x$ then QT = $(r + x)\delta\theta$ and PS = $r\delta\theta$, where $\delta\theta$ is the angle POS and the extension of the element QT is $(x\partial\theta/r\partial\theta) = x/r$ and the stress in QT is $Y(x / r)$. Consider ABCD to be a section of a beam and the line EF to be the neutral axis. Every infinitesimal, thick layer in ABFE is under varying degrees of extension and every layer in EFDC is in compression. The result of these strains is an effective couple which is attempting to straighten the beam. In order to find the couple, consider the force due to the layer of depth δx at a mean distance x from the centre line of EF. The stress in the layer will be $Y(x / r)\partial xb$, where b is the breadth of the beam. The total force over ABFE is equal to that over EFDC or else there would be translational movement. Hence, for a beam so thin that any variation in r over the section may be neglected, the two parts are symmetrical, and FE lies halfway between AB and CD.

The moment of the force over the layer for a layer δx located a distance x from the median is $(Yb / r)x^2\delta x$ and for the part ABFE, the total moment is:

$$(Yb / r)\int_0^{d/2} x^2\partial x$$

where d is the depth of the beam. Thus the moment is given by:

$$(Yb / r)\left[x^3 / 3\right]_0^2 = \left(Ybd^3\right) / 24r$$

and for the two halves, the restoring couple is equal to $= (Ybd^3) / 12r$. If the area A is defined by bd and $d^2/12$ is the square of the radius of gyration about the median EF of the section, then $(bd^3)/12$ is therefore Ak^2 and corresponds to the moment of inertia about EF of a plate of unit mass per unit area, which is essentially the moment of area, designated I, about EF. It should be noted that, whatever the shape of the cross-section of the beam, $\int_{-x}^{+x} x^2\partial x$ is the moment of area about the neutral layer, corresponding to the moment of inertia of a plate. The couple acting on

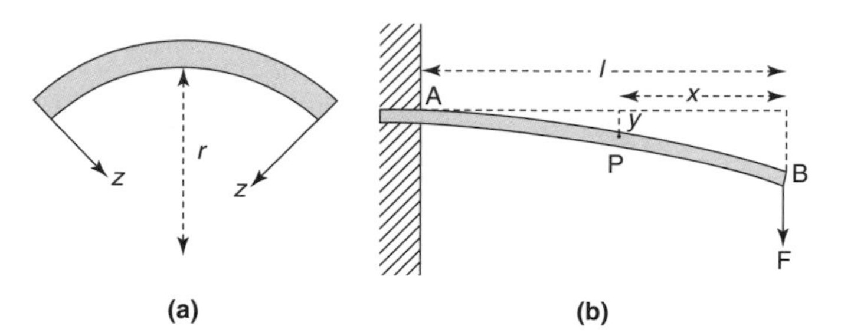

Figure 2.5 Beams: (a) bent beam; (b) cantilever beam.

any cross-section is YI/r and if the curvature at all points along the beam is the same, r is constant and the couple on every cross-section is the same. A thin beam bent by a couple z applied at each end (see Figure 2.5) will therefore be bent into an arc of a circle of radius r, where $z = (YI/r)$ or $r = (YI/z)$.

2.2.7 Cantilever

In certain test configurations, the piece to be tested is supported at one end and this is a cantilever (see Figure 2.5). If the end A is clamped and a load F applied at B, the beam will curve downwards to an extent which is dictated by its rigidity. It will be assumed that the thickness of the beam is small compared to the radius of curvature into which it is bent. The end of the beam, B, is depressed for two reasons. The beam bends and is also sheared. If Young's modulus were infinitely great, there would be no bending and AB would be a straight line. As B is depressed by an amount y_G due to shearing, the shearing strain is y_G/l and the stress is F/A. The rigidity G is then given by $G = (Fl)/(Ay_G)$ and therefore $y_G = (Fl)/(AG)$ and unless the beam is thick it is negligible in comparison with the depression of B due to bending.

To find the extent to which the beam is deflected, the origin is taken as the horizontal at A and the intercept is the projection of B back onto the horizontal. If the coordinates of the point P are x and y then the slope of the curve of the deflected beam is $\partial y/\partial x$ and the rate of change of slope, on passing along the axis of x, is $\partial^2 y/\partial^2 x$. Now the radius of curvature r is given by $(1 + (\partial y / \partial x)^2)^{3/2}/(\partial^2 y / \partial x^2)$ and since the slope of the curve will for rigid materials be very small, it follow that $(1/r) = (\partial^2 y / \partial x^2)$. The element of the beam at P is in equilibrium under the action of two couples, the couple Fx, known as the *bending moment,* and the couple $(YI/r) = YI(\partial^2 y / \partial x^2)$ due to the stresses over the cross-section at P. Therefore $YI(\partial^2 y / \partial x^2) = Fx$. On integration with respect to x: $YI(\partial y / \partial x) = (1/2)Fx^2 + C$, C being a constant of integration which can be determined by noting that at A $\partial y / \partial x = 0$ and $x = l$, therefore $C = -(1/2)Fl^2$ and $YI(\partial y / \partial x) = (1/2)F(x^2 - l^2)$. Integrating again, $FIy = (1/6)Fx^3 - (1/2)Fxl^2 + K$ and when $y = 0$ at A, where $x = l$ then $K = (1/2)Fl^3 - (1/6)Fl^3 = (1/3)Fl^3$ and $YIy = (1/6)Fx^3 - (1/2)Fxl^2 + (1/3)Fl^2$. This expression gives the depression at any point of the beam; and at the end B where $x = 0$, the depression is: $y = (1/3)(Fl^3/YI)$. For a beam of rectangular cross-section:

$$I = bd^3/12 \quad \text{and} \quad y = (4Fl^3)/(Ybd^3) \tag{2.14}$$

where b is the breadth and d is the depth of the beam. The ratio of the depression due to shear y_G to that due to bending is therefore:

$$\frac{y_G}{y} = \frac{Fl}{AG}\frac{3YI}{Fl^3} = \frac{3Tk^2}{Gl^2}$$

where $I = Ak^2$. For a thin beam, k^2 is small and the ratio k^2/l^2 is so small for such beams that y_G may be generally neglected and the deflection is determined by Equation (2.14).

2.2.8 Beam loaded in the middle

The cantilever is not suitable for taking measurements because it is difficult to support one end rigidly. Usual practice is to support the beam at both ends and load the test piece in the centre (see Figure 2.6).

The beam is supported on two knife edges at points A and B close to the edge of the beam and the force, the load applied at the centre of the beam C. The reaction at the knife edge corresponds to the force applied to the beam (see Figure 2.6). It must be remembered that the weight W applied at point C is twice the reaction at A and B. Again the length of the cantilever is half the length L of the beam AB between the knife edges. Making the substitutions in the equations we obtain $y = 1/3(W/2)(L^3/9)(1/YI) = (WL^3/48YI)$ or the depression of C can be used to calculate the Young's modulus:

$$Y = \left(WL^3/48yI\right) \tag{2.15}$$

The deflection y can be obtained for a given load and can be varied by altering the distance between the knife edges. In certain instruments, measurement of the deflection as L is varied at constant load W is used as a method of measurement. The alternative approach is to measure y at a series of different value of W.

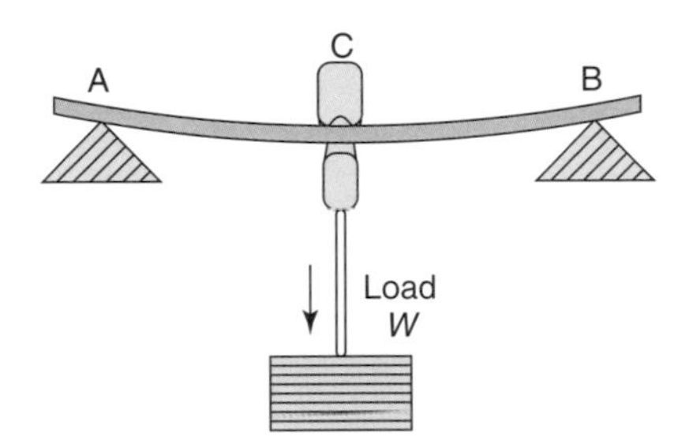

Figure 2.6 Beam loaded at its middle.

2.2.9 Poisson's ratio

Because of the nature of the structure of polymers and the different forces which may act between polymer chains, it is possible to observe different values of Poisson's ratio. Let us consider a thin beam made up of a series of layers; then on the convex side of the beam the filament (see Figure 2.4) undergoes a lateral contraction $\sigma(x/r)$, where x/r is the stretching strain, since the stretching at the surface is x/r, where x is half the thickness of the beam. The lateral contraction is therefore $\sigma(x/r)b$, if b is the uncontracted width of the beam; $(b - \sigma(x/r)b) = b(1 - \sigma(x/r))$ is the contracted width and the beam is curved (see Figure 2.7). If r_1 is the radius of curvature of the beam due to this lateral contraction, $(r_1 + x)\theta$ is the uncontracted width and $r_1\theta$ the contacted width, θx is therefore the actual contraction and $\theta x / (r_1 + x)\theta$ or since x is small compared with r_1 then x/r_1 is the coefficient of contraction and $\sigma(x/r) = (x/r_1)$ or $\sigma = r/r_1$.

If r_1 is the radius of curvature of the beam due to this lateral contraction, $(r_1 + x)\theta$ is the uncontracted width and $r_1\theta$ the contracted width; θx is therefore the actual contraction and $\theta x / (r_1 + x)\theta$ or since x is small compared with r_1 then x/r_1 is the coefficient of contraction and $\sigma(x/r) = (x/r_1)$ or $\sigma = r/r_1$. That is, the ratio of the longitudinal to the transverse curvature is equal to Poisson's ratio for the material. The neutral layer is thus a saddle-shaped surface having opposite directed curvatures in directions at right angles to each other. Such a surface is called an *anticlastic surface*.

2.3 Dynamic modulus

In many engineering applications, the load may be applied for a short period of time or may in certain situations vary in amplitude in a sinusoidal manner. The material will

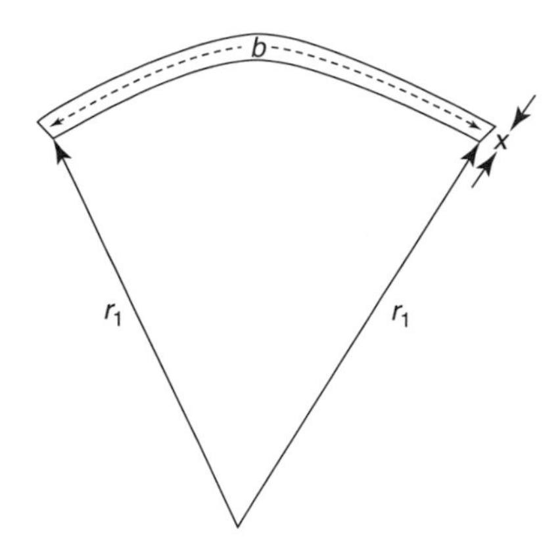

Figure 2.7 Thin beam subjected to bending at two points.

experience stresses which increase to a maximum value, then decrease and then increase once more (see Figure 2.8).

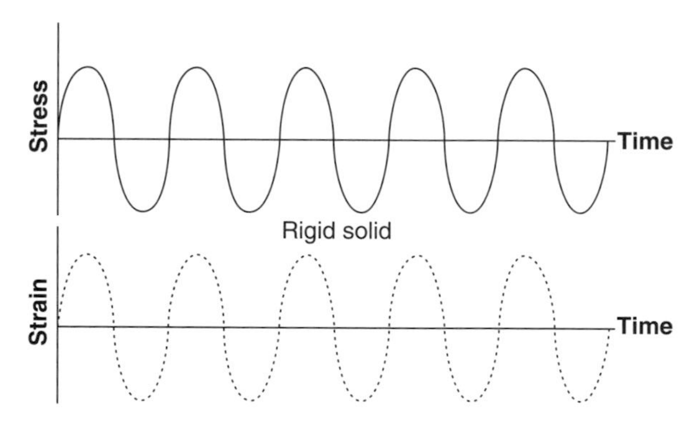

Figure 2.8 Variation of stress and strain for a rigid solid.

If the solid is simple and Young's modulus is applicable then the variations of the stress and strain stay in complete synchronisation and vary identically with time. However, if the material is able to show elastic properties, as is often found in the case of a plastic, then the strain may lag behind the stress and the curves are shifted in time. This shift in time is referred to as a *phase shift* and using the concept of an Argand

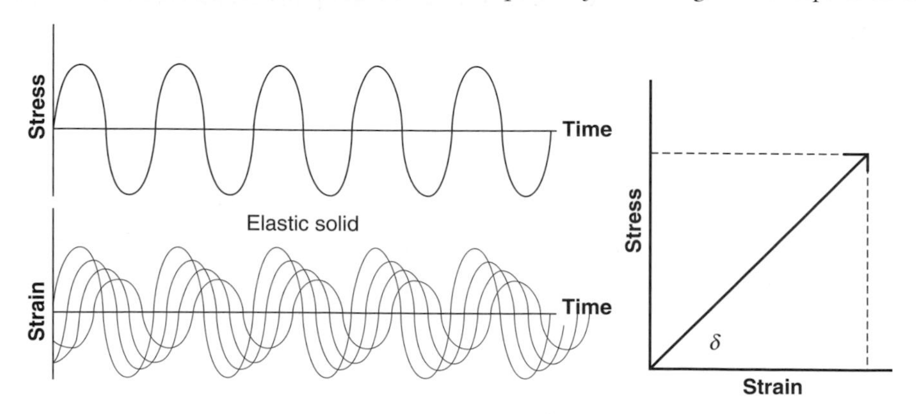

Figure 2.9 Strain responses for materials with increasing elasticity.

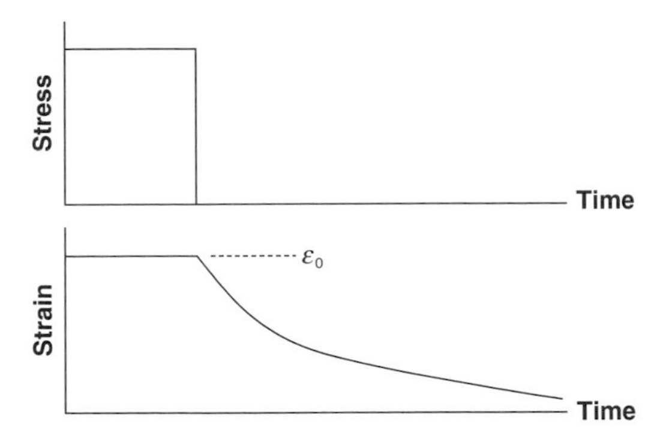

Figure 2.10 Decay of strain after removal of an applied stress.

diagram the phase angle is referred to as $\tan\delta$ (see Figure 2.9). The strain plot contains a series of responses for solids with increasing elasticity. If we take an elastic band and pull and relax it, the energy which is imparted during the extension is to some extent recovered during the contraction. However, if this process is carried out a large number of times, the rubber band is observed to heat up (see Section 7.3.4).

If the solid is simple then the response is assumed to be immediate and no energy is lost. However, if there is a time delay, as indicated by the shift on the time axis of the strain, then energy can be stored and dissipated as heat. The energy being dissipated is reflected in a decrease in the amplitude of the strain. The more elastic the material, the greater the shift along the time axis and the greater the reduction in the amplitude of the deformation produced. This type of behaviour implies a sensitivity of the material to the time involved in the application and removal of the stress.

If it is assumed that there is a time lag between the application of the strain and the response (see Figure 2.10) then the strain will not drop immediately, but will be reduced with a time constant which reflects the motion of the molecules in the solid.

The time required for a polymer to respond reflects the nature of the interactions between the polymer chains and is called the *relaxation time*, τ. The decay of the strain ε can have the form:

$$\varepsilon(t) = \varepsilon_0 \exp(-t / \tau) \tag{2.16}$$

where $\varepsilon(t)$ and ε_0 are, respectively, the value of the strain at any time t and the initial value of the strain, and τ is the characteristics relaxation time. More complete forms of Equation (2.16) are often used to better model real behaviour. If the stress is applied in the form of a sine wave then the modulus is considered to be a complex quantity. The real part of the modulus is essentially the static component and the complex component is the energy dissipation. The stress will have the form: $\varepsilon(t) = \varepsilon(0)\sin(\omega t)$ and the strain will vary as $\sigma(t) = \sigma(0)\sin(\omega t + \delta)$, where ω is the angular frequency and δ the phase lag.

The periodic stress can be written in terms of its components in phase with, and 90° out of phase with, the periodic strain:

$$\sigma(t) = \sigma(0)\sin(\omega t)\cos(\delta) + \sigma(0)\cos(\omega t)\sin(\delta) \tag{2.17}$$

which in terms of a complex modulus can be represented by an Argand diagram (see Figure 2.11), where $G^* = G' + iG''$, with $G' = (\sigma(0) / \varepsilon(0))\cos(\delta)$ and $G'' = (\sigma(0) / \varepsilon(0)\sin(\delta))$ Using complex variables and separating the components we have

$$G'(\omega) = G'(0)\left\{(\omega^2\tau^2) / (1 + \omega^2\tau^2)\right\}$$

$$G''(\omega) = G(0)\left\{(\omega\tau) / (1 + \omega^2\tau^2)\right\}\tan(\delta) = 1 / (\omega\tau)$$

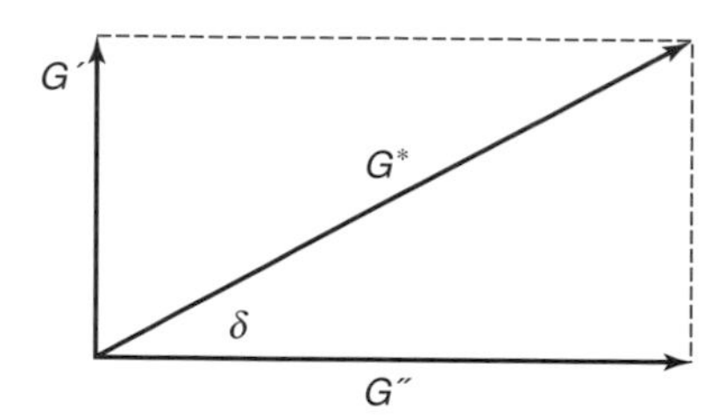

Figure 2.11 Argand diagram representation of complex modulus.

The phase shift between the applied stress and the corresponding strain can be measured experimentally and is a parameter which can be used to define when changes in a polymeric material are taking place.

2.4 Methods of measurement of mechanical properties

A variety of different methods have been devised to measure the moduli of solid polymers. Requirements for mechanical testing of polymers are specified in the relevant standards: ISO 10350–1, ISO 11403, and ISO 17282.

2.4.1 Tensile testing

The tensile test procedures are described in ISO 527–1/2 and ASTM D 638. A test sample is placed in a tensile testing machine, one end of the sample is clamped in a rigid base, B (see Figure 2.12) and the other attached to a movable cross-head which contains a load cell, A. The difference between the test methods is the recommended test speeds. For certain polymeric materials, the test speed can be important. As a general rule, the stiffer the material the lower the speed of testing; a rubber will be measured at 500 mm per minute, whereas a rigid composite may be measured at 1 mm per minute.

In the case of rubbery samples, it may be necessary to tension the sample and hold it for a period to remove the effects of hysteresis. The extension of the sample needs to be measured to

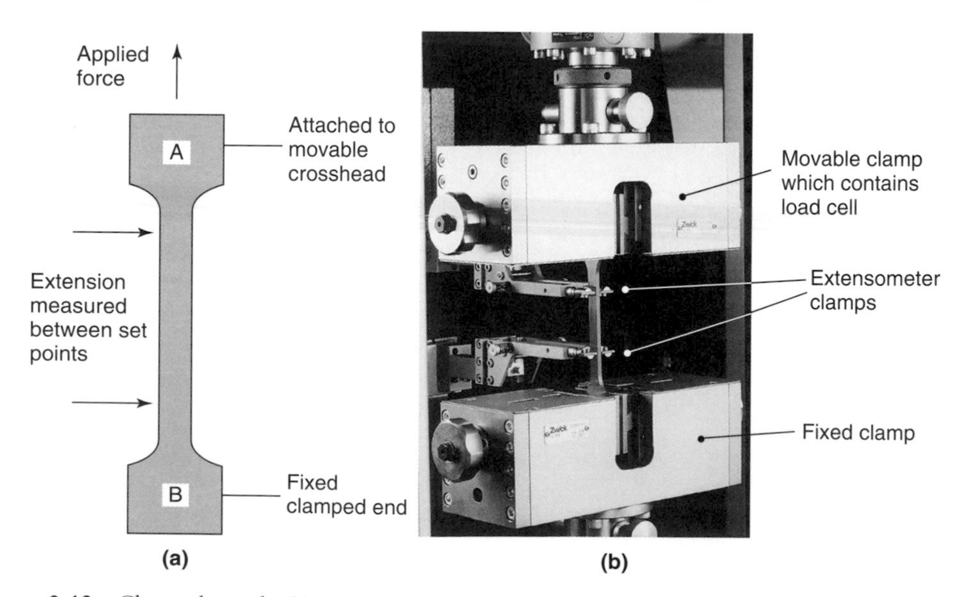

Figure 2.12 Clamped sample: (a) extensometer attached to sample in Zwick test machine; (b) diagram of sample. (Photo courtesy Zwick Testing Machines Ltd)

a high degree of accuracy; ±1 μm would be typical. The simplest tests assume that the movement of the clamps are an accurate reflection of the extension of the sample. Greater precision can be achieved using extensometers. The load cell used will depend on the material being investigated: for a rubber a 500 N capacity is appropriate, whereas for a more rigid material a 25 kN load cell may be necessary.

In the case of polymers, the samples can be deformed in the clamps and may slip, making the assumption invalid. A dumbbell is usually used as the test piece as this shape avoids problems with nonuniformity of the load distribution due to slight misalignment of the sample in the clamps. It is important to ensure that there are no visible defects in the thinner section of the sample as this will be the region where failure will occur. Defects can cause premature failure and unreliable test results.

2.4.2 Clamping of samples

The method used to clamp the sample is critical for obtaining reliable data and a number of approaches exist (see Figure 2.13). For tensile measurements, the simplest configuration is a bar, but there can be problems with the clamping, so it is not usually used if accurate measurements are required.

A dumbbell shape is the preferred configuration but the quality of the data can still depend on the clamping arrangement. The clamps are attached to the broad section of the dumbbell (see Figure 2.13(a)–(c)). To avoid slippage with a very difficult sample, holes may be drilled through the dumbbells and a pin introduced (see Figure 2.13(c)). This approach is only valid if the material is very stiff and the distortion around the hole can be neglected. Metal plates can be attached to provide a firm surface for clamping (see Figure 2.13(d)) but the adhesive used has to be stronger than the polymer being tested. In the case of composite materials this can be a problem. The principal problem with clamping is that when the load is applied to the plastic, contraction naturally occurs, reducing the force between the clamps and the material and thus allowing slippage. For slippage to be avoided the clamping pressure has then to be increased and clamps are designed to achieve this effect (see Figure 2.14). The gap between the jaws is reduced in proportion to the increasing load.

2.4.3 Problems with sample clamping

If the pressure of the grips on the specimen is not uniform along the clamped length, extension of the sample within the grips may take place and introduce an error in the data. Because the grip penetration is proportional to the load, the load–elongation curve may remain smooth and apparently be normal. The extent to which grip penetration has occurred can be determined by plotting the elongation against gauge length for a given applied force (see Figure 2.15).

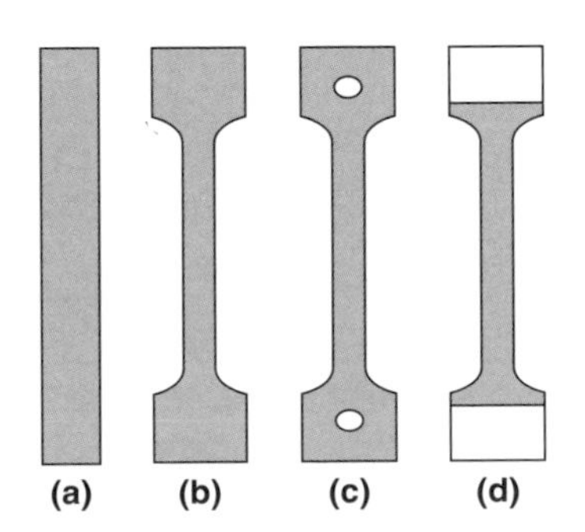

Figure 2.13 Test piece configurations for tensile testing: (a) simple bar; (b) dumbbell shape; (c) dumbbell with holes; (d) dumbbell with attached metal plates.

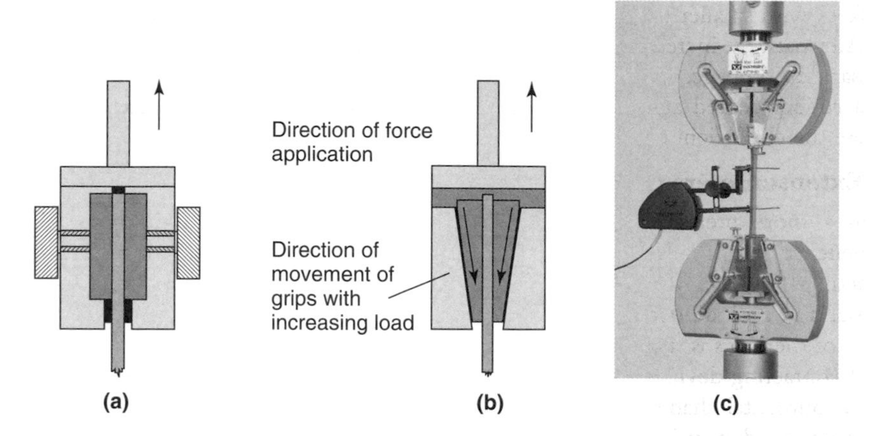

Figure 2.14 Clamps: (a) conventional clamp; (b) wedge-type clamp; (c) Instron wedge clamp. (Photo courtesy Instron, UK)

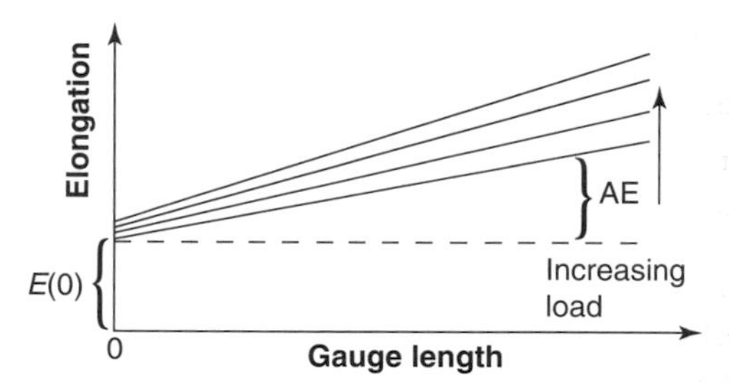

Figure 2.15 Elongation against gauge length for increasing load.

If the resulting line, when extrapolated to zero gauge length, does not pass through the origin, but gives a positive intercept on the elongation axis, then grip penetration may have occurred. These tests must always be carried out at the same strain rate since polymers are often sensitive to the strain rate, i.e. the longer the gauge length, the faster the required cross-head speed. The intercept gives a value $E(0)$ representing elongation within the grip at a specified load. The quantity AE represents the true elongation for corresponding gauge length or grip separation. When calculating elongation from a load–strain graph, the value $E(0)$ should be subtracted from the total elongation before dividing this value by the gauge length or grip separation.

To avoid slippage, the contact area between face and specimen should be as large as possible. To achieve a *good* break, both the load to slip and the load to jaw break should be higher than the value to break. The load required for slippage to occur is defined by:

$$\begin{aligned}\text{load to slip} &= (\text{friction force}) \times (\text{force applied to the faces})\\ \text{force applied to the faces} &= (\text{pressure})/(\text{contact area})\\ \text{friction force} &= (\text{friction coefficient}) \times (\text{contact area})\end{aligned}$$

Jaw breaks are caused by forces concentrated at points of contact in the jaws and are usually related to the use of serrated surfaces to increase the grip.

Several approaches can be used to eliminate specimen slippage:

- Use pneumatic or hydraulic grips. It is then possible to adjust the pressure on the specimen during the test procedure and avoid problems associated with purely mechanical clamping.

- Use faces with smaller contact area to concentrate the applied pressure over a smaller surface area. Alternatively, increase the area over which friction can act, thereby increasing the load necessary for slippage.
- Use a rubber-coated and serrated face on the clamps. Different combinations of faces can increase the coefficient of friction.

2.4.4 Extensometers and measurement of strain

To obtain a more precise measure of the extension, *extensometers* are used to measure the displacements in the sample under test (see Figure 2.12). Extensometers may be contacting or noncontacting. The optical extensometer allows the most precise measurements and comprises a small laser and a diode detector. Small metal spots are attached to the sample and the relative movement of the spots is tracked. Noncontacting optical extensometers offer many benefits over traditional contacting devices. They do not suffer problems with knife-edge slip and can be used within environmental chambers over a wider temperature range. With elastomeric materials, the extensions observed may be many hundred per cent and precise measurements are critical in order to obtain accurate data. Using high resolution digital cameras and advanced real-time image processing, it is possible to make very precise strain measurements. The imaging technology uses a set of grid lines on the surface of the sample and tracks their changes with time.

2.4.5 Properties measured

Mechanical properties are usually obtained from stress–strain curves (see Figure 2.16). The ultimate strength is defined as the force that is required to break the sample. Other important parameters are the strain to failure (break point) and the energy required to break the sample, measured as the area under the curve. The shape of the curve is a function of the rate at which the measurements are performed and temperature. The Young's modulus is obtained from the initial linear slope of the stress–strain curve. If the curve is no longer linear, Hooke's law no longer applies and permanent deformation occurs in the specimen and the 'elastic limit' has been exceeded. Beyond the elastic limit, the material responds plastically to further increases in load or stress and does not return to its original dimensions when the load is removed. The *yield strength* corresponds to the point at which plastic deformation starts to occur.

For some plastics, the departure from the linear elastic region cannot be easily identified. In these cases, an offset method discussed in ASTM D 638 is used. An offset is specified as a percentage of strain used. The *yield stress*, R, is determined from the intersection of a line shown dotted in Figure 2.16

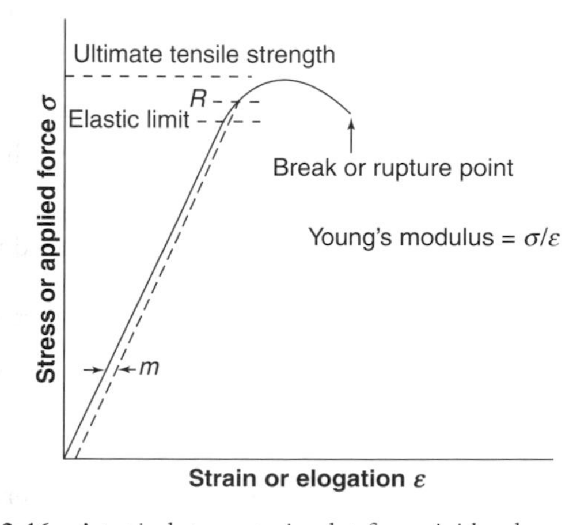

Figure 2.16 A typical stress–strain plot for a rigid polymer material.

parallel to the original line and with a slope displaced by a strain, m, which for a plastic would be typically 2%. Polyurethanes often exhibit hysteresis and require this approach to be applied.

2.4.6 Strain

The elongation, or strain, can be expressed as an absolute or relative measurement of the change in length. Strain can be expressed in two different ways. *Engineering strain* is the ratio of the change in length L to the original length L_0: $\varepsilon = (L - L_0 / L_0) = \Delta L / L$, whereas the *true* strain is the instantaneous length, L_i, during the test: $\varepsilon = \ln(L_i / L_0)$.

2.4.7 Ultimate tensile strength

The ultimate tensile strength (UTS) is the maximum load sustained during the test and may or may not equate to the strength at which the material breaks, depending on whether the material is brittle, ductile or has rubbery behaviour.

2.4.8 Flexural modulus

As indicated above, one of the methods used for determination of the modulus of a material involves bending a beam of the material (see Figure 2.17).

The three-point flexural test is specified in ISO 179 and ASTM D 790. A sample supported at its ends is subjected to deformation at its centre using a load cell. The deflections are usually four times higher than those observed in tensile tests and the crosshead movement is usually sufficiently precise to allow accurate measurement.

2.4.9 Shear modulus

The shear modulus for a solid can be measured using a conventional tensile test machine, but with samples which have been constructed to have a sandwich structure (see Figure 2.18). The material to be measured is bonded between two thick, rigid metal plates with dimensions such that when a force is applied to B there is no bending in the metal plates. The material has to be sufficiently

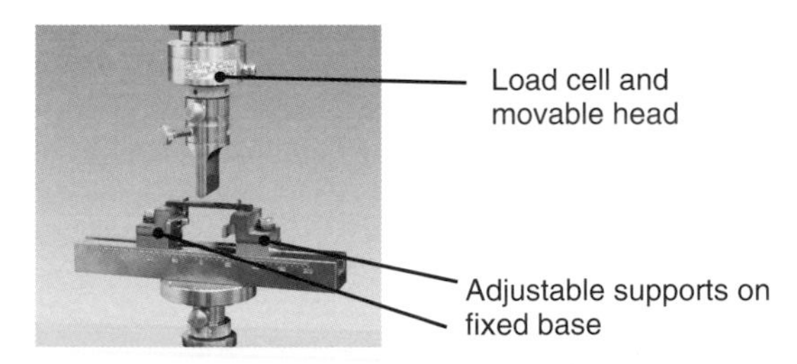

Figure 2.17 Typical configuration for measurement of bending modulus. (Photo courtesy Zwick Testing Machines Ltd)

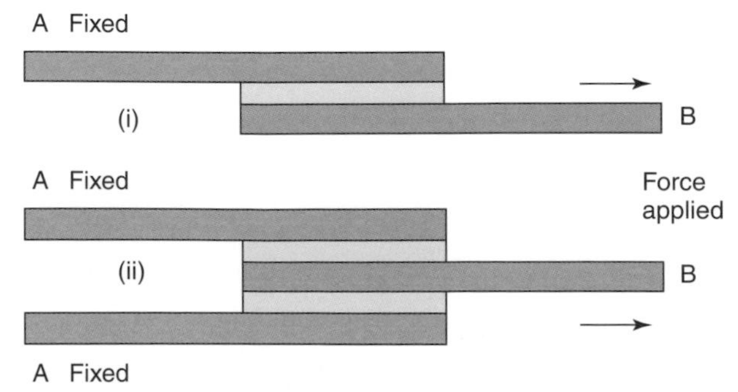

Figure 2.18 Two configurations used to measure shear modulus.

thick that it is subjected to a simple shear force. In configuration (i) there is the possibility of the sample being subjected to bending rather than simple shear. Configuration (ii) clearly avoids the bending problem but the sample is more complicated to create. The principle challenge with this test is to ensure that the bond between the material and the rigid substrate is sufficiently strong that it is not the limiting factor in the test.

2.4.10 Pendulum impact tests

Impact is a very important factor in governing the life of a structure. In the case of aircraft, impact damage can result from birds or debris hitting the plane during taking off or landing. Moulded-in stresses, polymer orientation, weak spots (e.g. weld lines), and geometry will affect impact performance. Impact properties change when additives, such as colouring agents, are added to plastics. Measurements usually involve a pendulum, held at a specific height, being released and hitting a sample, which breaks. The potential energy released by the swinging pendulum is absorbed by the sample and determines its impact strength. The two commonly used methods are the Izod and Charpy tests. The Izod test differs from the Charpy test in that the sample is held in a cantilevered beam configuration as opposed to a three-point bending configuration (see Figure 2.19). Both methods involve impacting a specimen of defined dimensions with a pendulum of well-defined mass and momentum and observing the height to which the pendulum hammer rises after impacting the test piece. The corresponding test methods are: Charpy impact tests (ISO 179–1, ASTM D 6110); Izod impact tests (ISO 180, ASTM D 256, ASTM D 4508) and unnotched cantilever beam impact (ASTM D 4812); tensile-impact tests (ISO 8256 and ASTM D 1822).

The Izod test method, according to ASTM D 256, always uses notched specimens (see Figure 2.19(a)). The ASTM 4812 describes the use of the 'Unnotched cantilever beam impact' method as in ASTM D 4812, which is similar to the Izod method, but specifies the use of an unnotched specimen. If only small specimens are available then the chip-impact test, ASTM D 4508, applies. According to ISO 179–1 the single-point data Charpy test (ISO 10350–1) is the preferred method (see Figure 2.19(b)). The test is typically carried out with an unnotched specimen and edgewise impact. The pendulum is raised to progressively increasing heights until failure occurs. If the specimen does not break then the test is repeated using a notched specimen. Notching of the sample to conform to the specification of the test method is often a source of great variability. The Charpy method has a larger range of applications and is better suited for testing materials exhibiting interlaminar shear fracture or surface effects. The Charpy method has the advantage that for low temperature tests the increased distance from the supports to the tests area reduces the effects of heat transfer. The best results are obtained when the pendulum is operated using 10–80% of the nominal energy deployed. It is best to use the highest energy pendulum available in order to minimise effects due to the deceleration of the pendulum during impact.

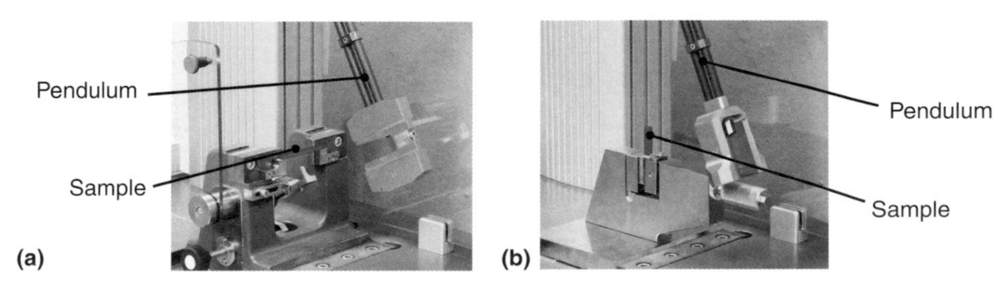

Figure 2.19 Configurations used for pendulum impact testing. (Photos courtesy Zwick Testing Machines Ltd)

2.4.11 Drop tests

As the name implies, a falling object impacts a clamped disc of the material (see Figure 2.20). The form of the falling object reflects the particular test method and is usually a ball or a loaded needle. The test involves dropping a weight in a vertical direction, with a tube or rails to guide it during the *free fall*. The potential energy is determined by the height and weight of the falling bob. Since the falling weight either stops dead at the test specimen, or destroys it completely, the test is usually a pass or fail.

The falling weight method can be used on moulded samples and is unidirectional with no preferred direction of failure. Failure originates from the weakest point in the sample. Samples exhibiting excessive deformation or delamination can be considered to have failed and do not have to break. A composite may suffer a high level of internal damage but exhibit little or no external damage.

2.4.12 Instrumented methods

Instrumented drop weight and pendulum testing is possible by monitoring the falling weight or pendulum with electronic sensors, allowing the load applied to be continuously monitored as a function of time and/or specimen deflection prior to fracture. Remote sensing using digital photography, laser tracking and simple measurements of the impact forces by attaching strain gauges to the sample improve the quality of data recovered. A complete picture of the failure process may require the use of different rates, heights and weights of impacting load. To make the tests realistic it is also useful to select the shape of the impact probe to mimic the type of impact incident which the material will commonly experience. The damage created by a needle-shaped load will be very different form that from a ball or cylindrical weight. Smooth rocks travelling down a pipe will create a very different impact from needle-shaped materials. Similarly a spanner falling on a composite aircraft wing will have a different effect from a bird impacting the same structure at high speed. It is often difficult to quantify the effects of the impact of falling or propelled objects, especially for composite structures where damage may occur within the mass of the material and not necessarily at the surface. The tests usually involve measurements at various rates and reflect different characteristics of the materials (see Figure 2.21).

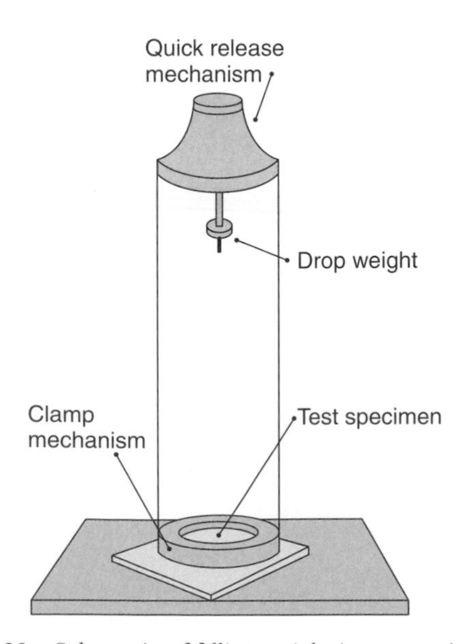

Figure 2.20 Schematic of falling weight impact testing machine.

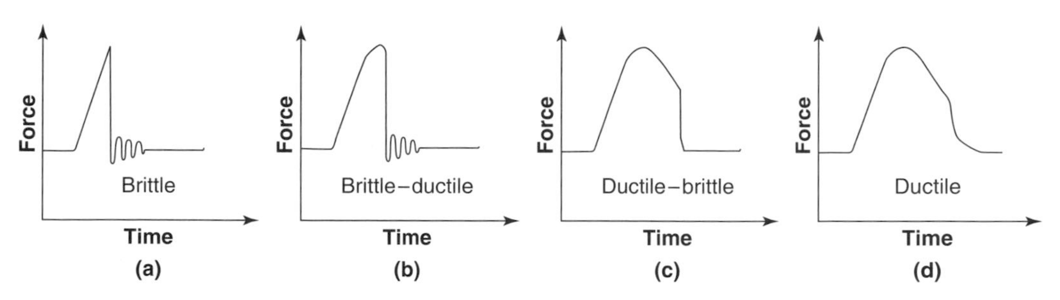

Figure 2.21 Typical force–time plots for different types of failure: (a) brittle; (b) brittle–ductile; (c) ductile–brittle; (d) ductile.

2.4.13 Ductile vs brittle fracture

In reality, most impacts are biaxial rather than unidirectional and are further complicated by the failure having varying degrees of brittle or ductile failure. Brittle materials, like untoughened window glass, require little energy to start to crack and little more to propagate to a shattering climax. Other materials possess ductility to varying degrees. Highly ductile materials fail by puncture in drop weight testing and require a high energy load to initiate and propagate the crack. Many materials are capable of either ductile or brittle failure, depending on the conditions used, rate and temperature, and can exhibit a ductile/brittle transition. The area under the force–time curve gives the energy required to achieve failure, the *impact energy*. Another parameter which is measured is the *impact velocity/strain rate*. In some materials, dropping a 5 kg weight from 1 m produces a very different result compared with that of dropping a 1 kg weight from 5 m, the material exhibiting strain rate sensitivity. Every material will behave differently depending on the geometry of the striker/bob (free-falling load), how the specimen is clamped, and the geometry of the clamping.

2.4.14 Creep properties

A common characteristic of plastic materials is that they can undergo *creep*, a slow deformation when held under constant load. Creep is temperature dependent and increases in rate as the melting or softening point is approached. It is measured by observing the extent of deformation, usually elongation of a dumbbell sample with time (see Figure 2.22).

Measurements involve either direct observation of the extension using contacting or non-contacting extensometers or the use of strain gauges attached to the specimen. The creep process can be divided into three stages. The first stage, or primary creep, starts at a rapid rate and then slows down with time. The second stage, secondary creep, occurs with a relatively

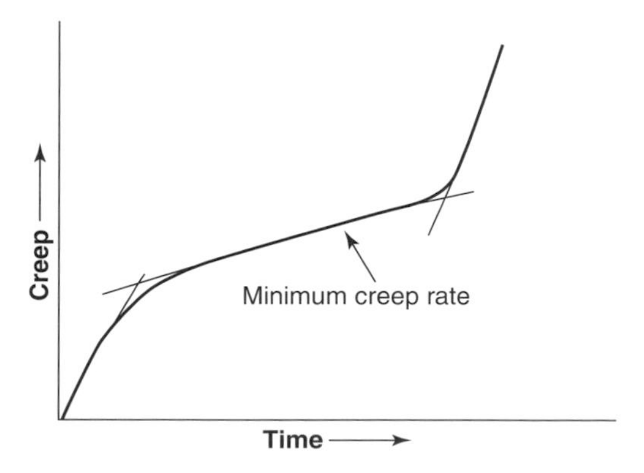

Figure 2.22 Creep–time plots for typical polymer material.

uniform rate. The third stage (tertiary) creep has an accelerating creep rate and is terminated by the failure of the material. If the specimen does not fail within the test period then creep recovery can be measured. Unloading the material allows it to recover. The difference between the original and recovered dimensions is known as the *permanent set*.

2.4.15 How to determine stress relaxation?

Stress relaxation is connected to creep and is measured by determining the rate at which the stress decreases with time for a specimen which has been deformed to a given amount (see Figure 2.23). The stress relaxation rate is the slope of the curve at any point. Stress relaxation is a consequence of the time-dependent characteristics of the polymer and can be associated with molecular conformational changes.

2.4.16 Fatigue testing

If a polymer is subjected to repeated loading and unloading, usually in a cyclic manner, it will ultimately fail. This process may require the cycle to be repeated millions of times and for several hundred times per second. This test simulates the conditions experienced when materials are subjected to vibration or oscillations and materials can behave differently under dynamic or static loads. Because the material is subjected to repeated load cycles in actual use, designers are often faced with predicting fatigue life, which is defined as the total number of cycles to failure under specified loading conditions.

2.4.17 Compression testing

The compression strength determines the behaviour of materials under a crushing load. Compression may be considered to be the reverse of extension, but due to Poisson's ratio it will cause an extension in the plane perpendicular to the applied force (see Figure 2.24). However, for isotropic behaviour to be retained, the area making contact with the plates must be increased. This condition requires that the faces through which the force is being applied must allow the material to slip freely in order to ensure that the perpendicular forces are equilibrated (see Figure 2.24).

If the surface becomes pinned then the perpendicular surfaces will become distorted as a result of shear forces created within the sample. Deviations of the stress–strain plots from linearity will reveal the effects of the anisotropy of the deformation. The *elastic limit, yield point and yield strength,*

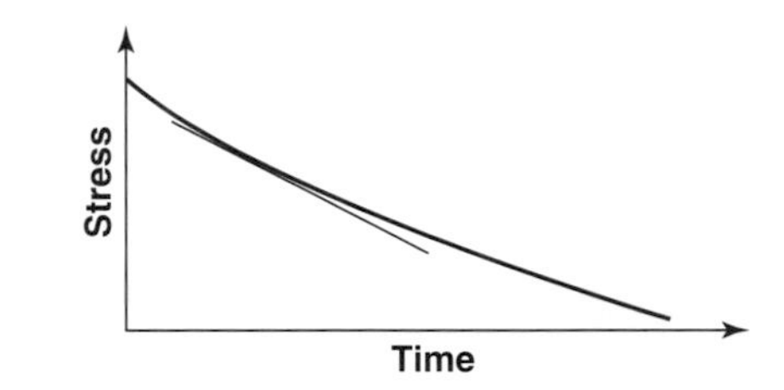

Figure 2.23 Stress relaxation plot for a typical polymer.

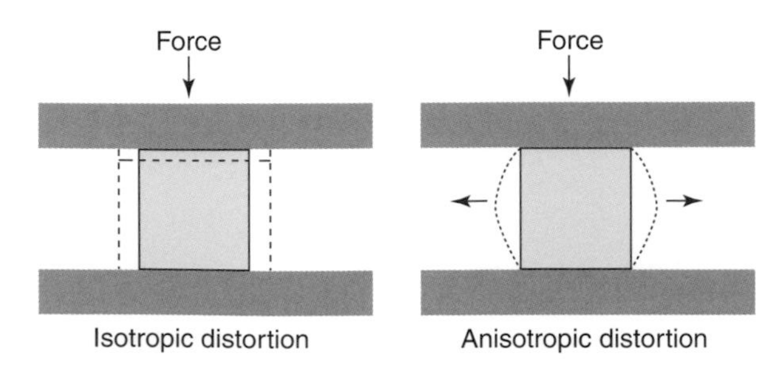

Figure 2.24 Compression testing.

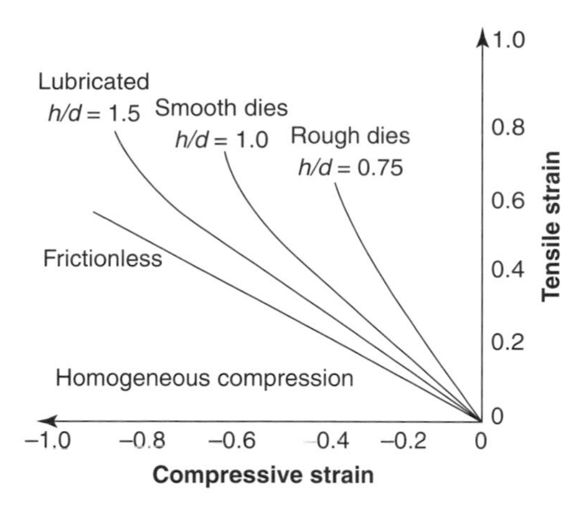

Figure 2.25 Variation of compressive strain against tensile strain for various sample plate interactions.

and *compressive strength* can be calculated from the *stress–strain* diagram. The *elastic limit* is defined as the greatest stress that can be applied to a material without causing permanent deformation.

The *yield point* is the stress at which strain increases without accompanying increase in stress. There are a few polymers which exhibit this type of behaviour. The *compressive strength* is the maximum stress which a material can sustain under crush loading. The compressive strength of a material that fails by shattering fracture can be defined within fairly narrow limits as an independent property. However, the compressive strength of a material that does not shatter in compression must be defined as the amount of stress required to distort the material by an arbitrary amount. Compressive strength is calculated by dividing the maximum load by the original cross-sectional area of a specimen in a compression test.

When performing compression tests, samples with a large height (h) over diameter (d) ratio should be avoided as these will be subject to high levels of shear distortion. The effects of surface roughness can be easily recognised in the stress–strain plots (see Figure 2.25). Axial compression testing is used to measure the plastic flow behaviour and ductile fracture limits of a material. Measuring the plastic flow behaviour requires frictionless (homogeneous) compression. Measurement of the ductile fracture limits takes advantage of the barrel formation and controlled stress and strain conditions at the equator of the barrelled surface when compression is carried out with friction. Axial compression testing is useful for measuring the elastic and compressive fracture properties of brittle or low ductility materials. The use of specimens having large h/d ratios should be avoided to prevent buckling and shearing modes of deformation.

2.4.18 Modes of deformation in compression testing

Figure 2.26 illustrates the modes of deformation in compression testing: (a) buckling, when $h/d > 5$; (b) shearing, when $h/d > 2.5$; (c) double barrelling, when $h/d > 2.0$ and friction is present at the contact

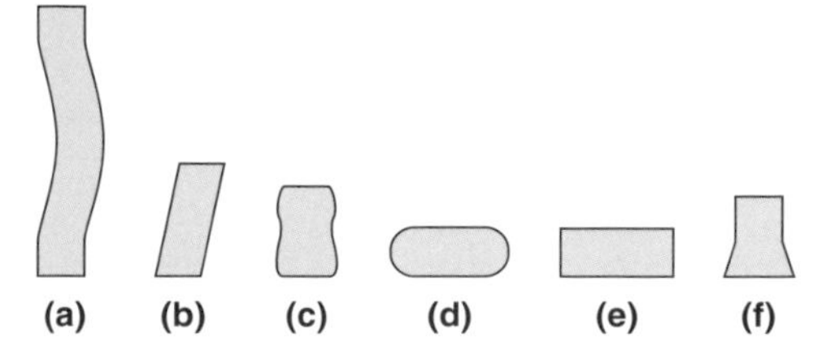

Figure 2.26 Modes of deformation of various shapes under compression.

surfaces; (d) barrelling, when $h/d < 2.0$ and friction is present at the contact surfaces; (e) homogeneous compression, when $h/d < 2.0$ and no friction is present at the contact surfaces; (f) compressive instability due to work-softening material. The most ideal shape for the measurement of the compression strength of materials is (e). For this ratio of h/d the deformation is close to the ideal situation.

2.4.19 Other mechanical property measurements in polymer systems

Engineers are used to measuring the modulus from stress–strain curves, however, this usually provides data at a single temperature. The modulus measured at room temperature, together with the bending modulus and ultimate elongation to breaking, is usually adequate to determine whether or not a selected material is fit for purpose. However, this information is usually inadequate for a polymer system, where the temperature dependence of the physical properties can be an important factor in selecting the correct material. Whilst data sheets will often quote a value of the tensile and flexural modulus, what is often required is data on the temperature dependence of these parameters. The *work range* for a plastic material will often need to be specified in terms of retaining the values of these parameters over the temperature range in which they will be used. For instance, in the automobile industry a typical range will be 20–40°C, this being the range which cars could experience across the world during a normal year. However, under the bonnet the working range may be −20°C to +120°C. Temperature dependence of the modulus data is available by using a technique known as *dynamic mechanical thermal analysis* (DMTA).

2.4.20 Dynamic mechanical thermal analysis

The test involves subjecting a small bar of material to a cyclic stress as discussed in Section 2.3 and varying the temperature. The DMTA apparatus consists of a device which is capable of applying a sinusoidal perturbation to the sample and a means of measuring the displacement produced. The force is applied via a force transducer of sufficient strength that it can induce a constant vibration in materials. The oscillation is maintained constant and independent of the temperature and can be represented by the displacement A in Figure 2.27.

The sample to be investigated is clamped to the force transducer and will be placed in tension. The force transducer is excited by an oscillatory frequency which is transferred to the clamped sample. The point A will then undergo oscillatory motion as shown in the top right-hand diagram of Figure 2.27. A transducer monitors the response of the sample to the forced oscillation (B). If the material is rigid then the displacements A and B will have similar amplitudes and will

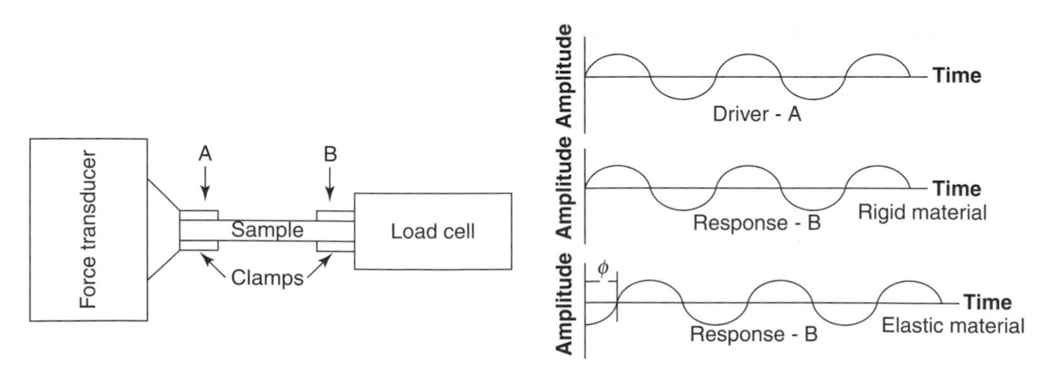

Figure 2.27 Schematic of dynamic mechanical thermal analysis apparatus working in tension. Typical responses for a rigid and flexible material are shown as amplitude of oscillation vs. function of time: Top is driver response; middle is for a rigid material; bottom is for a flexible rubber.

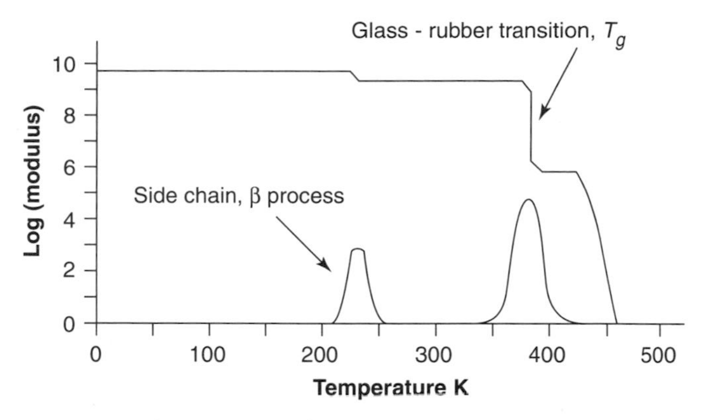

Figure 2.28 Typical DMTA trace for an amorphous polymer such as polystyrene.

follow the same oscillatory cycle, and will be in phase. However, if the material becomes flexible then the amplitude of the response B will be reduced in proportion to the ability of the sample to stretch and the oscillations will be phase shifted (ϕ) in proportion to the amount of energy that is stored when the material is elastically deformed (see Section 2.3). The apparatus is housed in a temperature-controlled cell that can typically be cooled to liquid nitrogen temperatures and heated to about 400°C.

Knowing the dimensions of the sample and appropriate calibration factors, it is possible to determine the temperature dependence of the appropriate stiffness modulus. The loss modulus can be calculated from the phase shift, ϕ. The loss modulus is a measure of the energy absorbed by the sample and dissipated as heat and is obtained from $\tan\phi$. An idealised DMTA trace for a polymer, such as polystyrene, is shown in Figure 2.28.

At low temperature, a peak in the $\tan\phi$ trace can be observed which is associated with the onset of motion of the pendant phenyl group. The temperature at which this motion occurs is controlled by the steric hindrance of neighbouring elements of the chain (see Figure 2.29).

This process is known as a *subglass transition* process. The drop in modulus from a value of ~10^9 N m^{-2} ~10^6 N m^{-2} reflects the onset of the motion of the chain backbone and is designated T_g, the *glass* transition point (also known as the glass–rubber transition point). The T_g is associated with a cooperative motion of the polymer backbone and resembles the motion of a crank shaft. The collective chain motion requires there to be volume available for it to occur and is also marked by a significant drop in the modulus. The modulus of a glass has a value of typically 10^9–10^{10} N m^{-2}, whereas that of a rubber is typically of the order of 10^6 N m^{-2}. Above the T_g, the modulus may appear to be constant for a temperature interval before dropping to values which are typical of viscous and free-flowing liquids. For the processing of a thermoplastic it is usual for the temperature to be raised to a value at which the viscosity is of the order of 100 Pa s.

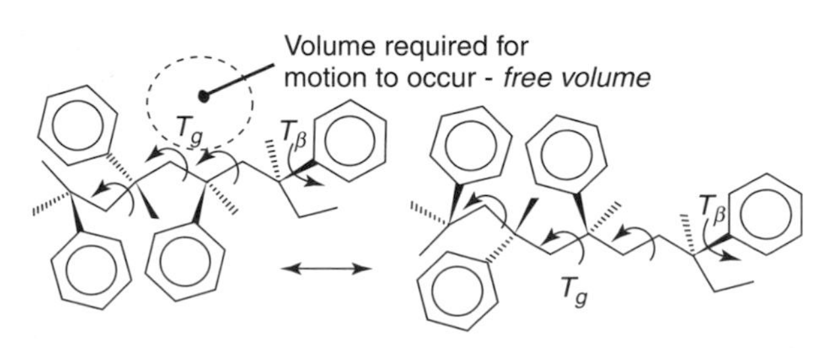

Figure 2.29 Motions of chain corresponding to side chain motion (β-process) and backbone motion (T_g).

2.5 Thermal expansion coefficient measurements

Some of the first observations of the T_g for amorphous polymers were made from measurements of the expansion coefficient of the solid. Nearly all materials will exhibit a linear expansion coefficient; this is also true of polymers in the glassy (amorphous) state (see Figure 2.30).

At the T_g, the expansion coefficient of the material changes and a higher value is observed. The change in the expansion coefficient reflects an increase in the volume occupied by the polymer and is associated with a greater separation between individual polymer chains (see Figure 2.29). The volume created by increasing the chain separation allows the elements of the backbone of the polymer chain to start to rotate as shown in Figures 2.29 and 2.31. The moving element of the polymer chain is shown in bold in Figure 2.31(b). The circle indicates the size of the volume which is necessary for the chain to undergo the segmental motion.

The motion of the chain backbone is not controlled by the chain acquiring thermal energy to overcome a potential energy barrier to rotation, as is the case in the motion of a side chain element, but is associated with there being sufficient volume for the backbone motion to occur. The *free volume* is a very important aspect of the process associated with the T_g and makes the process pressure dependent.

It is very important for engineers to understand that T_g is pressure dependent as it is possible that a material which is a rubber at ambient pressure, and so may have been used as a seal, may become

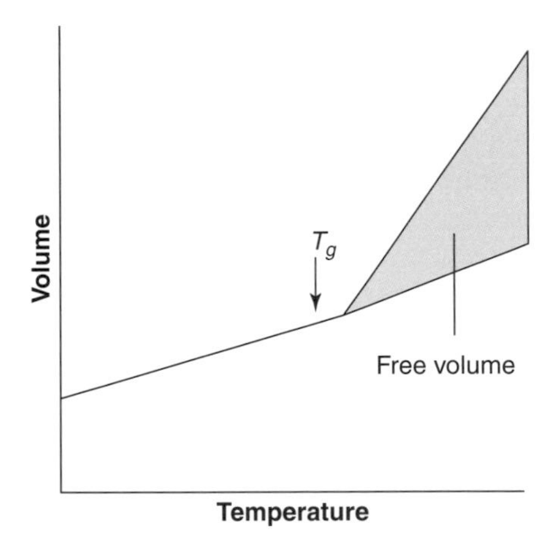

Figure 2.30 Variation in volume of polymer with temperature.

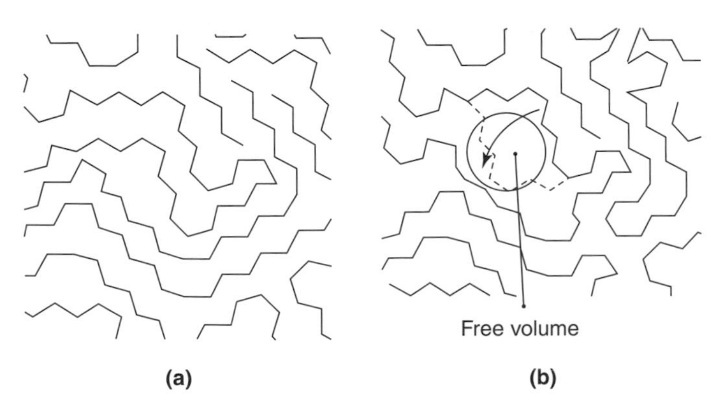

Figure 2.31 Schematic of packing of polymer chains: (a) below T_g; (b) above T_g. Conformational change indicated by arrow requires the volume shown as a circle. New position of chain is indicated by dotted line.

a glassy solid when pressurised. The application of pressure raises the T_g to above ambient and so the material will cease to act as a sealant. Such a problem was associated with the *Challenger* space shuttle disaster on 28 January 1986, when the failure of seals associated with the fuel tanks caused an explosion. The *Challenger* NASA space shuttle broke apart 73 s into its flight, leading to the deaths of its seven crew members. Disintegration of the shuttle stack began after an O-ring seal in its right solid rocket booster (SRB) failed at liftoff. The seal failure caused a breach in the SRB joint, allowing a flare to reach the outside and impinge upon an external fuel tank. The SRB breach flare led to the separation of the right-hand SRB and the structural failure of the external tank. Aerodynamic forces promptly broke up the orbiter. The designers had known that the SRBs contained a potentially catastrophic flaw in the O-rings since 1977, but they failed to address the problem properly and ignored warnings from engineers about the dangers of launching on such a cold day. The sad result was a major loss of life.

Increasing the pressure will reduce the free volume and as a consequence increase the value of T_g, which increases linearly with increasing pressure:

$$T_g(P) = T_g(0) + sP \tag{2.18}$$

where $T_g(P)$ is the glass transition temperature at some value of the pressure P and $T_g(0)$ is the value at atmospheric pressure. The coefficient s can be obtained from pressure–volume–temperature data, but for many polymers it is a simple linear function and has a value of 0.2 $\mathrm{K\,MPa^{-1}}$ for flexible aliphatic chains and a value of 0.55 $\mathrm{K\,MPa^{-1}}$ for semirigid aromatic chains. The effect of pressure can be important in some processing applications, such as injection moulding (see Section 5.4).

2.5.1 Molar mass dependence of glass transition point

As in the case of the melting point variation with chain length (see Section 1.10), similar effects are observed with the T_g. The chain ends will require essentially no *free volume* to move and are only restricted in their movement by the steric hindrance of neighbouring groups. Rotational motion is controlled purely by the potential energy surface describing the conformational change. However, as the chain length increases the motion has to involve more bonds moving synchronously to achieve the rotational motion and sufficient volume needs to be available to accommodate the motion that arises. The free volume restriction changes with molar mass and the proportion of the main chain to the ends in a polymer. Most polymers show a similar variation of T_g with molar mass which can be described by:

$$T_g(M_n) = T_g^{\infty} - K/M_n \quad K \sim 10^5 \tag{2.19}$$

where K is a constant characteristic of a particular polymer system but generally has a value of the order of 10^5, and T_g^{∞} is the characteristics value for that polymer type and is the asymptotic limiting high molecular weight value. A polystyrene sample with a molar mass of ~600 will be a liquid at room temperature, i.e. is well above its T_g. At about 2000 molar mass the polystyrene is a solid, but has a T_g of ~30°C. It is only once the molar mass is above 100,000 that a constant value of ~100°C is observed. Data on the glass transition temperatures of a variety of different polymers are listed in Brandrup *et al.* (1999).

2.5.2 Influence of chemical structure on the glass transition point

Change in the chemical structure will influence the value of the T_g^{∞} and reflects the contribution to the energy requirement dictated by the potential energy surface associated with the bond rotation (see Section 1.6.1). Summaries of T_g values for some common polymers are presented in Tables 2.1 and 2.2.

Polymethylmethacrylate and polystyrene, although chemically very different, have T_g values of ~100°C. Polycarbonate has a value of ~160°C, polyethylenetetraphthalate a value of ~75°C and polydimethylsiloxane (PDMS) a value of about −120°C. The increase in the bond length in changing from

Table 2.1 Glass transition temperatures for a series of polyalkyacrylates

Polymer	Structure	Glass–rubber transition (°C)
Polymethylacrylate	$-[CH_2-CH(COOCH_3)]_n-$	10
Polymethylacrylate	Head-to-tail	5
	Head-to-head	31
Polyethylacrylate	$-[CH_2-CH(COOC_2H_5)]_n-$	−24
Polypropylacrylate	$-[CH_2-CH(COOC_3H_7)]_n-$	−37
Polybutylacrylate	$-[CH_2-CH(COOC_4H_9)]_n-$	−54
Polypentylacrylate	$-[CH_2-CH(COOC_5H_{11})]_n-$	−57
Polyhexylacrylate	$-[CH_2-CH(COOC_6H_{13})]_n-$	−57
Polyheptylacrylate	$-[CH_2-CH(COOC_7H_{15})]_n-$	−60
Polyoctylacrylate	$-[CH_2-CH(COOC_8H_{17})]_n-$	−65

Table 2.1 (*Continued*)

Polymer	Structure	Glass–rubber transition (°C)
Polynonylacrylate	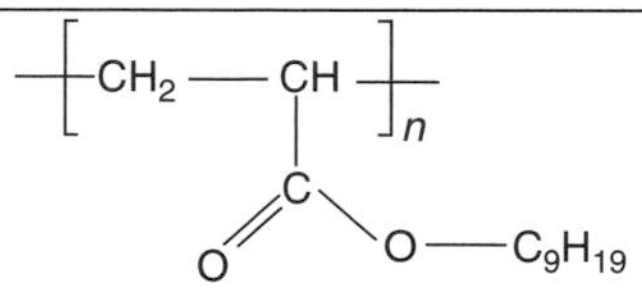 $-[CH_2-CH(-C(=O)-O-C_9H_{19})]_n-$	−58
Polydodecylacrylate	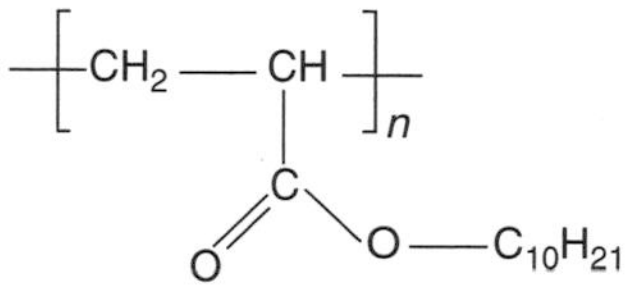 $-[CH_2-CH(-C(=O)-O-C_{10}H_{21})]_n-$	−3

Table 2.2 Variation of glass–rubber transition point T_g with chemical structure

Polymer	Structure	Glass–rubber transition (°C)
Polymethylmethacrylate	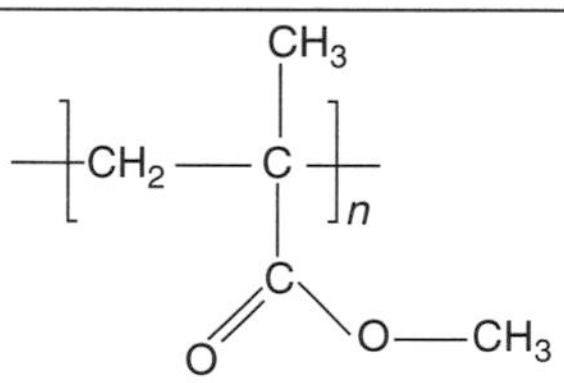 $-[CH_2-C(CH_3)(-C(=O)-O-CH_3)]_n-$	~100
Polyethylene (depends on sample, amorphous phase)	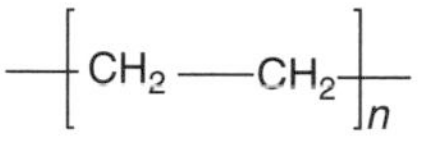 $-[CH_2-CH_2]_n-$	−78
Polypropylene	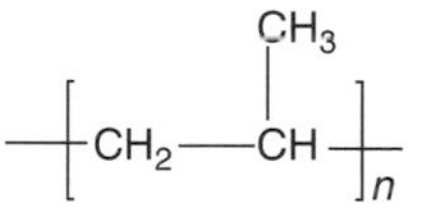 $-[CH_2-CH(CH_3)]_n-$	−10
Polyvinylchloride	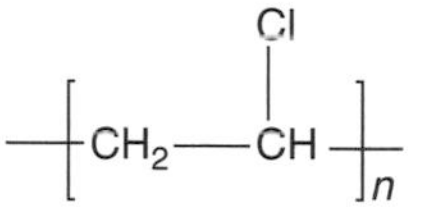 $-[CH_2-CH(Cl)]_n-$	86
Polyacrylonitrile	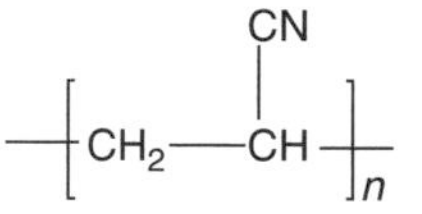 $-[CH_2-CH(CN)]_n-$	125
Polyvinylidenechloride	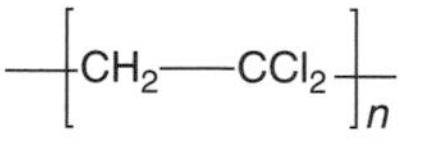 $-[CH_2-CCl_2]_n-$	
Polystyrene	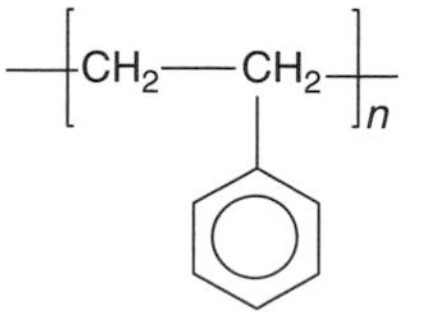 $-[CH_2-CH_2(C_6H_5)]_n-$	100

(*continued*)

Table 2.2 (*Continued*)

Polymer	Structure	Glass–rubber transition (°C)
Poly(α-methylstyrene)	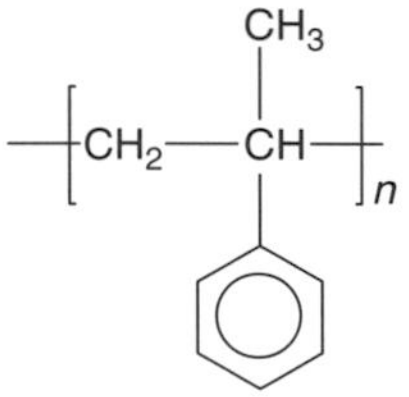 CH_3 $-[CH_2-CH]_n-$	105
Poly(2-methylstyrene)	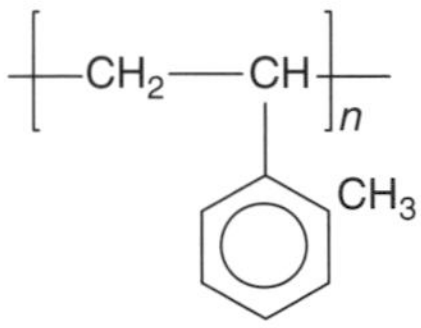 $-[CH_2-CH]_n-$ CH_3	136
Poly(3-methylstyrene)	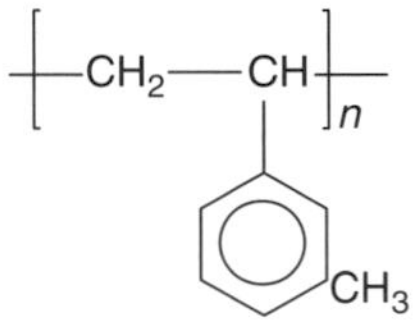 $-[CH_2-CH]_n-$ CH_3	97
Poly(4-methylstyrene)	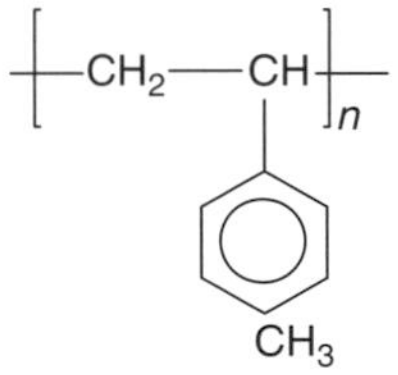 $-[CH_2-CH]_n-$ CH_3	109

a C–C bond to a Si–O bond significantly reduces the interaction of neighbouring groups and there is a concomitant reduction in the value of T_g. Polymethylacrylate has a T_g of ~10°C. If the polymerisation is carefully controlled and all the sequences are head-to-tail then the T_g is 5°C. However, if the structure is created so that the bulky groups go head to head then the value of the T_g is raised to 31°C. The conventional polymer is a mixture of head-to-tail and head-to-head structures and has a T_g of ~10°C. Changing the methyl to an ethyl group in the pendant ester leads to a lowering of the T_g to a value of −24°C. Although the ethyl group is larger than the methyl group and hence might be expected to be a bulkier group to rotate about the backbone, it will increase the separation between neighbouring chains and it is easier to create free volume. Increasing the length of the aliphatic chain in the ester to a butyl group in polybutylacrylate lowers the T_g to −54°C. The trend in the T_g continues to fall with increasing length of the alkyl chains until it reaches a minimum value of around −60°C. The alkyl chain is nonpolar and can significantly reduce the dielectric permittivity of the material and hence the strength of the interaction between the neighbouring polar ester groups. Reduction in the strength of the interaction between chains increases the free volume and lowers the T_g.

Increasing the alkyl chain length to 12 carbon atoms increases the T_g to −3°C. The longer alkyl chains start to order and create the interactions which lead to crystal formation in polyethylene. These interactions reduce the free volume and increase the energy required for bond rotation.

The effect of restricting rotation about the polymer backbone can be seen when the H atom next to the ester group is replaced by a methyl group. Polymethylmethacrylate can occur in three different tactic forms (see Figure 1.10). The atactic form of the polymer has a T_g of ~100°C; the isotactic form has a T_g of ~50°C and the syndiotactic has a T_g of ~105°C. The methyl group is severely restricting rotation about the backbone and there is a corresponding increase in the value of the T_g. The increase in T_g of polymethylmethyacrylate compared to polymethacrylate is ~90°C. In contrast, atactic poly(*sec*-butylacrylate) has a T_g of −20°C; syndiotactic has a T_g of −21°C and isotactic has a T_g of −23°C, which shows that with the longer side chain the structure of the backbone is less important in defining the energy and volume required to achieve free segmental rotation.

Polyethylene (see Section 3.1), tends to form crystalline phases, but the disordered amorphous phase has a T_g which is typically about −60°C. The introduction of a methyl group to produce polypropylene produces an amorphous phase which has a T_g of −10°C. The T_g of polyvinylchloride (PVC) at 86°C reflects the effects of the polar C–Cl bond tightening up both the polymer backbone and reducing the free volume through increasing the interaction between neighbouring chains. The highly polarisable cyano group in polyacrylonitrile further increases the interactions with the result that the T_g has now been raised to 125°C. In general, polymers which contain halogen atoms have significantly higher values of T_g than nonpolar polymers, however if the dipoles cancel one another out, as in the case of polyvinylidenechloride, then the effects are significantly reduced. Polyvinylidenechloride has chloride atoms symmetrically distributed about the backbone with the effect that the T_g is −23°C.

Introduction of a phenyl ring structure into the polymer backbone will usually lead to an increase in the barrier (see Table 2.2). The introduction of a methyl group into the backbone leads to an increase in the value of T_g to 105°C. If the substitution takes place in the phenyl ring rather than in the backbone then different effects are observed. The substitution in the 2 or *ortho* position produces a T_g of 136°C, which is a substantial increase, reflecting the effect of the methyl group hindering the rotation about the polymer backbone. Substitution in the 3 or *meta* position does not lead to an increase in the steric hindrance to internal rotation and the T_g has a value of 97°C. Substitution in the 4 or *para* position has a positive effect in increasing the energy for rotation and a T_g of 109°C is observed.

By change of the chemical structure it is possible to change the T_g of the polymer to a value that is appropriate for an application. For instance, PDMS is very useful as a sealant below 0°C, but is not very useful where a load bearing capability is required. The useful working temperature range for a polymer will depend on the application. Often the sub T_g peaks will determine the point at which the material is too brittle to be used and the T_g peak defines the point at which the material ceases to be able to take a load. The *working* temperature is often considered to be between the sub T_g and the T_g of a material.

Within a family of similar polymers, increasing the chain stiffness and interchain cohesion increases the glass transition temperature. The tighter the polymer chains can pack together, the higher the temperature will have to be raised before sufficient free volume will be created for rotational motion to occur. Most structures of polymer molecules are idealised and do not reflect the complexities which can be introduced by small variations in the synthetic method used for their production. Structural uncertainties can arise both from multiplicity of possible chemical reactions, down stream processing and from the structures of the polymers themselves.

2.5.3 Plasticisation

Mixing a low molar mass polymer with a higher molar mass material can change the glass transition temperature. The value of T_g that is obtained is described by a simple mixing law:

$$1/T_g(mix) = \left(x_1/T_g(1)\right) + \left(x_2/T_g(2)\right) \tag{2.20}$$

where $T_g(mix)$ is the glass transition temperature of the combination of a polymer with $T_g(1)$ present as x_1 (the volume fraction of the whole) and $T_g(2)$ is the value of the second material present as a volume fraction x_2.

A surprisingly broad range of polymer materials obey this simple mixing law provided that specific addition interactions are not introduced as a consequence of making the blend. The same equation describes the effect of the addition of a small molecule or another soluble polymer. The addition of the material with the lower T_g forces the chains apart, introduces additional free volume and leads to a lowering of the T_g of the blend.

Plasticisation is used to enhance PVC. Pure PVC has a T_g of ~75°C, the precise value depends on the method of synthesis. If PVC is used to cover electrical cable it has to be flexible and addition of plasticiser allows the T_g to be lowered to ambient temperature. Similarly, if used in automobile applications for seat covering, lowering the T_g plasticisation is essential to provide comfortable seats. However, rigidity and a high value of T_g are desirable when PVC is used for window frames or ducting. The plasticiser typically used with PVC is a low molecular weight aliphatic ester. These oligomeric species are soluble in the polymer but of sufficiently high molar mass to be none volatile or to diffuse easily through the matrix. Some of the earlier low molecular weight plasticisers were volatile and diffused to the surface of the plastic, giving it a greasy appearance and a distinctive 'new' smell. Loss of the plasticiser from the surface causes shrinkage and eventually cracking and crazing.

2.5.4 Examples of T_g calculations

First, we consider the case where the glass transition temperatures of two polymers were observed to be, respectively, 100°C and 80°C. The polymers were present in the mixture in the ratio of 1:2 and were mixed to form an intimate blend; the glass transition temperature was then determined. What was the glass transition temperature?

The most common mistaken made in these calculations is to forget that temperature has to always be given in degrees absolute. If the T_g of A is 100°C then it is 373K and the T_g of B is 80°C, which is 353K. By the rule of mixtures: $1/T_g = x_1/T_{g1} + x_2/T_{g2}$, $1/T_g = (0.33/373) + (0.66/353)$, $1/T_g = 0.000{,}884{,}7 + 0.001{,}869{,}6 + 0.002{,}754{,}3$, $T_g = 363\text{K} = 90°\text{C}$. The T_g value for the blend is approximately half way between the values of the two polymers.

Secondly, we consider the case where a low molar mass plasticiser was added to a polymer which exhibited a glass transition temperature of 80°C. The plasticiser was present in the system at a level of 10% by weight and has a glass transition temperature of –20°C. What is the glass transition temperature of the plasticised polymer?

T_g of A = 80°C = 353K. T_g of B = –20°C = 253K. Applying the rule of mixtures: $1/T_g = (0.9/353) + (0.1/253)$, $1/T_g = 0.002{,}549{,}5 + 0.000{,}395{,}2 = 0.002{,}944{,}7$, $T_g = 339\text{K} = 66.5°\text{C}$.

2.5.5 Other molecular mass effects

Examination of the DMTA plots for a series of polystyrene samples with a narrow molar mass distribution and with different molar masses (see Figure 2.32) indicates interesting variation in properties around and above the T_g.

The sample with molar mass ~2000 will have a T_g value of ~50°C. Increasing the molar mass to a value of 10,000 increases the T_g to ~70°C, and a further increase to a value of about 15,000 achieves the true polymeric value of ~100°C. In the case of the 2,000, 10,000 and 15,000 molar mass polymers, above T_g, the modulus falls rapidly to a low value corresponding to a free flowing liquid (see Figure 2.32). Above 30,000 molar mass a plateau region with a modulus of ~10^6 Pa is observed. As the polymer molar mass is further increased so the temperature range over which this plateau exists is increased. The *ceiling temperature* designates the temperature at which the polymer starts to unzip and degrade lowering the molar mass and leading to a loss in the physical properties of the article that is being formed. The ceiling temperature is usually considered as an upper guide to the temperature at which processing can be performed and assumes

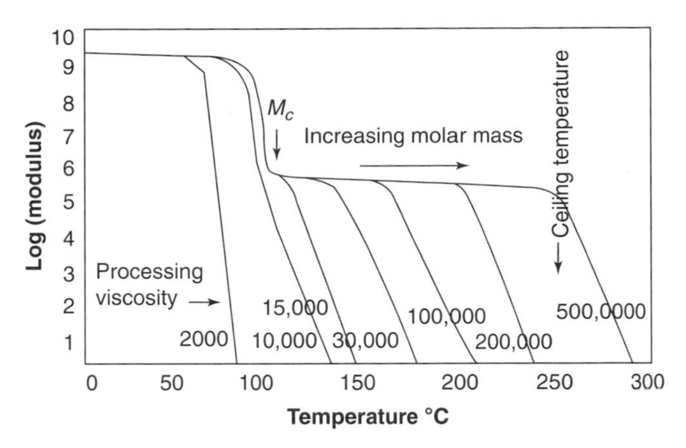

Figure 2.32 Schematic of DMTA traces for narrow molar mass distribution polystyrene materials.

that the polymer material will be exposed to this temperature for a very short period of time. The incorporation of long chains into the material has the advantages of helping to develop high elongation to break and other advantageous properties, but clearly presents a challenge in terms of being able to achieve the required low level of viscosity for good mould production.

2.5.6 What is happening at the critical molecular mass?

A plot of the variation of the melt viscosity against the number average molar mass, M_n (see Figure 2.33), shows a change of slope at the critical molar mass M_c. For short chain polymers with a molar mass below a critical value M_c, the viscosity, η, varies in proportion to the molar mass (see Figure 2.33). This type of behaviour is observed for all polymers. Once the molar mass is increased above M_c, the viscosity dependence can now be described by the relation $\eta \propto M_n^{3.5}$.

Theoretical and experimental studies of polymer melts have concluded that the reason for the higher power dependence is the entanglement of polymer chains forming a 'transient' network structure. If we take spaghetti, and cook it, the pieces will form a tangled mass that requires skill to disentangle. However, if we use a knife and cut the spaghetti we can produce short lengths that do not entangle and can be scooped up from the plate. There is a critical length for the entanglement of the spaghetti similar to that for a polymer chain. For the

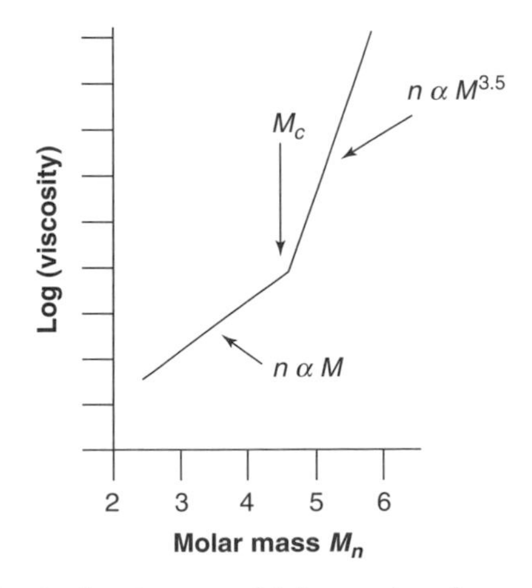

Figure 2.33 Plot of melt viscosity versus M_n for a series of narrow molar mass polymers.

polymer chains to disentangle they have to move like a snake. The theory indicates, however, that the chains form new entanglements as fast as they manage to disentangle from the original entanglements. The effect is that the melt appears like an entangled matrix and shows pseudo rubber-like behaviour, the modulus having a value of ~10^6 Pa.

2.5.7 Effects of chain entanglement on the mechanical properties of thermoplastics polymers

The plateau region in the case of the DMTA traces is indicative of the effects of entanglement of the long polymer chains. These physical entanglements are effectively like cross-links in thermoset materials and pin the polymer chains in space in the solid. When a thermoset is subjected to a load it will initially undergo elastic deformation and if there is a sufficient level of entangled polymer present will recover its original dimensions when the load is removed. The polymer is thus exhibiting the classical elastic behaviour of a solid. As the glass transition temperature is approached so the material may exhibit nonreversible deformation when loaded (creep) and the behaviour is rather like that of a very viscous liquid. The creep behaviour of polymers has been studied most extensively by engineers and physicists and is described under the general heading of *viscoelastic* behaviour. The term viscoelastic reflecting the fact that, in this temperature region, the behaviour of the polymer is neither that of a solid or a liquid but somewhere in between.

2.6 Viscoelastic behaviour

In practice, the modelling of the viscoelastic behaviour (Ward and Hadley, 2004) is carried out using sophisticated finite element analysis packages, which are based on the very simple theory now outlined. The theory considers the physical behaviour of the polymer to be represented by a combination of a *spring*, which models the elastic behaviour of the solid and a *dashpot*, which describes the flow behaviour of a viscous fluid. There are two approaches that can be adopted: the so-called Maxwell and the Kelvin–Voigt models.

2.6.1 Maxwell model

The Maxwell model considers the solid to be described in terms of a series combination of the elements of the spring and dashpot (see Figure 2.34).

The elastic component of the response is described by a spring with ξ as the elastic constant – Young's modulus of the material – and when subjected to a stress σ_1 it undergoes elongation according to the relationship:

$$\sigma_1 = \xi\varepsilon_1 \tag{2.21}$$

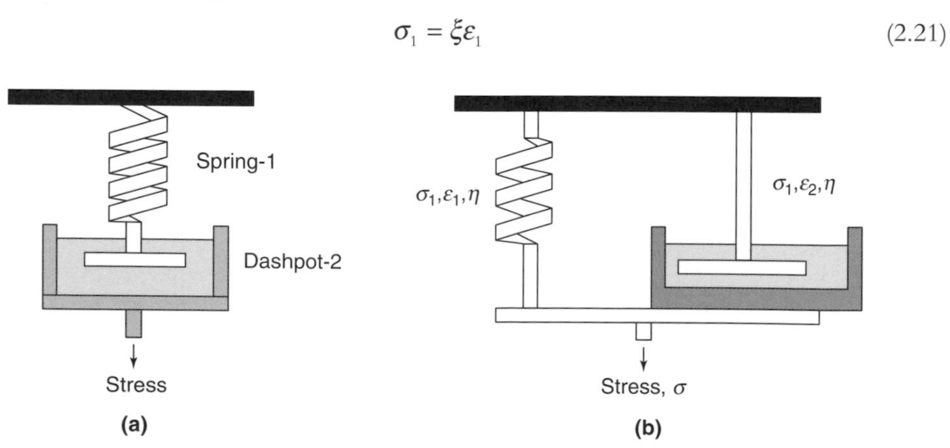

Figure 2.34 Models for polymer viscoelastic behaviour: (a) Maxwell model; (b) Kelvin–Voigt model.

where ε_1 is the strain in the sample. The dashpot describes the viscous component of the response and is proportional to the rate of change of the stress, σ_2, the proportionality constant being 'η', the 'effective' viscoelastic constant for the material. The components of the stress can be described as follows:

$$\sigma_2 = \eta d\varepsilon/dt \quad \sigma_1 = \xi\varepsilon_1 \tag{2.22}$$

where η is a constant for the material and can be related to the viscoelastic behaviour of the material. The constant, η, may be a function of molecular mass of the polymers and will be sensitive to the presence of high molar mass components. In the series arrangement it is reasonable to assume that the applied stress is equal on both components, i.e. an equilibrium condition exists and hence:

$$\sigma = \sigma_1 = \sigma_2 \tag{2.23}$$

The total strain, ε, will be equal to the sum of the strains in the spring and the dashpot and equal to:

$$\varepsilon = \varepsilon_1 + \varepsilon_2 \tag{2.24}$$

Combining Equations (2.20) and (2.21) and (2.23) we obtain:

$$\partial\varepsilon/\partial t = (1/\xi)(\partial\sigma_1/\partial t) + (1/\eta)\sigma_2 \tag{2.25}$$

but $\sigma_1 = \sigma_2$ therefore the above equation simplifies to:

$$\partial\varepsilon/\partial t = (1/\xi)(\partial\sigma/\partial t) + (1/\eta)\sigma \tag{2.26}$$

Let us examine what happens to the strain when a constant stress σ_0 is applied to the material. In the dashpot element, the strain increases at a constant rate with time defined by:

$$\partial\varepsilon/\partial t = (1/\eta)\sigma_0 \tag{2.27}$$

The initial application of the stress will cause a displacement of the spring which is a direct function of the spring constant and will be equal to:

$$\varepsilon_0 = \sigma_0/\xi \tag{2.28}$$

The strain at any time will therefore be given by:

$$\varepsilon(t) = \sigma_0/\xi + (\sigma_0/\eta)t \tag{2.29}$$

The first term indicates that on the application of the stress, the spring instantaneously undergoes an extension dictated by the spring constant; the dashpot does not move. The dashpot will elongate as a function of time, the rate being dictated by the viscoelastic constant, η. Equation (2.28) can be rearranged to give a time-dependent modulus $E(t)$, the so-called creep modulus:

$$E(t) = \sigma_0/\varepsilon(t) = (\xi\eta)/(\eta + \xi t) \tag{2.30}$$

Let us consider a situation where a constant stress of σ is applied to the Maxwell polymer for a period t_1 and then removed instantaneously (see Figure 2.35(a)). Equation (2.29) predicts the strain as having the form shown in Figure 2.35(b).

Many polymeric materials do not exhibit an instantaneous strain on application of a stress and the Maxwell model is not a good description of the observed behaviour. An alternative approach is to add the elements in parallel, the Kelvin–Voigt model.

2.6.2 Kelvin–Voigt model

Application of stress to the spring and dashpot in parallel requires that the total stress is distributed between the two elements:

$$\sigma = \sigma_1 + \sigma_2 \tag{2.31}$$

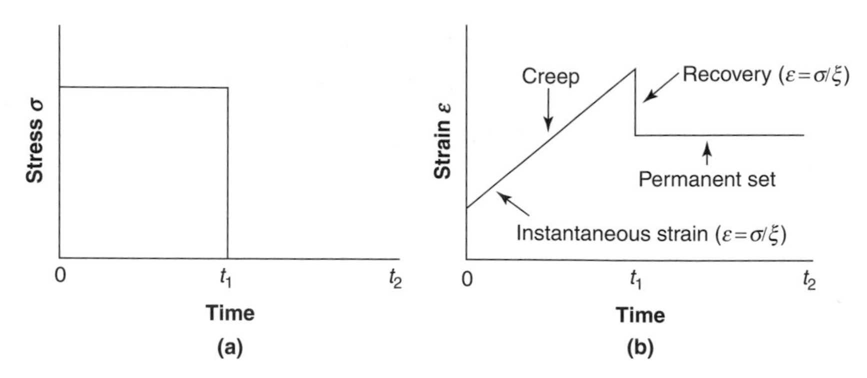

Figure 2.35 Stress and resultant strain which is generated in material.

and the total strain is equal to the strain in each of the elements:

$$\varepsilon = \varepsilon_1 = \varepsilon_2 \tag{2.32}$$

From Equations (2.21), (2.22) and (2.31) we have:

$$\sigma = \xi\varepsilon_1 + \eta(\partial\varepsilon_2/\partial t) \tag{2.33}$$

or using Equation (2.32):

$$\sigma = \xi\varepsilon + \eta(\partial\varepsilon/\partial t) \tag{2.34}$$

But ε_2 has an initial value of zero and will only grow with time. As a consequence the initial value of the strain must also be zero. This is the first difference between the predictions of the Maxwell and Kelvin–Voigt models. If a constant stress σ_0 is applied then Equation (2.34) becomes:

$$\sigma_0 = \xi\varepsilon + \eta(\partial\varepsilon/\partial t) \tag{2.35}$$

but Equation (2.32) requires that ε is initially zero. Rearranging Equation (2.35) the time-dependent strain can be described by:

$$\varepsilon(t) = (\sigma_0/\xi)\left[1 - \exp(-\xi/\eta)t\right] \tag{2.36}$$

This indicates that there is an exponential increase in strain from zero up to the value σ_0/ξ, which is a limiting value controlled by the elastic constant of the spring. Unlike the Maxwell model, removal of the stress leaves an 'internal' stress within the system, designated by the equation:

$$0 = (1/\xi)(\partial\sigma/\partial t) + (1/\eta)\sigma \tag{2.37}$$

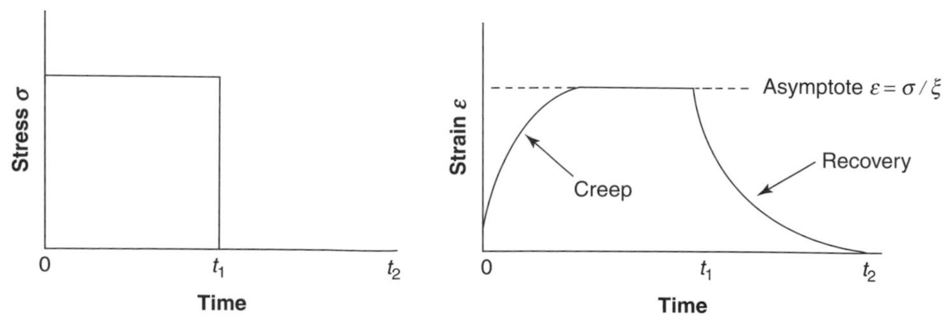

Figure 2.36 Stress and strain profiles with time for a Kelvin–Voigt model.

Solving this differential equation with the initial condition $\sigma = \sigma_0$ at $t = t_1$ then:

$$\sigma(t) = \sigma_0 \exp(-\xi / \eta)t \tag{2.38}$$

The internal stress will decay with a time constant η/ξ to zero, in other words the polymer will recover its original length but will require a finite time to do so. This type of behaviour is found in elastomeric polymers. The stress–strain profile described by the Kelvin–Voigt model is shown in Figure 2.36. This profile is different from that for the Maxwell model (Figure 2.35) and illustrates the way in which the different arrangement of the components influences the response profile.

2.6.3 More complex models

The Maxwell model gives an acceptable first approximation to elastic deformation and simple creep behaviour, but does not account for relaxation. The Kelvin–Voigt model can account for relaxation but provides a poor description of the elastic deformation and does not allow for permanent set. It is appropriate to combine the models to obtain a more accurate model for real data (see Figure 2.37). The total strain is a combination of the values in the various elements and can be written as:

$$\varepsilon = \varepsilon_1 + \varepsilon_2 + \varepsilon_3 \tag{2.39}$$

where ε_3 is the strain response of the Kelvin model and ε_1 and ε_2 are, respectively, the responses of the spring and dashpot elements of the Maxwell model. Using Equations (2.21), (2.22) and (2.39) we obtain:

$$\varepsilon(t) = (\sigma_0 / \xi_1) + (\sigma_0 t / \eta_1) + (\sigma_0 / \xi_2)\left[1 - \exp(-\xi_2 / \eta_2)t\right] \tag{2.40}$$

and the strain rate can be obtained as:

$$(\partial\varepsilon/\partial t) = (\sigma_0 / \eta_1) + (\sigma_0 / \eta_2)\exp(-\xi_2 / \eta_2)t \tag{2.41}$$

which is the sum of the effects of the creep for both of the previous models.

The response of this model to creep relaxation and recovery is the sum of the effects described for the previous two models (see Figure 2.38). It can be seen that although the exponential responses predicted by these models are not true representations for a real viscoelastic material, for many purposes, the picture described is an acceptable approximation to the actual behaviour for many polymer systems.

2.6.4 Standard linear solid: Zener solid

For some stiffer solids a better reproduction of their physical behaviour is the so-called Zener solid, a model consisting of elements in series and parallel attributed to Zener. It is known as the

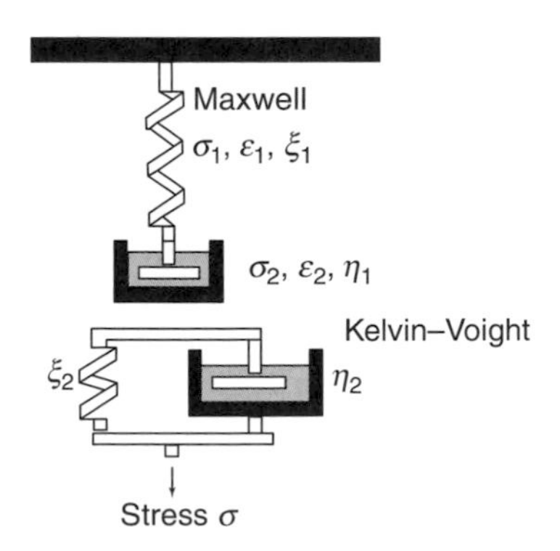

Figure 2.37 Series combination of Maxwell and Kelvin–Voigt models.

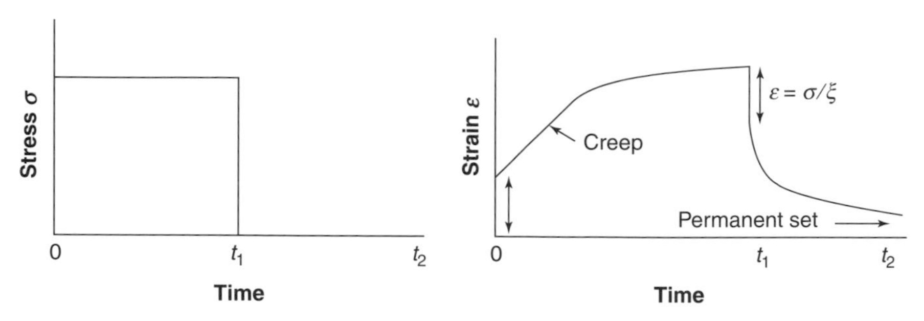

Figure 2.38 Stress and strain variation with time for a combined Maxwell–Kelvin–Voigt model.

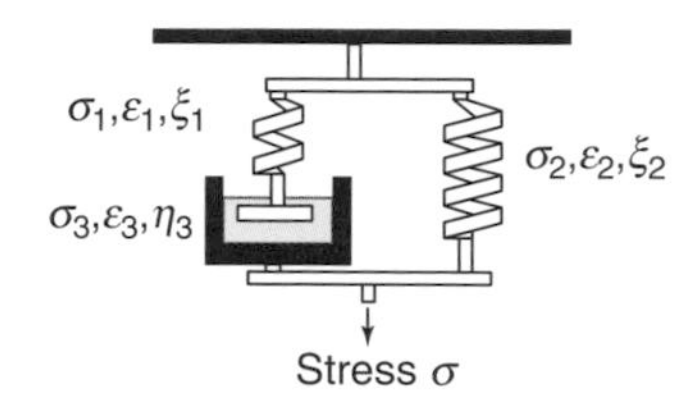

Figure 2.39 Zener solid.

standard linear solid model (see Figure 2.39). The governing equation may be derived as follows. For the Zener solid model the stress–strain relations relationships are:

$$\sigma_1 = \xi_1\varepsilon_1 \quad \text{small spring}$$
$$\sigma_2 = \xi_2\varepsilon_2 \quad \text{large spring} \tag{2.42}$$
$$\sigma_3 = \eta_3 \frac{d\varepsilon}{dt} \quad \text{dashpot}$$

If we examine the way in which the forces are distributed, we can see at equilibrium that:

$$\sigma = \sigma_1 + \sigma_2 \quad \text{and} \quad \sigma_1 = \sigma_3 \tag{2.43}$$

The spring and spring/dashpot being in parallel leads to the equation for the deformation, ε, given by:

$$\varepsilon = \varepsilon_2 = \varepsilon_1 + \varepsilon_3 \tag{2.44}$$

Differentiating Equation (2.44) with respect to time gives:

$$(\partial\varepsilon / \partial t) = (\partial\varepsilon_1 / \partial t) + (\partial\varepsilon_3 / \partial t) \tag{2.45}$$

but from Equation (2.43) we obtain:

$$(\partial\sigma_1 / \partial t) = (\partial\sigma / \partial t) - (\partial\sigma_2 / \partial t) \tag{2.46}$$

and

$$\sigma_3 = \sigma - \sigma_2 \tag{2.47}$$

which leads to:

$$\frac{d\varepsilon}{dt} = \frac{(\partial\sigma / \partial t) - \xi_2(\partial\varepsilon / \partial t)}{\xi_1} + \frac{(\partial\sigma / \partial t) - \xi_2(\partial\varepsilon / \partial t)}{\eta_2} \tag{2.48}$$

which on rearranging gives:

$$\eta_3\left(\partial\sigma/\partial t\right)+\xi_1\sigma=\eta_3\left(\xi_1+\xi_2\right)\left(\partial\varepsilon/\partial t\right)+\varepsilon_1\varepsilon_2\varepsilon \tag{2.49}$$

This is the governing equation for this model. As with the other models it is useful to describe the behaviour of certain types of polymers; in this case it is the stiffer materials that are best suited to the Zener model.

2.6.5 Use of the time-dependent modulus approach

The creep behaviour of many polymers can be described by a simple time-dependent modulus of the form $E(t)$, so that the creep strain has the form:

$$\varepsilon(t)=\left(1/E(t)\right)\sigma_0 \tag{2.50}$$

To analyse the effects of complex stress behaviour it is useful to apply the Boltzmann superposition principle. The principle is based on the assumption that it is possible to calculate the strain at any time by calculating the change in the stress that occurs over a defined time interval. Consider the situation in which the stress σ_0 was applied at zero time and an additional stress σ_1 applied at time u_1 then the Boltzmann superposition principle that says that the total strain at time t is the algebraic sum of the independent responses:

$$\varepsilon(t)=\left(1/E(t)\right)\sigma_0+\left(1/E(t-u_1)\right)\sigma_1 \tag{2.51}$$

The equation can be generalised for N step changes to the form:

$$\varepsilon(t)=\sum_{i=0}^{i=N}\sigma_i\left[\left(1/E(t-u_i)\right)\right] \tag{2.52}$$

where σ_I is the step change of stress which occurs at time u_i. This equation is very useful as it is very easily incorporated into finite element analysis (FEA) programmes and allows theoretical exploration of consequences of application of a particular type of stress to a material.

Worked examples of viscoelastic calculations

To help understand how the FEA models the behaviour of polymers it is useful to consider a few simple examples.

Example 1

An acrylic polymer is found to obey a Maxwell type of model with spring element constant ξ_1 equal to 2,800 MNm^{-2}. What is the instantaneous strain ε_1 if the stress is 14 MNm^{-2} and is purely an extension of the spring element? Use the data: $\xi_1=\sigma_0/\varepsilon_1$, $\varepsilon_1=\sigma_0/\xi_1=14/2{,}800=0.005$. If the slope of the creep curve is 1.167 × 10^{-6} h^{-1}, calculate the dashpot constant for the Maxwell element. Thus $\partial\varepsilon/\partial t=\sigma_0/\eta_1$, $\eta_1=14/1{,}167\times10^{-6}=1.2\times10^7\,\text{MNhm}^{-2}$.

Example 2

A styrene copolymer can be fitted by the Kelvin–Voigt model. If the initial strain is 0.5 × 10^{-2} and the final strain is 0.7 × 10^{-2}, what is the spring constant if the stress is 14 MNm^{-2}? The second value is the asymptotic value and hence we can relate the change in the strain to the stress: $\xi_2=\sigma_0/\varepsilon_2=14/(0.7-0.5)10^{-2}=7{,}000\ \text{MNm}^{-2}$.

The viscoelastic behaviour of a certain plastic is to be represented by spring and dashpot elements having constants of 2 GNm^{-2} and 90 GNsm^{-2}, respectively. If a stress of 12 MNm^{-2} is applied for 100 s and then completely removed, compare the values of strain predicted by the Maxwell and Kelvin–Voigt models after: (a) 50 s and (b) 150 s.

$$\varepsilon = \sigma_0/\xi + \varepsilon t = \sigma_0/\xi + (\sigma_0/\eta)t \; \varepsilon(50) = 12/2{,}000 + 12 \times 10^5/90 \times 10^9 (50) = 1.26\%$$

$$\varepsilon(100) = 12/2{,}000 + 12 \times 10^6/90 \times 10^9 (100) = 1.933\% \quad \text{but} \quad \sigma_0/\xi = 0.6\%$$

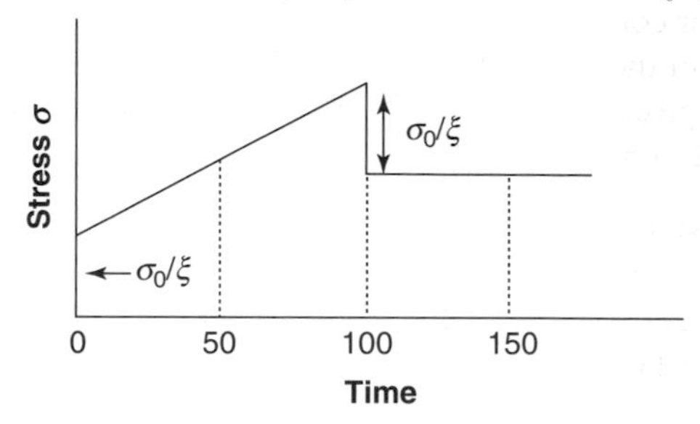

The spring will instantaneously recover at 100 s and therefore: $\varepsilon(150) = 1.933 - 0.6 = 1.333\%$. The Maxwell model predicts a permanent set and hence the value of the stress $\varepsilon(100)$ at 100 s is equal to the value $\varepsilon(150)$ at 150 s.

The Kelvin–Voigt model predicts a $(1 - \exp(-\xi t / \eta))$ growth in the stress over the first 100 s. The stress will then decay with a rate $\exp(-\xi t / \eta)$ from a value that is determined by the growth for the first 100 s.

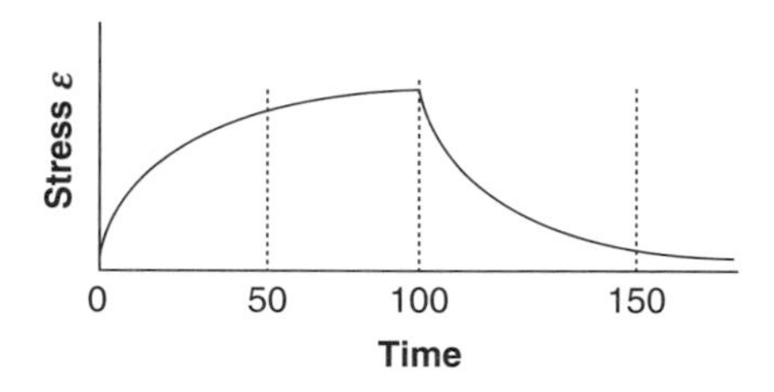

$$\varepsilon = \sigma_0 / \xi\left(1 - \exp\left(-\xi t / \eta\right)\right)$$

$$\varepsilon_{50} = 12/2{,}000\left(1 - \exp\left(-2 \times 50/90\right)\right) = 0.402\%$$

$$\varepsilon_{100} = 12/2{,}000\left(1 - \exp\left(-2 \times 100/90\right)\right) = 0.535\%$$

$$\varepsilon_{150} = \varepsilon_{100} \exp\left(-\xi t / \eta\right) = 0.535 \exp\left(2 \times 50/90\right) = 0.176\%$$

Example 3

Maxwell and Kelvin–Voigt models are to be set up to simulate the creep behaviour of a plastic. The elastic and viscous constants for the Kelvin–Voigt models are $2\,\mathrm{GN\,m^{-2}}$ and $100\,\mathrm{GN\,m^{-2}}$, respectively, and the viscous constant for the Maxwell model is $200\,\mathrm{GN\,s\,m^{-2}}$. Estimate a suitable value for the elastic constant for the Maxwell model if both models are to predict the same creep strain after 50 s: Maxwell strain (50 s) = Kelvin strain (50 s) $\sigma_0 / \xi_1 + (\sigma_0 / \eta_1)t = (\sigma_0 / \xi_2)(1 - \exp(-\xi_2 t / \eta_2))$, thus substitution and rearrangement gives:

$$\xi_1 = \left[1/2\left(1 - \exp\left(\xi_2 t / \eta_2\right) - t / \eta_1\right)\right]^{-1}$$

$$\xi_1 = \left[1/2 \times 10^9\left(1 - \exp\left(-2 \times 50 \setminus 100\right)\right) - \left(50/200 \times 10^9\right)\right]^{-1} = 15.1\,\mathrm{GN\,m^{-2}}$$

Example 4

Suppose a plastic that can have its creep behaviour described by a Maxwell model is to be subjected to the following stress history: initially a stress of $10\,\mathrm{MN\,m^{-2}}$ is applied for 100 s and then removed for 100 s and then a stress of 5 $MN m^{-2}$ is applied for a period of 100 s and then increased to a value of 15 $MN m^{-2}$ for a further 100 s at which point it is reduced to a value of 10 $MN m^{-2}$. If the spring and dashpot constants for this model are 20 $GN m^{-2}$ and 1,000 $GN s m^{-2}$, respectively, then predict the strains in the material after: 150 s; 250 s; 350 s and 450 s. For the Maxwell model, the strain up to 100 s is given by $\varepsilon(t) = \sigma / \xi + \sigma t / \eta$. The modulus for a Maxwell element may be expressed as $E(t) = \sigma / \varepsilon(t) = \xi\eta / (\eta + \xi t)$. Then the strains may be calculated as follows:

(1) at 150 s: $\sigma_0 = 10$ $MN m^{-2}$ at $u_0 = 0$, $\sigma_1 = -10$ $MN m^{-2}$ at $u_1 = 100$ s.
$\varepsilon(150) = \sigma_0[\eta + \xi(t - u_0) / \xi\eta] + \sigma_1[(\eta + \xi(t - u) / \xi\eta)] = 0.002 - 0.001 = 0.1\%$
(2) at 250 s: σ_0, σ_1 as above $\sigma_2 = 5$ $MN m^{-2}$ at u_2, $a = 200$ s.
$\varepsilon(250) = \sigma_0[(\eta + \xi(250 - 0)) / \xi\eta] + \sigma_1 + \sigma_2[(\eta + \xi(250 - 200) / \xi\eta)] = 0.003 - 0.002 + 0.000{,}5 = 0.15\%$
(3) at 350 s: σ_0, σ_1, σ_2, as above $\sigma_3 = 10$ $MN m^{-2}$ at $u_3 = 300$ s, so $\varepsilon(350) = 0.003 = 0.3\%$.
(4) and in the same way, e(450) = 0.004 = 0.4%.

Example 5

A plastic is stressed at a constant rate up to 30 $MN m^{-2}$ in 60 s and the stress then decreases to zero at a linear rate in a further 30 s. The time-dependent creep modulus for the plastic can be expressed in the form: $E(t) = \xi\eta / (\eta + \xi t)$.

Use the Boltzmann superposition principle to calculate the strain in the material after (i) 40 s; (ii) 70 s; and (iii) 120 s. The elastic component of the modulus is 3 $GN m^{-2}$ and the viscous component is 45×10^9 $N s m^{-2}$.

$$\varepsilon(t_1) = K_1 t_1 ((1/\xi) + (t_1 / 2\eta))$$

$$\varepsilon(40) = 0.5 \times 40((1/3{,}000) + (40/2 \times 45{,}000)) = 1.55\%$$

$$\varepsilon(t_2) = (K_1 T + K_2 T)((1/\xi) + (t_2/\eta) - (T/2\eta)) - K_2 t_2 ((1/\xi) + (t_2/2\eta))$$

$$\varepsilon(70) = (0.5 \times 60 + 1 \times 60)((1/3{,}000) + (70/45{,}000) - (60/90{,}000))$$
$$-70((1/3{,}000) + (70/90{,}000)) = 3.22\%$$

$$\varepsilon(t_3) = K_1 T / \xi\eta\{\eta + \xi t_3 - (1/2)\xi T\} - K_2 (T' - T) / \xi\eta\{\eta + \xi t_3 - (1/2)\xi(T'' - T)\}$$

which for $t_3 = 120$ s, $T = 60$ s and $T' = 90$ s gives $\varepsilon(120) = 3\%$.

2.7 What does the experimental data look like for a real polymer system?

Polyethylene is used for high pressure gas distribution pipes which could be subject to creep as a consequence of use in a constant stressed condition. Figure 2.40 shows strain–elongation and tensile creep data for a typical pipe-grade high density polyethylene (HDPE) measured at different temperatures.

HDPE is a paracrystalline material, implying that a large proportion of the material is crystalline but some of the material is disordered and amorphous. Application of a stress to the material causes it to creep through a reorganisation of the amorphous and crystalline phases as discussed in Section 3.6. The amorphous content of HDPE gives it classic viscoelastic behaviour. At low levels of stress (6–8 MPa) the behaviour is similar to that described by a Maxwell model: there is a small initial elongation reflecting the instantaneous elastic response of the material followed by a linear

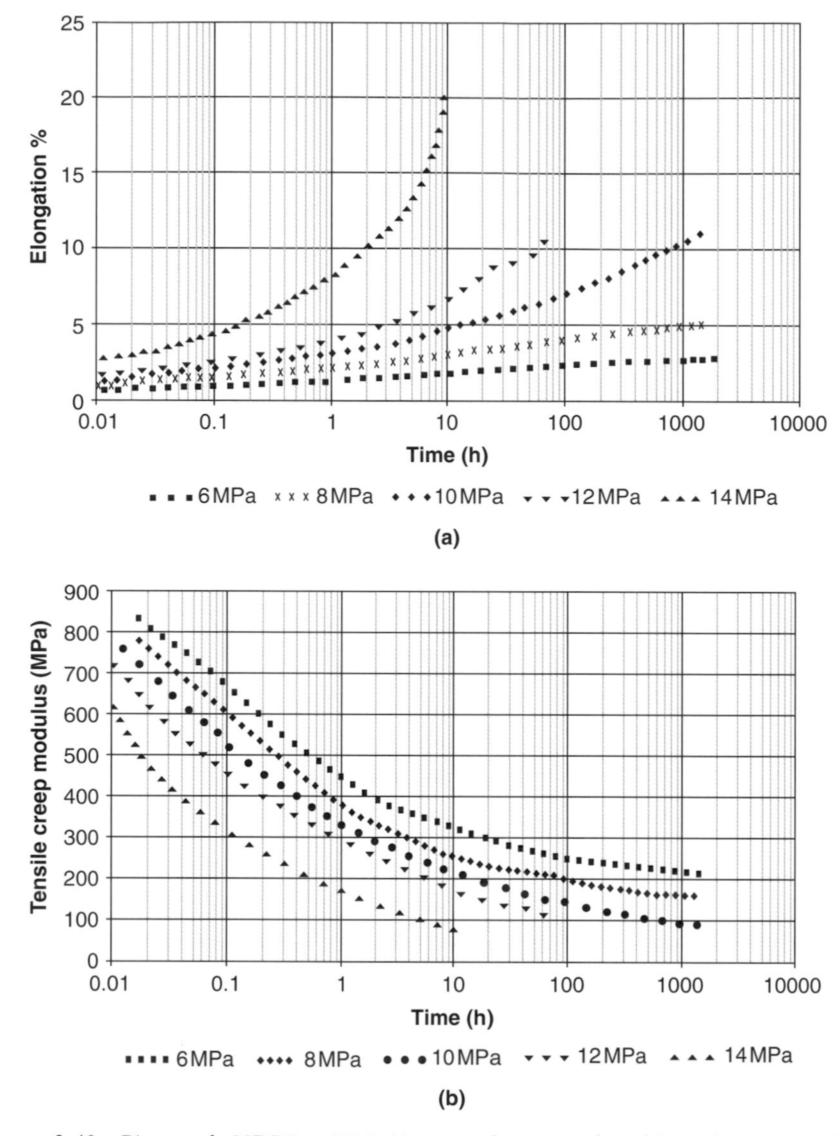

Figure 2.40 Pipe-grade HDPE at 25°C: (a) strain–elongation data; (b) tensile creep modulus.

creep with an almost constant slope. However, for higher stresses, deviations from simple Maxwell behaviour are observed, the creep curve changing dramatically with time. The data indicate that the material is failing, the tensile creep modulus falling to an unacceptably low value. Modelling these data would require the use of two Maxwell elements arranged in parallel.

The second Maxwell element would have a low spring constant and a very high viscoelastic constant. This combination would describe the large increase at long times, but make negligible contribution at short times. The tensile creep modulus is shown in Figure 2.41(b) and can be described in terms of Equation (2.30) for the Maxwell model or in a generalised form as Equation (2.50). For the low levels of the stress the initial drop is approximately linear with time, whereas at higher stress levels the curves show a high initial drop and an increasingly nonlinear dependence with time.

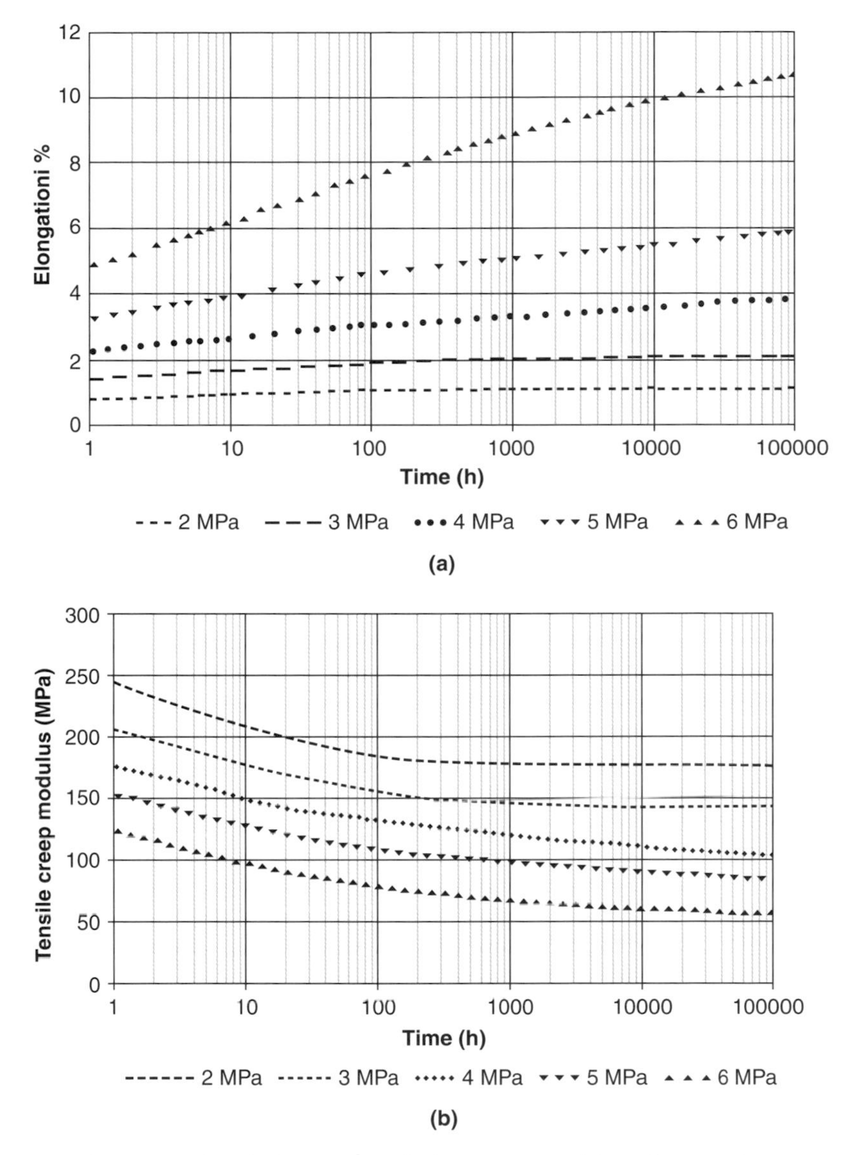

Figure 2.41 HDPE at 65°C: (a) elongation data; (b) tensile creep data.

Increasing the temperature will cause low melting fractions to become mobile and a consequent change in behaviour. The higher temperature data are obtained at significantly lower stress levels (see Figure 2.41(a)) and the shape of the curves are different from those obtained at the lower temperature.

The elongation–strain data at low levels can be described by the Maxwell model whereas at higher stress levels the data are clearly better described by a Kelvin–Voigt or possible a universal model. As at the lower temperatures there is an instantaneous elastic deformation, but at the higher temperatures the growth of the elongation is not linear and curves towards the time axis rather than away from it as in the low temperature data. The greater mobility of the chain elements has led to a reduction in the value of the modulus and created the possibility of the material failing after a long period of time at relatively low stress levels.

It should be apparent that to select a material which is fit for purpose it is essential to consider the temperature range over which it will be used, have creep data for that range and also understand which the critical criteria are for failure. The latter is usually defined in terms of a parameter, which in this case is the tensile creep modulus that is determined from the application.

2.8 Other mechanical properties of polymer systems

The DMTA trace for an idealised polymer thermoplastic material is shown in Figure 2.42. The stress–strain curve for the polymer (see Figure 2.42(b)) indicates that the material has a high ultimate strength, high stiffness and strain to failure. The polymer is initially stiff and will fail in a brittle manner. Changes in resin type will be reflected in the tensile strength and stiffness. The ultimate properties of a polymer will depend on the chemical structure of the polymer. The chains can either be rigid or flexible, long or short and will be influenced by the characteristics of the material. The toughness of a material is influenced by the energy that is required to destroy the material and is calculated from the area under the stress–strain curves.

Examples of the ways in which the stress–strain curves will change with temperature are shown in Figure 2.42(b). At low temperatures, [A], the material is a brittle glass which will show

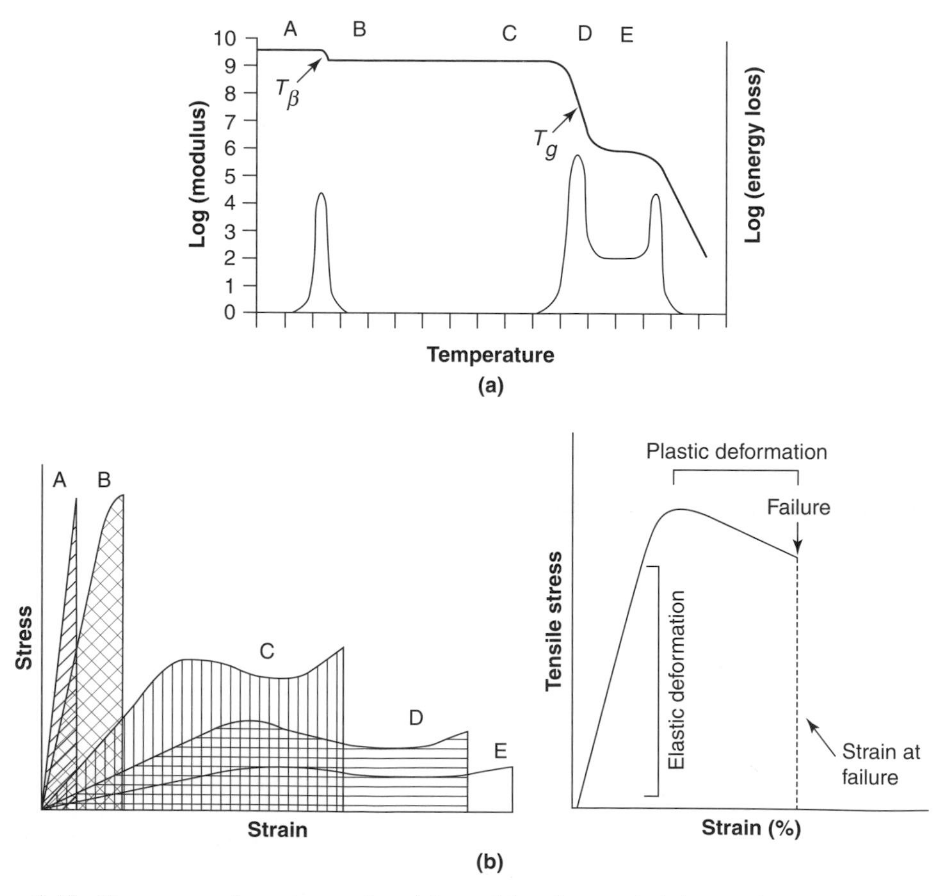

Figure 2.42 Temperature dependence of modulus and loss for a typical thermoplastic (a) and the stress–strain curves obtained at various temperatures (b).

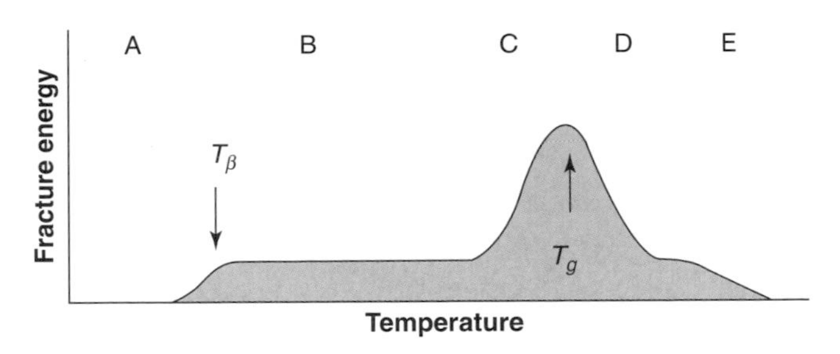

Figure 2.43 Plot of variation of fracture energy with temperature.

a high modulus indicated by the high initial slope of the stress against strain. The material will exhibit *brittle* failure. Brittle failure is characterised by the failure surface as seen by optical or electron microscopy as being almost glass smooth, indicating that the failure progressed very rapidly through the material.

The energy dissipated associated with brittle failure is relatively small, indicted by the small area under the stress–strain curve and the very small strain to failure. Once the temperature has been raised above the β transition temperature T_β, the matrix has the ability to absorb energy and the stress–strain behaviour is characterised by a slightly greater degree of strain before failure [B] (see Figure 2.42(a)). This type of ductile behaviour will be observed over the temperature range up to just below T_g. As T_g is approached, so the initial slope will decrease reflecting a drop in the modulus and the matrix will undergo a greater degree of elastic distortion [C]. The ability of the matrix to absorb energy will increase the energy that is required for failure to be achieved. The larger area under the stress–strain curve indicates that there is a large energy to failure. Further increasing the temperature allows the matrix to become more elastic and a high degree of strain is achievable before failure is achieved. The degree of elastic deformation before failure will depend on the molar mass of the polymer. If the molar mass is just above M_c, the extent to which the matrix can be stretched before failure will be limited. For molar masses well above M_c, the entanglements will allow a significant degree of elastic distortion before failure occurs. In the rubbery *plateau* region, the elastic distortion will be determined by the degree of entanglement in the rubbery plateau [E]. The strength of a plastic can be described by the variation of the free volume with temperature relative to the glass transition temperature:

$$\log\left[\sigma_{ultimate}(T)/\sigma_{ultimate}(T_g)\right] = \left[a(T-T)\right]/\left[b+(T-T_g)\right] \quad (2.53)$$

This shift with temperature of the data is often described by the Williams–Landel–Ferry (WLF) shift equation and reflects the effects of the chain dynamics on the ultimate strength. Fracture toughness, which is the integral under the stress–strain curve, changes with temperature and shows a maximum at the T_g as indicated in Figure 2.43.

2.9 Effects of water

Water can act as a plasticiser for polar polymers such as polymethylmethacrylate, polycarbonate, nylon, polyesters, epoxy resins and similar materials. The absorption of moisture will lower the glass transition point and can lead to loss of physical properties. Polymers can be divided into two classes according to their susceptibility to water.

- *Hydrophilic polymers*: These will have polar groups and include nylon and polyester. Epoxy resins will take up varying quantities of moisture depending on their ability to be swollen by moisture. For amine-cured epoxy resins typical uptake levels can be of the order of 2–3%. As a rule of thumb, 1% of water uptake will lower the T_g by 10°C and hence a material with a T_g of 100 can after exposure to moisture/water end up with a T_g of 70°C which may compromise its application in certain structures. Water ingress into nylon plus the repeated application of stress can distort the bristles of a tooth brush. The lesson to be learnt is that you should never put your tooth brush under warm water as this will aid the ingress of the moisture and shorten its life.

- *Hydrophobic polymers*: These are generally nonpolar and include polystyrene, polyethylene and polypropylene. They do not take up significant amounts of moisture and hence do not suffer plasticisation by water. It is possible for these polymers to absorb small amounts of moisture by the water diffusing into microvoids, but the level of uptake is usually of the order of 0.1% or less.

2.10 Environmental stress crazing

Environmental stress crazing (ESC) (the term 'crazing' can be used interchangeably with the term 'cracking') is the formation of external or internal cracks in a plastic caused by tensile stresses less than its short-time mechanical strength, when such strength has been reduced by ageing or exposure to certain environmental conditions. ESC may be defined as the acceleration of stress cracking by contact with a liquid or vapour without chemical degradation. The mechanism is purely physical. The interactions between the fluid, the stress, and the polymer include local yielding, fluid absorption, plasticisation, craze initiation, crack growth, and fracture, without irreversible chemical change (i.e. without change in molecular weight, substitution or abstraction). Plasticisers can soften the surface of a polymer and promote swelling (see Figure 2.44).

The swelling of the polymer will create stresses due to the expansion of the polymer, however, the swollen region is constrained by the unplasticised material in the bulk. If the plasticisation is effective then the stresses aid diffusion and the material softens without failure. However, in many cases the plasticisation is less effective and the stresses plus external stress can lead to dramatic failure of the material. The stress relaxation causes the chains to move so as to lower the energy and there will be a tendency to want to increase the surface area. The creation of microcracks increases the area and hence lowers the stresses in the surface and lowers the surface energy, resulting in the formation of surface crazes or cracks (see Figure 2.44). The ultimate effect of the ESC is for the item under test to fail. ESC tests are always carried out under stress and often the time to failure is reported for a particular material.

ESC is the most common cause of failure in plastics. It is responsible for ~25% of serious failures (see Figure 2.45). About 90% of these failures involve glassy amorphous thermoplastics

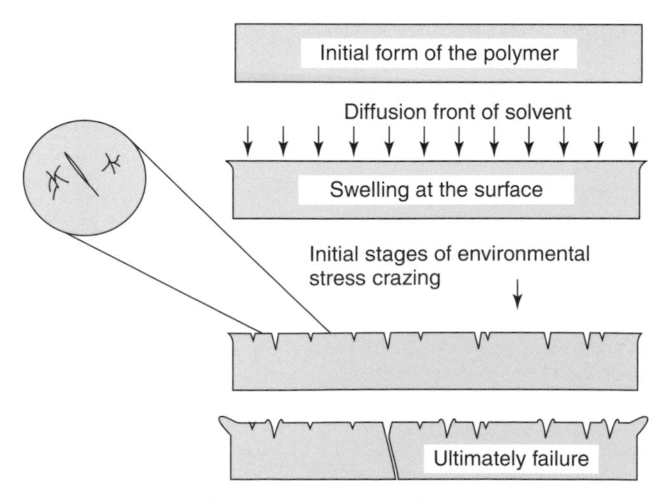

Figure 2.44 Stages of ESC.

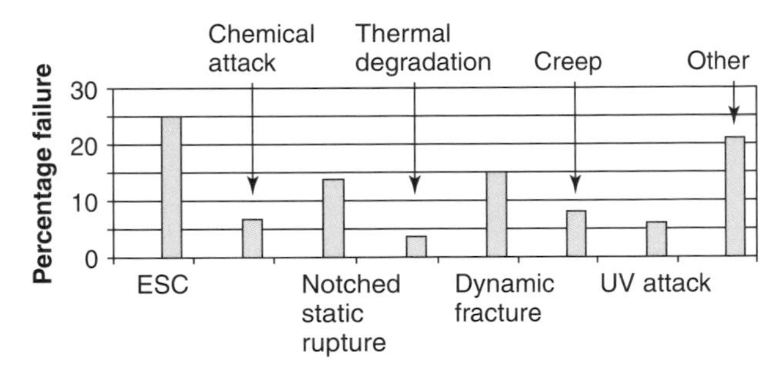

Figure 2.45 Distribution of typical types of failure in plastics.

in contact with 'secondary fluids'. Such fluids include: paints, adhesives, cleaning agents, lubricants, plasticisers, inks, aerosol sprays, antirust agents, leak detection fluids, lacquers, fruit essences and vegetable oils. ESC failures due to contact with 'primary fluids' i.e. those that are purposely contained by the plastic product or those into which the product is purposely immersed, are comparatively rare. In the clearly defined situations faced by the industries that handle bulk fluids, materials can be exhaustively tested to avoid the problem. This is a very important topic and will be considered again as a series of case studies (see Section 7.2).

Brief summary of chapter

- Unlike metals and ceramics, polymers exhibit complex mechanical behaviour. They can behave as crystalline or glassy material at one temperature but have the properties of a rubber at another.
- Mechanical properties: The molar mass, chemical and geometric structure can all influence the physical properties. However, there are some simple relationships which allow us to predict the molar mass variation of the T_g, and viscosity. The free volume plays a vital role in determining T_g and its potential pressure sensitivity.
- Viscoelasticity: The ability of polymers to simultaneously show characteristics of solids and liquids is unique to these materials. Viscoelasticity is closely connected to the creep and failure of plastics. In the limit of low stresses it can be usefully adapted for damping vibrations etc.
- Modelling viscoelasticity: Simple spring and dashpot models can provide the basis for more complex finite element models for creep and related phenomena. It is important to recognise the implied relationships between the models selected and the physical properties of the polymers. Recognition of that connection enables a more confident prediction of the life expectancy for a particular material in particular circumstances.

References and further reading

Brandrup J., Immergut E.H. and Grulke E.A. (Eds.) *Polymer Handbook*, Wiley Interscience, New York, NY, USA, 1999.

Ebewele R.O. *Polymer Science and Technology*, CRC, New York, NY, USA, 2000.

Griskey R.G. *Polymer Process Engineering*, Chapman and Hall, New York, NY, USA, 1995.

Starling S.G. and Woodall A.J. *Physics*, Longmans, London, UK, 1950.

Ward I.M. and Hadley D.W. *An Introduction to the Mechanical Properties of Solid Polymers*, Wiley, Chichester, UK, 2004.

3

Crystallinity and polymer morphology

3.1 Introduction

As indicated in Chapter 1, polymers can be either crystalline, partially ordered or totally disordered (amorphous). Whether or not an ordered crystalline structure is created depends on the regularity of the polymer backbone and the strength of the polymer–polymer interactions. In the melt, the polymer chain is flexible and will adopt a number of higher energy *gauche* conformations. On cooling, the lower energy *trans* form becomes predominant and chains prefer a more extended structure. The all-*trans* sequences produce linear sections of chain that can interact with other chains and grow crystals. The extent to which the polymer eliminates the higher energy conformations on cooling will dictate its ability to crystallise. For a polymer to exhibit crystallinity it must either have a regular backbone structure, possess strong interchain interactions or alternatively adopt a specific chain conformation. The presence of chain stereochemical defects and/or chain branches at high concentrations make it impossible for the polymer chain to form a close-packed structure and an *amorphous* structure results. Large side chains will usually inhibit packing and promote the formation of an *amorphous* structure. However, long side chains can assist the polymer to crystallise. Specific interactions, such as hydrogen bonds or strong dipole–dipole interactions, may promote order between neighbouring polymer chains and facilitate crystallisation.

In general, polymers that are symmetrical undergo favourable interactions and create highly ordered crystalline phases (see Figure 3.1). All these polymers all have a high degree of regularity in the polymer backbone and pack together to form stable crystalline structures.

X-ray scattering studies of polymers show that, in addition to the expected shape peaks characteristic of an ordered solid, there will be a broad scattering pattern due to the presence of disordered amorphous regions. Crystalline polymers are therefore composed of crystalline regions dispersed in an amorphous matrix.

3.2 Crystallography and crystallisation

To help us understand the factors that influence crystal formation in polymers, it is appropriate to consider particular polymer systems.

Polyethylene

The X-ray scattering pattern for polyethylene corresponds to that of a simple unit cell based on close packing of the monomer units. The polymer chains that adopt the lowest *trans* conformation align with the *c*-axis. However, polyethylene shows two different crystal forms at normal pressure, indicating that the all-*trans* chains may pack in more than one way. The extended polyethylene chains will create a 'sheet' of molecules with a one-dimensional repeating structure. The next stack coming alongside the first will 'see' small differences in the potential energy surface, leading to a slightly displaced form of the normal packing of the chains. Due to this misalignment the potential surface contains only small differences in energy between similar states and more than one form is possible. The existence of more than one crystal form for a specific

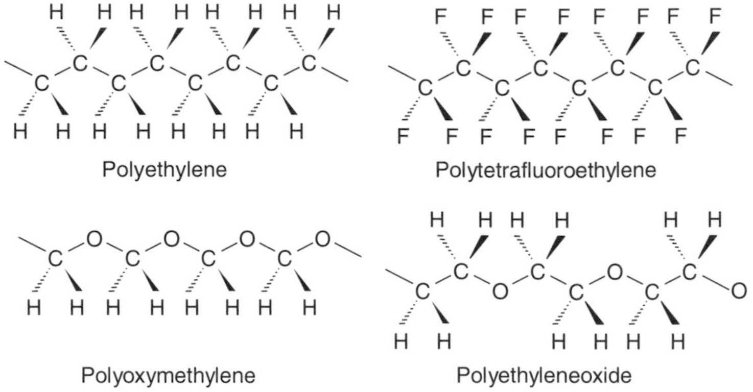

Polyethylene

Polytetrafluoroethylene

Polyoxymethylene

Polyethyleneoxide

Figure 3.1 Some polymers which exhibit a high degree of crystallinity.

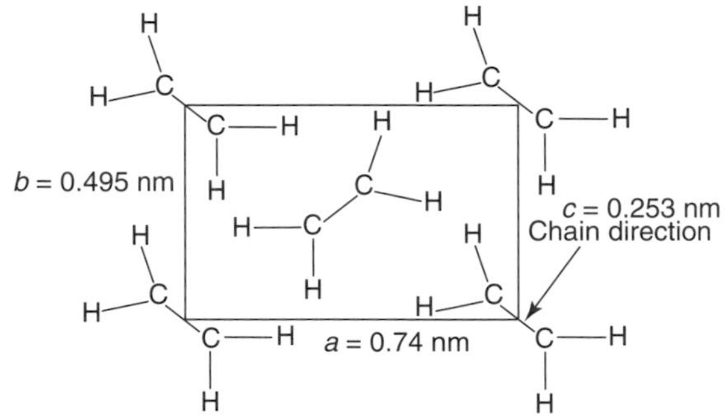

$b = 0.495$ nm

$c = 0.253$ nm
Chain direction

$a = 0.74$ nm

Figure 3.2 View along c-axis: along chain direction, for orthorhombic polyethylene crystal.

compound is referred to as *polymorphism*. For polyethylene, the orthorhombic structure (see Figure 3.2) is the most stable.

Polypropylene

The methyl groups in the *isotactic* polypropylene would be expected to inhibit the packing of the chains; however, these groups impose a helical twist on the backbone and crystallisation can occur (see Figure 3.3).

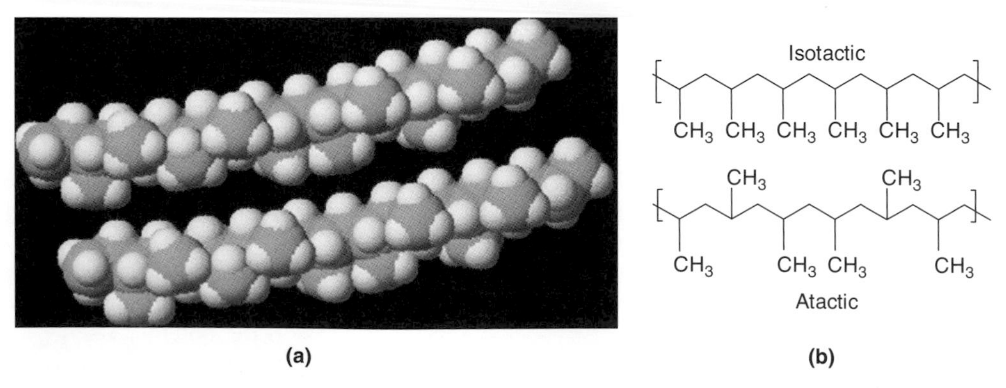

Figure 3.3 (a) Space-filling model of isotactic structure of polypropylene; (b) symbolic structures for polymers.

X-ray studies indicate that the fibres have a monoclinic unit cell structure containing four chains and 12 monomer units, their cell parameters are: a = 6.65 Å, b = 20.96 Å, c = 6.50 Å, $\beta = 99°26'$. The chains form a helical structure and pack to form a regular crystal matrix. *Isotactic* polypropylene is used to fabricate hot water pipes. In contrast, *atactic* polypropylene cannot form a regular helical structure and does not crystallise. It is a soft, flexible solid which can be used as an additive in lubrication oils.

In the *isotatic* polypropylene crystal, there are four possible arrangements of the helix: a right-hand helix pointing upwards and another with alternate helices pointing downwards and a complementary pair of left-handed helices. Although at first sight the upward and downwards helices would appear the same, closer inspection indicates that they are different. *Isotactic* polypropylene exhibits a number of polymorphs, known as the α-, β- and γ-forms. Packing adjacent chains with opposite senses of the helices creates the α form of *isotactic* polypropylene. In general, the packing helices of an opposite form, helices pointing in an opposite direction, are better than for the same form. The γ-structure is associated with the high-pressure crystallisation and the β-form is associated with a spherulitic structure produced when a nucleating agent is present. *Isotactic* polypropylene can exhibit a so-called *smectic* or mesomorphic phase on rapid cooling which has a density of 880 kg m^{-3}, compared with 850 kg m^{-3} for the fully amorphous polypropylene.

Polyoxymethylene

Replacement of one of the methylene groups of polyethylene by an oxygen creates polyoxymethylene. The lone pair of electrons on the oxygen will interact with the hydrogen atoms on the adjacent carbon atoms and the preferred conformation is the nearly all *gauche* structure which produces a stable trigonal form (I) and a less stable orthorhombic form (II). The unit cell contains chains with the same handedness, left- and right-handed molecules appear in different crystal lamellae.

Polyethyleneoxide

In the crystalline state, the polyethyleneoxide chains for a structure has seven monomer elements: -$(CH_2CH_2O)_n$- and two helical turns per unit cell. The chains have dihedral symmetry two-fold axes, one passing through the oxygen atoms and the other bisecting the carbon–carbon bond. The chain conformation is assigned to internal rotation about the -O–CH_2-, -CH_2–CH_2- and -CH_2–O- bonds is, respectively, *trans*, *gauche* and *trans*. Although polyethyleneoxide is essentially very similar to polyethylene, the complexity of its chain conformation in the solid form is quite surprising and reflects the dominance of the local interactions in determining the unit cell structure.

Other polymer systems

Many *isotactic* forms of polymers or simple linear chains, such as polytetrafluoroethylene and polyvinylidenechloride, have helical conformations. The pitch of the helix is determined by influence of the nonbonding short range interactions between adjacent atoms on the polymer backbone. Helical structures are frequently observed and reflect the subtle effects of these interactions. The controlling factor is the enthalpy of the melt process. Table 3.1 summarises values for some common polymer systems.

Although polystyrene, with its bulky phenyl side group, is normally considered to be an amorphous polymer, the *isotactic* form has a helix structure and can crystallise. Usually the density of the crystalline form is higher than that of the amorphous solid, however, in the case of 4-methylpentenene-1 the reverse is true. The crystal of 4-methylpentenene-1 is helical and occupies a large volume, leaving a hollow cylinder down the centre of the coil. Although nylon 6 and nylon 6,6 have similar melting temperatures, they have significantly different enthalpies that reflect the differences in the hydrogen bonding in the two polymers and explain the significant difference in their densities and susceptibilities to moisture uptake.

Table 3.1 Heats of fusion, melting points and densities for some common polymers

Monomer unit	Enthalpy of fusion, ΔH ($J\,g^{-1}$)	Melt temperature (°C)	Density ($g\,cm^{-3}$)	
			Amorphous	Crystalline
Ethylene (linear)	293	141	0.853	1.004
Propylene (isotactic)	79	187	0.853	0.946
Butene (isotactic)	163	140	0.859	0.951
4-Methylpentene-1 (isotactic)	117	166	0.838	0.822
Styrene (isotactic)	96	240	1.054	1.126
Butadiene (1,4 polymer) (*cis*)	171	12	0.902	1.012
Butadiene (1,4 polymer) (*trans*)	67	142	0.891	1.036
Isoprene (1,4 polymer) (*cis*)	63	39	0.909	1.028
Isoprene (1,4 polymer) (*trans*)	63	80	0.906	1.051
Vinyl chloride (syndiotactic)	180	273	1.412	1.477
Vinyl alcohol	163	2165	1.291	1.350
Ethylene terephthalate	138	280	1.336	1.514
Ethyleneoxide	197	69	1.127	1.239
Formaldehyde (oxymethylene)	326	184	1.335	1.505
Nylon 6	230	270	1.090	1.190
Nylon 6,6	301	280	1.091	1.241

Although the unit cell indicates the local order, the mechanical properties of the polymer are influenced by the longer range effects of the structure. Polymer crystals are never perfect and the unit cells do not infinitely duplicate through space. A variety of defects can be created as a consequence of chain ends, kinks in the chain and jogs (defects where the chains do not lie exactly parallel).

The presence of molecular defects can be seen by studying the expansion of the unit cell for branched and linear chain polyethylene. The *c*-parameter which is the measurement along the polymer backbone remains constant but the *a*- and *b*-directions are expanded for the branched polymer crystals. Both methyl and ethyl branches induce expansion, whereas larger pendant groups, propylene or longer homologues are largely excluded from the crystals.

3.3 Single crystal growth

Our understanding of how polymer chains form crystals has been obtained from studies of crystallisation carried out in dilute solution. The crystals formed are unimpeded by the presence of other polymer molecules and the structures created are as close to ideal as can be achieved. Examples of an electron micrograph of a single crystal are shown in Figure 3.4.

The polyethylene (PE) crystals are rather like platelets. At first sight we would expect the extended PE chains to pack together to form a perfect crystal structure, but this does not allow

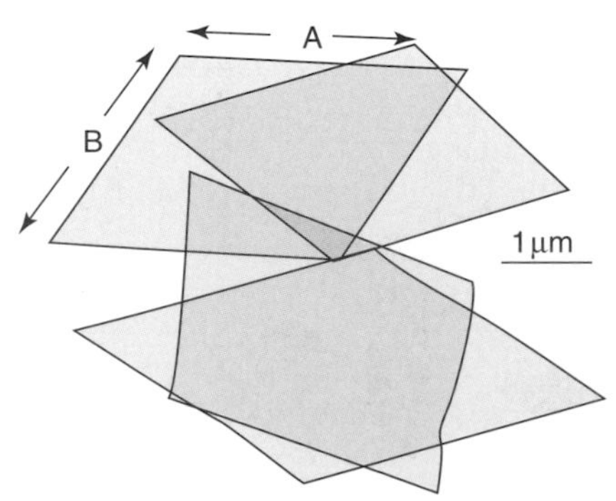

Figure 3.4 Transmission electron micrograph of solution-grown single crystals of linear polyethylene showing platelet structures.

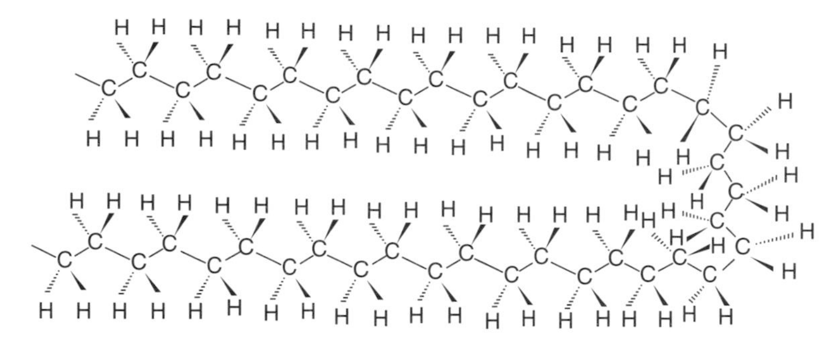

Figure 3.5 Loop structure in a polymer chain.

for the effects of thermodynamics. At the melt temperature there is a balance between the enthalpy and entropy. As the melt cools entropy has to be lost and is achieved by chains looping back on themselves. As a consequence, the length of the linear section of a chain is limited by the requirement for it to contain a series of loops. The picture of the structure which emerges is represented schematically in Figure 3.5.

The chain elements in the loop area will have *gauche* conformations and have a higher energy and entropy than those in the extended all-*trans* regions of the chain. It is easier to visualise the polymer chains in the crystalline solid if we omit the hydrogen atoms from the picture of the backbone and use a zig-zag structure to represent the chain. The chains are held together by weak Van der Waals forces and hence the enthalpy of interaction comes from a large number of separate interactions arising from individual methylene units.

The crystals are formed by the chains folding back on themselves and allowing the linear sections to optimise their interactions (see Figure 3.6(a)). The thickness of the crystal is defined by the occurrence of the folds (see Figure 3.6(b)).

At a molecular level, the polymer chains are packed into well-defined crystal forms, but at a larger scale they show platelets, fibres or other structural forms. A sequence mismatch can produce a distortion of the lattice and resulting roof top-type structures (see Figure 3.6(b)). Crystals grown from the melt exhibit similar structure with a roof top-shape. This has implications for the way the crystalline polymers align when the melt is drawn below its melt temperature.

The interface between the crystals within the solid will contain a number of disordered polymer chains, and contains the amorphous polymer often observed in 'crystalline' polymeric materials (see Figure 3.7). As a consequence of the chain folding the thickness of the crystalline regions is limited to about 8–10 nm and this dictates the *lamellae* thickness. The *lamellae* may extend in space for hundreds of nanometres and in certain cases can stretch for distances of micrometres. The space between the lamellae will contain some polymer chains that are connecting the lamellae but the majority of the materials will be in a rather disordered state that

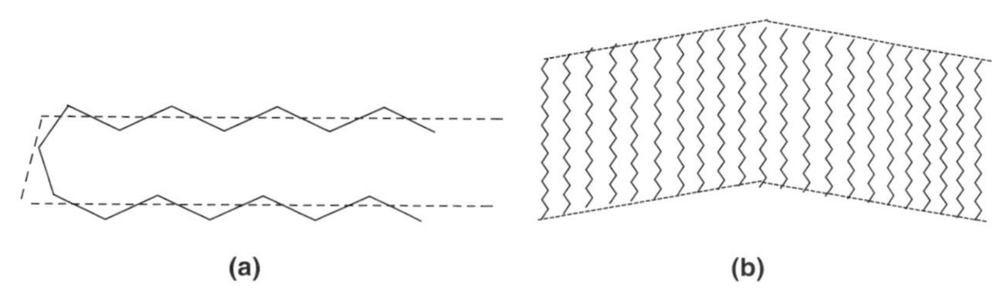

Figure 3.6 Chains packing to form pyramidal shaped single crystal of polyethylene: (a) close up of (110) chain fold with chain axes parallel but slightly displaced and resulting inclined edge; (b) chain alignment within crystal.

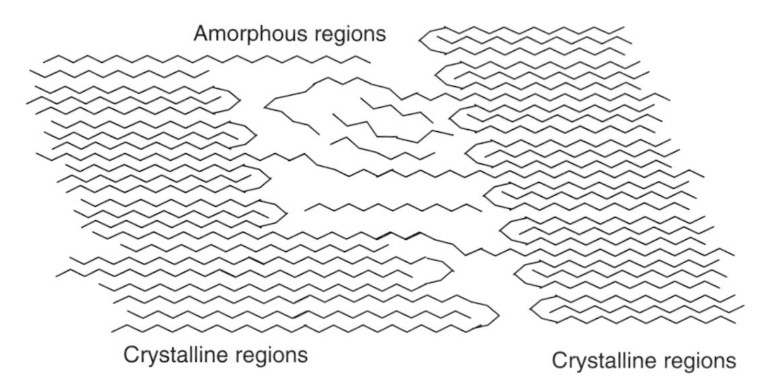

Figure 3.7 Schematic of a section of structure of polyethylene.

resembles the amorphous state. Occasionally some long polymer chain bridge between lamellae and these molecules may have a significant influence on the ultimate mechanical properties of this type of polymer.

3.3.1 Habit of polymer crystals

The temperature at which crystallisation occurs will influence the *habit* of the crystal formed. The habit is the macroscopic shape which the crystals exhibit at micrometre rather than nanometre scale. In the polymer system, the nucleus from which the crystals grow is formed by the alignment of sections of all-*trans* or similar chain structures. The conformational entropy of the polymer molecule in the melt will influence the length of these low energy sections. The higher the temperature the more frequent the occurrence of the *gauche* sequences and the smaller the B/A ratio (see Figure 3.4) The more *gauche* sequences, the shorter the value of A and hence B/A will decrease and the crystal will become more like a ball than a fibre. At the melt temperature of 130°C, only a very small part of the periphery of the crystal lamellae is faceted, the remaining parts are rounded. The variation in the shape, or *habit*, of the crystals is a direct reflection of the influence of chain folding on the growth mechanism. Despite the apparent complexity of the growth behaviour of polyethylene, a number of characteristic shapes have been identified for particular polymer systems (see Table 3.2).

Most of the polymer crystals will exhibit facets and some like polyoxymethylene form hollow pyramids. The occurrence of the hollow pyramid is a direct consequence of the constraints on the chain folding. The smooth surfaces observed for many crystal systems are evidence of regular chain folding, but are not proof that this occurs.

Table 3.2 Characteristic shapes for some dilute solution-grown polymer crystals

Polymer	Characteristics shape of crystal
Polyoxymethylene	Hexagonal hollow pyramid
Poly(4-methyl-1-pentane) (isotactic)	Square-based pyramid
Polytetrafluoroethylene	Irregular hexagonal platelets
Poly(1-butene)	Square or hexagonal platelets
Polystyrene (isotactic)	Hexagonal platelets
Poly(ethyleneoxide)	Square or hexagonal platelets
Poly(ethyleneterephthalate)	Flat ribbons of ~30 nm width
Polyamide 6	Lozenge shaped lamellae
Polyamide 6,6	Irregular hexagonal platelets tending to flat ribbons
Polypropylene (isotatic)	Lath shaped platelets

3.4 Crystal lamellae and other morphological features

Polymers are long chains and so far in the discussion there has not been any clear evidence of the effects of chain length on the crystal structure, as chain folding removes the obvious effects of the chain length. However, *sectorisation* is the first effect that can be related to the length of the polymer chains. It is observed as a surface texture in which the lamellae are divided into discrete regions bounded by a growth face. This phenomenon is a direct consequence of chain folding along the growth face and would not occur if the chain formed prefolded blocks before attachment to the crystal. *Sectorisation* arises because the folding of the chains transforms a long molecule into a pleated sheet which can extended across several successive lattice planes but essentially lies along the relevant growth surface denoted as the *fold* plane. The so-aligned chain breaks the symmetry and slightly distorts the repetitive unit of chain packing within the lamellae. The fold plane in a given sector differs from nominally equivalent ones along which there is no folding, thus transforming a single lamella into a multiple twin. The size of the sector is governed by its growth face, which can be very small if it is dendritic growth, where each individual facet has its own microsector with dimensions as small as 20 nm in width. The consequence of this sectorisation is that the platelets do not necessarily stack in a regular fashion, thus making the lamellae nonplanar. Folding along the growth faces makes nominally equivalent planes in the subcell become unequivalent, adopting slightly different spacings.

3.5 Melt crystallised lamellae

The bulk material is full of lamellae that often have different profiles. In a typical crystalline polymer, such as polyethylene, spherulitic growth is responsible for the spatial variation in physical properties. The structures that are observed are the results of the growth of dominant lamellae that branch and diverge. Except in the case of very low molecular weight materials, the space created between the lamellae is filled by *subsidiary* or *infilling* lamellae. This type of growth will produce lamellae with different characteristics for two reasons. First, fractional crystallisation, in which the chains with different molar mass-chain lengths (molar mass is simply the monomer molar mass multiplied by the degree of polymerisation, i.e. the number of monomer units in the polymer; the chain length is the length of a bond multiplied by the degree of polymerisation) segregate from one another, allows the shorter chains to form lamellae which have a lower modulus and melting point. Secondly, the different orientations of the lamellae help to develop isotropic properties, a proportion of the lamellae being in the direction of the applied force and others being perpendicular with a statistical distribution in all other directions. If the material is cold drawn then the applied stresses will develop physical characteristics in the material that are the result of the alignment of the dominant lamellae and the development of the enhanced modulus in the draw direction. The natural draw ratio is influenced by the 20° alignment of the subsidiary lamellae to the dominant structures. Studies of melt grown crystals have revealed terrace-like structures associated with the regular stacking of the lamellae (see Figure 3.8).

3.6 Polymer spherulites

The supermolecular structure exhibited by many polymers has features that are in the range of 0.5 μm to several millimetres and are best observed using polarised light optical microscopy. The common structure exhibited by many polymers is the spherulite, which is a circular structure. The first spherulites were found in igneous rocks and reflect the ball or globe nature of the structure. Two unique refractive indices may be determined, namely the tangential (n_t) and radial (n_r) refractive indices. In crystalline polyethylene, the polymer is uniaxially birefringent, with the unique direction (largest refractive index) along the chain axis down the *stem*.

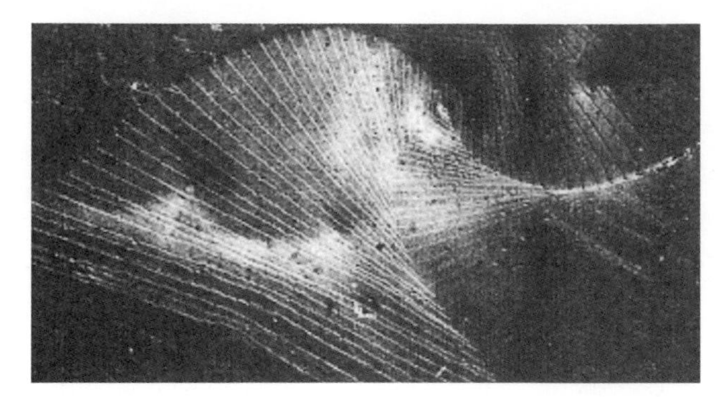

Figure 3.8 Electron micrograph of melt grown multilayered crystal of polyethylene displaying regular rotation of successive growth terraces.

Negative spherulites with $n_t > n_r$ have a higher proportion of the chains in the circumferential planes than along the radius of the spherulites. The direction of growth of polyethylene spherulites is always in the plane of the lamellae (see Figure 3.9).

Other polymers exhibit similar spherulitic structures. The size of the spherulites is controlled by the nucleation process which is almost invariably heterogeneous, i.e. growth starts from extraneous material such as particles of dust or residual catalysts. It is generally agreed that spherulitic structure is controlled by the dominant lamellae growth and usually adopts a circular format. Starting from an individual lamella, the growth progression is to first create the dominant lamellae that then creates a multilayered axialite, a parallel organised set of lamellae. Several fast growing lamellae that may splay and which present a sheaf-like appearance down the principle axis of the splay (see Figure 3.9) will emerge from the axialite. Lamellae, which will eventually be at right angles to the original axis, as well as growth occurring parallel to the original structures (see Figure 3.10), will emanate from these splays. The axialite is a nonspherical and irregular superstructure. Axialites are primarily found in low molar mass polyethylene at essentially all crystallisation temperatures and in intermediate molar mass polyethylene crystallised at higher temperatures at undercooling by less than 17K. The gross morphology is a consequence of the intersection of the growth fronts from

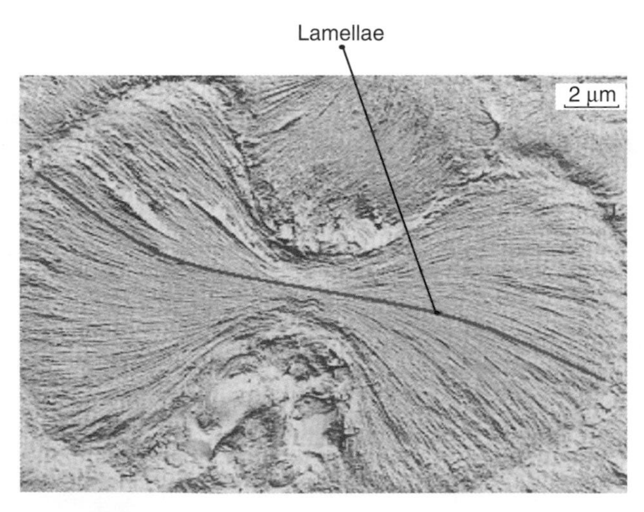

Figure 3.9 Sheaf-like lamellar aggregates crystallised from melt at 125°C in a blend of linear and low density polyethylene.

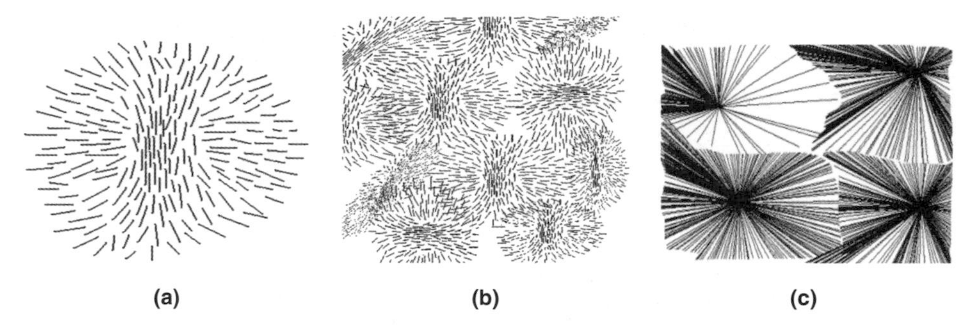

Figure 3.10 Growth of spherulites: (a) schematic of lamellae growth from an axialite to a spherulite; (b) packing into a solid; (c) schematic diagram of spherulites viewed with polarised light.

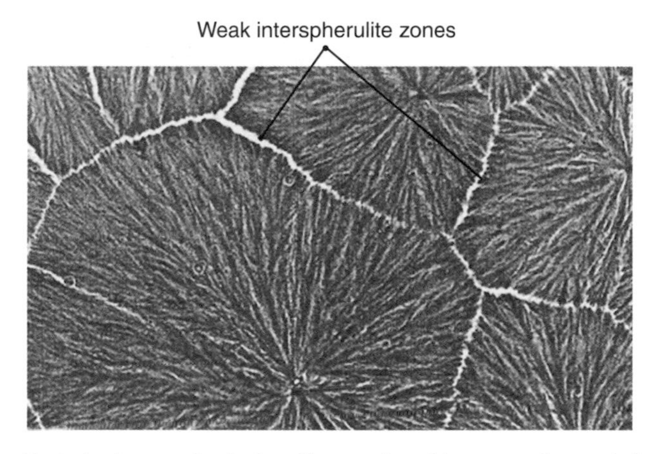

Figure 3.11 Optical micrograph of spherulites produced by a propylene–ethylene copolymer.

the various nucleation centres and boundaries are often hyperbolic reflecting the intersection of the spherical structures. It must always be remembered that these are three-dimensional structures and apparently distorted shapes are often structures viewed at a different angle. All the structures in Figure 3.10(b) were created by rotation or skewing of the original spherulite (see Figure 3.10(a)); rotation and skewing are slightly different processes, the latter involving translation as well as rotation. When viewed with polarised light, a structure like a Maltese cross is usually observed and the spherulites appear to have different shades. This behaviour is exemplified in the case of the optical microscopic images for ethylene–propylene (see Figure 3.11).

A scanning electron micrograph of a fracture surface of polypropylene shows the characteristic rings of an underlying spherulitic structure (see Figure 3.12). The spherulitic crystal structure can be subjected to low temperature drawing. During this process the lamellae remain intact but are rotated into the line of the applied stress. As a consequence, the lines of weakness that are the points at which the spherulites intersect open and form voids between the developing fibres. The spherulites elastically respond up to ~30% strain, followed by permanent inhomogeneous changes which involve combinations of slip, twinning and phase changes.

If a crystalline polymer is heated to just below its melting temperature, it is relatively easy to align the lamellae in the direction of an applied stress (the *draw* direction). This process is used to produce polyethylene string that is widely used to secure computers, TVs and DVDs and other electrical goods. Polyethylene drawn to form fibres produces material that can be used to

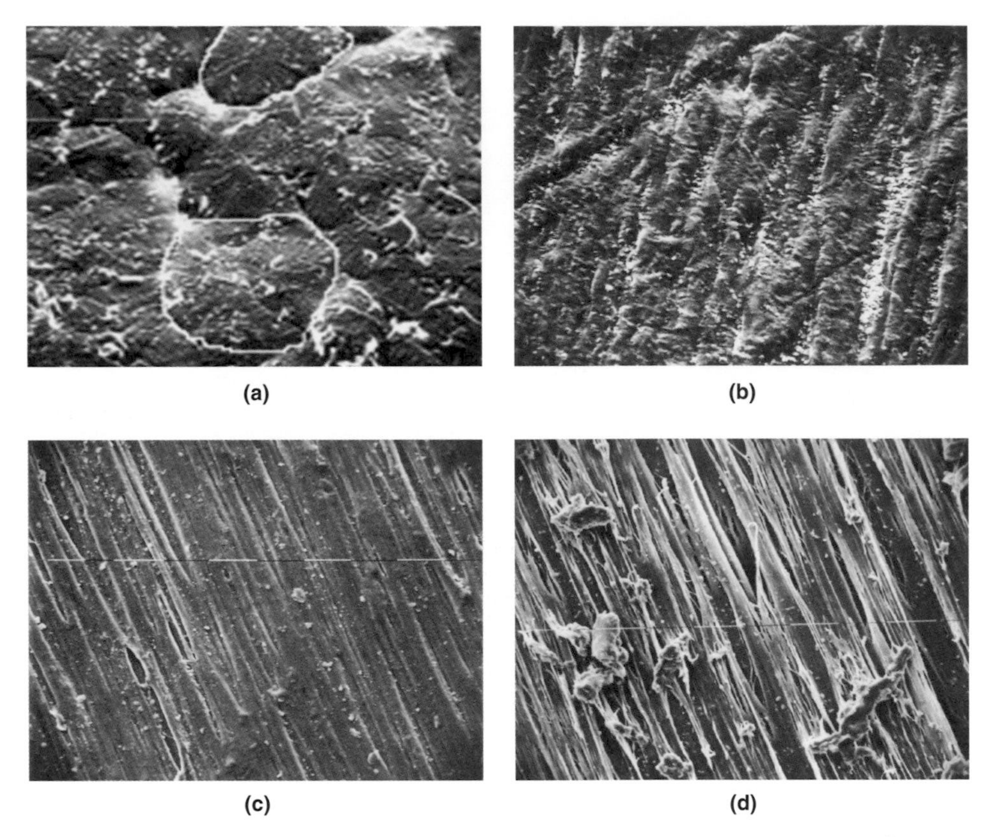

Figure 3.12 Scanning electron micrographs of fracture surfaces: (a) undrawn surface, $\lambda = l/l_0 = 2.6$; (b) drawn surface, $\lambda = l/l_0 = 2.6$; (c) $\lambda = 6$; (d) $\lambda = 14$; where l is drawn length and l_0 is undrawn length of sample.

produce useful composites. The processes which occur are easily seen by examining the changes which occur to the solid using scanning electron microscopy (see Figure 3.12).

The draw ratio (*DR*) is the ratio of the length after the material is drawn, l compared with the length before it was drawn, l_0, thus $DR = l/l_0$. At $DR = 2.5$, a significant proportion of the lamellae in the spherulites have been orientated into the draw direction with the results that the modulus in that direction has been enhanced. However, there will be a significant proportion of the lamellae that are aligned transverse to the draw direction and the material retains a significant strength transverse to the draw direction. Increasing the value of *DR* to 6 causes a significant increase in the degree of alignment of the lamellae and the weak interfaces that correspond to the points at which the spherulites meet start to appear as voids in the structure. Increasing the draw ratio to a value of 12 further increases the development of fibres associated with the alignment of the crystallites and clear voids appear in the structure. The 'debris' in the surface is the residue of the small spherulites that have been stripped from the structure. The variation of the modulus with draw ratio is shown in Figure 3.13. The development of the modulus often follows a two-step process. At low draw ratios the lamellae are aligned with the draw direction but there will still be a significant number that retain the transverse modulus. At the natural draw ratio significantly greater alignment is achieved and the material will often neck and may change optically. Polyethylene will change from an opaque solid to a transparent film. The change in optical characteristics reflects the creation of structures that are submicron in dimension, the fibres will have

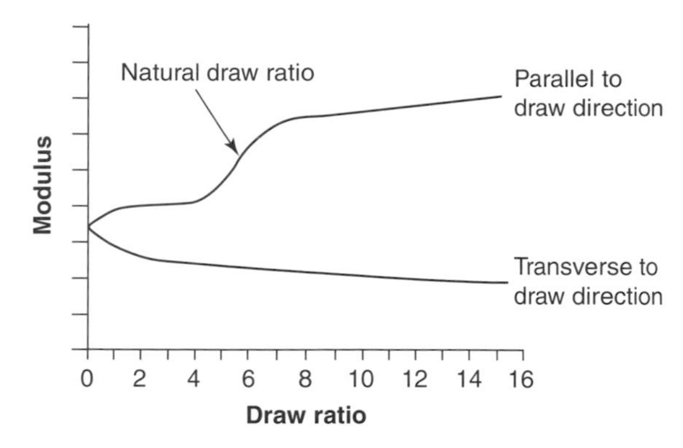

Figure 3.13 Variation of modulus in draw direction and transverse to it for a crystalline polymer.

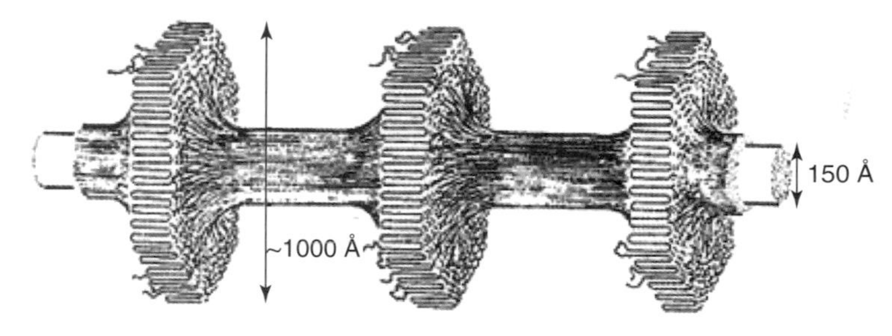

Figure 3.14 Schematic of a section through a shish-kebab structure showing extended core with radial lamellae growth.

a length of many millimetres or greater. The original material has spherulites with dimensions in the micrometre scale which can scatter light and so the material appears to be opaque.

The morphologies discussed so far have all been formed from isotropic melts. In practice, melts are often subjected to shear forces and can induce alignment in the chains, which aids crystallisation in particular directions. The orientation of the melt causes an increase in the free energy and this itself constitutes an important factor in practical processing. Shish kebabs are formed from solutions that are subjected to elongational flow, which induces orientation of the solute molecules (see Figure 3.14).

A central core of orientated bundles of fibres is formed at first as a direct consequence of the orientation. The shish kebab consists of a central group of highly orientated fibrils from which lamellar crystals can grow and form the kebab-like structures shown in Figure 3.14. The central fibrils are formed from high molar mass material. Similar structures have been obtained during extrusion/injection moulding of the melt when extreme conditions are used and the melt is subjected to high elongational flow in combination with a high pressure or a high cooling rate. The orientated melt solidifies in a great many fibrous crystals from which lamellae overgrowth occurs. The radiating lamellae which nucleate in adjacent fibrils are interlocked, a fact that is considered to be important for the superior stiffness of melt extruded fibres. The fibrillar structure is present in ultra-orientated samples. It consists of highly orientated microfibrils. These microfibrils are sandwiches of alternating sequences of amorphous and crystalline regions. A great many taut interlamellar tie chains are present and the resulting mechanical properties are excellent. The fibres are formed from stacks of the lamellae and have a very high strength consistent with the crystal structure of the polymer. At high draw ratios the fibres part and a significant void structure can be observed between the fibres. Studies of highly linear low molar mass metallocene synthesised polyethylene has shown a tendency for this material to grow a sheaf-like morphology (see Figure 3.7).

3.7 Differential scanning calorimetry

There are slight differences in the methods of making the actual measurements in differential scanning calorimetry (DCS) which depend on the manufacturer of the instrument, but the information obtained is essentially the same. If we raise the temperature a sample of material heat will be absorbed in proportion to its specific heat, C_p. If, however, as the temperature is raised the sample undergoes a melt transition then extra heat will need to be provided to the sample to achieve the same heating rate. The experiment is usually carried out using two pans, the term differential indicating that the measured heat flow is the difference in heating the unknown in reference to some standard material. A typical trace for a crystalline polyethylene is shown in Figure 3.15.

The initial heating up to approximately 50°C appears as a linear traces, as would be expected if the heat capacity of the material were constant. The first peak at ~77°C is indicative of the melting of small crystals which are often found around the larger crystals. The main melting feature occurs at ~112°C and corresponds to the melting of the bulk of the crystals which are all of a similar size. However, there are a few larger, slightly more stable, crystals and these melt at 116.4°C. The area under the peak is proportional to the total volume of crystalline material present and allows the % of crystallinity to be determined provided that the enthalpy for the perfect crystalline material is known.

If we consult the technical literature looking for a polyethylene, a variety of different grades are available which have different densities and crystallinities and are referred to as low (LDPE), medium (MDPE) or high density (HDPE) polyethylene. The principle difference between these grades of PE is the number and distribution of branches in the side chain. Either as part of the synthesis or as a consequence of random backbiting reactions, side chains can exist in the PE structure (see Figure 3.16(b)). Essentially, the highest density material will have no branches and is able to form a near-ideal packed crystalline structure. The medium- and low-density materials will have increasing branched chain content. If the polymers are made using a radical initiated process then there will be a prevalence of ethyl and butyl side chains with an occasional methyl side chain. Manufacturers will sometimes introduce a longer chain vinyl terminated monomer into the polymerisation and it is possible to find C_8H_{17} or C_9H_{19} groups in the lower density materials. The side chains will inhibit the packing of the chains together and hence increase the 'amorphous' content of the polymer.

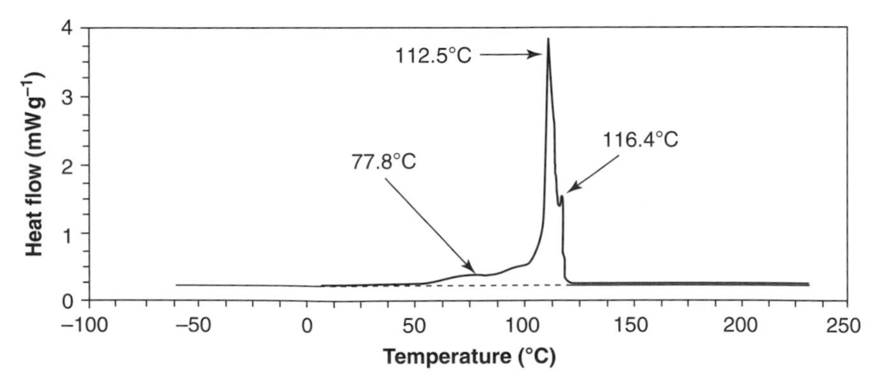

Figure 3.15 DSC trace for a high density linear poly(ethylene) sample.

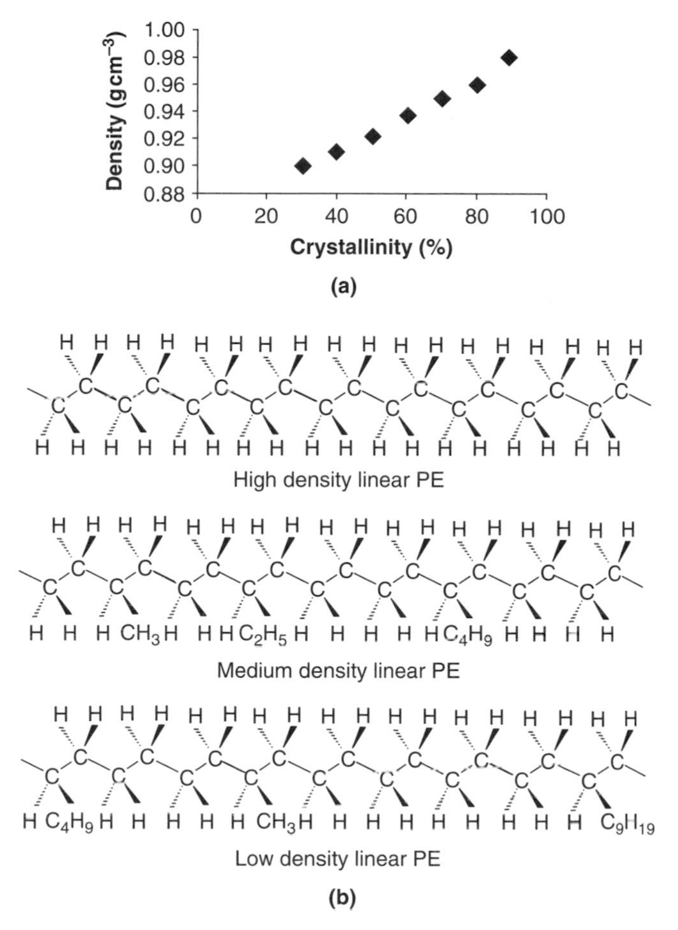

Figure 3.16 Samples of PE: (a) plot of density versus % crystallinity; (b) variations on basic polymer structure.

3.8 Polytetrafluororethylene

Substituting the hydrogen atoms in PE with fluorine atoms produces a very crystalline material (93–98% crystalline) with a melting point of ~350°C. In order to make polytetrafluororethylene (PTFE) usable the pure PTFE is sintered in an inert nitrogen atmosphere. The sintering process at 350–400°C leads to the bonds in the backbone of the polymer breaking and C_4,C_6 side chains being formed (see Figure 3.17).

The branched chain structure, as in the case of PE, inhibits crystal packing and results in a reduction in the crystallinity and a softer, more pliable material results. The polymer structure can be thought of as being a copolymer of PTFE with a comonomer of C_6F_{12}. A 100% crystalline material is almost impossible to process and the usual PTFE tape used in plumbing and electrical repairs is only about 60% crystalline.

3.9 Other types of morphology in semicrystalline polymer systems

Hydrogen bonding, created by the interaction of a hydrogen atom with a polarisable chemical bond such as C–*X*, where *X* is a halogen (chlorine [Cl], bromine [Br], iodine [I] or fluorine [F]), or a group such as a hydroxyl (–OH) or a carbonyl (C=O) can have a major influence on the

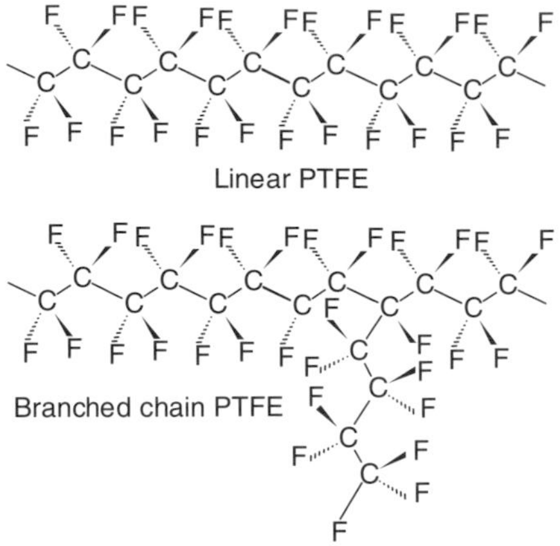

Figure 3.17 Structure of linear and branched PTFE.

morphology. The classic example of hydrogen bonding is DNA and the analogous interactions in synthetic polymers are found in nylon. The bonding in nylon is related to that found in DNA and has the form of an amide linkage (NH–C=O–CH_2-). Nylon is the generic name given to polymers which have the chemical structure: -NH–$(CH_2)_x$–NH–CO–$(CH_2)_y$–CO- where x and y are integers. For example: nylon 6,6 will have $x = 6$ and $y = 4$. This is called nylon 6,6 because it has six carbon atoms in each of the component elements, the second element which contains the carbonyl groups has six carbon atoms if we add the two carbonyl atoms to the four methylene carbon atoms. Nylon 6 has the structure –NH–$(CH_2)_6$–CO- , the repeat unit contains only one amine (NH) and one carbonyl (C=O) group (see Figure 3.18).

At first sight the two polymers would appear to look very similar, but when we look at the implications in terms of the crystal structure (see Figure 3.18) we see that they are different and these differences make the materials suitable for slightly different applications.

In nylon 6,6, the carbonyl and amine groups line up in adjacent polymer chains to form a well-ordered and regular crystal structure with a characteristic well-defined shift of ~20° in the crystal plane which reflects the periodicity of the chain structure. In nylon 6, a match in the carbonyl and amine spacing occur every other repeat unit and there is no shift in the crystal plane. The N–H and C=O groups which do not match can be rotated out of the plane and form hydrogen bonds in the z-direction. In the case of nylon 6,6 the amide bonds will naturally lie in the x,y plane, producing a layered structure. The mechanical properties in the z-direction will rely on the weak Van der Waals interactions and will not be reinforced by hydrogen bonding. These differences in crystal structure are reflected in the mechanical properties of the materials, nylon 6 being a tougher and stronger material than nylon 6,6. There are a variety of different nylon materials each having slightly different characteristics and properties. The development of a three-dimensional network in nylon 6 allows the material to be used in contact with water whereas nylon 6,6 shows susceptibility to swelling in contact with water.

3.10 Copolymers and phase separation

So far the discussion has centred on homopolymers, those polymers that contain only one type of monomer unit. There are many important polymer systems which are created using two or more monomers: these are called copolymers. Many of the monomers which are used to create the copolymers are thermodynamically compatible to a greater or lesser extent. At a molecular level the monomers will attempt to segregate so that chain-like elements are clustered together in space. The clustering or phase separation is only possible if the number of segments that form

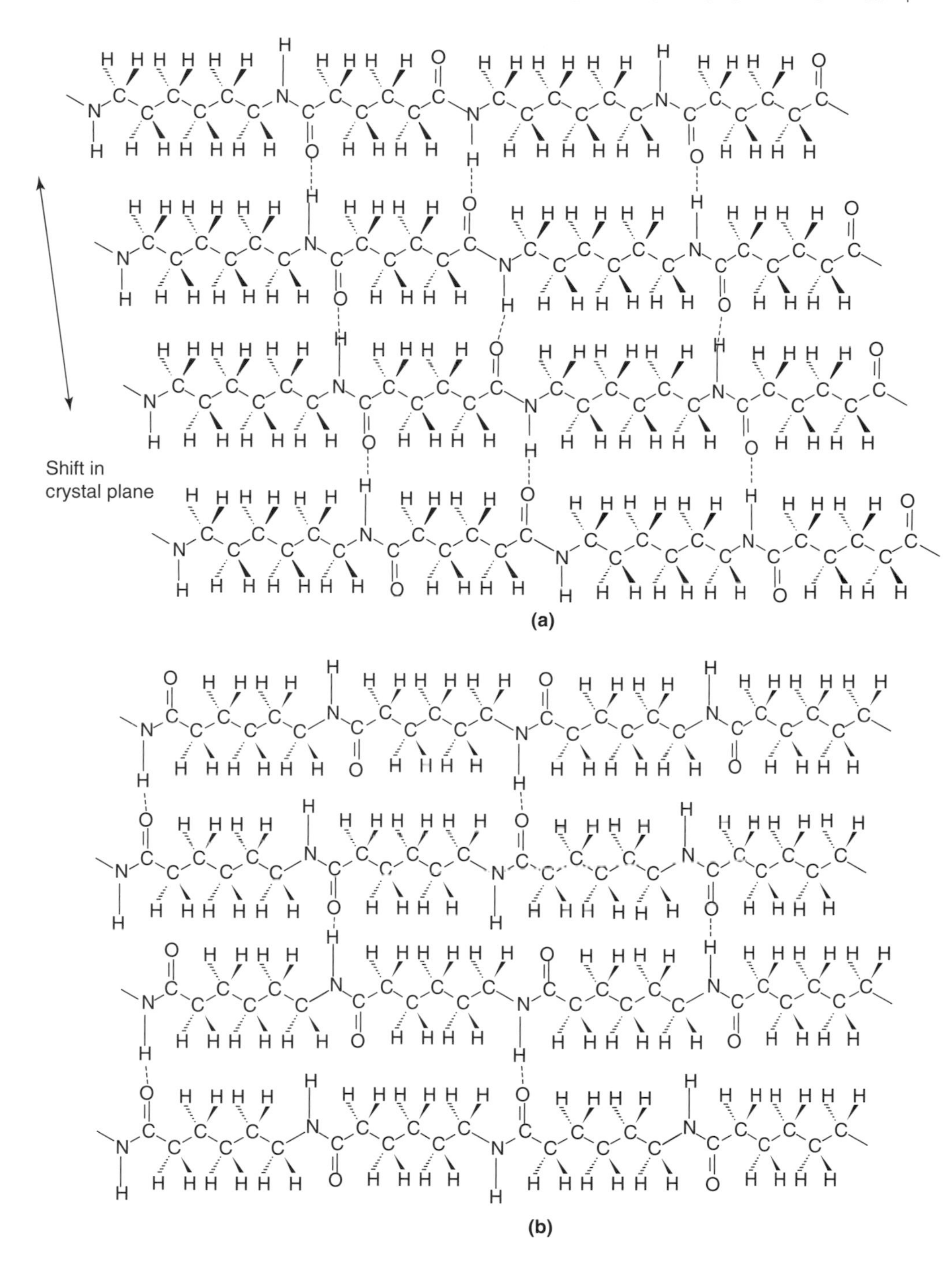

Figure 3.18 Schematics of crystal structure of (a) nylon 6,6 and (b) nylon 6.

a continuous sequence is of sufficient length for the thermodynamic driving force to be large enough to drive the elements into different regions of space.

3.10.1 What are the implications of phase separation?

A number of different copolymer structures were identified in Section 1.7. If we examine their physical properties we find that they exhibit characteristics that reflect the extent to which phase separation occurs in the system.

3.10.2 Alternating, random copolymers and blends of two polymers

The copolymer materials will show behaviour that resembles that of the blend, but can exhibit a distinct glass transition point (T_g) that reflects the internal plasticisation of the material, and the dynamic mechanical thermal analysis (DMTA) traces look like those of a simple polymer with a T_g value between the extreme values for the homopolymer. The DMTA behaviour of this type of system is shown schematically in Figure 3.19.

3.10.3 Block copolymers of incompatible monomers

If the monomers are incompatible then phase separation will occur. The phase structure will depend on the type of polymer, how it is being processed and whether or not it was cast from solution or from the melt phase. The types of morphology that can be observed will depend on the two phases that are present. Styrene–butadiene is a good example of a block copolymer system. The morphology can be observed by electron microscopy if the butadiene phase is stained with osmium tetroxide. The osmium tetroxide can add to the butadiene double bonds and stains the polymer phase having a high electron contrast to the only carbon-containing phase. The attachment of the heavy atom makes these bonds scatter electrons and they appear black in the electron micrograph (see Figure 3.20).

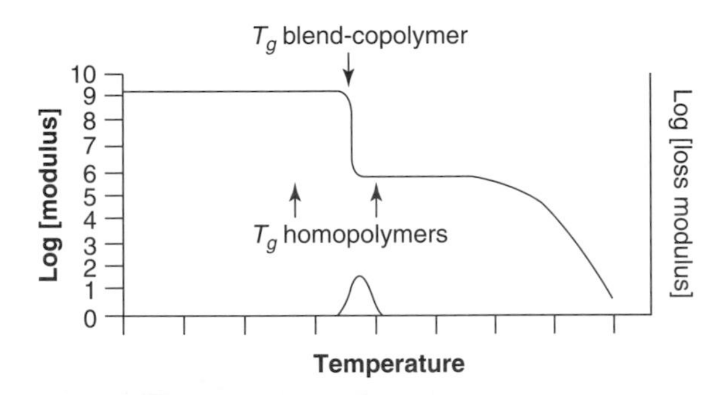

Figure 3.19 Schematic of DMTA traces for a compatible blend of homopolymers or random or alternating copolymers – (a) and incompatible blend.

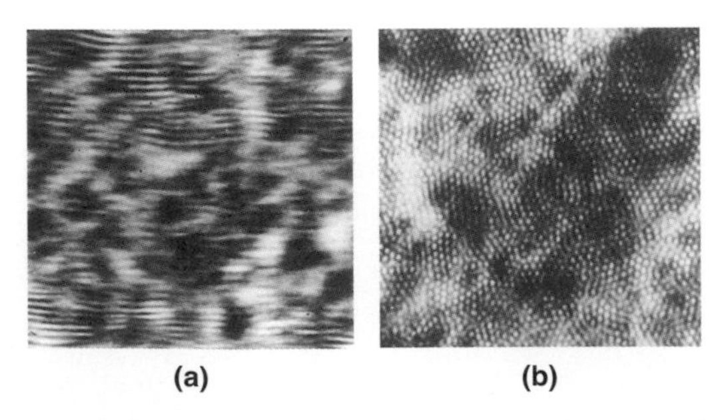

Figure 3.20 Electron micrographs of osmium tetroxide-stained styrene–butadiene–styrene triblock copolymer: (a) in direction of extrusion; (b) transverse to direction of extrusion.

The possible morphologies are:

- Isotropic morphology made up of spheres of styrene.

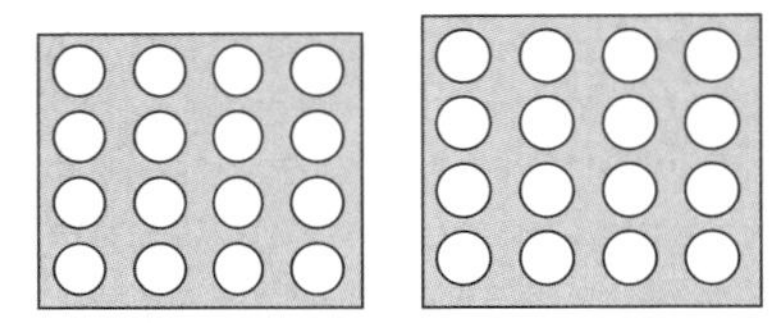

- Morphology in extruded materials depends on direction: cylinders in the extrusion direction and spherical transverse to extrusion direction.

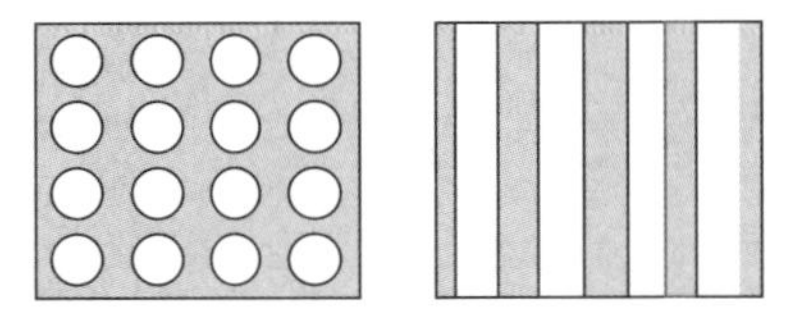

- Lamellar structures obtained in some solvent cast materials.

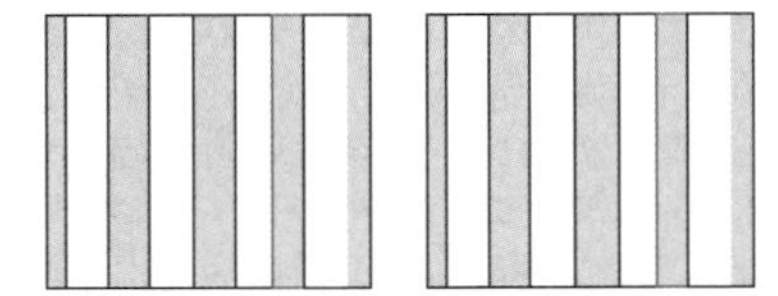

The styrene phases are high modulus phases with a value of 10^{10} N m^{-2} at room temperature, whereas the value for the butadiene phase is 10^6 N m^{-2}. The butadiene phase will exhibit elastic properties and can be deformed whereas the styrene phase cannot be deformed. The above structures lead to distinctly different mechanical properties in these materials. The aligned cylindrical structure is generated when the polymers are extruded and the resultant cylinders are orientated in the direction of extrusion.

Anisotropy of mechanical properties

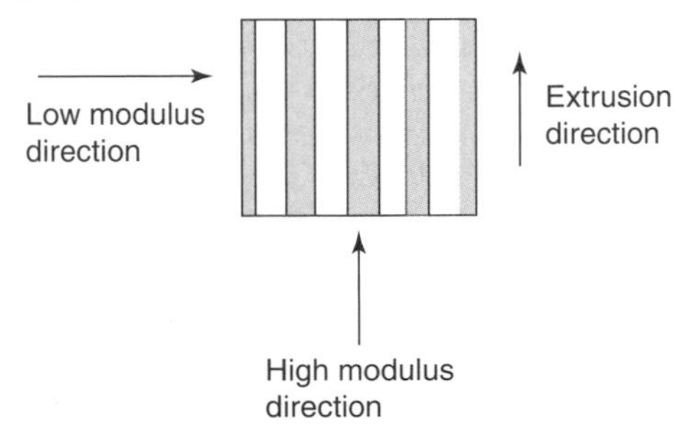

If the dark areas are polystyrene and are aligned in the direction of extrusion, the modulus will have a value of the order of $10^{10} \times$ N m^{-2}, whereas transverse to the extrusion direction the value will be of the order of $10^6 \times$ N m^{-2}. In the extrusion direction, the styrene columns control the modulus, whereas in the transverse direction it is the rubbery butadiene phase which is important. These materials are used in the generation of car tyres. The phase structure will influence the temperature profile of the modulus. The cylindrical morphology will be developed when the melt is subjected to shear (see Figure 3.20).

Diblock copolymer

The typical DMTA trace for a diblock copolymer is shown in Figure 3.21. The first transition is that associated with the T_g of the butadiene phase. Once these chains are mobile because they are only anchored at one end, the solid begins to lose its mechanical properties and the material becomes almost completely mobile at the T_g of polystyrene. Because the butadiene chains are mobile above the T_g of the polymer it is possible for flow to occur and the extent to which this occurs will depend on the molar mass of the butadiene polymer. If the butadiene has molar mass above M_c then the chains will be entangled and in effect the butadiene phase will act like a viscoelastic liquid and its modulus will be maintained to high temperatures. If the molar mass is below M_c the chains will become very mobile and flow can occur to a very significant extent.

Triblock copolymer

In the case of the triblock copolymer, the polymer chains are anchored at each end in the styrene phase and as a result the modulus stays almost constant with temperature between the T_g of the butadiene and the T_g of the styrene phase (see Figure 3.22). Above the styrene phase the phase separation will maintain the viscoelastic character of the material since the polymer chains will have a combined molar mass that will be above M_c.

Above the T_g of styrene the material will be viscoelastic and the length of the rubbery plateau region will depend on the molar mass of the polymer and the degree to which it is entangled. Usually the molar mass of the constituent polymers are sufficiently high for entanglement to exist and a rubbery plateau to be observed (see Figure 3.23).

3.10.4 Varying the styrene–butadiene ratio in triblock copolymers

By changing the molecular architecture it is therefore possible to control the modulus of the temperature profile for polymers and hence develop properties which make them suitable for various uses. In the case of the styrene–butadiene–styrene triblock copolymers, DMTA traces of

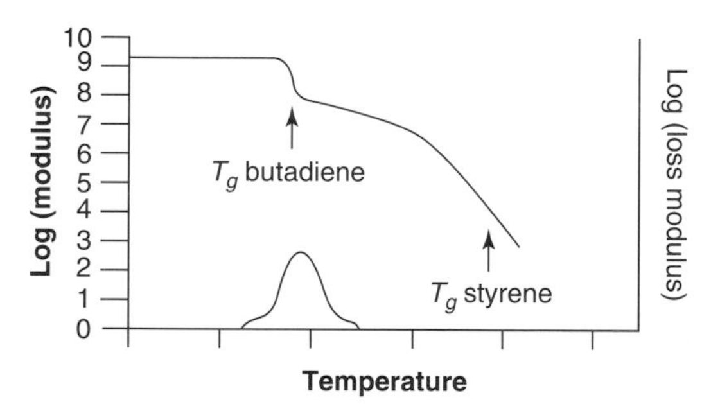

Figure 3.21 Schematic of DMTA trace for a styrene–butadiene diblock copolymer.

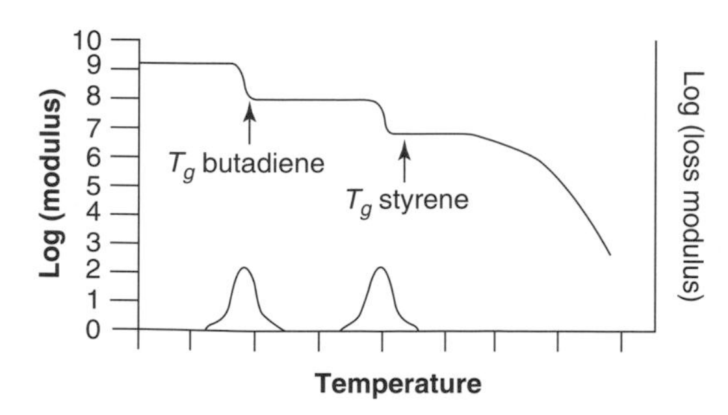

Figure 3.22 Schematic of DMTA trace for a styrene–butadiene triblock copolymer.

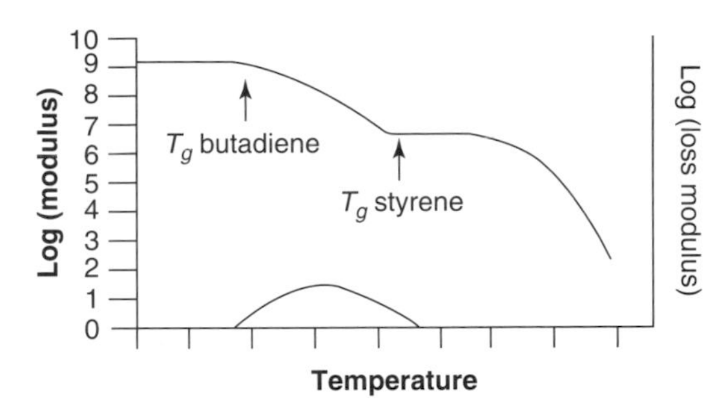

Figure 3.23 DMTA plot for a blend of styrene and butadiene.

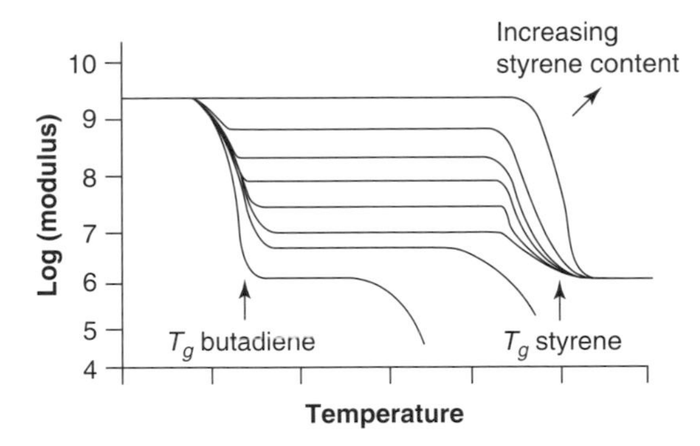

Figure 3.24 DMTA traces for styrene, butadiene and styrene–butadiene–styrene triblock copolymers with varying styrene content.

the copolymer show a regular variation of the modulus with composition, the value being essentially constant between the T_g of butadiene and that of styrene (see Figure 3.24).

The modulus of the plateau region between the two T_g values can be approximated by a simple mixing law:

$$\log\left[M_{copolymer}, T\right] = x_1 \log\left[M_{styrene}, T\right] + x_2 \log\left[M_{butadiene}, T\right] \tag{3.1}$$

where $\log[M_{styrene}, T]$ will be the value of the glass ~10^9 N m^{-2} and $\log[M_{butadiene}, T]$ will be the value of the rubbery phase ~10^6 N m^{-2} and x_1 and x_2 are their respective volume fractions. Whether or not a change in modulus at the T_g of styrene is seen for the low styrene content polymers will depend on whether the molar mass of the butadiene is above M_c and the extent to which phase separation has been achieved.

3.11 Why do we need to be able to change the modulus of polymeric materials?

A car tyre is a good example of the selective use of materials with different values of their modulus to achieve a particular load bearing characteristic. In the tyre we require stiffness in the walls that are attached to the rim, yet they need to be sufficiently flexible to give a good ride. The surface in contact with the road has to be hard wearing yet conformable to give the required level of friction with the road surface. To match the different flexibilities and modulus characteristics of the materials used in the running surface and the tyre walls other matching materials are used to help maintain shape and structure. Seven or eight different materials may be used in a typical high-performance car. Using different materials it is possible to match the performance of the tyre to the weather conditions and allow selection of 'dry' and 'wet' tyres which can have very significant influence in Formula 1 car racing. See Section 7.3.3 for further discussion of this topic.

3.12 Polyurethanes

A very important class of polymers that are tailored for their application are materials based on polyurethane (PU) and polyurea. Polyurethanes are often available as thermoplastic or alternatively thermoset materials. The main difference between these materials is the functionality of the monomers. If only difunctional isocyanates and diols are used then the material is a thermoplastic. However, if higher functional diols or isocyanates are used then a three-dimensional matrix is created and the material is a thermoset. There are two types of isocyanate commonly used: 4,4′methylenediphenyl diisocyanate (MDI) and mixtures of 2,2′- and 2,4-toluene diisocynate (TDI) (see Figure 3.25).

The TDI is more difficult to handle and is usually only used by foam manufacturers. MDI is usually used in the construction industry. PUs are used in a variety of different applications, which include the soles of shoes, conveyor belts, doctor blades in photocopiers and specialist tyres. They are also widely used for many engineering applications.

The PUs are formed by the reaction of the isocyanate with a diol. The diol will usually be polymeric and may be a polyether or polyester (see Figure 3.26). By changing the nature of the polyether or polyester the susceptibility of the material to oil and water can be changed. Polyethers can absorb water whereas polyesters with large aliphatic chains between the ester groups are less susceptible to moisture uptake. If water is present in the isocyanate or the polyol then a side reaction can occur (see Figure 3.27).

As with nylon 6 the hydrogen bonding does not all occur in the plane and hence matching one set of hydrogen bonds promotes the set on the other end of MDI to seek to form hydrogen

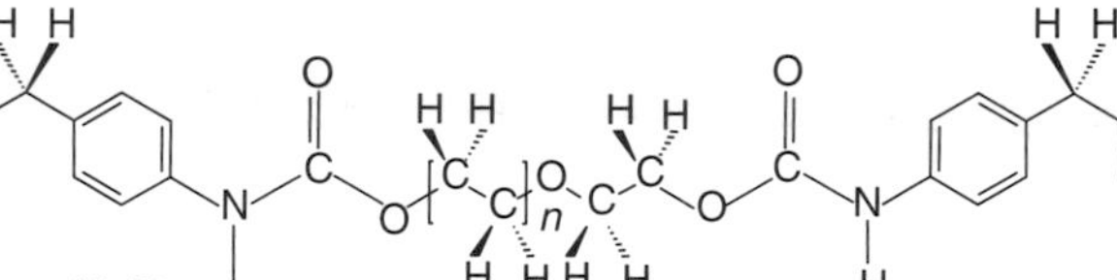

OCN NCO CH$_3$ NCO OCN
MDI TDI

Figure 3.25 Structures of MDI and TDI.

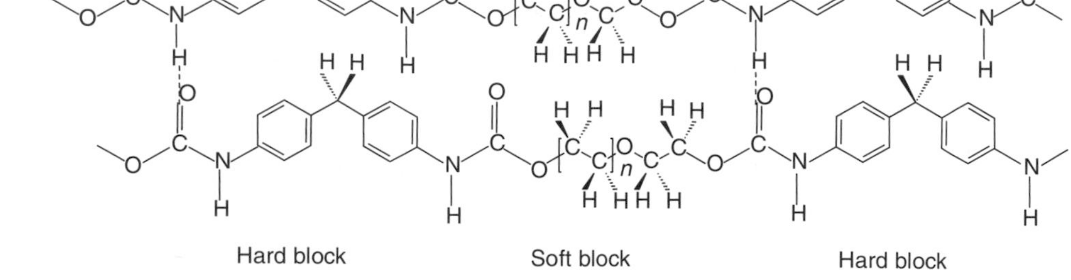

Hard block Soft block Hard block

Figure 3.26 Schematic structure of a polyurethane containing a diol and MDI.

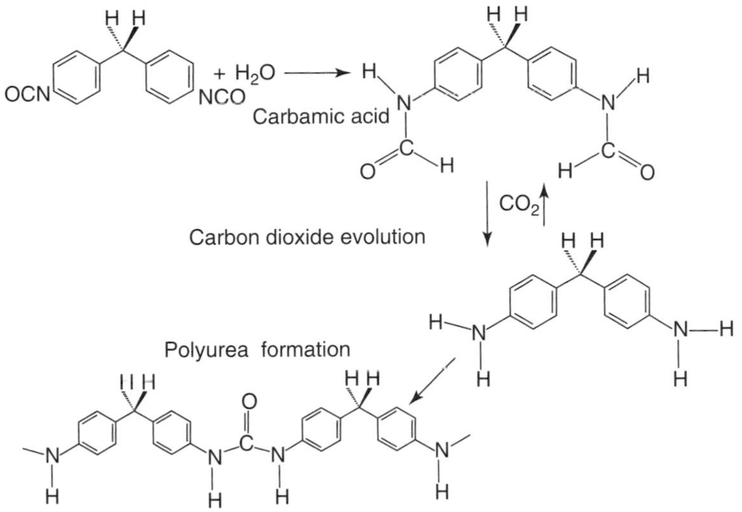

Figure 3.27 Schematic for formation of urea and generation of carbon dioxide.

bonds out of the plane. The MDI hydrogen-bonded sections tend to phase separate and form rigid phases within the matrix. The melting point of these phases is ~150°C and this gives the stability which is desired in these materials. The T_g of the polyethyleneoxide phase depends on the chain length; short chains have low values of the T_g which can be as low as –20°C. As the chain length is increased sufficient lengths of polyethyleneoxide chains exist for this 'soft phase' to have crystalline zones and the melting point is typically of the order of 40°C. The properties of the polymer can easily be changed by varying the amount of the MDI and the molar mass of the polyethyleneoxide. Polyurea is created by the reaction of an isocyanate with an amine or can be created when water is present in the system (see Figure 3.28). As in the case of nylon 6, both groups are available to form hydrogen bonds in the plane and at right angles to the plane and the resulting material is hard and strong.

The reaction of water with the isocyanate first forms a carbamic acid; this then decomposes with the generation of carbon dioxide and the production of an amine. The amine can then react

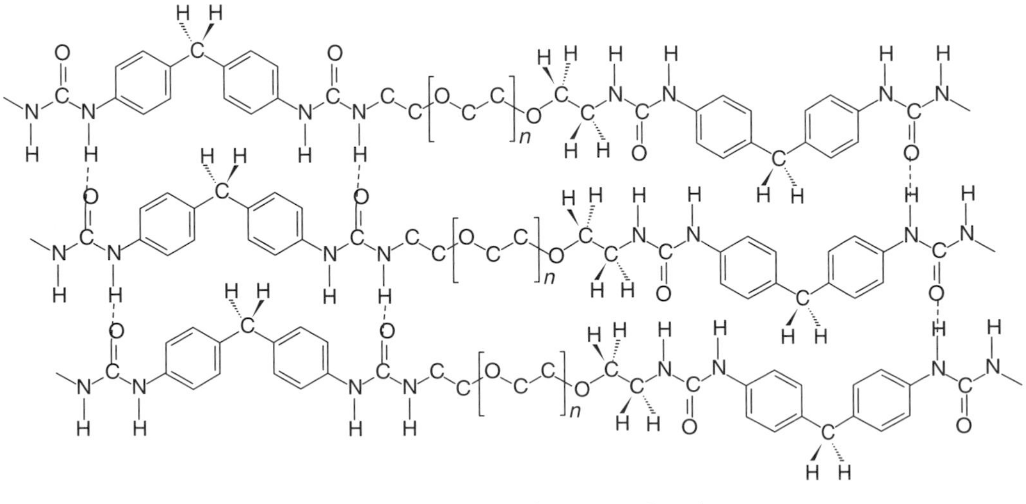

Figure 3.28 Schematic of structure of a polyurea.

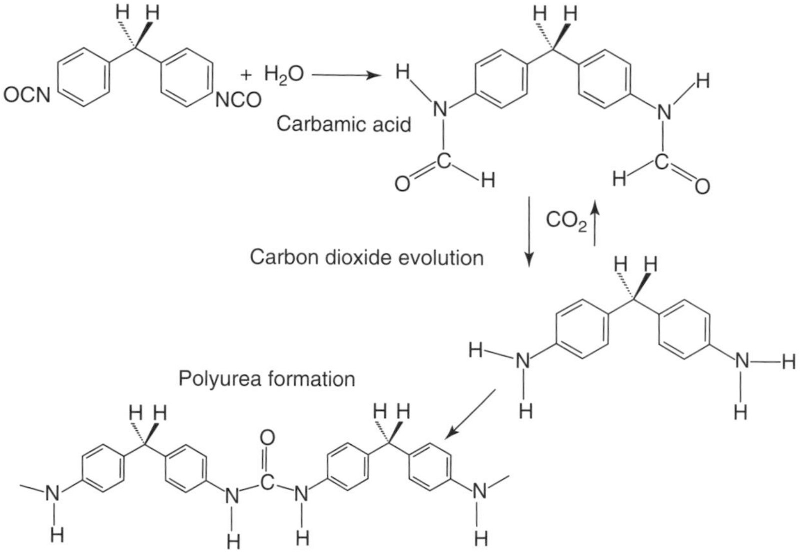

OCN NCO + H_2O → Carbamic acid

CO_2 ↑ Carbon dioxide evolution

Polyurea formation

Figure 3.29 Schematic for formation of urea linkages and generation of carbon dioxide.

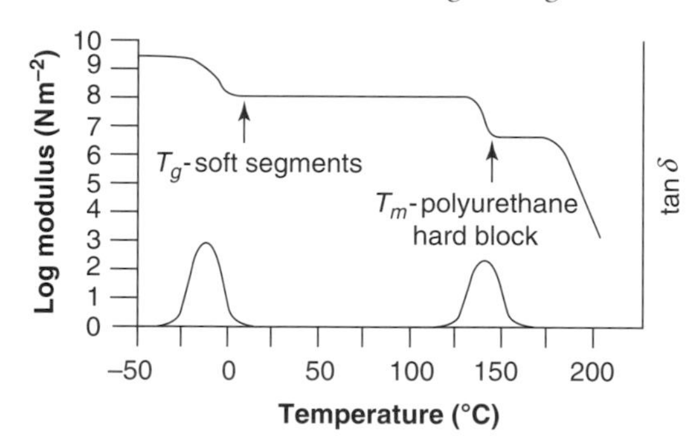

Figure 3.30 Schematic for DMTA response of a typical phase-separated polyurethane elastomer.

with isocyanate to form a urea linkage. In a typical foamed material water is added to the polyol and the evolved carbon dioxide creates the foam (see Figure 3.29). A typical DMTA trace for a thermoplastic polyurethane is shown in Figure 3.30.

Depending on the type of polyether or polyester so the extent of phase separation may vary, as will the value of the T_g. The more phase separated the polyether or polyester, the more likely that it will exhibit a sharp transition and the location will usually be close to that for the soft block as a pure material. If the material is well phase separated then the modulus of the material will remain fairly constant between the T_g of the polyether or polyester (*soft block*) and the melt transition for the hydrogen bonded isocyanate/urea–urethane phase. The phase separated urea–urethane phase is known as the *hard block*. The apparent temperature independence of the modulus makes these materials ideal for fabricating shoes, conveyor belts and, in general, materials where elasticity is required in addition to a degree of load bearing capability.

3.13 High-temperature polymers

Certain polymers are classed as high-temperature materials. The so-called engineering polymers fall in this group and include: polyether ether ketone (PEEK), polyethersulfone, polysulfone and polycarbonate (see Figure 3.31).

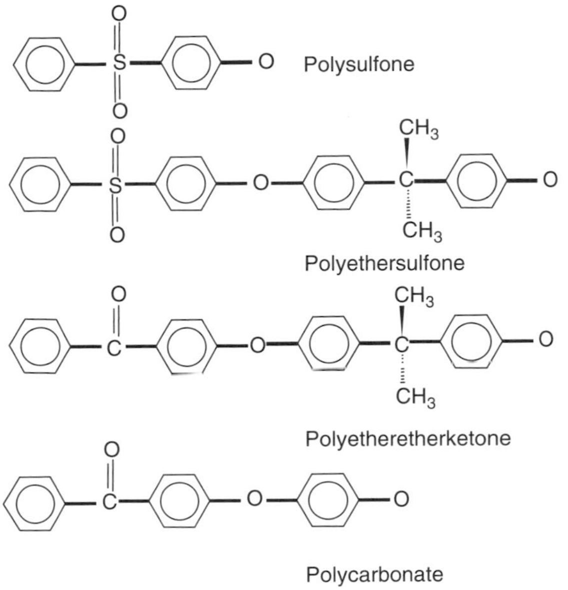

Figure 3.31 Structures of polysulfone, polyethersulfone, polyetherketone and polycarbonate.

These polymers can be heated to above 250°C without substantial degradation in their properties. The polyimides are, however, very useful for high-temperature applications and can be heated to 420°C before they suffer significant degradation. In fact, their modulus at 450°C is only slightly lower than the value at 350°C and is 2/3 of its value at 250°C. Polyimides are extensively used for high-temperature and high-modulus applications (see Figure 3.32). The polyamidic acid is a flexible material which on heating forms a rigid and inflexible polymeric form which has applications as fillers, where high-modulus and high-temperature performance are required.

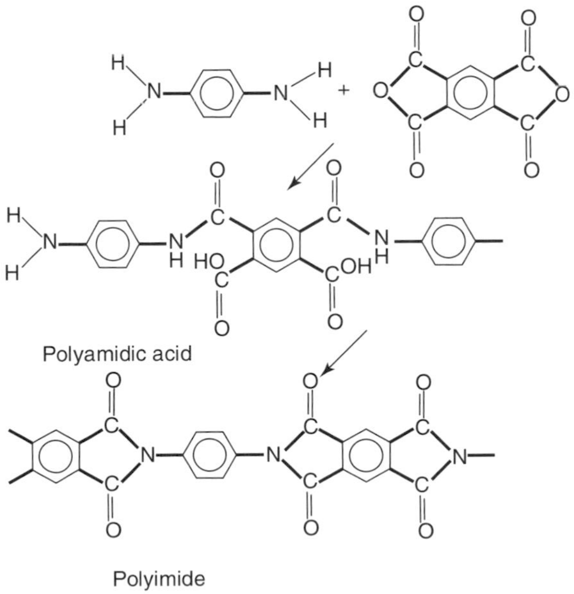

Figure 3.32 Structure and synthesis of polyimide.

Brief summary of chapter

- Crystalline polymers can form a range of organised structures which dramatically influence the resulting physical properties. The chemical structure of the polymer backbone dictates whether or not a particular polymer will crystallise.
- Even if polymers crystallise they can exhibit a variety of different morphologies which are influenced by the conditions used to solidify the material.
- Applying a stress just below the melting point can enhance the morphology and anisotropy may be created.
- Copolymers can phase separate. Their mechanical properties differ from those of homopolymers.

Additional reading

Basset D.C., Olley R.H. and Vaughan A.S. *Techniques for Polymer Organisation and Morphology Characterisation*, Wiley, Chichester, UK, 2003.

Pethrick R.A. *Polymer Structure Characterisation: From Nano to Macro Organisation*, Royal Society of Chemistry, London, UK, 2007.

4

Chemistry of polymer processing

4.1 Introduction

As indicated in Chapter 1, polymeric materials can be divided into two classes: *thermoplastics* which can be reshaped by heating and *thermosets* which cannot. Thermoplastics are formed from linear polymers and the monomers have a functionality of two, whereas thermosets are created from monomers with a higher functionality and each monomer becomes attached to a number of other molecules to form a three-dimensional structure. The challenges in processing these materials are very different and create very different physical characteristics in the final polymer.

4.2 Processing thermoplastic materials

Heating a thermoplastic will lead to the formation of a free-flowing liquid. The extent to which a polymer requires to be heated will depend on the molar mass of the polymer and the viscosity associated with the process. The processes usually used are:

- rotational moulding
- injection moulding
- compression moulding
- solution/melt casting
- plastisol moulding

The technology of each of the above processes is discussed in more detail in Chapter 5.

4.3 Thermosets: elastomers

A thermoset structure can be formed in a variety of different ways. Thermoset materials are created by cross-linking pseudothermoplastic materials. For instance, a thermoset can be created by the vulcanisation of rubber.

4.3.1 Rubbers and vulcanisation

Rubbers (elastomers) are used in applications where the ability to both stretch and carry a load is important. Traditionally, processed natural rubber has been the material of choice, but more recently materials such as the polyurethanes have been used. Natural rubber and gutta percha are artificial polyisoprene (see Figure 4.1).

Replacement of the methyl group by chlorine gives polychloroprene which has characteristics that are similar to those of natural rubber but are superior in some respects. A related material is chlorinated rubber that is produced by treating rubber with chlorine gas in hot carbon tetrachloride. The reaction produces free radical species which cause substantial substitution of hydrogen by chlorine to give a mixed chlorinated product which has properties which retain a rubber-like characteristic but also resemble polyvinylchloride.

In the early 1820s Mackintosh discovered how to vulcanise natural rubber using sulfur to produce a material with high durability. Goodyear refined this vulcanisation process in 1836

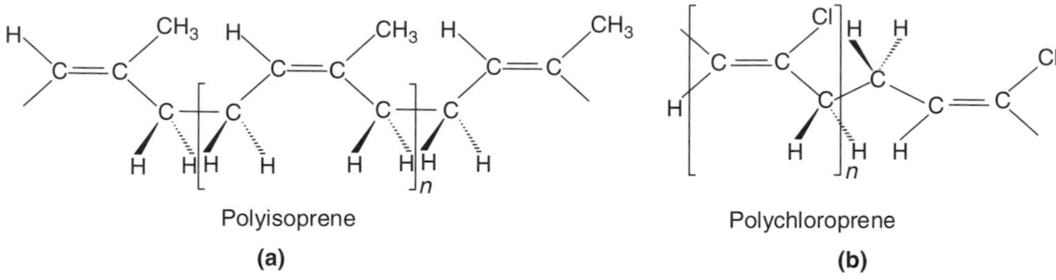

Figure 4.1 Structure of elastomers: (a) polyisoprene; (b) polychloroprene.

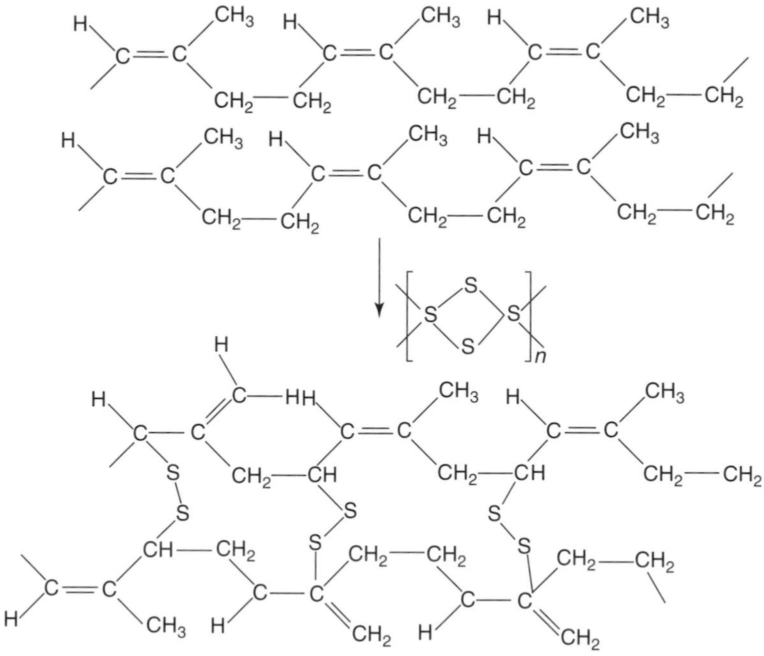

Figure 4.2 Schematic of cross-linking–vulcanisation process for polyisoprene.

to create flexible products that eventually evolved into the products used in the modern tyre industry. The chemistry is complex but can be represented by the diagram shown in Figure 4.2.

The vulcanisation process is complex and can be accelerated by the use of various sulfur containing compounds such as mercaptobenzothiazole (MBT) or diphenylguanidine (DPG) (see Figure 4.3), together with zinc oxide and surfactants for compatibility. In most applications,

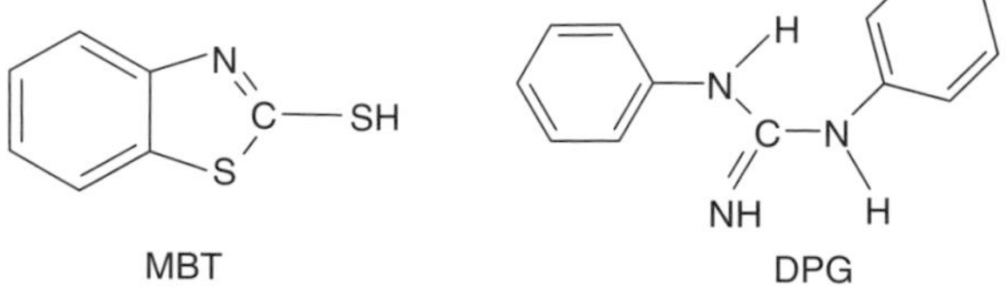

Figure 4.3 Schematic of accelerators used for vulcanisation of rubbers.

carbon black is added to rubber to act as a filler. If the density of the cross-linkages is very high then a very tough material, known as ebonite, is created.

In the case of polychloroprene, cross-linking is achieved by the addition of zinc and magnesium oxides without the use of sulfur. The chemistry is complex but rearrangements involve the removal of chlorine and the creation of ether linkages. This process is more acceptable in applications where the release of smelly thio compounds is deemed unacceptable.

4.3.2 Siloxanes

A very important class of elastomers is based on siloxanes. Polydimethylsiloxane (PDMS), is a very versatile polymer which is liquid at ambient temperatures and even the very high molar mass materials show viscoelastic rather than solid behaviour. The long Si–O bond and large distance between the methyl groups makes the backbone very flexible. The polydimethylsiloxane polymer is a thermoplastic, with a simple linear chain polymer with terminal OH groups at each end of the chain. The glass transition point (T_g) of the backbone is about –90°C; however, the siloxane chains can form a crystalline phase at about −60°C, but still retain their rubbery characteristics down to the T_g. PDMS can be converted into a stable solid and is used as a sealant. The process of cross-linking the PDMS can be achieved by the addition of a hydrolysable tetrafunctional silane (see Figure 4.4). The hydrolysis process liberates silanol which rapidly condenses with the terminal hydroxyl group on the end of the PDMS to form a cross-linked structure. This process is known as *room temperature vulcanisation* (RTV) and is used to form the common sealants used for baths and showers and other applications where gaps are to be filled with a water impervious, flexible material.

The cross-linking process will liberate CH_3COOH (acetic acid, the chemical name for vinegar) and this can sometimes be detected as the RTV systems cure. An alternative cross-linking agent is based on tetraethoxysilane $(C_2H_5O)_4Si$ and the generation of the silanol $[Si(OH)_4]$ liberates C_2H_5OH (ethanol). This latter cross-linking agent is preferred as the smell is more acceptable than that of acetic acid.

A more stable matrix can be created using a process known as *high-temperature vulcanisation* (HTV). The polymer used for this process contains a proportion of vinyl substituted silane units of the type shown in Figure 4.5. The number of vinyl groups in a polymer chain can be varied and this will influence the physical properties of the material being created. The larger the number of vinyl groups, the greater the extent to which a carbon-based cross-linked structure is created. Since the carbon-based chain is resistant to hydrolysis, the HTV materials have

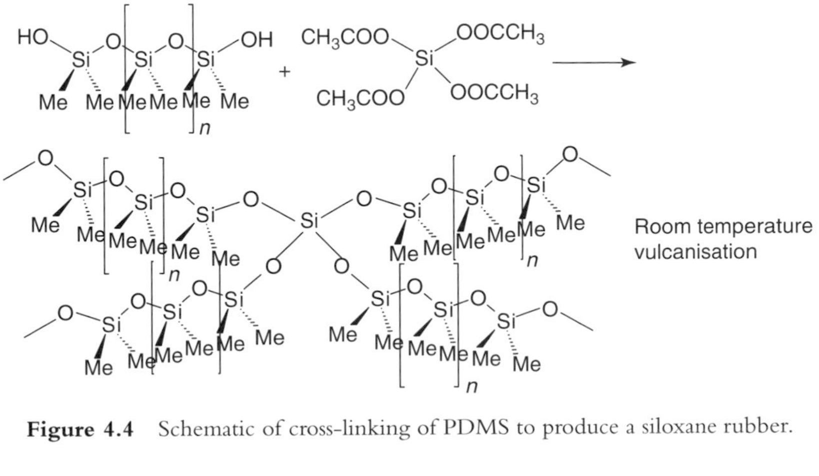

Room temperature vulcanisation

Figure 4.4 Schematic of cross-linking of PDMS to produce a siloxane rubber.

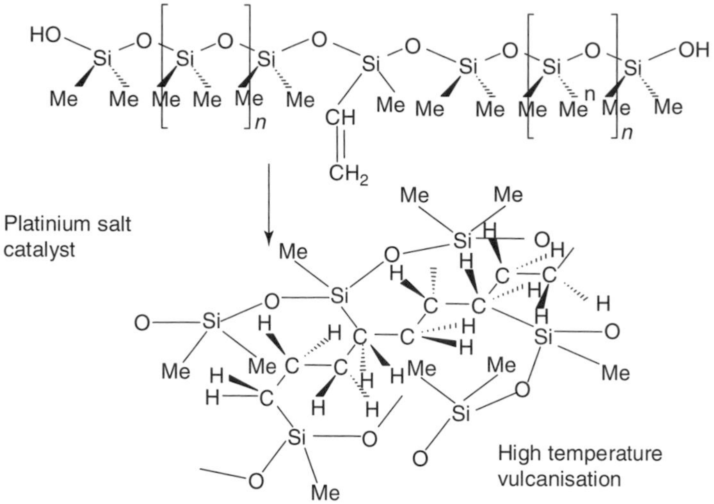

Platinium salt catalyst

High temperature vulcanisation

Figure 4.5 Schematic of high-temperature vulcanisation of vinyl substituted siloxanes.

superior resistance to alkaline attack compared with comparable RTV materials. The HTV is achieved by chain growth polymerisation initiated using a platinum salt that is heated to a temperature above 100°C. The resulting matrix has both Si–O and C–C chains and very good thermal and chemical stability. HTV siloxane polymers are used as a rubber coating for fuser rollers in photocopiers and other applications where operation to temperatures in excess of 150°C may be desirable.

Both RTV and HTV materials contain fumed silica as an active reinforcing agent. By adding 5–35% of fumed silica, it is possible to increase the hardness, modulus and tear strength of the material. An unfilled silicone rubber has very little tear strength and will readily snap if subjected to stress. However, the addition of fumed silica dramatically increases the tear strength and makes the material very useful as a sealant and filler. The surface of the fumed silica will contain SiOH groups which can undergo condensation reactions with silanol created by the hydrolysis process and incorporate the particles into the cross-linked matrix.

4.3.3 Rubber elasticity

We will discuss the theory of rubber elasticity in detail in Section 9.1. The siloxane, isoprene and butadiene materials exhibit behaviour which is close to that of a theoretically ideal rubber. If the chains between the cross-links are completely flexible then the effective modulus, E, of the material is directly related to the number of cross-links by the simple formula:

$$E = 3\mathrm{k}TN_c \qquad (4.1)$$

where N_c is the effective cross-link density of the material, k is the Boltzmann constant and T is the temperature. Cross-links are chemical and unlike the physical entanglements or hydrogen bonds found in thermoplastics cannot be broken. The net result is that the high temperature modulus does not show the expected drop above the values of T_g found with thermoplastics (see Figure 4.6). The high temperature drop in the modulus is associated with polymer chain degradation and results in the creation of monomer or small molecular species by chain scission. In certain systems, the degradation processes will lead to char formation and the modulus will increase as would occur if filler were incorporated into the rubber.

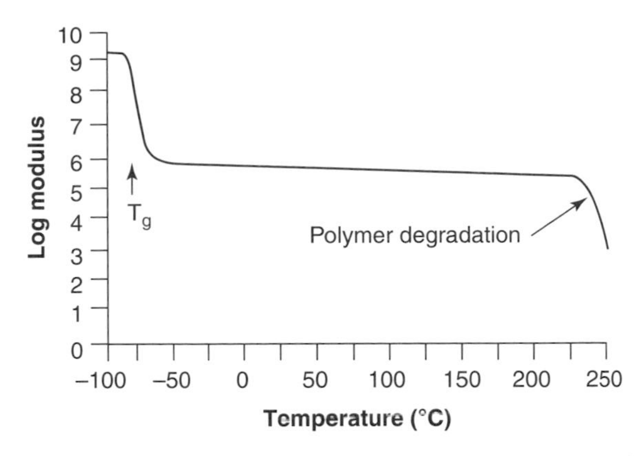

Figure 4.6 Schematic of DMTA trace for a cross-linked rubber.

4.4 Thermoset polymers: rigid materials

Rigid thermoplastic materials are produced predominantly from high molar mass linear polymers. In contrast, thermosets are formed from low molar mass starting materials and have to be created in the form of the final structure. The creation of a thermoset uses the same chemistry as the creation of thermoplastic material, the only difference being that one or more of the component monomers must have a functionality that is greater than two.

Thermoset materials are usually used where very good thermal stability is essential. If we are constructing an aircraft wing, we do not want it to change its shape when it is thermally cycled or after it has been exposed to a variety of climates. Typical materials used in thermoset chemistry include the following:

- *Epoxy resins*: These are the highest performance resins that are currently available. They will generally outperform most other resin types in terms of their mechanical properties and are resistant to environmental degradation. Epoxy resins are extensively used in aircraft fabrication. As a laminating resin, their increased adhesive properties and resistance to water degradation make these resins ideal for use in applications such as boat building. Epoxy resins may be cured with a difunctional amine which has *four* reactive hydrogen atoms and hence forms four bonds to epoxy groups (see Section 1.53), forming a cross-linked three-dimensional network. Epoxy resins are readily cured at any temperature in the range 5–150°C, depending on the choice of curing agent, and exhibit low shrinkage during cure, minimising fabric 'print-through' and internal stresses. High electrical insulation and good chemical resistance complement high adhesive strength and good mechanical properties to make these very useful resins.
- *Epoxy resins with anhydrides*: cationically initiated reactions and imadazole cured materials. The dominant reaction is the opening of the epoxy ring and a linear material bristling with epoxy groups is initially formed which is essentially a multifunctional macromonomer. Subsequent reaction of these pendant epoxy groups forms a very tight cross-linked resin and so they are excellent potting compounds for applications where high voltage electrical breakdown is a major concern.
- *Polyesters*: These are widely used and relatively cheap, but have only moderate mechanical properties. Their disadvantages include high styrene emissions in open moulds, and a large amount of shrinkage during cure. Polyester resins are widely used in the marine industry with a number of different acids, glycols and monomers being available. Two types of polyester resin are in use: orthophthalic polyester resin is the standard, cheapest resin, whereas isophthalic polyester resin is preferred when water resistance is desirable (see Figure 4.7).

$R = H, CH_3$

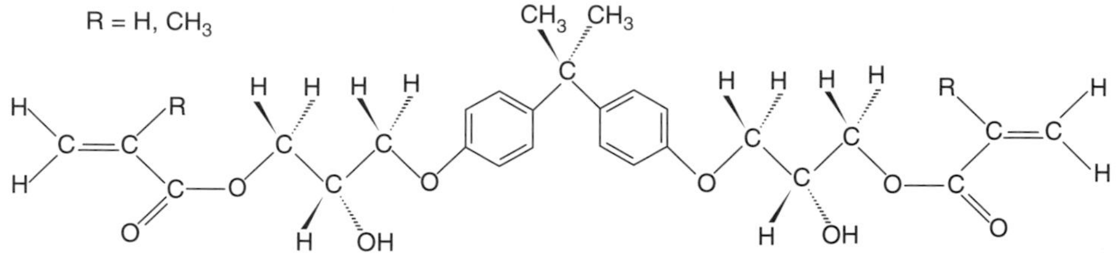

Bisphenol-A epoxy-based (methyl) acrylate vinyl ester resin

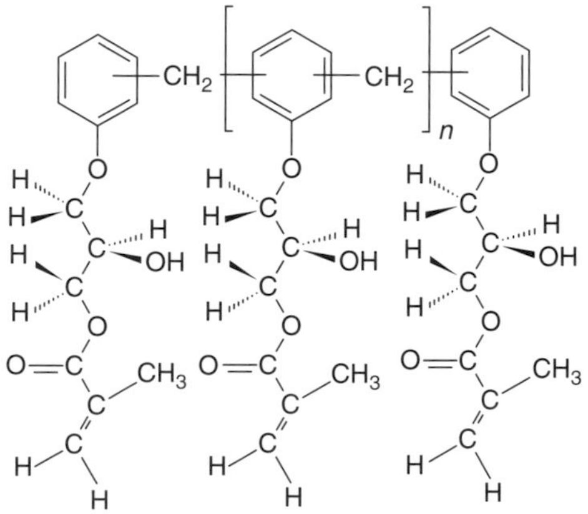

Phenolic–novolac epoxy-based (methyl) acrylate vinyl ester

$n = 3–6$

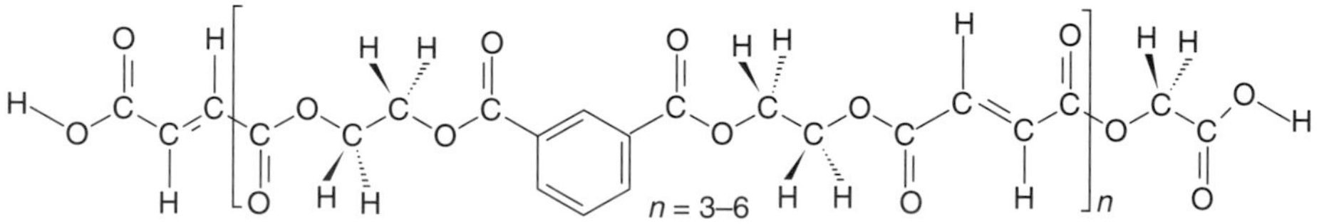

Figure 4.7 Idealised structure for a polyester resin.

The monomers are dissolved in styrene monomer and radically polymerised to produce a cross-linked structure in which polystyrene chains are bridged by the polyester groups which act as cross-links. The ester groups (CO–O–C) increase the reactivity of the vinyl –CH=CH– bonds. The most common resins are based on bisphenol A, fumaric acid or urethane groupings (see Figure 4.8). Polyester resins are viscous, pale coloured liquids which contain up to 50% styrene. These resins can be moulded without the use of pressure and are called 'contact' or 'low pressure' resins. Small quantities of inhibitor are usually added during resin

$R_1 = H, CH_3$ U = Urethane R_2 = Bisphenol A

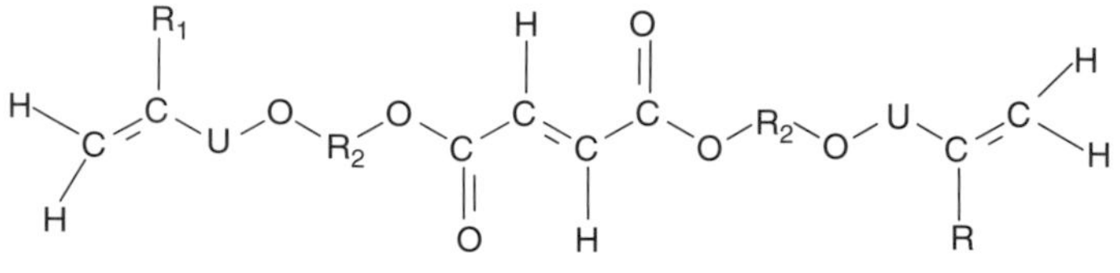

Urethane bisphenol–A fumaric acid-based polyester

Figure 4.8 Chemical structures of commonly used polyester resins.

manufacture to slow the gelling process. These resins are very important for engineering applications and form the basis of glass-reinforced plastic (GRP) composites.

- *Vinyl esters*: These resins are similar in their molecular structure to polyesters, but differ primarily in the location of their reactive sites, being positioned only at the ends of the chains (see Figure 4.8). The whole length of the chain is available to absorb shock, making vinyl ester resins tougher and more resilient than polyesters. The vinyl ester resin contains fewer ester groups that are susceptible to degradation by water and other chemicals. These resins are used for pipelines and chemical storage tanks and as a barrier or 'skin' coat for polyester laminates, such as boat hulls, which are immersed in water. Resins having isophthalate groups cure more slowly than resins without these functionalities.
- *Phenolics*: These are based on resole or novolac structures and are cured with a source of methylene radicals, usually hexamethylene tetramine. Phenolics are used where high fire-resistance is required. The condensation nature of their curing process tends to lead to inclusion of voids and surface defects unless pressure is used during moulding. The resins tend to be brittle.
- *Cyanate esters*: These are primarily used in the aerospace industry and have excellent dielectric properties that make them useful for the manufacture of radomes. These resins exhibit temperature stability up to 200°C.
- *Polyurethane*: This can be moulded to form a tough material, sometimes hybridised with other resins. Mixed PU/polyester materials are used for certain coating applications.
- *Bismaleimides*: These are used in aircraft composites where operation at higher temperatures (230°C wet/250°C dry) is required, e.g. engine inlets and flight surfaces of high-speed aircraft.
- *Polyimides*: These resins can operate at higher temperatures than bismaleimides and can stand up to 250°C wet/300°C dry. Typical applications include missile and aero-engine components. Polyimides tend to be difficult to process due to the condensation reaction emitting water during cure and are relatively brittle when cured.

4.5 Cure of thermoset resins and time–temperature transformation diagrams

The cure process in all these thermoset resins involves conversion of a low viscosity fluid into a hard solid (Pethrick, 2002). The cure process can involve either a step-wise addition or a chain reaction depending on the nature of the monomer. In the case of a monomer that involves a step-wise reaction, the viscosity of the resin stays very low for much of the curing process. It is only when cure has advanced well into the polymerisation reaction that the viscosity starts to increase and does so in an almost exponential manner. The viscosity–modulus plotted against time (see Figure 4.9), shows a break at the point at which *gelation* occurs. At *gelation* a three-dimensional

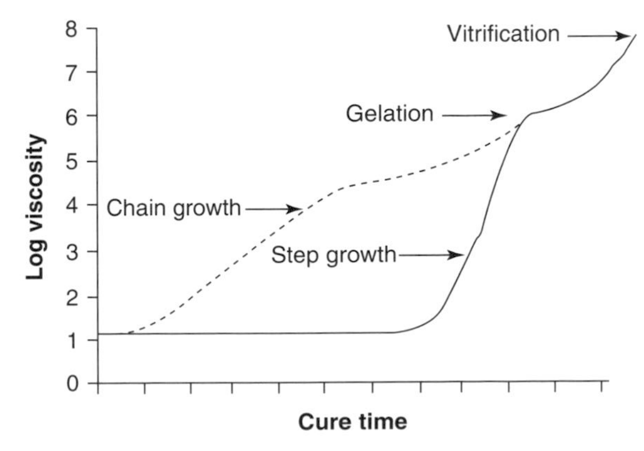

Figure 4.9 Typical viscosity against cure time plots for chain and step growth processes.

network is formed and the final shape of the article is fixed. Further reaction will convert the gel into a rigid solid (see Figure 4.9).

Epoxy resins are very mobile initially but after a small degree of cure can be converted into a solid at room temperature. A mixture of an aromatic amine and an epoxy resin can form a solid phase at room temperature with a very low reactivity. On heating to a high temperature, the cure process (see Figure 4.9) will occur. This mixture of amine and epoxy is called a *one-part resin* and forms the basis of much of composite fabrication technology. The alternative is a two-part resin process, in which each of the reactants is mixed at the point at which the reaction is required. The *one-part resin* can be very stable when stored at low temperature and can have a useful life for several months and in certain conditions up to a year. On heating the resin mixture to temperatures of the order of 70–120°C, rapid reaction occurs and the liquid is transformed into the solid. The significant increase in viscosity occurs once the degree of reaction is of the order of 60–70%. A quick rule of thumb for the accelerating effect of heat on the rate of cure of a resin is that an increase of 10°C in temperature will roughly double the reaction rate. Therefore if a resin gels in a laminate in 25 min at 20°C it will gel in about 12 min at 30°C, provided that additional exothermal reactions do not occur. Curing at elevated temperatures is often required to ensure that the reaction is complete and many resin systems will not reach their ultimate mechanical properties unless the resin is 'postcured'. The postcure involves increasing the reaction temperature after the initial cure has been completed. Postcuring increases the extent of reaction and increases the density of the cross-linkages in the material. Depending on the structure of the material, a three-dimensional gel phase will be formed and after further reaction a glass is created. Once the system has passed into the vitrified or glassy state, the reaction will cease. However, it is possible that groups will remain that are capable of further reaction. Increasing the temperature will transform a glassy solid into a rubbery phase, which will then undergo further reaction. These phase changes can be represented in a time–temperature transformation (TTT) diagram (see Figure 4.10). It is sometimes desirable to leave the matrix slightly undercured, residual

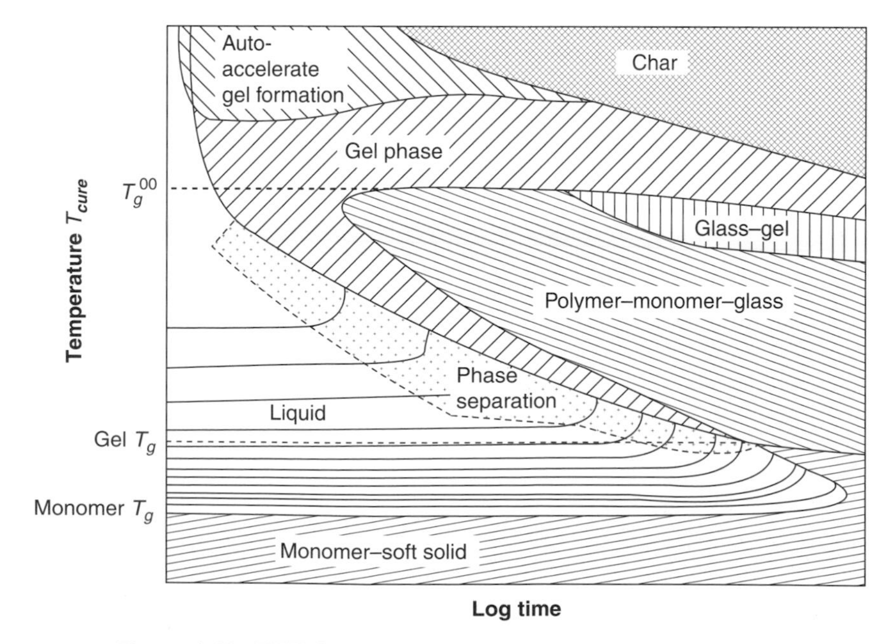

Figure 4.10 TTT diagram for a phase-separating epoxy resin system.

uncross-linked trails of chains effectively plasticising the matrix and improving the mechanical impact strength.

The TTT diagram describes the path by which the cure process proceeds at various temperatures and is typical of the behaviour of a one-pack epoxy resin system (see Figure 4.10). The characteristics features of the TTT diagram are as follows:

- At low temperatures, the *monomer* mixture forms a soft solid rather like butter. The monomer components will react only very slowly and it is possible to store this mixture for a long period of time without compromising its curing behaviour.
- If the *monomer* mixture is heated, it can be transformed to a liquid. The lines on the diagram represent the values of viscosity profiles during curing. For much of the period of the reaction, the viscosity is constant but rises dramatically as gelation is approached (see Figure 4.9). Increasing the temperature of the liquid will lower the viscosity, as reflected in the increased spacing between the lines. The rate of reaction will increase with the increase in temperature and this is indicated by the increase in viscosity decreasing in time with increasing temperature.
- In many systems, *phase separation* can occur. Because epoxy resins are intrinsically brittle it is common practice to toughen the material with either the addition of a rubbery phase or a thermoplastic. The point at which phase separation occurs depends on the solubility of the second phase in the epoxy mixture and the effect of the increasing molar mass on the solubility of the second phase in the matrix. In some systems, phase separation occurs at a relatively early stage in the cure, for other systems it is a processes that occurs just before gelation.
- As the polymerisation processes proceeds, a point is reached at which the network being formed effectively fills the whole reaction volume and this point is designated the *gelation* point. The gel will contain a significant amount of unreacted monomer and will have rubbery, elastomeric characteristics.
- Further reaction of the gel will lead to the unreacted monomer being incorporated into the matrix and the material is slowly converted into a glassy solid. The solid contains monomer yet has glassy characteristics and is designated a *polymer–monomer–glass* solid. The network will contain unreacted monomer and it will have a glass transition temperature (T_g) which is lower than that of the completely cured system, designated T_g^{∞}.
- Heating the solid to the point at which it once more regains elastomeric characteristics produces a *glass gel*. The matrix will be fully formed but not all the groups may have been reacted into the matrix. This phase can also be produced as a consequence of chain degradation leading to the creation of short chains which will lower the T_g of the resin allowing the formation of a gel phase.
- At higher temperatures the fully formed resin network can be transformed into an elastomeric phase, known as the *gel phase* if the temperature is raised above the T_g of the fully cured system (T_g^{∞}). In principle, this is a stable elastomeric phase but it is clear from the phase diagram that if the material is held at these temperatures for a long period of time degradation can occur and a *char* will be formed.
- The *char* phase represents the creation of a degraded resin structure in which some elements of the structure will become very highly cross-linked and have very brittle characteristics whilst others may become very open and exhibit elastomeric or even liquid-type properties. A *char* is the same state which would be achieved if an object were burnt.
- If the curing temperature is raised to a very high value then it is possible for certain systems to exhibit *auto-accelerated* cure processes which are often accompanied by the premature formation of a gel network. High temperatures can be created if the heat generated during the cure process is not effectively dissipated. The heat created can lead to hot spots being generated which lead to volatilisation of the unreacted monomer and can produce an explosion! This type of problem

can be encountered in the cure of heavy composite structures and must be avoided at all costs. Very often the temperatures created will lead to the degradation of the resin and char formation.

- The glass transition point, T_g, of the mixture designated *monomer* T_g will be the temperature at which there is sufficient fluidity in the mixture for curing to effectively start to occur. The *gel* T_g corresponds to the lowest temperature at which a three-dimensional matrix has been formed. At this point there will be large quantities of monomer present which will plasticise the resin and the value will be relatively low. The T_g achieved at a particular cure temperature T_{cure} will usually be slightly higher than the cure temperature used and reflects the influence of the unreacted monomer on the resin.

4.5.1 How do we effectively cure resins?

The objective of the cure process is to achieve a material in which almost if not all of the monomer has been converted into polymer. The most obvious approach would be to raise the reaction mixture to a value close to the ultimate T_g^{∞} value. This could have several effects. First, it is possible that a phase-separated structure might not be created and this could have consequences for the impact properties. Secondly, the material would gel very quickly making the processing very difficult. Thirdly, in the case of the resin being the matrix for composite manufacture, the viscosity may have been so decreased that it flows away from the fibres, forming a resin-rich region and leaving fibres that are no longer surrounded by resin. Fourthly, the thermodynamic state of the gel structure created will correspond to the temperature at which the cure process is carried out. As we have seen before, the chemical structure will adopt conformations that reflect the temperature the material is at. In the case of the cure process, the gel network structure will retain a memory of the temperature at which it is formed. When cure is completed, the object is cooled and the conformations trapped at the stage of gel formation may not be able to adopt the appropriate lower thermodynamic state corresponding to the lower temperature. This frozen-in thermodynamic strain can be very large and may be sufficient to lead to chain scission. On an object that has thick sections, it has often been observed that the use of a high temperature for the cure can lead to major cracks being formed when the object is cooled. Thus, it is normal to start the cure process at a relatively low temperature where the reaction is sufficiently slow for the resin to be easily handled and the gel is formed with a structure which leaves the object with minimal strain when cooled to room temperature. However, the T_g of the object will be rather low and it is normal to *postcure* the object. *Postcuring* may be part of the overall cure cycle and is achieved by slowly heating the article once gel has been formed to a value that approaches T_g^{∞}. An alternative approach is to hold the article at some elevated temperature that is below the value of T_g^{∞}, for a longer period. Which approach is used depends on the circumstances under which the article is fabricated.

4.5.2 Thermoset cure resins

Polyesters, vinyl esters and epoxy resins account for ~90% of all structural composites. Any resin system for use in a composite material will require the following properties:

- good mechanical properties
- good adhesive properties
- good toughness properties
- good resistance to environmental degradation

4.5.3 Mechanical properties of the cured resin system

The stress–strain curve for an 'ideal' resin system (see Figure 4.11) indicates that the material has a high ultimate strength, stiffness and strain to failure. Changes in resin type will be reflected in the tensile strength and stiffness. Figure 4.12 compares the values for typical polyester, vinyl ester and epoxy resin systems cured at 20°C and 80°C. After a curing period of seven days at room

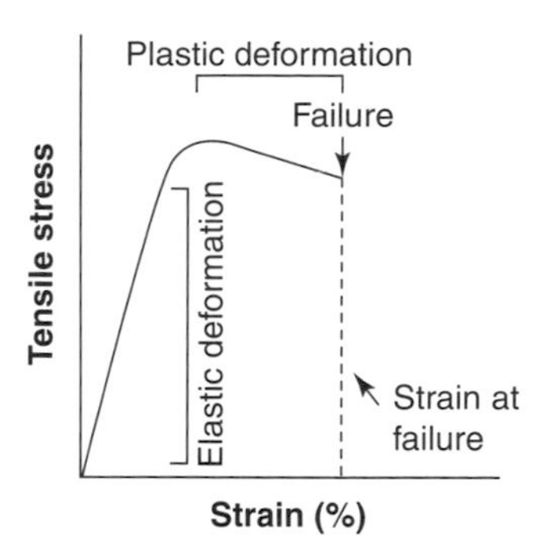

Figure 4.11 Stress–strain plot for an ideal resin.

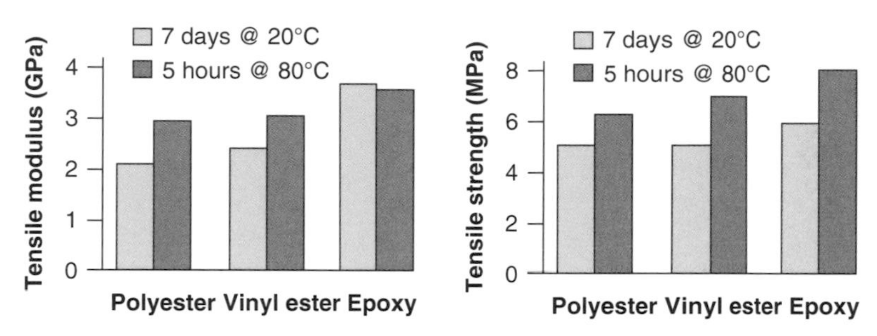

Figure 4.12 Typical mechanical properties for polyester, vinyl ester and epoxy resins obtained using different times and temperatures of cure.

temperature, a typical epoxy will have higher mechanical properties than a typical polyester and vinyl ester for both strength and stiffness. Postcuring at 80°C leads to significant enhancement of the mechanical properties.

4.6 Commercial thermoplastic polymers

The composition of a commercial polymer will depend on its intended application. Most systems will contain: polymer 40–100%, fillers 0–40%, plasticisers 0–60%, antioxidants 1–6% and processing agents 0–5%. All the above may influence the physical properties of the material. Fillers may be especially beneficial and can increase the modulus of a plastic but can also lead to a reduction in some physical properties.

4.6.1 Antioxidants

These are organic molecules which are added to the mixture to stop the depolymerisation reaction and inhibit the reduction in molar mass which occurs when the polymer approaches its ceiling temperature. At the ceiling temperature a polymer will spontaneously decompose to produce radicals. Many antioxidants are hindered amines which are able to scavenge the free radicals and suppress the degradation processes.

4.7 Fillers

A variety of fillers are used with polymers. They can be divided into two types: reinforcing and nonreinforcing fillers. They differ in their interaction with the polymer. Hence, it is possible for a material to be a reinforcing filler in one situation and a nonreinforcing filler in another. Typical fillers are: quartz, fused silica, sand, talc, calcium carbonate, carbon black, carbon fibres, glass fibres, kevlar (polyimide), polyethylene fibres, wood fibres and sawdust.

4.7.1 Carbon black

Carbon black is commonly used to produce black plastic products. It is produced by the partial combustion of various organic media, usually high molar mass hydrocarbon materials. The process generates graphitic materials that have dimensions of typically 1–20 μm. A close examination of the particles indicates that they are porous and are created by the fusion of smaller particles. Using similar techniques, it is possible to create Buckminster fullerene (C_{60}), or nanotubes that are the cylindrical equivalent of the spherical C_{60} structure. The typical carbon black is a less perfect fusion of graphite platelets and structures which resemble the more exotic spherical and tubular forms of carbon (see Figure 4.13).

There are a variety of different types of carbon black, their properties depending on the conditions used in their formation. Conducting blacks will contain larger graphitic structures and may have additional functional groups that reflect the nature of the gas stream used in their formation. Acid blacks are usually postoxidised using nitric acid, nitrogen oxides (N_2O_4, N_2O_5, NO) or ozone. The types of surface chemistry found on carbon blacks are illustrated in Figure 4.13.

The main physical differences between various carbon blacks are their surface area and porosity. Surface areas can typically range from ~100 $m^2\ g^{-1}$ to over 600 $m^2\ g^{-1}$. The pore size will influence whether or not polymer molecules can enter the internal structure of the carbon black and hence its ability to reinforce the matrix. Some carbon blacks will *not* reinforce the matrix whereas others can. However, it is possible that a nonreinforcing carbon black is capable of reinforcing a thermoset, where penetration of the monomer is possible.

There are at least seven different ways of producing carbon black:

- *Lamp-black process*: Produces carbon black from burning rubber or paint. The particle size can vary in the range 50–120 μm. The carbon black is produced by depositing the product of combustion in a flue.
- *Thermal process*: The process uses two furnaces that run alternately for about five minutes; one is heated with a mixture of natural gas and air and the other has a stream of 100% natural gas. A coke oven produces thermal decomposition and the carbon black particles have sizes in the range 120–500 μm.
- *Channel process*: This process burns natural gas through a large number of small luminous flames impinging on slowly moving cooled iron channels, forming fine particle blacks of ~5–30 μm.

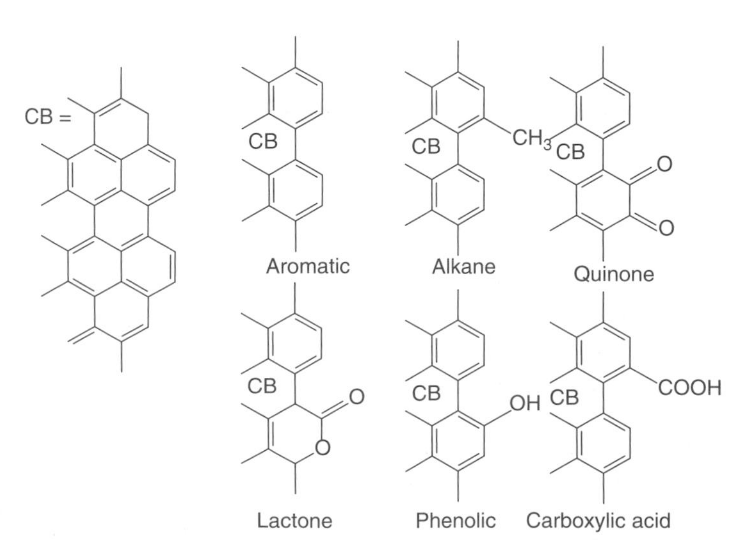

Figure 4.13 Types of chemical structure found on the surface of carbon black (=CB).

The product is removed mechanically from the channels and collected on a conveyor. The yield is only 3–6% of the theoretical maximum and channel blacks are expensive.

- *CK black process*: This is similar to the channel black process except that oil is used instead of gas. The average size of the particles is 10–30 µm and they are used as pigments.
- *Electric arc process*: The process is used to produce acetylene by hydrocarbon decomposition and yields large quantities of carbon black. The particle size is 35 µm, however, the particles are laminar or flaky, not spherical. This process also yields nanotubes and nanocarbon blacks.
- *Furnace process*: This differs from the channel process in that combustion of oil raw materials is carried out with a single large flame in a refractory lined furnace. The particle sizes are in the range 10–80 µm.
- *Acetylene black process*: The process is operated at 800°C and acetylene is decomposed into carbon and hydrogen in a self-sustaining exothermic process and the carbon is separated from the hydrogen gas stream. The particle size is typically 10–30 µm and contains nanomaterials.

Each carbon black has a different application depending on the media in which it is used and the extent to which conductivity is to be imparted to the object.

4.7.2 Quartz, silica and clay fillers

Silica in the form of sand, fumed silica and exfoliated clay is used to reinforce plastics. Many of the materials will have been derived from a geological source and are not perfectly defined from either the point of view of their chemistry or their physical properties. Two of the most abundant elements are silicon and aluminium and these combine to give a wide range of materials. Clays have been traditionally used as fillers for various plastics, and if treated with organic modifiers can be dispersed (exfoliated) to the level of the primary platelets and substantially enhance their physical properties. Clays can be classified according to their chemical composition, shape and size.

4.7.3 What is the structure of a clay?

The primary building block for a clay structure is a *tetrahedral sheet* composed of individual tetrahedrons based on silicon or aluminium which share three out of four oxygens. The primary units are arranged in a hexagonal pattern with the basal oxygens linked and the apical oxygens pointing up/down. The resultant sheet composition is T_2O_5, where T is the common tetrahedral cation of Si, Al and sometimes Fe^{3+} and B. In many clays there are *octahedral sheets* in which individual octahedrons share edges composed of oxygen and hydroxyl anion groups with Al, Mg, Fe^{3+} and Fe^{2+} typically serving as the coordinating cation. The octahedral structures can be subdivided into dioctahedral (gibbsite) $Al_2(OH)_6$ and trioctahedral (brucite) $Mg_3(OH)_6$. If cation substitution occurs in the sheet structure, charge imbalances result and cations add into the layers to satisfy the charge imbalance or exchangeable cations (K, Ca, Mg, Na and many others). Variants on the structure of the layers arise from substitution of Al^{3+} and Fe^{3+} for Si^{4+} in the tetrahedral layer. In octahedral layered structures, the cations are usually Al^{3+}, Mg^{2+}, or Fe^{2+}, but such structures may also contain Fe^{3+}, Ti, Ni, Zn, Cr and Mn. In the octahedral layers, the anions are oxygen and a hydroxyl and some of the hydroxyls may be replaced by F or Cl ions.

Clays with an octahedral and a tetrahedral layer, e.g. kaolinite, which has a platelet dimension of 7 Å, are designated 1:1 clays. Clays which have an octahedral layer plus an interlayer are known as 2:1 clays, e.g. chlorite, which is trioctahedral and has a 14 Å basal spacing. The various clays are characterised by various charges. The neutral lattice structures are: talc (2:1), pyrophyllite (2:1), kaolinite (1:1), chlorite (2:1+1), all of which have a net charge of zero and their sheets are bound by van der Waals-type bonds. The high-charge 2:1 structures have charges in the range 0.9–1.0 created by ionic substitution, which is compensated for by an interlayer cation (K held firmly, diocatahedral). Low charge 2:1 structures have a 0.2–0.9 charge compensated for by loosely held ions which can easily be exchanged and have a tendency to swell. Dioctahedral clays (1:1 clays) include kaolinite $Al_2Si_2O_5(OH)_4$, dickite and halloysite $Al_2Si_2O_5(OH)_4$. The 2:1 clays include micas,

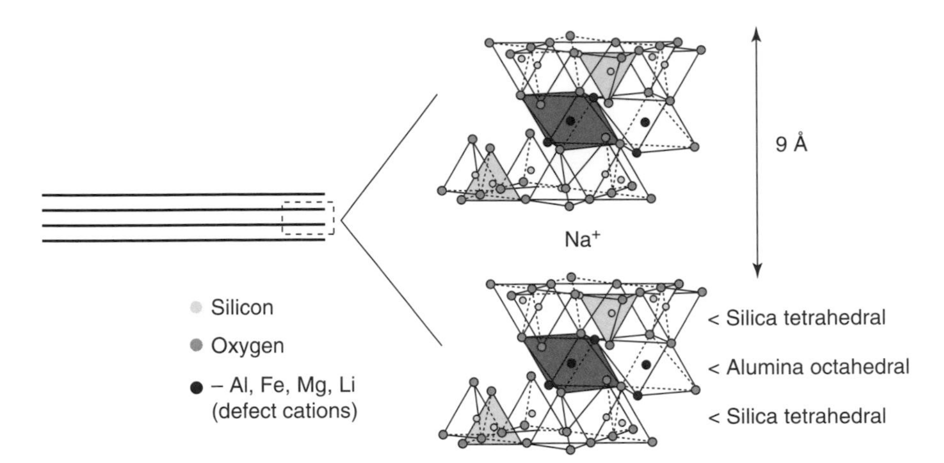

Figure 4.14 Schematic of platelet structure of montmorillonite.

pyrophyllite $Al_2Si_4O_{10}(OH)_2$ and muscovite $KAl_2(Si_3Al)O_{10}(OH)_2$. Smectites include beidellite $X^+_{0.33}Al_2(Si_{3.67}Al_{0.33})O_{10}(OH)_2$ and montmorillonite $X^+_{0.33}(Al_{1.67}Mg_{0.33})Si_4O_{10}(OH)_2$. Vermiculites include altered smectite and illite $K_{0.8}Al_2(Si_{3.2}Al_{0.8})O_{10}(OH)_2$. A typical layered structure is that of montmorillonite (see Figure 4.14). The silica and alumina elements fit together to produce a layered structure which is about 1 nm thick and will extend as sheets for distances of the order of one or more micrometres. In nature, the charge imbalances are matched by sodium hydrated ions which hold the platelets together. The exfoliation process is the dispersion of these platelets as individual sheets. To aid this exfoliation process organic cationic surfactant molecules are added to replace the sodium, increase the platelet separation and ease the exfoliation process.

These platelets have a large surface area and can reinforce the plastic raising the glass transition temperature by 10–20°C for the addition of 2–3% of exfoliated clay. The large sheets can overlap and are able to introduce barrier properties by increasing the percolation path for a gas moving through the material. Other clays have a ribbon structure and are useful as thixotropy index modifiers.

4.8 Plasticisers

These are usually low molecular weight polymers or small molecules. The necessary criteria are that they should be soluble in the polymer and this usually requires that they are chemically very similar to the polymer in which they are dispersed. Plasticisation has been discussed more fully in Section 2.5.3.

Brief summary of chapter

- Polymers can be broadly divided into two classes: thermoplastics and thermosets.
- The production of articles from these materials involves very different approaches to be adopted for the creation of useful structures.
- This chapter has summarised the chemical–physical issues associated with polymer fabrication.

Additional reading

Pethrick R.A. Cure monitoring. In: Kulshreshta A.K. and Vasile C. (Eds.) *Handbook of Polymer Blends and Composites, Vol. 1*, RAPRA, Shrawbury, UK, 2002, Chapter 10.

Pinnavaia T.J. and Beall G.W. *Polymer–Clay Nanocomposites*, Wiley, Chichester, UK, 2000.

5

Polymer processing: thermoplastics and thermosets

5.1 Introduction

In the previous chapters, the chemical aspects of processing thermoplastics and thermosets were considered. It is now appropriate to look at the engineering issues associated with the production of polymeric artefacts.

5.2 Processing thermoplastics

Processing of thermoplastics involves heating the polymer to an elevated temperature and forcing it into a die/mould. Raising a polymer to high temperature will allow attack by oxygen and possible degradation. It is therefore desirable that the temperature used in the process is as low as possible and *antioxidant* molecules, which are usually hindered amines, are added to suppress degradation (see Sections 4.6.1 and 8.9).

A variety of methods are available: which is appropriate depends on what is being made. The processes which will be considered are: rotational moulding, compression moulding, injection moulding, solution/melt casting and plastisol moulding.

On heating, a thermoplastic will form a viscous liquid which can be poured into the mould. The temperature to which the polymer must be heated will depend on the characteristic viscosity associated with the particular process. Some processes, in which fine features are to be created, may require a lower viscosity than others. Rotational moulding can be used to produce hollow sealed cavities and involves powder being melted within a rotating mould. In this process, fairly high viscosities are desirable to stop the molten polymer flowing around the mould. Compression moulding can handle fairly viscous materials, the molten polymer being forced by an external force into the desired shape. Injection moulding involves the flow of molten polymer into small cavities and requires low viscosities. Melt and solution casting require, respectively, temperatures for flow or evaporation of the solvent. The plastisol process is complex, but requires a relatively low temperature at or just above the glass transition temperature, for gelation to be achieved. Each process requires slightly different conditions and hence different grades of polymer.

5.3 Rotational moulding

Rotational moulding (Pethrick and Hudson, 2008) is used to produce hollow sealed cavities such as buoys, tanks and cylindrical structures (see Figure 5.1). Rotation moulding allows the creation of a seamless 'one piece' construction and involves the rotation of a heated mould containing the thermoplastic (see Figure 5.2). When heated, the free-flowing polymer powder melts and sticks to the mould. On cooling, the plastic shrinks and becomes detached from the mould. In rotational moulding, the possibility of the polymer flowing from the point where it melts must be suppressed as this would lead to an uneven coating of the mould.

When all the powder has been melted, the cooling process can take place. The rate of cooling will influence the growth of the crystalline structure and hence the mechanical properties of the

Figure 5.1 Typical type of structure created by rotational moulding.

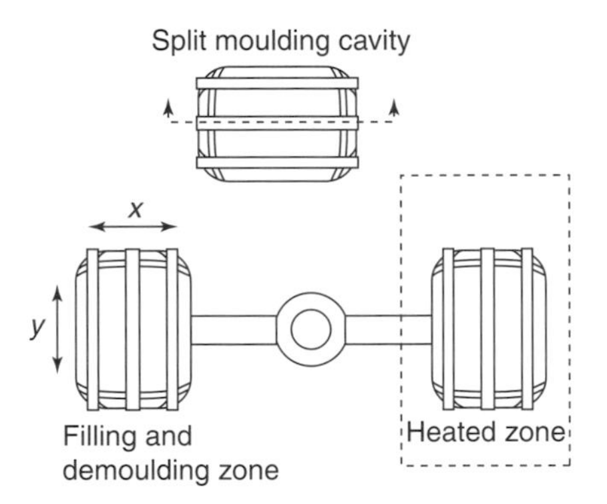

Figure 5.2 Schematic of a rotational moulding process.

final material. Rotational moulding usually involves large machines and the zone heated can have a volume of 10 m^3 or even larger.

The moulds are usually heated with gas burners and constructed from thin steel plate to achieve efficient heat transfer to the inner surface of the mould, whilst retaining rigidity of the mould. The inner surface of the mould is treated with a mould release agent to ensure release of the moulded article on cooling.

5.3.1 Moulding process

The required amount of powder is placed in the split mould. The moulding machine will often use two or more moulds to balance the load on the supporting system (see Figure 5.2). The connection arm of the mould to the machine allows the cavity to be rotated independently about *two* axes (x, y). In order to achieve a uniform coverage of the inside of the mould, ratios of the rates of rotation about the x- and y-axes are selected to be nonintegers and typically ~3, depending on the shape. A good moulding requires optimisation of the heating and cooling processes. The various stages in the moulding process are:

- The initial tumbling of the powder ensures that there is good coverage of the surface of the moulding cavity.
- Heat transferred from outside of the mould will cause the powder in contact with the heated surface to melt and stick to the walls. If the temperature is too high then flow can occur and wall thinning arises.

- The amount of powder used is critical, as this will determine the thickness of the wall of the moulded article. The larger the amount of powder used, the more energy required and the longer some of the polymer will be in the melt. A long time in the melt phase or the use of high temperatures can lead to polymer degradation, a reduction in chain length and loss of physical properties.
- If the wall is too thick, the molten polymer can become detached from the mould when it reaches the top of the arc of rotation, with a resulting distortion of the moulding. In the molten phase, gas is released and densification results. Without this densification process, the mechanical strength is not developed. For effective gas release to occur, the melt viscosity and surface tension have to be sufficiently low for bubble formation to take place.
- The cooling process has to be sufficiently slow to allow optimum development of the crystalline structure.

5.3.2 Theory of the rotational moulding process

Rotational moulding can be modelled by considering a simple heated cylinder rotating with a constant angular frequency, ω_1. In practice, the mould will also be rotated about a second axis at an angular frequency, ω_2. The ratio of the relative speeds about the axes, ω_1/ω_2, dictates the rotational ratio. This ratio for a cylinder would be ~8:1. The cylinder mould cavity (see Figure 5.3) has a radius R, the final wall thickness of the moulded polymer article is h and a is a segment of the wall. Four segments are identified; a, b, c and d as regions which will be consider in the subsequent mathematical modelling of the process.

5.3.3 Powder deposition

The molten polymer film is created by fusion of the powder particles at the heated mould wall (*sintering*) followed by densification. During the fusion process, air between the particles is expelled and a homogeneous molten polymer film is formed.

5.3.4 Powder melting in contact with the heated surface

The powder deposition process (see Figure 5.3) involves cascading powder within the mould. The powder adheres to the surface at the point at which it melts. The total mass of powder will define the final wall thickness of the moulding. Initially, the powder will be located in the base of the mould and only sticks to the wall when it melts. The amount of energy required is a combination of the energy required to heat the metal mould, $C_p(mould)$, and the powder heat capacity $C_p(powder)$. The heat capacity of the mould can be assumed to be a simple linear function of the

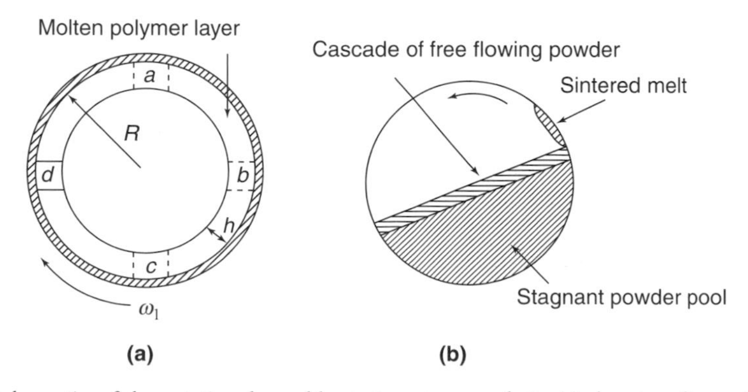

Figure 5.3 Schematic of the rotational mould rotating at a speed ω_1: (a) showing dimensions; (b) cascading powder in various states.

temperature and the total energy required will be:

$$E(mould) = \int_{T_1}^{T_2} C_p(mould)\,dT \tag{5.1}$$

where T_1 is the initial temperature and T_2 the final temperature during the heating process. The amount of energy required to raise the powder to the temperature T_2 will be:

$$E(power) = \int_{T_1}^{T_1^s} C_p\,(solid)dT + \Delta H(fusion) + \int_{T_2^1}^{T_2} C_p(melt)\,dT \tag{5.2}$$

where $C_p(solid)$ and $C_p(melt)$ are to a first approximation independent of temperature and $T_1^s(powder)$ and $T_2^1(powder)$ are, respectively, the end of the solid and the beginning of the liquid range, i.e. the melt temperature of the polymer. The total energy required will be:

$$E(total) = E(powder) + E(mould) \tag{5.3}$$

The ΔH (fusion) is the enthalpy of melting and heating the polymer is described in terms of an effective thermal diffusivity ($\alpha_{effective}$) and reflects the rate of heat transfer from the mould to the powder. A thermal penetration thickness, δ, has the form:

$$\delta = (24\alpha_{effective}t)^{1/2} \tag{5.4}$$

The temperature of the powder layer thickness, L, will depend on the mould surface temperature and the thermal diffusivity through the powder:

$$T_1 = at\left[1 - (L/(8\alpha_{effective}t)^{1/2}\right]^3 \tag{5.5}$$

where a has the form between T_1 and T_1^s:

$$a = C_p(mould) + C_p^s(powder) \tag{5.6}$$

At the melting point of the polymer, T_1^s, it has the form:

$$a = C_p(mould) + \Delta H(powder) \tag{5.7}$$

and above T_1^1 the value becomes:

$$a = C_p(mould) + C_p^1(melt) \tag{5.8}$$

Because of the large value of the ΔH(powder), the heating process will slow down at T_1^s. The thickness of the polymer layer at T_1^s has the form:

$$x_1 = \delta\left[L - (T_1^5 / T(t)_s)^{1/3}\right] \tag{5.9}$$

The surface temperature $T_s(t)$ is dictated by the heating profile for the mould. The surface area of the mould will be the area of the cylindrical part and the ends; $2\pi RL + 4\pi R^2$, where L is the length of the cylinder. The wall thickness with the polymer as a powder will be x_t. So that the volume of powder added V_p will be:

$$V_p = (2\pi(R - x_t)Lx_t + 4\pi R^2 x_t) \tag{5.10}$$

which for large moulds and thin walls becomes:

$$V_p = (2\pi(R)L + 4\pi R^2)x_t \quad (5.11)$$

5.3.5 Bubble removal

The rate of bubble formation is influenced by the viscosity and decreases in diameter as the melt temperature increases. The high viscosity at the melt temperature prevents movement of the bubbles and a further increase in the temperature is necessary for the bubbles to be released. Oxygen has about twice the solubility of nitrogen in polyethylene. At high temperatures, the oxygen is further depleted by direct oxidation reactions with polyethylene. The depletion of oxygen reduces the bubble diameter. The laws of surface tension dictate that the pressure inside the bubble has to increase as the diameter decreases. The increase in pressure forces the nitrogen to dissolve in the polymer, thus the bubble diameter is further reduced until the bubble disappears. As the moulding time increases, the size and quantity of bubbles decrease. However, for long heating times, the impact strength is lowered because of the effects of oxidative degradation. There is a critical bubble size above which the gases will not dissolve, regardless of temperature or time, because the surface tension forces cannot generate enough bubble pressure to help dissolve the gases inside the bubble. As a consequence of the densification process the thickness of the moulded part will reduce from the value given by the following equation:

$$V_m = \rho_{powder} / \rho_{melt}(2\pi(R)L + 4\pi R^2)x_t \quad (5.12)$$

where ρ_{powder} and ρ_{melt} are the densities of the powder and the consolidated melt, respectively. The rate of coalescence of adjacent spheres under the action of surface tension is given by:

$$x^2 / r = 3/2(\gamma_p / \eta_0)^t \quad (5.13)$$

where x is the neck radius associated with the fusing of the particles (see Figure 5.4), r is the radius of the particles, γ_p is the surface tension, η is the viscosity and t is the time. Combining with the viscosity, Equation (5.13) becomes:

$$x^2 / r = 3/2\left(\gamma_p t / \eta_0\left(1 - e^{(-t/\tau)}\right)\right) \quad (5.14)$$

where η_0 is the viscosity of the melt and τ is an apparent relaxation constant to account for the viscoelastic nature of the material.

In polymeric materials, deviations from Newtonian behaviour are observed for systems which contain molecules with a molar mass above the critical entanglement value, M_e. For good bubble release both η_0 and τ should be reduced. The initial size of the bubble influences the rate at which it dissolves, the surface area to volume ratio at which it dissolves being inversely proportion to the diameter:

$$(\varphi / \phi_0)^2 = K_1 - K_2 t + K_3 t^2 \quad (5.15)$$

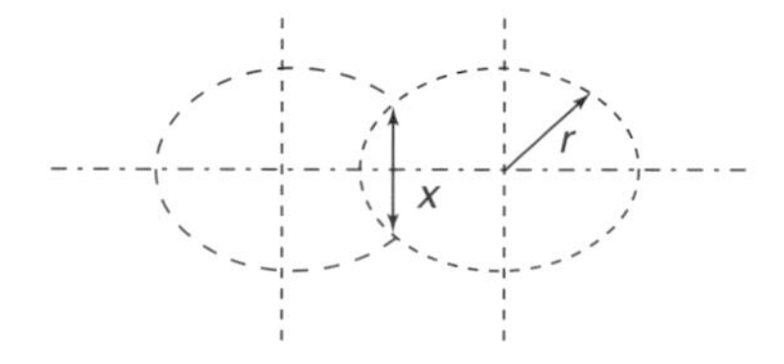

Figure 5.4 Schematic of two polymer powder particles fusing.

where K_1, K_2 and K_3 are constants and ϕ and ϕ_0 are, respectively, the diameter of the bubble at a point in time and the original diameter of the bubble. Equation (5.15), implies there is a limiting size of bubble that will not collapse. Combining equations:

$$\left(\phi / \phi_0\right)^2 = K_1 - 3/2\varphi\left(\gamma_p / \left(\eta_0\left(1-e^{-t/\tau}\right)\right)\right)t \tag{5.16}$$

where φ is the shape factor for the bubble and will be a constant for a particular system. In practice, air will diffuse out of a bubble in polyethylene when the polymer viscosity has been reduced to a value of 3,000–4,000 Pa s, and defines the limit of the working viscosity for the system. Polymers with low viscosities should be better at bubble release than those with higher viscosity, but may not produce articles with good mechanical properties.

5.3.6 Behaviour of the polymer melt

The flow of a liquid in a rotating cylinder is essentially the same as that of a lubricant and viscosity balances the effects of gravity, with inertia and surface tension effects being negligible, leading to the following analysis. The number of independent dimensionless groups reduces to two, viz. the fill ratio F, i.e. the ratio of total volume of liquid to the volume of the cylinder, and $\alpha \equiv (\Omega v/gR)^{1/2}$ where Ω is the angular velocity of the rotating cylinder, R is the radius, v is the kinematic viscosity of the liquid and g is the gravitational constant. If $F \ll 1$, the film thickness profile $h(\theta)$ where θ refers to the angular of rotation coordinate (see Figure 5.5), is small compared to R hence the equations of motion are simplified using standard lubrication theory. The solution of the corresponding momentum equation along the direction of rotation θ has the form:

$$\partial^2 v / \partial z^2 - \cos\theta = 0$$

giving the first approximation to the velocity profile as:

$$V = 1 - \left(\eta z - \left(z^2/2\right)\right)\cos\theta \tag{5.17}$$

where v is the velocity component along θ rendered dimensionless with Ω/R. Also:

$$\eta = h / R\alpha \quad \text{and} \quad z = \left(R - r\right) / R\alpha \tag{5.18}$$

where r is the radial coordinate. This expression for v satisfies the condition for no slip at $z = 0$ and to first-order the zero shear stress condition at $z = \eta$. In this analysis, it is assumed that the

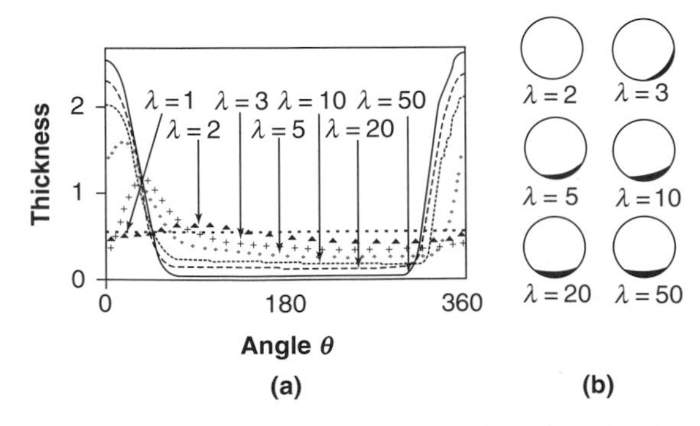

Figure 5.5 Range of solutions showing effect of variation in λ: (a) interface height plotted as a function of θ using Cartesian axes; (b) interface height plotted using plane polar coordinates.

fluid is Newtonian and the possibility of a time-dependent term is not considered. Integration of the velocity v across the film leads to:

$$q = \eta - \left(\eta^3 / 3\right)\cos\theta \tag{5.19}$$

where:

$$q = \int_0^\eta v \partial z = Q / \left(\Omega R^2 \alpha\right)$$

with Q being the total volumetric flow rate within the film per axial distance. At steady state, the dimensionless flow rate q is independent of θ but its value is unknown. The thin film approximation can be introduced:

$$F = 1/\pi R \int_{-\pi}^{+\pi} h(\theta)d\theta = \alpha\beta \quad \text{where } \beta = \frac{1}{\pi}\int_{-\pi}^{+\pi} \eta d\theta \tag{5.20}$$

These equations show that the initially independent parameters, F and α combine into one, viz. $\beta = F/\alpha$, the value of which implicitly determines q and therefore the dimensionless film thickness profile $\eta(\theta)$. This analysis showed that the stability of the fluid flow depends on the numerical value for the mass flux q, the value of η being independent of θ for values of $0.3 > q > 0$. No solutions to the equations are possible for value of q greater than 0.65. If q is small, the problem can be simplified and yields:

$$\left\langle \bar{h} \right\rangle < \rho\Omega^2 R^4 / 45\gamma \quad \text{if } \left\langle \bar{h} \right\rangle << \sqrt{v\Omega R/g} \tag{5.21}$$

Thus increasing the velocity of rotation or the cylinder's radius strengthens instability, whereas increasing surface tension or thickness of the film weakens the instability and the hydrostatic pressure gradient does not affect the stability. Surface tension is an important factor in consideration of film stability. We define two dimensionless parameters, $\lambda - A^2\rho g R/\mu\Omega$, where A is the filling fraction of fluid inside the cylinder, ρ is the density, g is the gravitational constant, R is the radius of the cylinder, η is the viscosity and Ω is the rate of rotation. Three regions can be defined: $0 < \lambda < 2$ (fast rotation), $2 < \lambda < 5$, (moderate rotation) and $\lambda > 5$ (slow rotation). There are restrictions on the Bond number: $\beta = \rho g R^2/A\gamma$ for surface tension effects to be negligible. For the condition $\lambda > 5$, the region of large curvature where the film is extracted from the powder pool leads to a capillary pressure gradient that restricts the fluid flux in the film. Detailed analysis leads to a prediction of the film thickness that scales as $0.798(\eta\Omega R)^{2/3} A^{-1/3}\gamma^{-1/6}(\rho g)^{-1/2}$ in the limit $\beta^{-3/5} << A << 1$ or $0.731(\eta\Omega)^{2/3}R^{7/15}\gamma^{-1/15}(\rho g)^{-3/5}A^{-2/5}$ in the limit $A << \beta^{-3/5} << 1$, depending on the geometry. When surface tension is negligible then the capillary number must be large. If the surface tension term is significant (β finite), two distinct steady-state solutions exist when appropriate defined capillary numbers are small and the film thickness scales with rotation rate and material parameters and describes the situation depicted in Figure 5.5(b). There is an upper limit to λ corresponding to the film becoming so thin that it is unstable and dewets. Numerical solutions of these equations for λ in the range 1–50 are presented in Figure 5.5. In all cases, $\beta = 100$, $A = 0.1$. As λ increases from small to large, an evolution is observed from a nearly uniform to a deep pool at the bottom of the cylinder which is then connected to a thin film. If the surface tension and first-order gravity effects were neglected entirely then no solutions would exist for $\lambda > 5$. The film varies from a nearly uniform thickness characteristic of higher rotation speeds (λ small) to highly nonuniform shapes that exhibit sharp gradients at lower rotation speeds (λ large). In the above, the filling parameter

Table 5.1 Summary of results in zero and nonzero surface tension cases, divided according to value of λ (gravitational parameter)

	Surface tension and first-order gravity neglected	**Surface tension and first-order gravity included**		
		$A^2\beta/\lambda >> 1$ $\beta^{-3/5} << A << 1$	$A^2\beta/\lambda << 1$ $\beta^{-3/5} << A << 1$	$A^2\beta/\lambda << A''^2\beta^{6/5} << 1$ $A << \beta^{-3/5} << 1$
$5 < \lambda$	No steady solution	Surface tension unimportant	Surface tension important	
$2 < \lambda < 5$	Discontinuous	Region of shape variation, larger gradients reduced by first-order gravity and surface tension		
$\lambda < 2$	Small gravitational perturbation to uniform film, surface tension is unimportant			

has been held constant. However, if the filling factor is varied, it is found that it is not the actual value but rather its ratio to material and geometrical parameters that is important. For a flat pool to form at the bottom of the cylinder the radius must be large compared with the capillary length that leads to the condition: $R >> A^{-1/3}\left(\gamma / \rho g\right)^{1/2}$. The cases of zero and nonzero surface tension in relation to the gravitational parameter can be described by Table 5.1.

The above theory does not include any consideration of possible viscoelastic effects on the fluid flow. The flow will be further complicated if the mould contains protrusions or weirs that would enhance polymer build-up in these areas and lead to uneven wall thickness. The analysis can be extended to include viscoelastic effects by an expansion of the viscosity term in the above expressions:

$$\eta(\Omega) = \left(\left(\eta(0) - \eta(\infty) / \left(1 + \Omega^2 \iota^2\right)\right)\right) + \eta(\infty) \tag{5.22}$$

Equation (5.22) indicates that the effective viscosity depends on the rotation rates. The parameter τ in the equation is indicative of the intrinsic relaxation time of the fluid. For many fluids, this will be effectively very short in which case Equation (5.22) ceases to have any significance. However, in polymer systems, the value of τ can become comparable with the rotation rate and then the form of the equation can become important in defining the film forming properties. Equation (5.22) predicts that at low rotational rates the values of the effective viscosity will be larger and can compensate for the gravitational effects.

5.3.7 Degradation effects on the melt

In all processing of thermoplastic materials the problem of holding the material at a high temperature, which induces oxidative degradation, arises. The oxidation process can have two effects: coupling, which will increase the molecular weight, and degradation, which will decrease the molecular weight. The viscosity of a polymer is lowered as the temperature is increased because the chains coil up to form smaller structures that will have less resistance to flow (see Section 8.9).

5.3.8 Solidification

The final stage in the rotational moulding process is the conversion of the melt to a solid. The rate at which the melt is cooled will influence the size and structure of the crystalline structure that is formed.

5.3.9 Moulding cycle

The above equations can help define the moulding cycle (see Figure 5.6). The precise profile will depend on the material and the type of moulding being created.

5.4 Injection moulding

As the name implies, the process involves the injection of molten polymer into a closed cavity. The temperature to which the polymer has to be heated is dictated by the ability to achieve

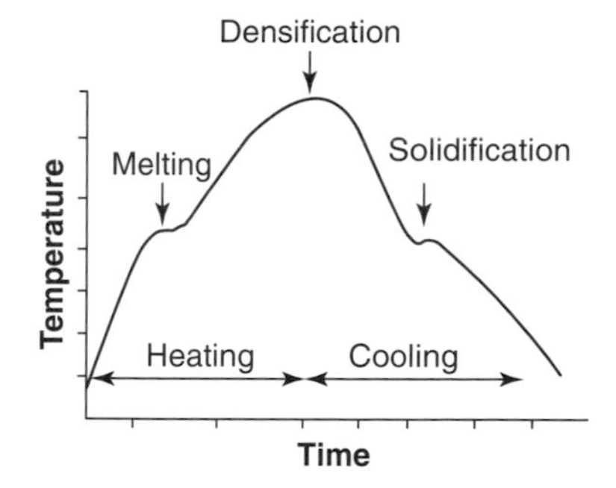

Figure 5.6 Schematic of a typical thermal profile for rotational moulding (rotolog trace).

flow into the mould. Whilst there are a number of variants of injection moulding machines, a *screw extruder* is usually used either as part of the process or in the compounding of the moulding powder. The compounding process involves blending various components: polymer, fillers, dyes, antioxidants, mould release agents, etc. to produce the pellets which are used in the injection moulding process.

5.4.1 Extruder

An extruder is essentially a barrel, which can be heated or cooled and contains a screw that helps compress the powder into a homogeneous liquid. Schematically, a screw extruder can be considered to have several zones (see Figure 5.7). The first stage ensures that thorough mixing has occurred between the various components that are added to the charge. Usually the components

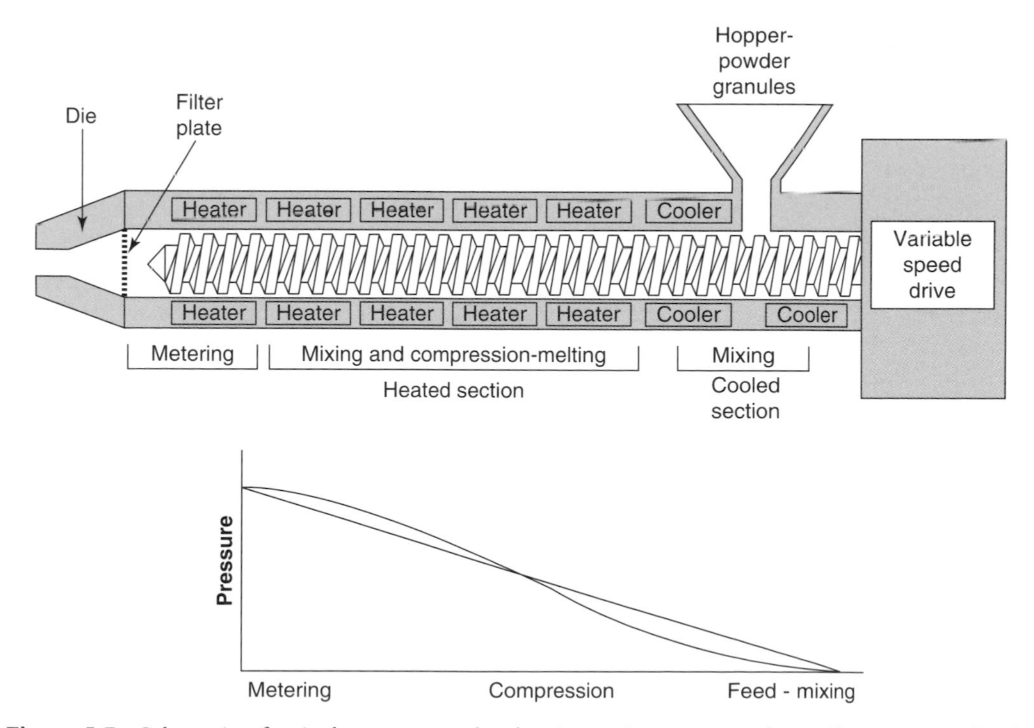

Figure 5.7 Schematic of a single screw extruder showing various zones and overall pressure profile for barrel.

will be dry blended in a tank prior to being poured into the barrel; however, true mixing only occurs when the materials are in the melt phase. The second phase is compression and melting, the final stage involves homogenising the mixture.

5.4.2 Feed or mixing zone

In the *feed* or *mixing zone*, the cold powder is thoroughly mixed and the barrel may be cooled to avoid the mixture undergoing premature melting. To aid the mixing, the screw profile may be from the regular thread to an interrupted thread, or screw of different pitch (see Figure 5.8). In the *parallel interrupted mixing flight* the screw thread has a shallow angle that leads to the powder being moved quickly along the barrel into the region where there are no screw threads. The change in screw profile helps to build up the amount of material in this area and thus helps to churn the powder before it is moved to the next section. The same effect is achieved by the use of the *undercut spiral barrier type* of screw section. The varying angle of the screw will cause uneven flow and hence a churning effect will develop. The *ring type barrier mixer* has a similar effect, the ring stops the forward movement of the powder and promotes the churning. A barrier to the free flow can also be created by the radial insertion of pins at a point in the barrel as in the use of *mixing pins*. For some very delicate fillers, needles or hollow spheres that are fragile, the mixing process may take place in the melt phase and the *cavity mixer* is placed at the end of the extruder barrel just before the melt is ejected.

5.4.3 Compression zone

Once the powder has been mixed it is then compressed and heated. The energy required to melt the powder comes from a combination of shear heating, thermal transfer from the barrel and convection from material that has already been melted. The compression process is assumed to produce a steady increase of pressure with distance down the barrel and is linear with distance. The molten polymer is then compressed into a free-flowing liquid that has viscoelastic characteristics but also will contain entrapped gas which has to be expelled from the melt. If the melt viscosity is sufficiently low and the transfer rate is not too high then gas bubbles will be formed and released. However, if the viscosity of

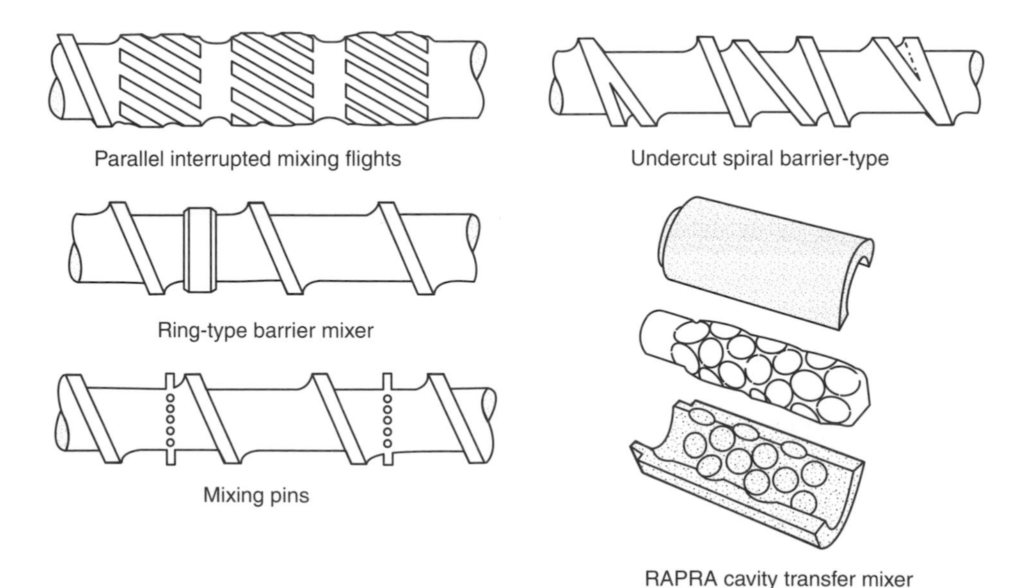

Figure 5.8 Schematic drawings of variations in screw structure to achieve efficient mixing with a screw extruder (RAPRA is the Rubber and Plastics Research Association).

the liquid or the surface tension is high then bubble formation may be difficult (see Section 5.3.4). If the gas is not naturally expelled then it may be necessary to introduce a breather zone.

5.4.4 Metering zone

The metering zone ensures the homogeneity of the mix. The length of each of the zones will depend on a number of factors, which will include: the thermal characteristic of the polymer, the imposed temperature gradient and the rate at which the screw is turned. If we consider some typical polymers, we can see how different factors come into play:

- Nylon is a crystalline polymer and as such melts very quickly, so that compression of the powder and melting can occur within one pitch of the screw.
- Polystyrene is an amorphous material and will slowly be transformed from a hard solid to a viscoelastic liquid by heating. Further heating will be required to achieve sufficient mobility for moulding. The reduction in viscosity will take place over several rotations of the screw and the number of turns will depend on the molar mass of the polymer.
- PVC is an amorphous polymer but it is sensitive to heat and can be easily degraded. It is necessary when moulding PVC to keep the compressive zone narrow and minimise the heat transferred to the polymer.
- Polycarbonate and some other polar polymers have the ability to rapidly absorb water and will expel water vapour in the melt, undergo hydrolysis and thermal degradation. A number of hydrophilic polymers must be carefully dried before they are extruded.

The typical extruder has a length that is between 20 and 30 times the screw diameter. The variables in the operation of the extruder are: the pitch of the screw, the number, size and settings of the zone heaters, and the rate at which the screw is operated.

5.4.5 Analysis of flow in extruder

The operation of a screw extruder has been studied extensively and the total flow can be considered in terms of three distinct effects:

- *Drag flow* in which the polymer is pulled along the barrel by the rotation of the screw.
- *Pressure flow* describes the pressure gradient created by molten polymer in the barrel stopping the forward movement of material behind it.
- *Leakage* of the polymer past the screw will occur as a consequence of the pressure gradient down the barrel.

Each of these factors will be considered separately.

5.4.6 Drag flow

Consider the polymer element defined by the volume, dx, dz and dy (see Figure 5.9). The screw pitch is defined by $\pi D\tan\phi$, and the screw is width e, diameter D with a pitch of angle of inclination ϕ and a rotation velocity of V_p which is equal to πDN, where N is the number of rotations (see Figure 5.9).

The gap between the screw and the barrel is δ and changes as the screw wears. The gap dictates the amount of polymer that will leak back in the opposite direction to the general flow. Assume that the fluid element is subjected to a shear gradient between the moving screw and the static barrel. Consider a small element ABCD, which is located in a channel of depth H (see Figure 5.10). The volume flow rate dQ is given by:

$$dQ = Vdydx \tag{5.23}$$

Assuming the velocity gradient is linear, then:

$$V = V_d[y/H] \tag{5.24}$$

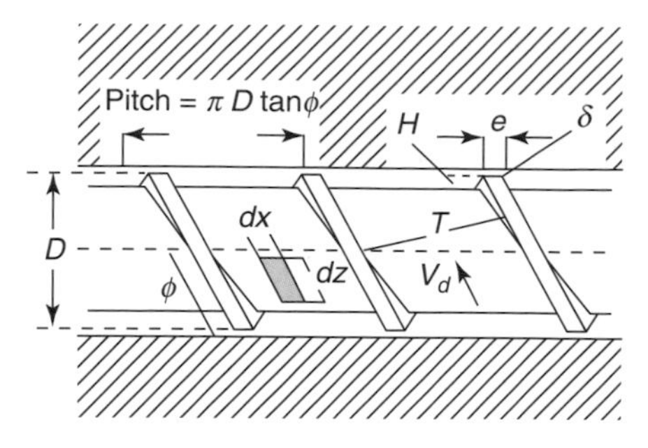

Figure 5.9 Schematic of a screw extruder.

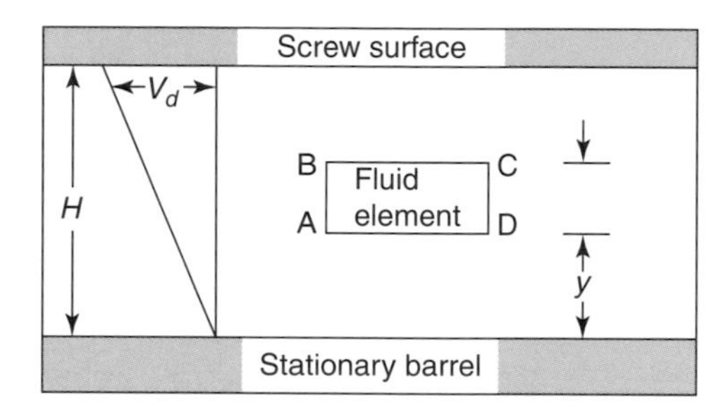

Figure 5.10 Schematic of a fluid element subject to drag flow.

In Equation (5.24), V_d is the velocity parallel to the screw pitch and will be different from the velocity of rotation of the screw V_p that is defined as being $V_p = \pi DN$, where D is the diameter of the screw and N is the number of revolutions per minute.

The axial velocity is V_0. Integrating the total drag flow Q_d is:

$$Q_d = \int_0^H \int_0^T \frac{V_d}{H} dydx = 1/2THV_d \tag{5.25}$$

If the screw speed is N then $V_d = \pi DN \cos\phi$. The distance T can be defined in terms of the screw pitch as follows:

$$T = (\pi D \tan\phi - e)\cos\phi \tag{5.26}$$

where e is the width of the screw. Substitution in Equation (5.25) gives:

$$Q_d = 1/2(\pi D \tan\phi - e)(\pi DN \cos^2\phi)H \tag{5.27}$$

In most cases the thickness of the screw is significantly smaller in comparison with ($\pi D\tan\phi$) so that Equation (5.27) can be reduced to:

$$Q_d = 1/2\pi^2 D^2 H \sin\phi \cos\phi \tag{5.28}$$

Equation (5.28) shows that *drag flow* is determined by the diameter of the screw, the depth of the channel and the angle that the screw makes with the horizontal axis.

5.4.7 Pressure flow

The fluid element is dragged along by the screw but will also be subject to pressure from the liquid ahead and behind the element (see Figure 5.11). The two sides of the element will be

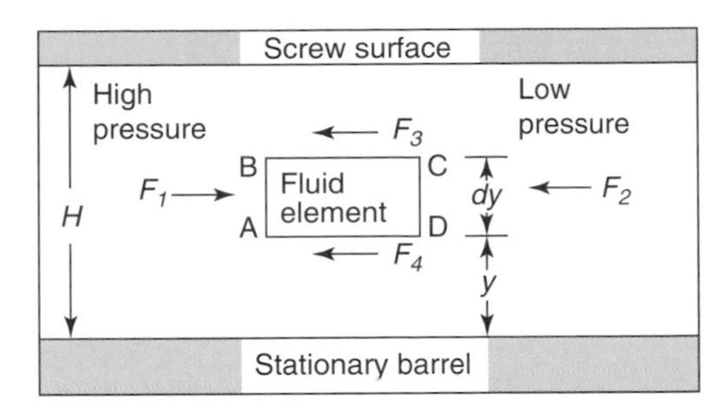

Figure 5.11 Schematic of fluid element subjected to pressure flow.

subjected to shear and if the element is thin i.e. dy is small then F_3 approximates to F_4. The forces on the element can be balanced according to the following equations:

$$F_1 = \left(P + \partial p / \partial z dz\right) dy dz, \quad F_2 = P dy dx, \quad F_3 = d\tau dz dx, \quad \text{and} \quad \int_0^v dV = 1/\eta \partial P/\partial y \int_{H/2}^{y} dy$$

$$\text{and} \quad V = 1/\eta dP/dz\left(y^2/2 - H^2/8\right) \tag{5.29}$$

The volume flow rate dQ is given by $dQ_p = VTdy$. Integration gives the pressure flow Q_p:

$$Q_P = \int_0^H 1/\eta \partial P/\partial z T\left(Y^2/2 - H^2/8\right) dy; \quad \text{integrate } Q_P = -1/12\eta \partial P/\partial z TH^3 \tag{5.30}$$

Assuming that e is small, then $T = \pi D \tan\phi \cos\phi$, and $\sin\phi = dL/dz$ so that $\partial P/\partial z = (\partial P/\partial L)\sin\phi$, where L is the length of the screw. The pressure flow is thus described by:

$$Q_P = \left(\pi D H^3 \sin^2\phi\right)/12\eta\left(\partial P/\partial L\right) \tag{5.31}$$

where dP/dL is the pressure gradient down the barrel and is assumed to be linear with distance down the barrel.

5.4.8 Leakage flow

The screw will not be a tight fit in the barrel and wear will increase the gap between the barrel and the screw thread (see Figure 5.12). The gap allows the melt to move *back* down the barrel as a result of the pressure gradient. The analysis of the *leakage* flow is essentially the same as that for the pressure flow, except that the element is now in the gap between the screw and the barrel. If the gap is δ and a pressure difference ΔP exists across the screw then the above model can be used by substituting $H = \delta$. In a typical extruder, the value of δ will be of the order of 0.2–0.5 mm. The leakage flow may be considered as flow through a wide slit which has depth d and a length ($e\cos\phi$) and a width ($\pi D/\cos\phi$). The pressure gradient across the gap is defined by: $\Delta P/e\cos\phi$ where ΔP is defined by $\Delta P = \pi D \tan\phi \cos^2\phi(\partial P/\partial L)$ which leads to an expression for the flow rate being:

$$Q_L = \left(\pi^2 D^2 \delta^3\right)/12\eta\left(\tan\phi\left[\partial P/\partial L\right]\right) \tag{5.32}$$

Equation (5.32) has the same form as that for pressure flow except that the dimensions are very different: Q_L is dependent on the third power of the gap and hence very sensitive to wear of the screw and its thickness, e. In practice, the screw will not be the same shape as the barrel and a factor should be introduced to allow for the eccentricity of the barrel–screw profile. In a real extruder, the total contribution to the flow from leakage can be as much as 20% of the total and

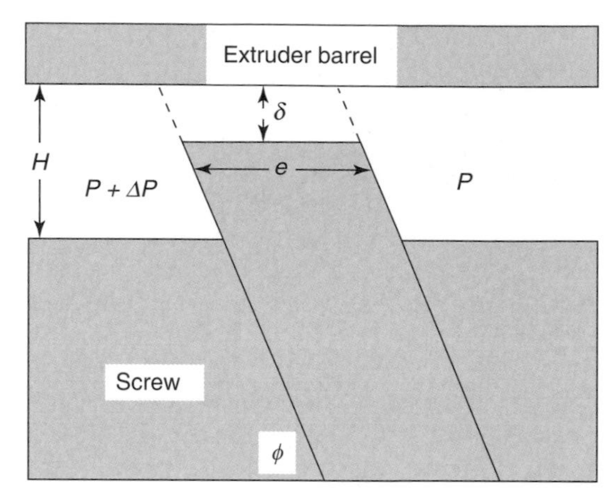

Figure 5.12 Leakage flow through gap between screw and barrel.

accounts for a significant amount of recirculation of the molten polymer within the barrel. If we assume that the pressure gradient is linear then $\Delta P = \pi D \tan\phi\ (\partial P/\partial L)$. Combining the three components to the total flow one obtains:

$$Q_{Total} = \frac{1}{2}\pi^2 D^2 NH \sin\phi\cos\phi - \frac{\pi DN^3 \sin^2\phi}{12\eta}\frac{dP}{dL} - \frac{\pi^2 D^2 \delta^3}{12\eta e}\tan\phi\frac{dP}{dL} \tag{5.33}$$

The three components to the total flow depend on the pitch of the screw, the rate at which it rotates, the viscosity of the molten polymer and the screw dimensions; the diameter D, thickness e and angle ϕ. Although in practice the pressure profile is not linear, it is usual to assume that a linear approximation is valid, i.e. $\partial P/\partial L = P/L$. The value of P will depend on the way in which the extruder is used. As a moulding machine it has to force the molten polymer through a gap into the mould and the dimensions of the gap will define the necessary pressure for the process. Alternatively, the extruder may be used to compound material and produces a thread of molten polymer and the relevant pressure is one atmosphere, since the material is freely discharging from the slit die.

The above theory does not consider the interaction of the polymer with the barrel, which can be introduced as a friction coefficient, f_b. The heat generated per unit of the barrel surface q_b is then given by:

$$q_b = f_b \pi ND\left[\sin\phi / \sin(\phi + \phi')\right]\left[(P_2 - P_1)/\ln(P_2/P_1)\right] \tag{5.34}$$

where $P_2 - P_1$ is the pressure differential and ϕ is the angular drift as the element moves across the channel as a consequence of the screw rotation. Assuming that conduction occurs only in the y-direction and bulk flow is in the z-direction, the differential equation for the temperature distribution is then:

$$\rho_b C_p v_p \left(\partial^2 T_p / \partial z\right) = k_p \left(\partial^2 T_p / \partial y^2\right) \tag{5.35}$$

where ρ_b is the density, v_p is the velocity of the fluid element down the barrel, C_p is the heat capacity of the polymer and k_p is the thermal conductivity. T_p is the temperature of the fluid element. The polymer melt will be able to transmit energy to the rest of the molten melt and is controlled by the equation:

$$\left(\partial T_p / \partial t\right) = \alpha_p \left(\partial^2 T_p / \partial y^2\right) \tag{5.36}$$

where α_p is the thermal conductivity of the polymer melt. Using these equations it is possible to create a thermal profile for the polymer down the barrel. In practice, the heaters attached to the barrel will offset the energy that is imparted to the barrel from the frictional heating by the polymer and may initially provide energy to achieve the transformation of the solid into the melt. It is fairly obvious that the detailed profile will depend on the heat capacity–temperature profile for each individual polymer.

5.4.9 Free-flow condition

In the extreme condition that the output of the extruder is into a wide bore pipe, then the pressure gradient is zero and the throughput of the extruder is given by the limiting case:

$$Q_{Total} = Q_{Max} = 1/2\pi^2 D^2 NH \sin\phi\cos\phi \tag{5.37}$$

In this equation, the pressure terms are placed equal to zero and it is assumed that the only controlling features are the rate at which the screw is rotated and its pitch angle ϕ.

5.4.10 Flow into a mould or die

The mould or die will create a restriction on the flow that will need to be overcome for the polymer to move down the barrel, and this pressure will usually depend on flow through a narrow channel. The flow will be defined by $Q = KP$ where K is a constant, which reflects the flow characteristics of the die and for a capillary of radius R, and length L_d has the form:

$$K = \pi R^4 / 8\eta L_d \tag{5.38}$$

The output of the extruder, if we ignore the effects of leakage, will then be given by:

$$Q_{total} = \frac{1}{2}\pi^2 D^2 NH \sin\phi\cos\phi - \frac{\pi DH^3 \sin^2\phi}{12\eta}\frac{P}{L} = \frac{\pi R^4}{8\eta L_d}P \tag{5.39}$$

The output is therefore directly related to the pressure that is created by the moulding cavity or the die. Solving the above equations gives an optimum value for the pressure and allows for the effects of the die or the mould:

$$P_{optimum} = \left(\left(2\pi\eta D^2 NH \sin\phi\cos\phi\right) / \left(\left(R^4 / 2L_d\right) + \left(DH^3 \sin\phi\cos\phi\right)\right)\right) \tag{5.40}$$

The above equations indicate that as the pressure is increased the amount of material flowing, Q, decreases. The optimum condition for the operation of the extruder is when the pressure in the barrel matches the pressure required for flow through the die or to fill the mould (see Figure 5.13).

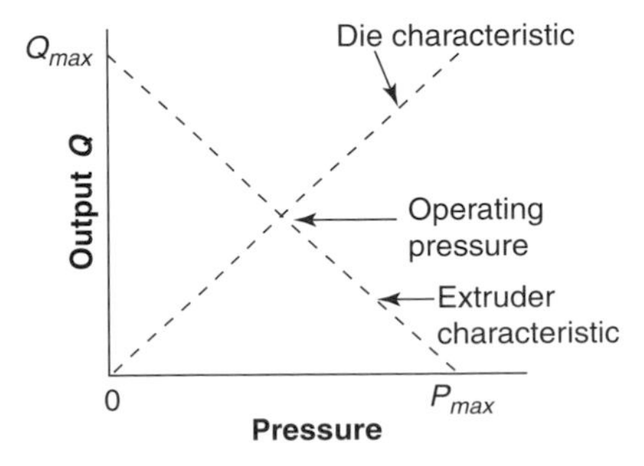

Figure 5.13 Extruder and die characteristics for determination of operational temperature.

5.4.11 Extruder volume efficiency

The efficiency of the extruder can be calculated as follows. For an axial velocity Va = pitch × screw speed = $\pi DN \tan\phi$. The velocity V_d is expressed by: $V_d = (V_a / \sin\phi) = (\pi DN \tan\phi)/\sin\phi$. The ideal output Q_{ideal} is given by $Va \times$ cross-section of screw flight:

$$Q_{ideal} = (\pi DN \tan\phi)/\sin\phi(\pi DH \tan\phi \cos\phi) = \pi^2 D^2 HN \tan\phi$$

The volume efficiency of a screw is defined as the maximum value Q_{max}, that is the value with no pressure restriction relative to the ideal value.

$$Q_{max}/Q_{ideal} = 1/2 \cos^2 \phi.$$

The ideal value for the angle ϕ is when Q_{max}/Q_{ideal} is equal to 1 and has a value of 42°. The pressure at the die will vary with the type of application; typical values are summarised in Table 5.2.

5.4.12 Power requirements

If we neglect the issues associated with frictional heating then the power requirement is equal to the peripheral speed×peripheral force. The peripheral velocity is $V_p = \pi DN$ and the peripheral force F_p is $F_p = F_s / \cos\phi$ where F_s is the shear force acting on the fluid element: $F_s = \tau\pi D \sin\phi dz = \tau\pi DdL$, where τ is the shear stress, given by $(\eta(dV/dy))$ for a Newtonian liquid. The power dE is given by:

$$dE = \eta(Dv / dy)\pi^2 D^2 NdL \tag{5.41}$$

where dV/dy is the shear rate at the barrel wall and integration from 0 to L gives the total power requirement, E.

5.4.13 Location of melt front

When operating an extruder, it is useful to be able to estimate at which point in the barrel the solid powder will be converted to a liquid. The processes involved are shown in Figure 5.14 and reflect the powder being in contact with the heated screw or being separated from it by a thin molten film.

Combination of a series of equations which describe the hydrodynamics, heat transfer and melting behaviour yield following expression:

$$X / W = \left(1 - \left[C_2 Z / \left(2\rho_s V_{sz} HW^{1/2}\right)\right]\right)^2 \tag{5.42}$$

where X is the width of the solid bed at any helical length Z, H being the channel depth and H_s is the height of the solid phase in the channel. V_{sz} is the velocity of the solid bed down the screw, the cross-channel velocity being V_x, and V_r is the resultant relative velocity. The density of the solid and melt are, repectively, ρ_s and ρ_m and Z_m is given by:

$$Z_m = \left(\left(2\rho_s V_{sZ} H_s W^{1/2}\right) / C_2\right) = 2\pi V_{sZ} H\left((2\lambda W) / (C_1 \rho_m V_s)\right)^{1/2} \tag{5.43}$$

where:

$$C_1 = \left[\left(\mu_t V_r^2\right)/2 + k_m\left(T_B - T_m\right)\right] \quad \text{and} \quad C_2 = \left(\left(\left(C_1 \rho_m V_x\right)/2\lambda\right)\right)^{1/2}$$

Table 5.2 Typical values of die pressure for different applications

Product	Die pressure (MPa)	Product	Die pressure (MPa)
Blown film	6.9–34.5	Cast film	1.4–10.4
Sheet	1.5–10.4	Pipe	2.8–10.4
Wire coating	6.9–34.5	Filament	6.9–20.7

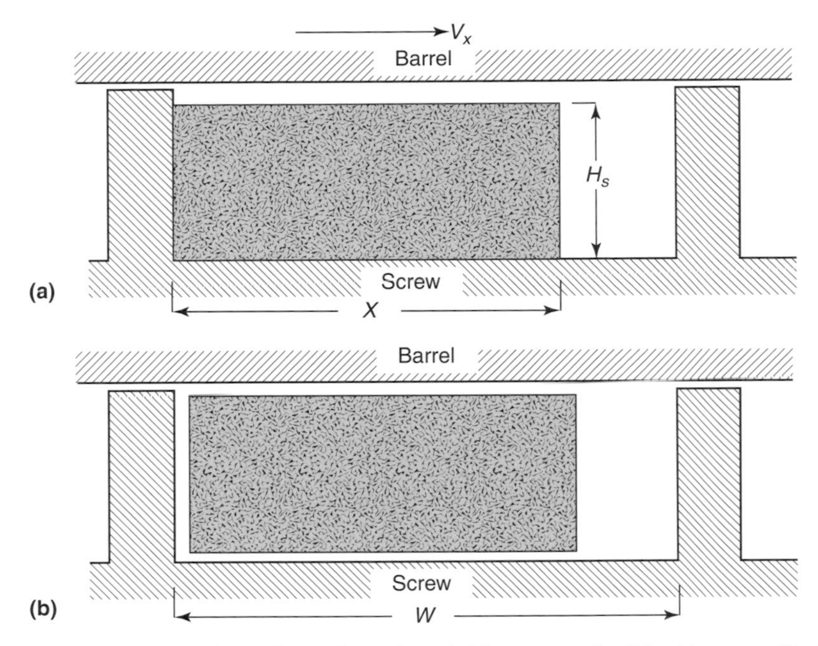

Figure 5.14 Schematic of melting of powder in barrel: (a) contact of solid with screw; (b) thin molten film of polymer separating solid from barrel and screw.

where T_B and T_m are, respectively, the barrel and melt temperature for the polymer, μ_1 is the film thickness between the solid and the screw, and λ is the latent heat of fusion. By combining these equations, the location of the melt front can be estimated.

The above theory is a very approximate model of the extrusion process. However, it captures the important elements of the extrusion process. In practice, modelling of the extruder performance can be best carried out using finite element methods and a number of computer packages based on more detailed analysis of the flow behaviour have been developed and allow visualisation of the various stages of the process and its practical optimisation (O'Brian, 1992). The computer models allow fine details such as the effects of the hanger die, the nonlinear characteristics of the shear dependence of the melt viscosity, heat transfer characteristics of the polymer melt, etc. to be included in the predictions for the optimisation of the extruder operation. The variables which are usually available for change are the screw rotation rate and the temperature of the zones (Tucker, 1989).

5.4.14 Twin-screw extruders

Before leaving the extrusion process it is important to consider the *twin-screw extruder*, a variant of the basic instrument that is often used in practice in preference to the simple single-screw instrument. The barrel has a figure of eight cross-section with the centres of the axes of rotation of the two screws being less than the sum of their radii (see Figure 5.15). The shape of the screw may vary both in its detail and in the extent to which it overlaps with its neighbouring screw. The typical arrangements are shown in Figure 5.15.

The cavity in which the screws are located can be opened up to allow the screws to intermesh as shown in Figure 5.16. The screws can either rotate towards one another (corotating) or be counter-rotating, and can be intermeshing or nonintermeshing. The closer the two screws are brought together the greater the shear gradient that is created where they overlap.

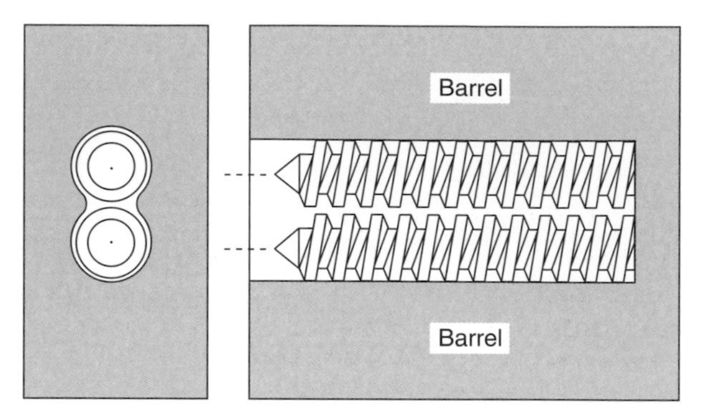

Figure 5.15 Schematic of a twin-screw extruder showing end view and section through barrel.

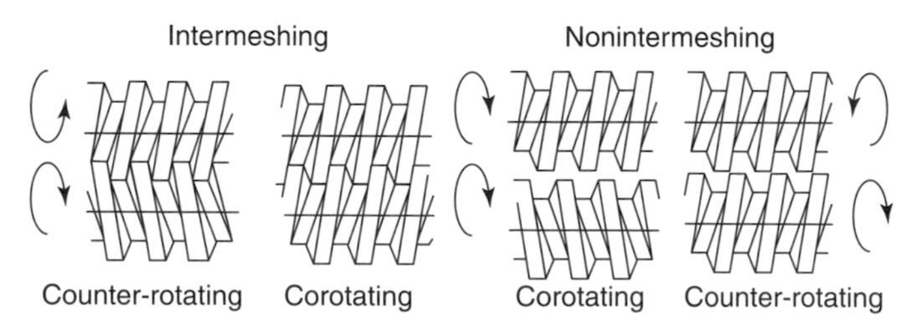

Figure 5.16 Schematic of screw arrangement for a twin-screw extruder.

The attraction of the twin-screw extruder is the extent to which the mixing can be controlled. For sensitive materials the fluid layer between the two screws balances the rotation of the screws and there is less tendency for the screw to run against the walls of the barrel.

5.4.15 Use of vented barrel

Many plastics will absorb moisture or contain other volatiles that need to be expelled before a homogeneous fluid can be created for moulding. The extruder is essentially the same as the single-screw extruder, but has a section inserted which contains a vent so that gas can be released (see Figure 5.17). The vent will usually exit into a gas ballast tank that allows control of the release of the gas and ensure that the liquid stays in the barrel.

The powder is taken through the usual compression, melting and mixing (metering zones) and then enters a decompression zone as it passes the vent. After the vent, the liquid is once more compressed, allowing the required pressure to be achieved for the molten polymer to flow through the die or enter the mould. To aid the transfer of the powder down the barrel, the screw is often asymmetric (see Figure 5.18).

The tapered edge is set at an angle of about 120° to avoid the creation of a *dead* space or alternatively a void region. If the leading edge is set at 90°, it provides the maximum contact with the fluid as the screw rotates.

5.4.16 Simplest use of extruder

The simplest application of the extruder is to compound materials for subsequent processing into a final moulded product. The extruder in this application is used to mix the components into a homogeneous mixture which is palletised for subsequent use. The process involves first drying the mixture of components, which has usually been ground to a fine powder (see Figure 5.19). The grinding process may need to be carried out in a cooled container to

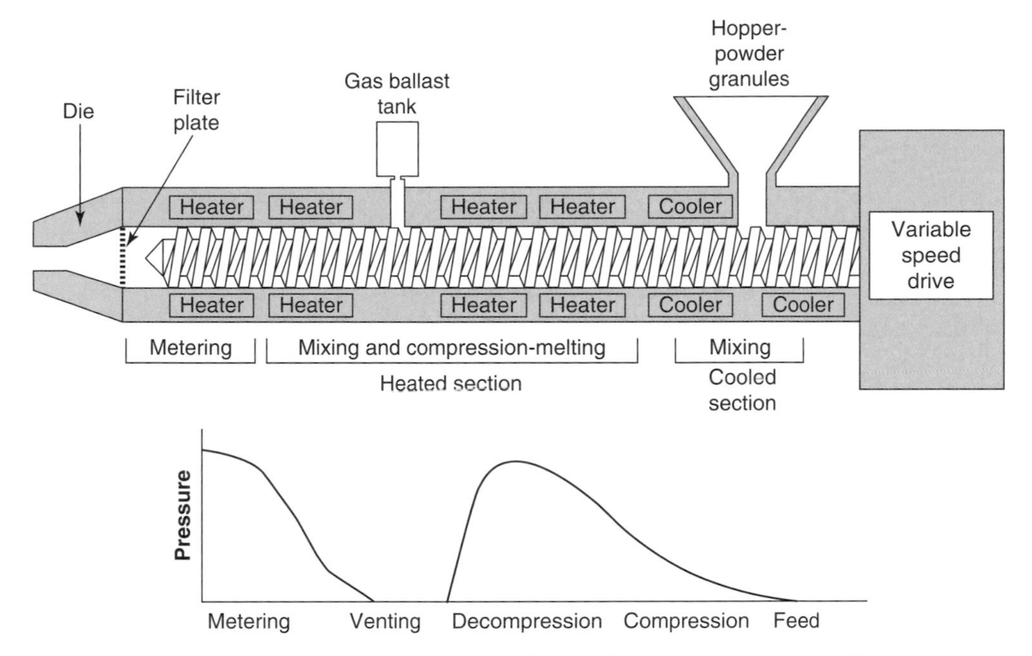

Figure 5.17 Schematic of an extruder in which a vent is incorporated.

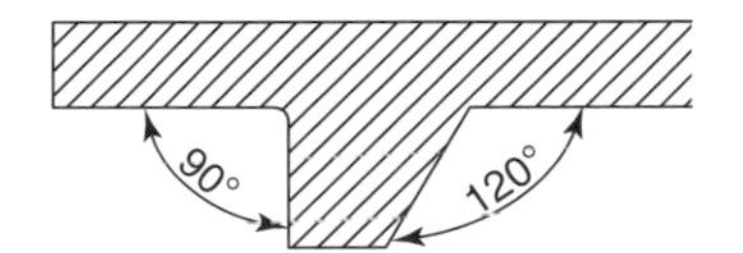

Figure 5.18 Cross-section of a screw blade.

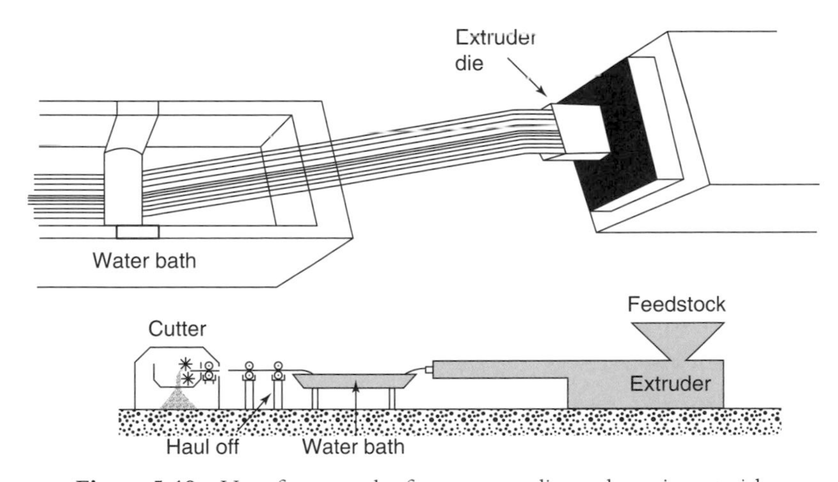

Figure 5.19 Use of an extruder for compounding polymeric materials.

avoid any of the components being prematurely melted. The mixture is then passed through the extruder where the screw will thoroughly mix the powder and then will produce a homogeneous melt. The melt is then extruded through a die which contains a number of holes, allowing a series of streams of molten material to be produced. The *laces* are quenched by passing through a water bath and then chopped into pellets. In this configuration, the effective pressure at the die is atmospheric and hence the extruder can be operated at its maximum throughput. In practice, to ensure that adequate mixing is achieved the extruder will be operated with a slight pressure at the die. The slight pressure gradient will be created by the constraints which are imposed by the filter and lace die.

5.4.17 Fabrication of simple, continuous profile materials

The exit die at the end of the extruder can be designed to create various profiles and allow the continuous production of sheets, tube structures, sections for civil engineering applications, window frames, etc. The typical configuration is illustrated for the formation of polymer sheet (see Figure 5.20).

As it leaves the extruder the molten polymer is confined to a pipe and the molten stream spread out into a uniform, broad profile before it is allowed to solidify. To achieve uniform flow across the width of the die, a so-called *coat hanger die* is used. The arch of the coat hanger is a tapered section so that the pressure across the exit to the die is the same at all points. The exit of the die has small screws which allow the gap (nip) to be altered and ensure that a perfect profile is achieved. Melt entering at the top of the coat hanger will have a long section of relatively narrow die through which to flow, whereas the broader but tapering channel will allow relatively easy flow to the edges of the die. The design of the die is carefully profiled to ensure that the exit pressure is constant across its width (O'Brian, 1992). To aid adjustment of the pressure profile, the gap can be varied either by using a variable lip or choker bar (see Figure 5.20(a)).

Another form of common product is the hollow tube. To create the hollow internal void, the die has inserted into either a solid rod or a hollow tube down which may be passed either a coolant or a gas stream (see Figure 5.21).

The polymer melt will often swell on exiting the die and a second *sizing die* is used to constrain the melt and define the outer diameter of the pipe. The sizing die will be cooled, thus transforming the melt into a solid. The molecular weight of the polymer has to be such that it will not flow too easily and can sustain the shape of the required pipe form on exit from the first die.

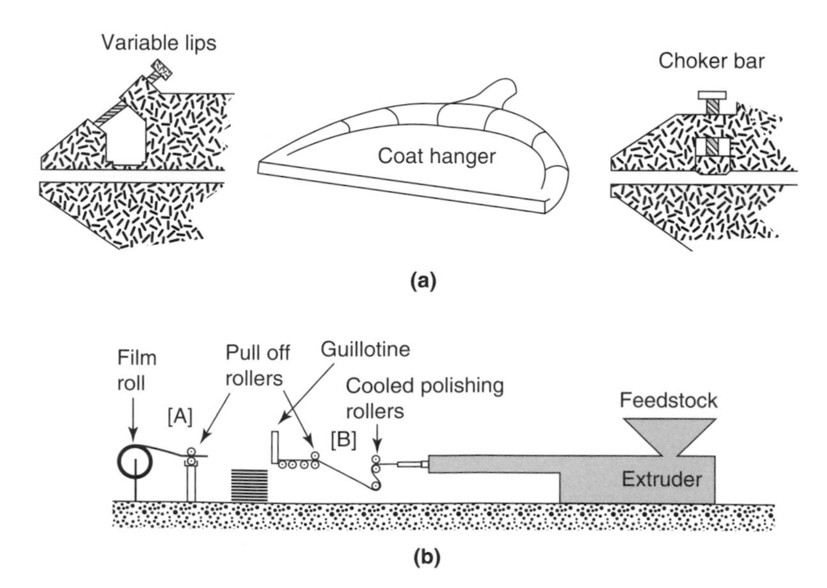

Figure 5.20 Schematic of extruder used to create polymer film.

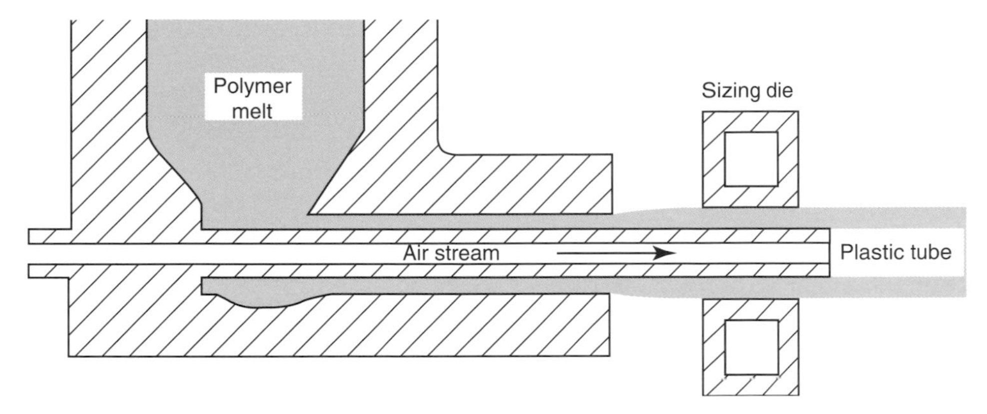

Figure 5.21 Die configuration for production of plastic pipe.

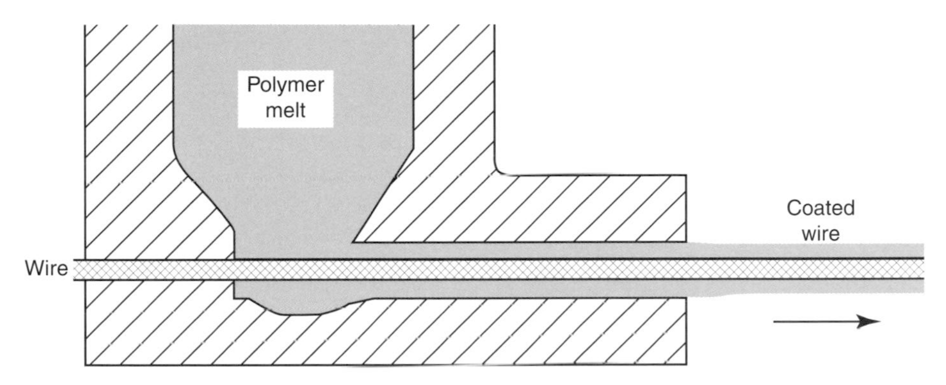

Figure 5.22 Die configuration for production of a coated wire product.

Coating wire can be achieved using a similar approach except that the wire to be coated is fed through the hole where the air stream would normally be located and there is no need for a sizing die as the coating thickness is determined by the velocity at which the wire is fed through the die (see Figure 5.22). The velocity of the wire exiting the die can also be used to reduce the thickness of the final coating. This relies on the melt having a small degree of viscoelasticity.

5.4.18 Polymer-coated products

Polymers are sometimes used as coatings for fabrics and to create simulated leather products. In this case the extruder will discharge the polymer melt into a heated slot die (see Figure 5.23). The slot die feeds a trough formed between a tensioning roller and a cooled roller. The fabric to be coated passes through the trough and solidifies on contacting the chilled roller. This process produces fabric which is coated on one side with a plastic coating and is used to produce fabric which may be converted into the uppers for shoes, handbags, etc.

5.4.19 Blow moulding

Large volumes of plastic are used in the food packaging industry: as shopping bags and as foil for wrapping goods. The film-forming process is similar to that used to create a plastic pipe, except that the molten polymer is subjected to a high pressure and blown through an annulus (see Figure 5.24).

The process of expansion of the tube of molten polymer takes place during the cooling of the melt to the solid. As a consequence the polymer is subjected to a *cold* drawing process.

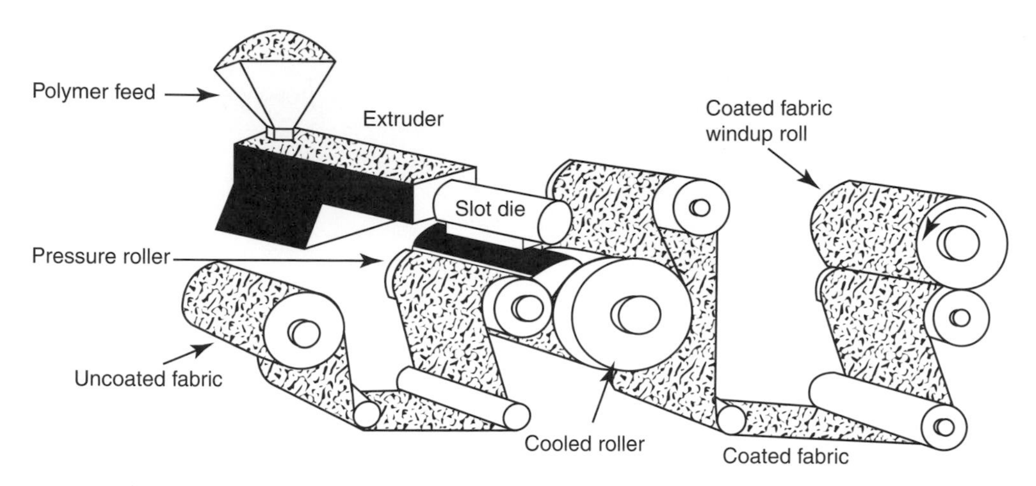

Figure 5.23 Schematic of a coating configuration using a slot die and chilled roller.

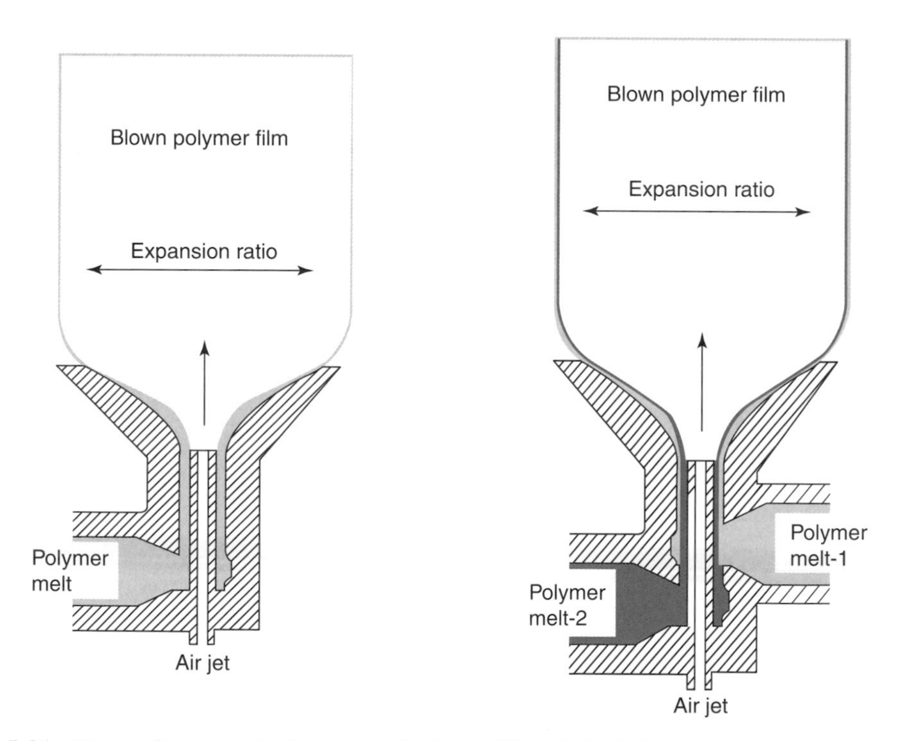

Figure 5.24 Die configuration for formation of polymer film: (a) single-layer operation; (b) production of double-layer film.

If the polymer is able to crystallise, as in the case for polyethylene or polypropylene, the expansion will orientate the lamellae and lead to enhancement of the mechanical properties. The expansion ratio is determined by the rate at which the film is extruded from the annulus and the pressure which is generated in the air stream. Stretching the paracrystalline polymer film will enhance its physical properties. An expansion ratio of the order of four or greater is easily

Table 5.3 Gas transmission characteristics for various plastics

Polymer	Layer distribution (μm)	Density ($kg\,m^{-3}$)	Oxygen ($cm^3\,m^{-2}$ for 24 h)	Water vapour ($g\,m^{-2}$ for 24 h)
LDPE	1000	920	140	0.5
HDPE	1000	960	60	0.3
Polypropylene	1000	910	60	0.25
Polyvinylchloride	1000	1390	5	0.75
PET	1000	1360	1	2
Polystyrene/ethylenevinyl alcohol/polyethylene	825/25/150	1050	5	1.6
Polypropylene/ethylenevinyl alcohol/polypropylene	300/40/660	930	1	0.25

achieved during blow moulding. You may have experienced using a shopping bag which does not take the load of shopping and stretches alarmingly, whereas an apparently similar bag easily takes the load and does not stretch. The difference between the two bags will be the extent to which the film has been subjected to *cold* drawing. Careful control of the rate at which the film is formed is needed. This will dictate the draw ratio in the direction in which the film is travelling. The pressure which determines the expansion ratio can create a biaxially drawn film which can have excellent mechanical properties.

Many polymer films are made from several layers of polymer film. If we consider the barrier properties in terms of the rates of diffusion of gases through materials (see Table 5.3) we see that certain polymers can be very good as barriers for one gas but not very good for others.

A hydrophobic polymer such as polyethylene or polypropylene will have good barrier properties for water, but will not necessarily stop the diffusion of oxygen. In packaging food, it is often important to control the ingress of oxygen, if one wishes to keep the food fresh for a period of time. More polar polymers, such as PVC, polyethyleneterephthalate or polyethylene-vinyl alcohol, will have better barrier properties with regards to oxygen but are less effective as barrier materials for moisture. For certain applications, these materials may be used as the pure materials, however, if they can be combined together either as a bilayer or alternatively as a triple-layered material. The appropriate layered structure can be created using a combination of two or three extruders feeding into a common die. These multilayered films allow balancing the properties of hydrophilic and hydrophobic polymers and achieve films with very good barrier characteristics (see Table 5.2). These multilayered films are used with microwaved products and the choice of a particular outer layer may be dictated by its compatibility with the substrate to which it is thermally bonded.

5.4.20 Moulding of bottles

A major use of plastics is in the bottle industry (see Figure 5.25). Bottles can be divided into two types: those which are used to contain simple liquids and those which contain carbonated products. The former can be created by direct moulding and the polymers are not necessarily subjected to any *cold* drawing in production, whereas the latter are subject to very significant cold drawing to achieve the barrier property enhancement required to retain the carbon dioxide in the drinks. The bottles which contain sauce, detergent or still water do not require the barrier to carbon dioxide required for those containing carbonated drinks but may need to have resistance to other materials, such as vinegar.

Simple bottle formation

The creation of a simple polyethylene, polypropylene or PVC bottle starts as a process similar to that of making a tube. As the tube (*parison*) is extruded from the die (see Figure 5.26) it passes into a cavity mould. At the lower end is a gas nozzle over which the tube is placed. The closing

Figure 5.25 Selection of plastic bottles.

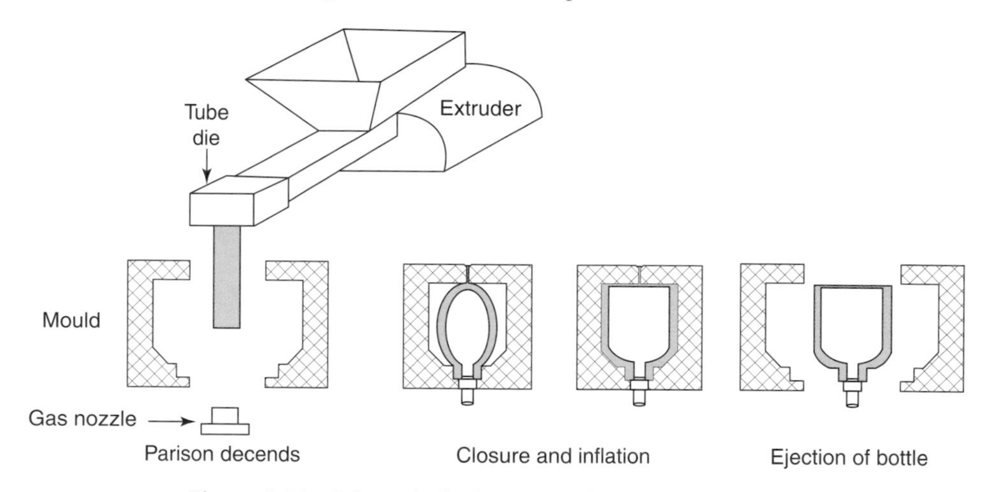

Figure 5.26 Schematic for formation of a simple plastic bottle.

of the mould traps the upper end of the tube causing it to close and be sealed. The injection of gas through the nozzle into the sealed tube causes it to inflate and take the shape of the cavity. The polymer will be in a viscoelastic state and only becomes a solid when it comes in contact with the mould. This inflation process will thin the walls of the original tube and in so doing may start to enhance the physical properties. However, since most of the shaping is taking place in the semimolten state the enhancement produced is usually considered to be minimal. The bottle is then ejected and the cycle starts again. Using a rotating mould system, it is possible to operate the extruder on a continuous basis. This process is used to produce a wide range of bottles: milk containers, detergent bottles, etc. The common characteristic of such bottles is that they are liquid containers and do not necessarily have to contain gas or be pressurised. However, many food applications may require the use of more than one polymer to fabricate such bottles, the outer layer being modified to allow printing of the labels, etc. In this case, multiple extruders may be used to produce the co-extruded *parison* allowing multilayer structures to be created. Even apparently simple bottles may, for various reasons, be created from more than one polymer material. For instance, to achieve the desired barrier properties and to avoid the fluids contained in the bottles extracting residual monomers from the plastics multilayered bottles are created. A bottle for tomato sauce is fabricated from no less than six layers!

Bottles which are pressurised or contain carbonated drinks

The bottles which are used for soft drinks need to have significantly better mechanical and gas barrier properties than those used for other liquids. A typical soft drink, carbonated water, may be pressurised well above atmospheric pressure. To achieve the pressurisation of the bottle, the plastic has to have a high modulus, and to ensure that the carbon dioxide is not lost, it also has to have good barrier properties.

Polyethyleneterephthalate (PET) is the polymer which best fulfils these criteria and is the most widely used plastic for this application. PET is usually considered to be a fairly amorphous polymer, however it can form very small ordered regions. These ordered regions can be aligned to dramatically increase the barrier properties of the polymer for the diffusion of carbon dioxide. The requirement to achieve a cold drawing process in the fabrication of the bottle leads to the following process. As in the case of the simple bottle a *preform* is created (see Figure 5.27). The *preform* is then expanded to the form of the final bottle (see Figure 5.27). The second stage involves the *preform* being heated to a temperature just below melting point and being stretched. The *perform* contains infrared absorbing agents which aid effective heating and the final shaping takes place in a mould where the bottle has been subjected to stretching in both the axial and radial directions (see Figure 5.28). Stretching is usually achieved by the application of pressure but may be assisted mechanically by the use of a metal rod to assist axial stretching. The preform will undergo an axial and radial extension of typically three times its original dimensions. The cold stretching of the PET, which causes the microcrystallites to be aligned parallel to the wall, will produce a percolation path that will inhibit the diffusion of the carbon dioxide through the walls.

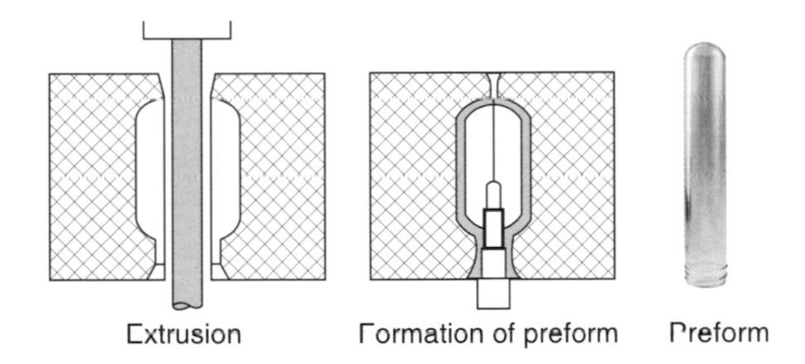

Figure 5.27 First stage in bottle formation: creation of perform.

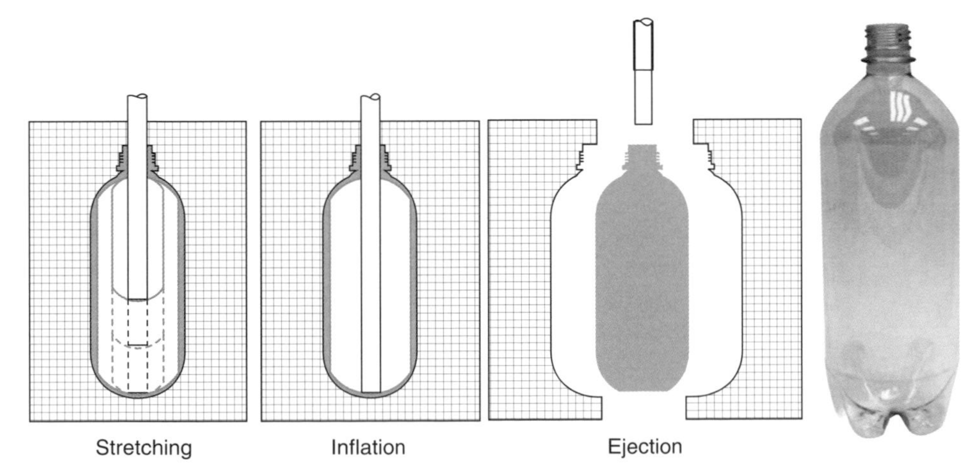

Figure 5.28 Stages in expansion of preform to create drinks bottle.

The barrier properties are directly related to the extent to which the alignment has been created by cold drawing. As a direct consequence, the small bottles (circa 500 ml) will not hold the bubbles for as long as the larger (2 L) bottles which can retain a significant pressure of carbon dioxide for 6–9 months. The PET which is used for bottle production contains a small amount of isophthalate moieties, which aid the creation of the ordered regions in the bottle walls.

5.4.21 Tensar process

The tensar process uses the cold drawing process to convert a low modulus pliable sheet of polyethylene or polypropylene into a rigid plastic network which can be used instead of wire netting to stabilise soil on the banks of motorways. The plastic netting is buried and stabilises the movement of the soil, which might be subject to water erosion or slippage. The advantage of tensar netting is that it will not be subject to corrosion as metal netting would be. The process involves taking a sheet of extruded plastic (see Figure 5.29) and subjecting it to cold drawing.

The drawing process is usually carried out in three stages. First, the sheet is punched with regularly spaced holes. The plastic which is discarded in forming the holes can be remelted and used to form another set of sheets. The second stage is the cold drawing in the longitudinal direction, where the sheet is gripped at the edges and the speed of the take up is increased. Ideally, the drawing process should induce a draw ratio in the range 3–4. The partially drawn sheet is then drawn in a transverse direction. The final *tensar* product has mechanical characteristics which are similar to those of wire netting but has the advantage of being corrosion resistant.

5.5 Compression moulding

Possibly one of the simplest methods of polymer processing is *compression moulding*, which involves heating the polymer powder in a cavity and consolidating the melt under pressure. The process allows the formation of articles with polymers that have a long rubbery plateau and are not easily processed by injection moulding. Cavity moulding, a variant of the process, can be used to process thermoset materials. In this case, the monomer is prepolymerised to a state where flow will be limited and gelation is achieved fairly rapidly.

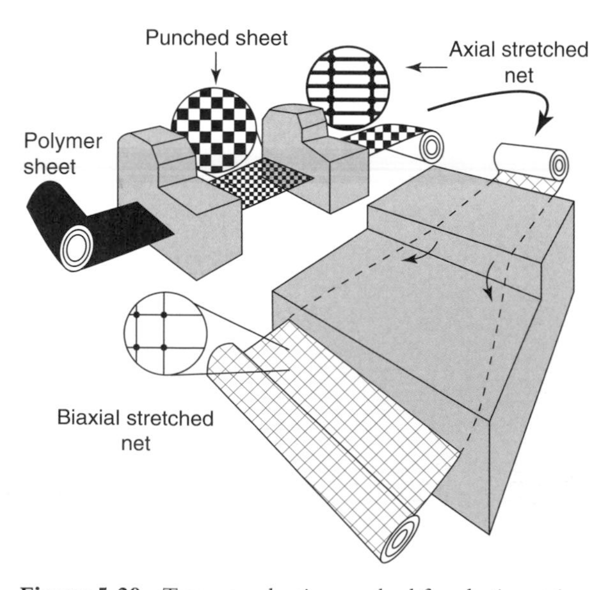

Figure 5.29 Tensar production method for plastic netting.

5.5.1 Vacuum and compression forming

Vacuum or compression forming processes are used to produce large mouldings relatively cheaply. Both processes start with a preformed plastic sheet or slab or polymer which is then subjected to heating. Under the influence of either a vacuum or with the assistance of an applied force the material is then converted into the shape of the final objects (see Figure 5.30).

The plastic sheet is securely clamped around its edge. The mould may be constructed from relatively lightweight material. To save money, wood may be used for large structures. The main requirement is that the mould should be capable of holding a modest vacuum. Heating of the plastic sheet is usually achieved by using infrared heaters. In the forming process, the plastic sheet will undergo a significant amount of cold drawing and the mechanical properties of the thinned sheet will often be superior to those of the thicker plastic sheet from which the moulding has been produced.

5.5.2 Pressure forming process

The pressure forming process is similar in concept to the vacuum forming process except that pressure rather than a vacuum is used to achieve the forming of the final moulded product (see Figure 5.31).

First, the plastic sheet which is clamped to the mould by the top ring is heated. The top ring contains a plunger, which is initially in its raised position and a series of inlets is arranged radially around the ring to ensure that an even pressure can be applied to the softened plastic sheet. The softened sheet can be forced into the mould and pressure is then applied to finish the moulding process. The gas used to pressurise the mould will aid the transfer of heat. This process is used to

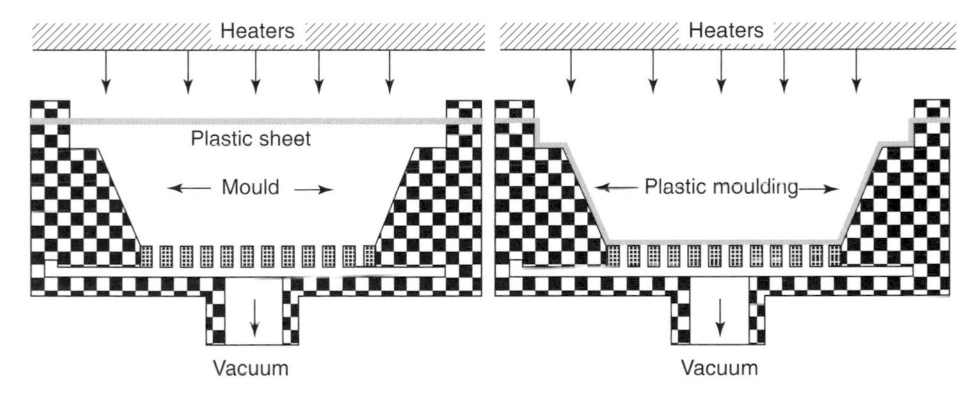

Figure 5.30 Vacuum forming of large plastic objects.

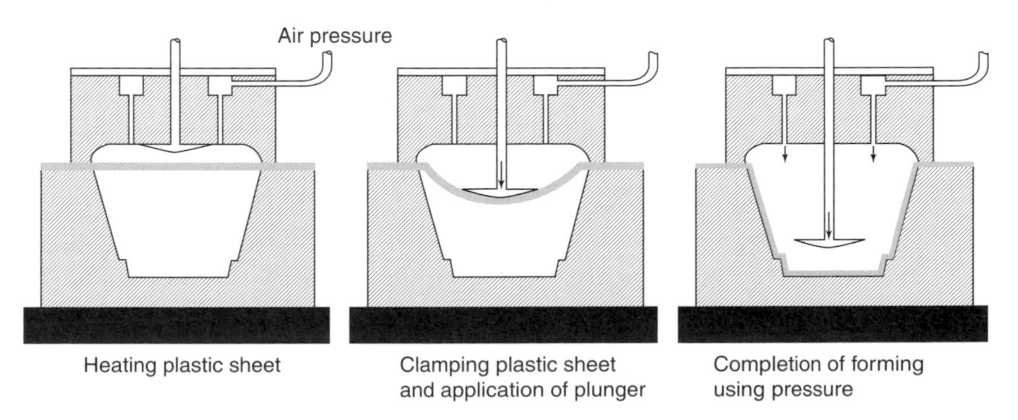

Figure 5.31 Operation of a pressure assisted forming process.

form wastepaper bins, bowls, and similar objects where the same basic form is being reproduced in large quantities and the creation of a mould can be justified. The vacuum and pressure forming processes are useful to create large volumes of simple objects, but are limited by the thickness of plastic sheet available and the dimensions which can be achieved during the forming process. There are variants of these processes which start from blocks of polymer and force the material to flow within the cavity once it is molten. In this case little or no cold drawing occurs and there is no enhancement of the physical properties.

5.6 Injection moulding

Perhaps the most important application of polymers is in the production of moulded products: cases for computers and other electrical goods, plastic containers, cups and other utensils used in cooking, etc. The injection moulding process involves melting the plastic and forcing it into a mould. The moulding process can use either a plunger-type of injection moulding machine or an adapted extruder.

5.6.1 Plunger-type injection moulding machine

The machine is relatively simple and combines the mixing of the components and transfer of the molten polymer into the mould (see Figure 5.32). The powder to be processed may be a simple plastic pellet, a composite of various components or a dry blend of powder, filler, dye, anti-oxidants, processing aids and other components. The powder is contained in a hopper and a metered quantity of the powder is fed into the machine and propelled by a hydraulic driven ram through to the mould. The material is melted as it enters the cylinder and passes through the torpedo, which helps to mix the components to achieve a homogeneous mixture prior to being forced into the mould. The flow of the melt into the heated mould is constrained by the nozzle. When the filled mould has been cooled, it can then be opened and the moulding can be discharged.

5.6.2 Extruders used for injection moulding

For the extruder to be used in injection moulding it is necessary for the screw to be capable of being translated backwards and forwards within the barrel and it is sometimes simpler to combine an extruder with a plunger-type delivery system. The molten polymer is translated into the mould by the action of a valve located at the end of the screw (see Figure 5.33). The simplest form of valve is a ring which is constrained to move on a rod attached to the

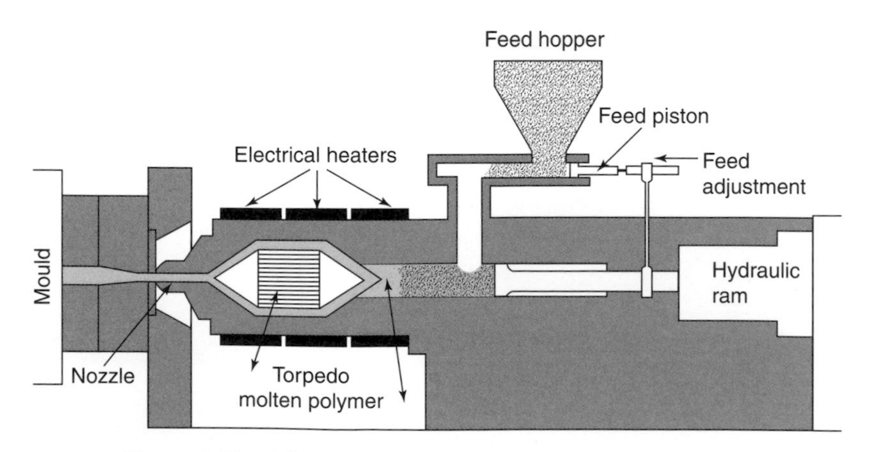

Figure 5.32 Schematic of a simple injection moulding machine.

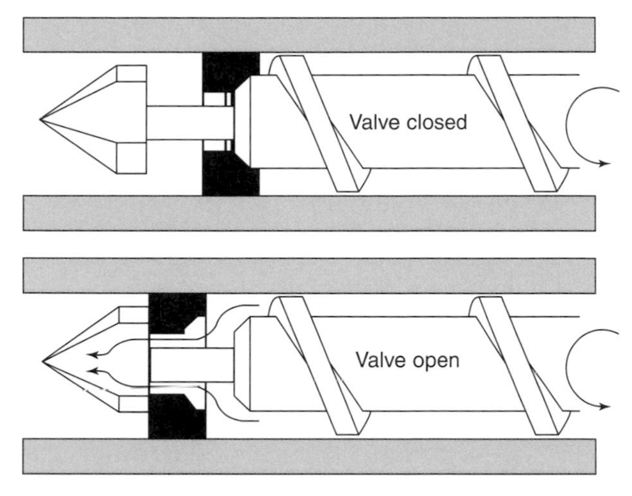

Figure 5.33 Valve attached to end of screw.

end of the screw. The ring is closed when it is pushed hard up against the screw and is open when located against a fluted nozzle attached to the end of the rod.

In its open state, the valve will allow the molten polymer to flow past the ring and fill the barrel ahead of the screw. When a sufficient volume of molten polymer has been transferred to the void ahead of the valve, the screw is advanced and the valve is closed. The closed valve allows pressure to be applied to the molten polymer and thus transferred to the mould. Once the mould has been filled the screw is withdrawn and the valve is once more opened, allowing the polymer to flow past the valve. The process is summarised in Figure 5.34.

In the first stage (Figure 5.35(a)), the valve is in its closed position and the molten polymer is moved ahead of the valve down the barrel towards the nozzle and the die. The pressure gradient and the flow of the molten polymer into the die will be controlled by the size of the nozzle. Once the die has been filled, the screw will pressurise the mould and ensure that no gas pockets or bubbles are formed in the plastic. Under pressure a small amount of residual gas will be dissolved in the molten polymer. In the second stage (see Figure 5.35(b)) the screw is withdrawn, thus breaking the fluid column. The mould is then drawn back from the nozzle to close the valve. The mould is then split to allow the mouldings to be extracted. The ejected moulding will cool and the short length of plastic which connected the main moulding to the nozzle, known as the *runner*, will then be cut from the moulding and can be discarded. Enthusiastic model builders will recognise the runner as the section of plastic to which the components of the model are attached. It is possible to introduce nozzles into the mould so that these *runners* are kept warm and are retained in the mould when it is opened. This is the so-called hot runner approach and is usually used to minimise the amount of waste produced by the process. In the third stage (Figure 5.35(c)) the screw is retracted to its full extent. The valve is closed once the screw starts to advance towards the die. The advancing screw will fill the die once more and the moulding cycle is completed.

5.6.3 Selection of plastics for extruder applications

Ideally, the polymer melt used in extrusion processes should have as low a viscosity as possible. The influence of the molar mass on the flow viscosity was discussed in Section 2.5.6 and a critical molar mass M_c was identified. For polymers with molar mass lower than M_c the viscosity of the melt is Newtonian and proportional to the molar mass. For molar masses above M_c, the effects of chain entanglement enhance the viscosity and the observed non-Newtonian behaviour is a direct consequence of chain entanglement. From the point of view of ease of processing, the polymer

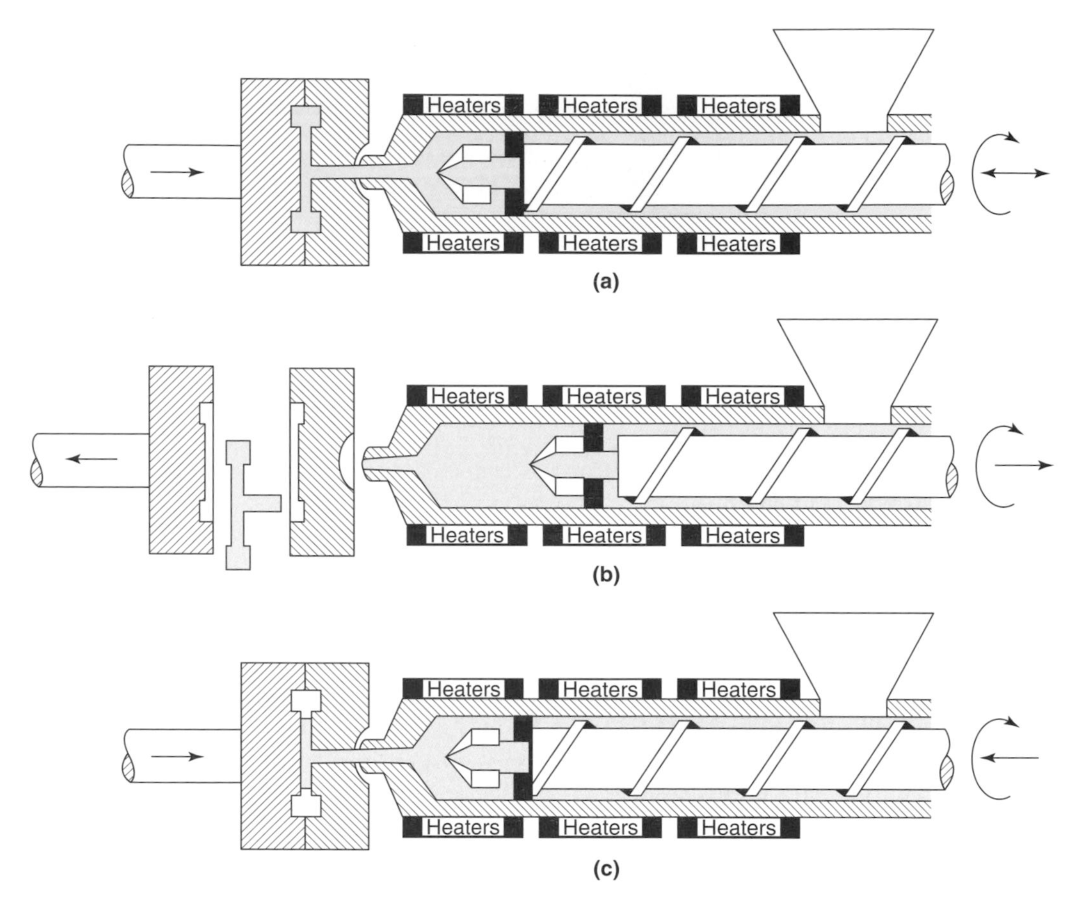

Figure 5.34 Summary of stages in filling of a mould using an extruder.

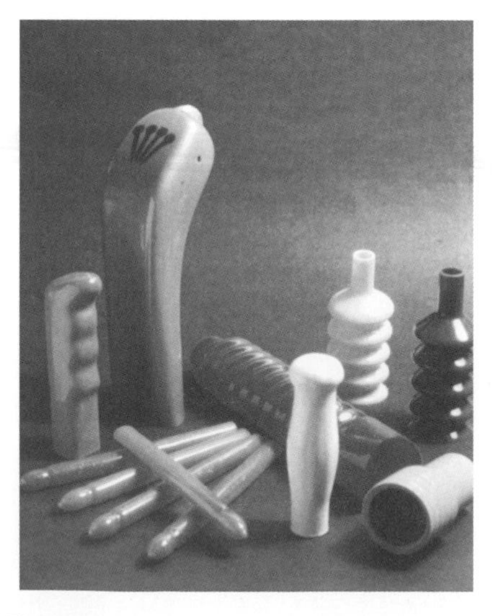

Figure 5.35 Examples of plastisol products.

should have a molar mass which is below M_c; however, many of the advantageous physical characteristics of polymers are only achieved by the use of high molar mass materials. Properties such as: high glass transition temperatures, high elongation to break, environmental stress crack resistance, barrier properties, etc. are all a function of the molar mass and depend on there being a critical number of high molar mass molecules present in the final material. Manufacturers of polymers sell the materials as different grades. These grades of materials are usually blends of different molar mass polymers adjusted to achieve the desired balance of ease of processing and physical properties in the final product. In a typical blend, a proportion of the material may have a molar mass well above M_c, but when blended with material with a molar mass below M_c creates a melt which will have minimal non-Newtonian flow characteristics and hence can easily be processed. The final solid, however, will benefit from the presence of the higher molar mass component and have superior performance when compared with low molar mass materials. It is not unusual to find a supplier having six or seven different grades of the same polymer available.

5.7 Plastisol processes

The plastisol process can be used to clad steel plate, produce moulded products such as the grips of tools, flexible tubing, protective covers for electrical insulations, ducting, car steering wheel covers, gear change grips, etc. (see Figure 5.35).

A plastisol is a dispersion of a thermoplastic powder, usually PVC, in a plasticising solvent. The powder is dispersed at room temperature and does not dissolve in the solvent. On heating, the powder is swollen, takes up the plasticising solvent and forms a gel. This gel is deposited on the former or article to be coated. On cooling the article can be detached from the former or may remain attached as a protective plastic coating. The process relies on the ability of the solvent to plasticise the polymer and cause the chains to swell. The swelling process is promoted by plunging a heated object or former into the dispersion, which is maintained at room temperature. Heating encourages diffusion of the solvent into the polymer particles and swelling occurs. The sticky particles fuse together and the glass transition temperature is lowered, producing a flexible material. The PVC powder may be pigmented and modified to give the desired colour to the final product. Use of different types of PVC can produce matt or glossy finishes. The PVC used will have been created from either emulsion or bulk polymerisation of PVC and needs to be ground and classified to achieve an appropriate distribution of particle diameters, typically in the range 10–50 μm. The first step in the dip moulding process (see Figure 5.36) is to heat the object to be moulded. The object or a former with the desired shape is then immersed or 'dipped' in a tank of liquid *plastisol.*

The heat in the object activates the *plastisol* and a gel is formed around the object. The thickness of the coating created will depend on the heat capacity of the mould and the time it is left in the plastisol. The hotter the mould and the longer the dip, the thicker the gel coating that is created. After the gel coating has been formed the object is removed and passed to a curing oven where the plastisol material is fused. Lastly, the moulded part is cooled and stripped from the mould. Since the parts that are created are elastic, it is relatively easy to create fine details on the objects. The thickness of the coating depends on the dipping time and can be controlled to an accuracy of 0.1″/min. If less flexible protective coatings are required, then a slightly different procedure is followed. The heated object is dipped and a gel coat formed which is then further heated to fuse the PVC. This process is used to create coatings inside bottles or containers.

Blisters and other irregularities can be created on the surface if the plastisol is contaminated by water, air or solvents. It is important to ensure that air has been removed from the plastisol by using a vacuum during the initial dispersion of the polymer powder in the plasticiser. The process used to produce the polymer will influence the molar mass distribution, the ability to swell the particles and the viscosity which is developed. Different grades of PVC exhibit different abilities to swell in the plasticiser.

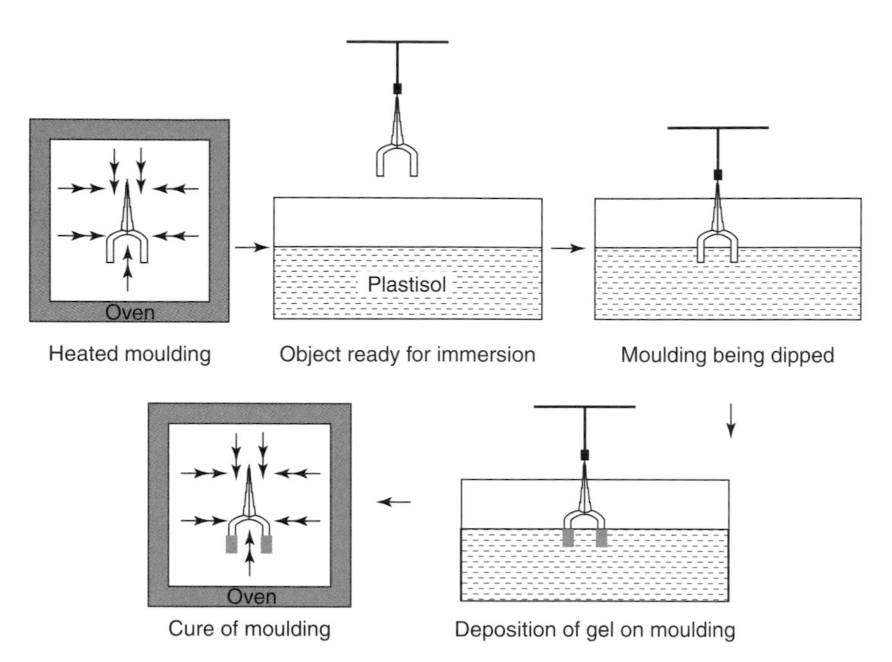

Figure 5.36 Dip moulding process.

5.8 Thermoset processing

The main problem encountered with the creation of thermoset structures is that the starting monomer has a low viscosity. Thermosets are often glass or carbon fibre reinforced and typically involve laying down the fibre in the form of woven, knitted, stitched or bonded fabrics.

5.8.1 Hand lay-up process

Liquid resin is applied to the dry fibre and consolidated using a roller, brush, or nip-roller impregnators to help force the resin into the fabric (see Figure 5.37). Laminates are left to cure under standard atmospheric conditions. The resins used will typically be epoxy, polyester, vinyl ester, or phenolic, together with a range of fibres. Heavy aramid fabrics can, in practice, be hard to wet out by hand. The process has been widely used for many years and is inexpensive. The quality of the structure formed depends on the skills of workers. Health issues can arise from the volatility of

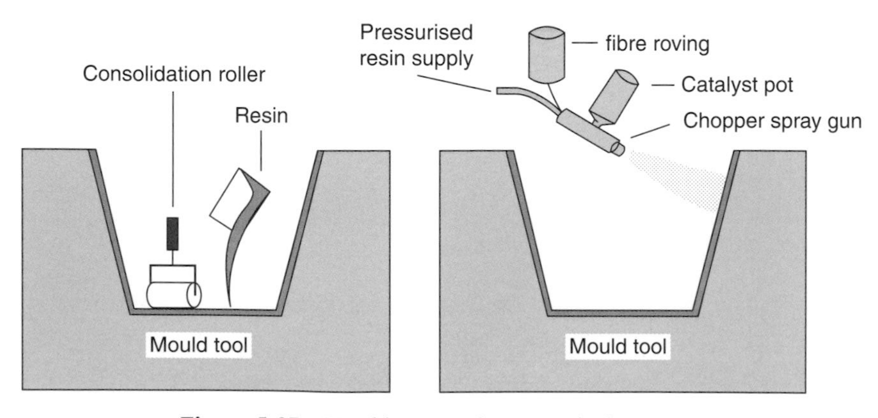

Figure 5.37 Hand lay up and spray applied processes.

the low molecular mass resins. The hand lay-up approach is widely used for wind-turbine blades, production boats, and architectural mouldings.

5.8.2 Spray lay-up method

An alternative approach is to spray the resin (see Figure 5.37). In this method, the fibre is fed to a gun which allows the short chopping fibre to be mixed with the resin and catalyst. Manual or remote control operation of the gun reduces problems of exposure of the operatives. Cure is achieved at ambient temperatures by adjusting the amount of catalyst. Laminates formed by this method tend to be resin-rich and therefore excessively heavy. Because the gun can only handle short fibres and low viscosity resins, the mechanical properties of the final material may be limited. The spray lay-up method is used to fabricate lightly loaded structural panels, e.g. caravan bodies, truck fairings, bathtubs, shower trays and some small dinghies.

5.8.3 Vacuum bagging

The main problem with the creation of a thermoset laminate is the elimination of voids within the final product. The voids are usually a consequence of air being entrapped, either around the fibre or dissolved in the resin during the mixing process with catalysts. In order to achieve a well-consolidated material, vacuum bagging is used. The process is the same as that used for hand lay-up except that the moulding is carried out in a vacuum bag (see Figure 5.38). Pressure is applied to the laminate and consolidation is achieved. The air under the bag is extracted by a vacuum pump which applies a pressure of up to one atmosphere to the laminate. The method is used for epoxy and phenolic resins: polyesters and vinyl esters are more difficult as the application of the vacuum may extract styrene from the resin.

The vacuum bag method allows high levels of fibre to be incorporated into the final laminate and a significant reduction in the void content of the laminate. The method is used in the production of structural members for marine, transportation and aerospace applications.

5.8.4 Resin transfer moulding

In order to improve the quality of the moulded product, a series of variants of the simple lay-up method have been developed. Resin transfer moulding (RTM) (see Figure 5.39), uses a double mould tool. Fabrics reinforcement is laid up in the moulding tool, and sometimes prepressed to the mould shape, and held together by a binder.

These 'preforms' are then more easily laid into the mould tool. A second mould tool is then clamped over the first, and resin is injected into the cavity. A vacuum is applied to the mould and resin introduced under pressure. This process is known as vacuum assisted resin injection (VARI). The flow of the resin into the mould is controlled by the ability to permeate the dry reinforcement. Both injection and cure can take place at either ambient or elevated temperature. Most resins can be used although high temperature resins, such as bismaleimides, require the use of elevated processing temperatures. To help achieve a uniform distribution of the fibre, layers of

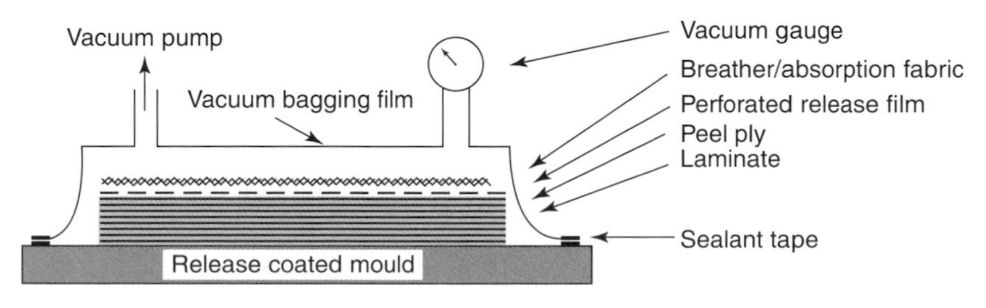

Figure 5.38 Configuration used for vacuum bag preparation of composite laminates.

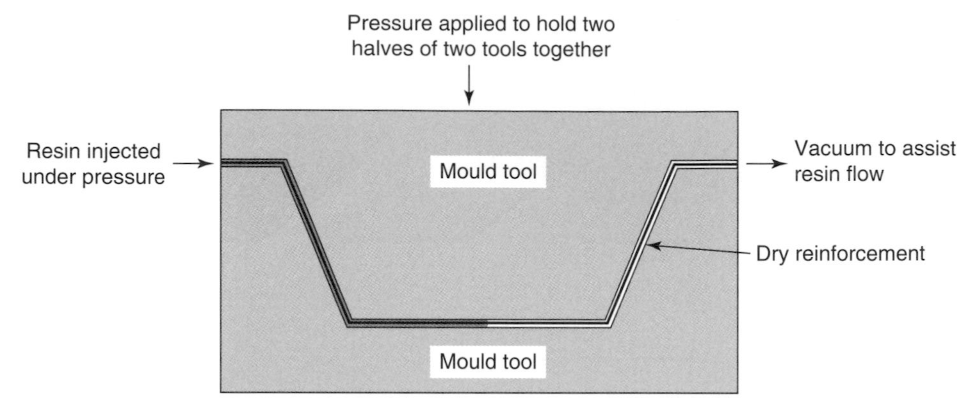

Figure 5.39 Resin injection moulding process.

material stitched together may be used. High fibre volume laminates can be obtained with very low void contents; however, the creation of matched tools can be expensive. The method is used for the fabrication of small complex aircraft and automotive components, train seats, etc. where the cost of mould fabrication is offset against the volume produced.

5.8.5 Resin infusion processes

The resin infusion process is analogous to the vacuum bagging process, but involves a process which resembles the RTM process. The resin infusion process (see Figure 5.40) (SCRIMP, RIFT, VARTM) involves the dry fabrics being laid up as a dry stack of materials as in RTM. The fibre stack is covered with peel ply and a knitted nonstructural fabric and the dry stack vacuum bagged and the resin allowed to flow into the stack. The process is used for semiproduction small yachts, train and truck body panels.

5.9 Composite fabrication

As discussed in Section 5.8, there are many different forms of composite which can be generated using thermosetting and thermoplastic resins. For composite structures used in structural engineering applications, which require quality, precision and durability, it is usual to follow the *prepreg* route. A pre-impregnated (*prepreg*) fibre resin laminate is obtained by taking fibres arranged in a continuous parallel fashion and feeding though a slot die so that they are coated

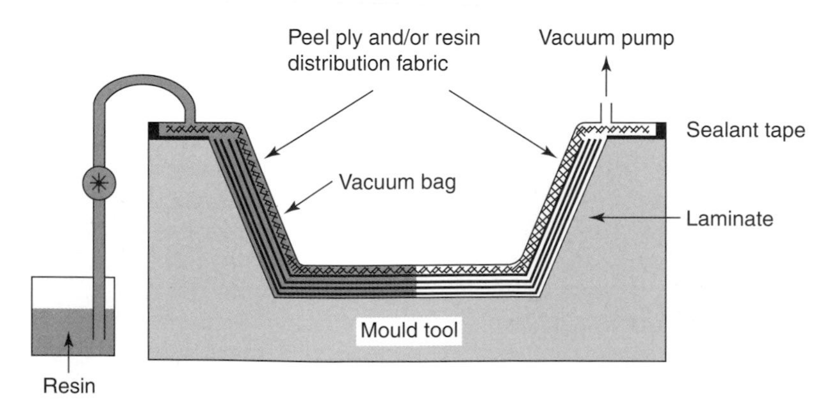

Figure 5.40 Resin infusion process.

or dipped in a bath of resin. The result is a roll of material approximately 1 mm thick which contains fibres, which are usually uniaxially orientated, surrounded by resins. The basic resin may not have sufficient viscosity to form a solid and so it will have been necessary to have increased its molar mass to achieve a semisolid material for the prepreg. The prepreg can have a variety of different forms, the most common being fibre laminated sheets, which are layered, one upon another. The orientation of the fibres can be adjusted so as to achieve the desired mechanical properties. The *prepreg* is manufactured from a partially cured resin in which the fibre reinforcements are dispersed. These sheets are usually wrapped into a roll and separated by a siliconised paper backing. The roll is stored at low temperatures to avoid the material being prematurely cured and will have a shelf life of about a year at 5°C. The process is often carried out in an autoclave.

5.9.1 Autoclave prepreg moulding

Fabrics and fibres are pre-impregnated with resin and orientated unidirectionally. The resin is a soft, sticky solid at ambient temperatures. The prepregs are laid up by hand or machine onto a mould surface, vacuum bagged and then heated to typically 120–180°C. The resin will reflow on heating and eventually cure. A pressure of up to 5 atmospheres is applied to consolidate the structure. The resins typically used are: epoxy, polyester, phenolic, polyimides, cyanate esters and bismaleimides. Optimum physical properties are achieved by using a high fibre–resin ratio and ensuring that all the air is removed before curing. A vacuum bagging technique is adequate for simple structures (see Section 5.8.3), but more complex structures may require the use of an autoclave (see Figure 5.41). The typical autoclave will allow controlled heating or cooling and application of pressure. The autoclave will be equipped with safety interlocks to avoid accidental release of hot gas or the burning of the operators. Safety devices should be fitted so the autoclave cannot be pressurised without the door being fully locked. The autoclave will allow controlled heating and cooling, and application of pressure. Temperature control is essential in order to achieve a good product.

Autoclaves use either direct or indirect heating. Indirect heating uses external heat exchangers, whereas direct heating is achieved using electrical heating in the autoclave. Control systems are designed to allow a temperature profile to be created that is made up of heating gradients, dwells and cooling gradients. A typical temperature profile is shown in Figure 5.42.

Dynamic control of the temperature is desirable as the part being created can self-heat as a consequence of the reaction exotherm. This problem can be encountered where thick laminate sections are being created or when the part is very large. In the ideal situation, the temperature

Figure 5.41 Views of autoclaves used to fabricate large composite items. (Photos courtesy ASC Process Systems, USA)

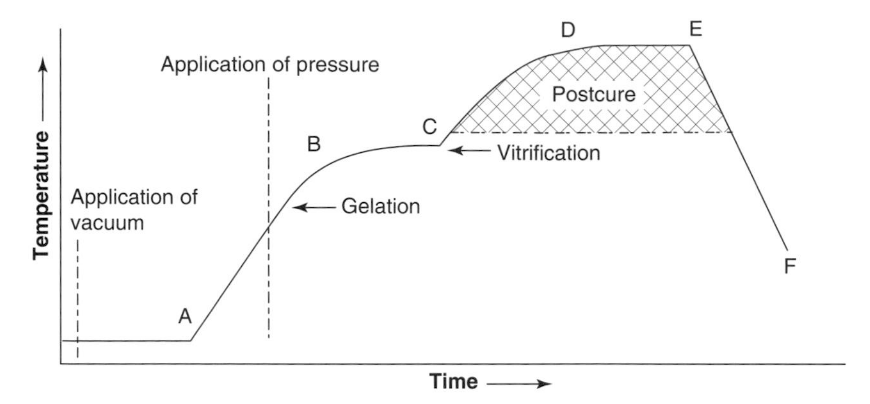

Figure 5.42 Typical operational cycle for curing a thermoset laminate using an autoclave.

follows as closely as possible an ideal cure cycle. The TTT diagram was discussed in Section 4.5. For a component to have low residual stresses, it is desirable to cure the item at a low temperature and then perform a postcure process. Gelation at low temperature creates a matrix which will have lower stresses than if it were created at high temperature and therefore there is less likelihood of subsequent stress cracking. In the temperature profile shown in Figure 5.42, the initial temperature ramp, A–B, allows the prepreg to gain a high degree of mobility and ensures adequate wetting and the opportunity for gas to escape from between its layers. The component to be created will have been vacuum bagged using the type of arrangement shown in Figure 5.38. The first step is to apply a vacuum to the bag to ensure that subsequent processing does not give rise to air entrapment between the prepreg layers. When the system has been allowed to settle, heating will take place from A to B. At B the prepreg resin will have started to gel and at this point the structure of the component is fixed and cannot be changed. Prior to the temperature reaching B, pressure will be applied to the structure to aid consolidation. It is essential that the consolidation process takes place before gelation, but not so early that the resin is forced from between the fibres in the structure. During the hold period at B, the component will have gelled and may have vitrified. A second temperature ramp, from C to D, allows the effective postcuring of the material and the final temperature in this cycle can influence the glass transition temperature of the final product. The temperature will be held at D for a period to achieve the desired degree of cure. The cooling from E to F is a very important part of the cycle and has to be carried out in a controlled manner to avoid cracking due to stress created from the contraction of the resin matrix. Autoclaves of the type shown in Figure 5.41 use data logging and cure monitors to ensure that the unit operates efficiently.

5.9.2 Filament winding

An important process for the construction of pressure vessels and pipe/tubular structures is filament winding (see Figure 5.43). The process is relatively simple and involves wrapping resin coated fibres around a mandrel. The resins used are often epoxy, polyester, vinyl ester or phenolic and the glass is pre-organised on a creel. To enhance the cure process, the resin-coated glass fibres are heated so that they consolidate effectively when they are applied to the mandrel. The structure often has a relatively thick section and the cure exotherm can be used to enhance the thermoset reaction processes. Fibre tows are passed through a resin bath before being wound onto a mandrel. The orientation of the fibres is controlled by the fibre feeding mechanism, and rate of rotation of the mandrel (see Figure 5.43). The process is ideal for the creation of a cylindrical structure, however the cost of the mandrel can be high. The resin content can be controlled by metering the resin onto each fibre tow through nips or dies. The fibre cost is purely that of the

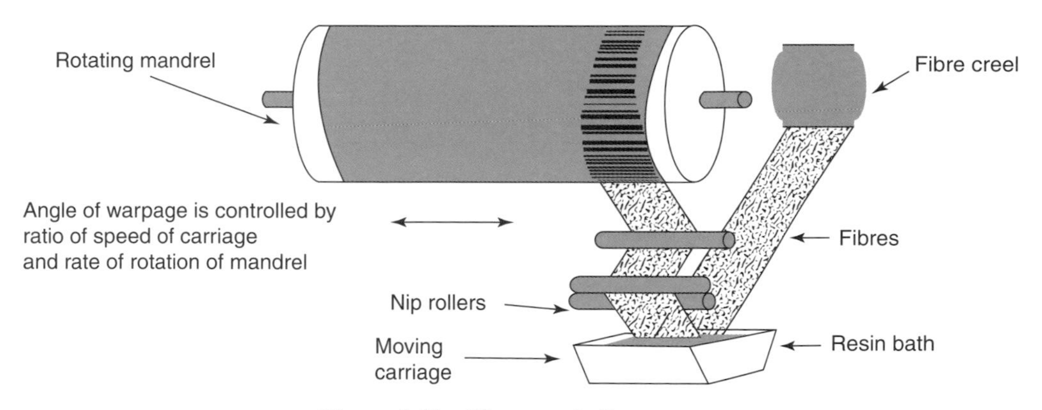

Figure 5.43 Filament winding process.

raw materials since there is no secondary process involved in preparing the fibres prior to their use. The structural properties of laminates can be very good since straight fibres can be laid in a complex pattern to match the applied loads. It is easy to control the radial mechanical properties and therefore the hoop strength but more difficult to influence the longitudinal properties since this system does not create fibres which are orientated along the axis of the cylinder. The process is used to create chemical storage tanks, pipelines, gas cylinders and fire-fighters' breathing tanks.

5.9.3 Pultrusion

As the name implies, the process of *pultrusion* involves pulling a laminated structure through a heated mould (see Figure 5.44(a)). As in the case of the filament winding, the fibres are coated with resin by being passed through a bath of liquid. Although in the simple illustration fibres are used from a creel, in practice woven fibres and other arrangements of reinforcement may be assembled to achieve the final laminated structure.

The pultrusion process will produce a continuous product and it is usual for the laminate to be cut into sections after the pull-off has taken place. The pull-off apparatus is usually a suitably located hydraulic ram which clamps the structure and pulls it out of the mould. The pull-off apparatus has to be able to apply the required force without damaging the newly formed laminate. It is therefore important that the laminate is fully cured when it leaves the mould.

The various stages in the cure of the laminate are summarised in Figure 5.43(b). Once the fibres have passed through the nip rollers they will be surrounded by a controlled amount of resin. The resin is usually sufficiently viscous that it will be retained on and between the fibres at room temperature. The resin must be able to effectively wet out the fibres. The wetted fibres then enter the first zone of the mould (A). The temperature is slowly increased to ensure that the cure process is increasing the viscosity and to avoid the resin becoming highly mobile and running off the fibres. At a point in zone B, the resin will pass through its gel point and the structure formed is now sealed. At this point, further heating causes the structure to expand and as a consequence pressure builds up in the mould. The pressure is initially at atmospheric pressure and the gelled resin forms a moving seal within the mould.

The increase in pressure will help to consolidate the moulding and helps to avoid the creation of internal voids and entrapment of air bubbles. If the temperature rise is too large then pressures can be excessive and will lead to resin being pushed back out of the mould and poor mouldings will result. The pressures which can be achieved have been known to be sufficient to shear the stainless steel bolts holding the mould together! If the heating is too slow then an undercured product may be obtained and may be damaged during the pull-off process. In practice, pressure probes may be used to monitor the location of the gel line in the mould and this can be used as

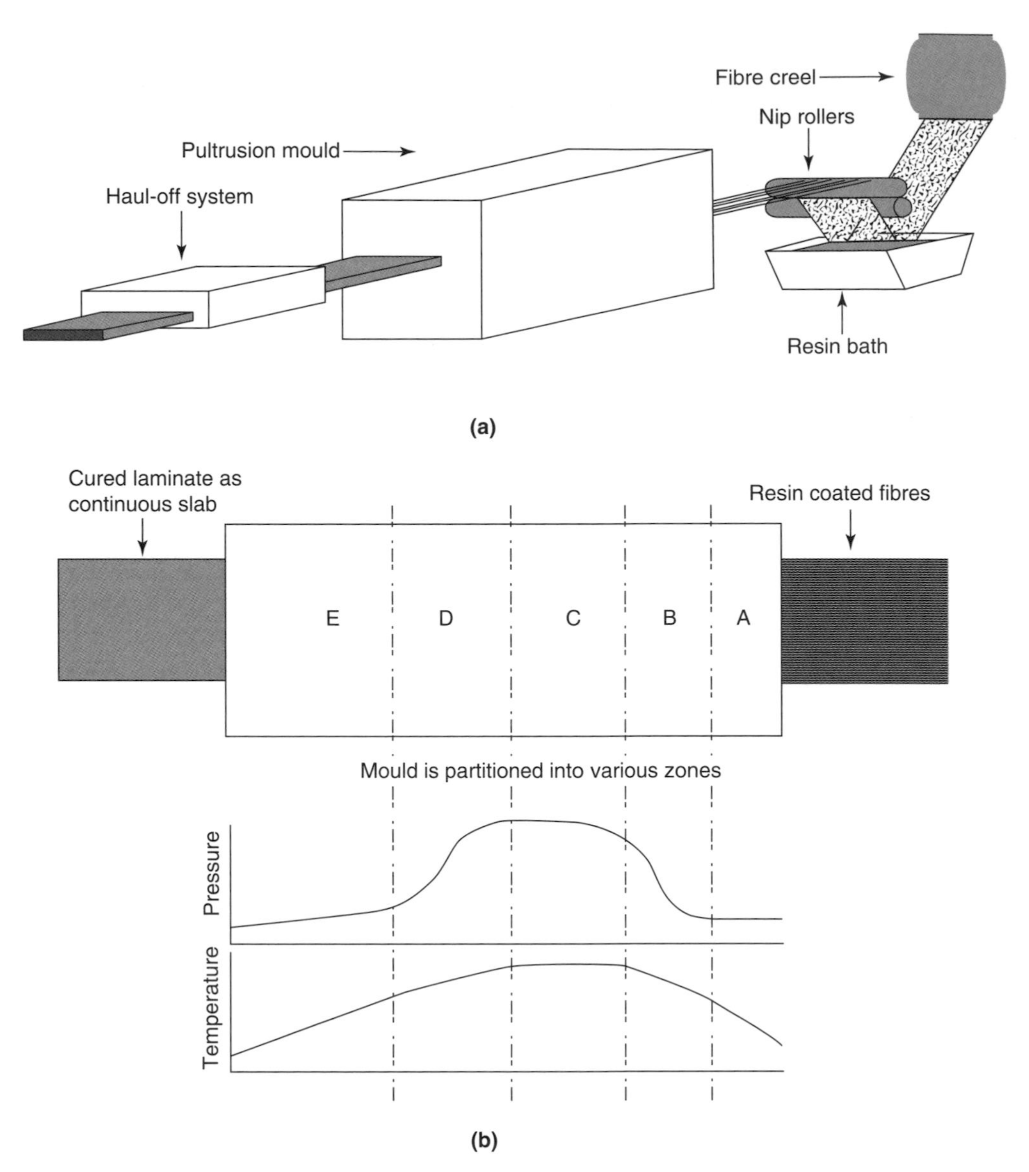

Figure 5.44 Pultursion process: (a) overall pultrusion process; (b) temperature and pressure changes during processing.

a process monitor. The temperature is raised to a point at which a full cure can be achieved and is then held constant in zone C. As the cure proceeds, so the pressure may continue to rise but to a much smaller extent than in B. The time for which the laminate is at the highest temperature needs to be sufficient to complete the cure. Clearly the time available will be determined by the haul-off rate and hence is a controllable variable in the process. Once the cooling of the laminate occurs it will shrink away from the mould and the pressure is reduced. The rate of cooling has to be controlled so that the resin is allowed to release the stresses which are created during the curing process, which occurs in zones D and E. The pressure will now have returned to an atmospheric value and the moulding cooled to about 30–40°C. The moulding will have to be sufficiently tough to be able to withstand the not insignificant forces which will be applied by the haul-off equipment. The resins used are usually: epoxy, polyester, vinyl ester and phenolic. The typical applications for this type of product are beams and girders used in roof structures, bridges, ladders and frameworks.

5.10 Cure monitoring

A liquid thermosetting prepolymer is turned into a rigid solid by the process of *curing*. This process is accompanied by a gradual and then sudden rise in the viscosity of the resin. The time at which this sudden viscosity rise is noted coincides with the *gelation* of the resin and indicates the formation of a three-dimensional molecular network. Further reactions tighten up this network, increasing its stiffness up to a point where no more reactions can take place at a given temperature. The network is then said to have *vitrified*. The regions of *gelation* and *vitrification* in the resin are of great practical significance in the processing of fibre-reinforced composites. Two types of microsensors are used reasonably widely to monitor the real-time state of cure: dielectric and optical fibre sensors. The first question which has to be asked when setting up the autoclave is what profile should be used. The profile in Figure 5.44 represents an optimised cure. When starting with an unknown resin, it is appropriate to use differential scanning calorimetry to determine the cure exotherm for a material. The energy released as a function of temperature indicates the point at which the cure is initiated, where it reaches a peak and when it is completed. Dielectric monitoring uses the intrinsic mobility of polar groups to allow the development of the thermoset matrix structure to be probed. Initially the monomer and ionic impurities are able to easily respond to an applied electric field and large dielectric losses and permittivity values are observed (see Figure 5.45).

Initially, the low frequency dielectric permittivity and loss for the uncured resin are both very high, reflecting ease of charge and dipole mobility. As the cure precedes so the unbound charges, which are impurities left over from the synthesis of the resin, are unable to move as quickly and the low frequency peaks in the dielectric permittivity decrease as does the loss. This general reduction indicates an increase in the viscosity of the curing resin. The appearance of an apparent plateau region in the dielectric permittivity can be associated with the occurrence of *gelation*. A shoulder is observed in the dielectric loss which merges into the low frequency loss and the overall profile is significantly reduced. A corresponding decrease is observed, a second drop in the dielectric permittivity. This second process is the freezing of the dipoles of the resin and is associated with vitrification. By mapping the changes at a selected frequency it is possible to identify the transition points and build a TTT diagram (see Section 4.5).

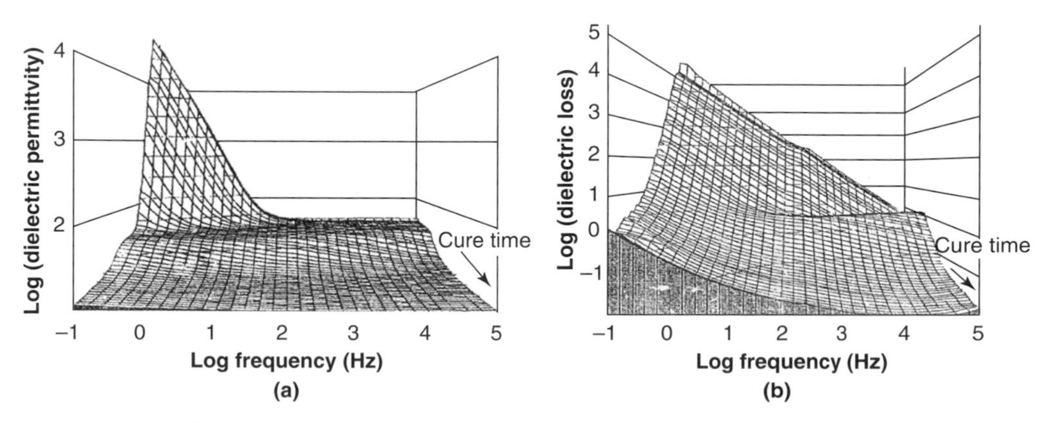

Figure 5.45 Dielectric permittivity and loss changes with curing time.

5.11 Repair of composite parts

Composite structures which have been damaged can readily be repaired (see Figure 5.46). Whilst the nature of the damage is obvious in some cases (see Figure 5.46(a)), in many cases structurally critical problems may be hidden within the component and are associated with delamination of the layered structure.

In carbon fibre composites, this has been called *spanner damage*, since it can be a consequence of a local impact to the outer surface of the component. Unfortunately spanner damage will not necessarily be evident at the surface of the component and hence can be difficult to detect until it has generated a major disbanded region (see Figure 5.46(c)). A related and frequently encountered problem is the hidden delamination of the skin from honeycomb composite structures. Low velocity impact can cause disbands which leave little evidence of their existence on the surface (see Figure 5.46(d)). Hidden damage may have occurred and this includes manufacturing defects that may not appear as visible features in the product or after exposure to moisture as blisters. For example, a vertical tail part of an aircraft may be designed to withstand hailstone impact but it may not able to resist damage from being dropped during shipping or removal for inspection.

5.11.1 Basic repair process

The repair of a damaged composite structure will usually involve some of the following steps:

- Inspect the composite part to assess the extent, type and degree of damage.
- Remove damaged material. The extent to which material is removed depends on the nature of the repair which will be undertaken.

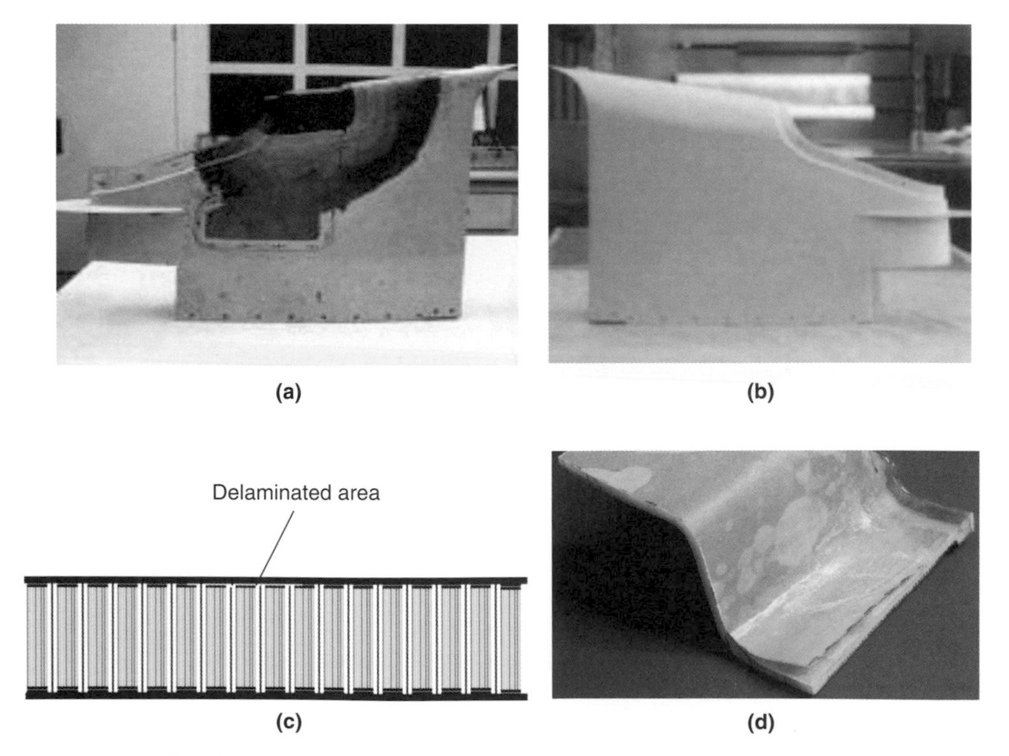

Figure 5.46 Defects found in composite structures: (a) damaged; (b) repaired; (c) delamination in honeycomb structure: (d) delamination in a thick composite.

- If the composite has been exposed to excessive heating (see Figure 5.46(a)) or there has been a high level of moisture ingress (see Figure 5.46(d)), it may be necessary to remove the contaminated material.
- For the repair to be successful it is usual to have to carefully prepare the surface of the composite adjacent to the repair to ensure good bonding and a satisfactory result.
- The repair which is undertaken depends on the nature of the damage as indicated below.
- It is critical that the repair is properly inspected to ensure that maximum physical properties are being developed in the repaired area.
- In certain repairs it may also be necessary to restore a surface finish. If this involves painting etc. the cure of the resin must be completed before it is exposed to solvents contained in the paint etc.

5.11.2 Types of repair

The type of repair can be classified according to the level of damage to be addressed and the extent to which structural integrity is to be restored.

Cosmetic

A superficial, nonstructural filler is used to restore a surface to keep fluids out until a more permanent repair is made (see Figure 5.47a) This type of repair will not regain any strength and is used only where strength is unimportant. Due to high shrinkage, cosmetic repairs may start to crack after a relatively short time in service.

Resin injection

Where the delamination is restricted to one ply it is possible to inject resin and execute a repair. Not much strength is regained, however the repair is quick and cheap. This type of repair will slow the spread of delamination but is not permanent.

Semistructural plug/patch

This type of repair often involves a major element of the composite being replaced with a mechanically fastened plug (see Figure 5.47(b)). This type of repair regains some strength and can be especially effective where thick solid laminates are used.

Structural mechanically fastened doubler

Full structural repairs using bolted doublers are usually used in heavily loaded solid laminates (see Figure 5.48(a)). However, such repairs are not aerodynamically smooth and may cause 'signature' problems in smart structures where low radar is desirable. This type of repair leaves the original damage and simply attempts to transfer loads around the damage. Stress concentrations will also be created at the corners and edges of the doubler. A doubler is a localised area of extra layers of reinforcement, usually used to provide stiffness or strength for fastening, or other abrupt load transfers.

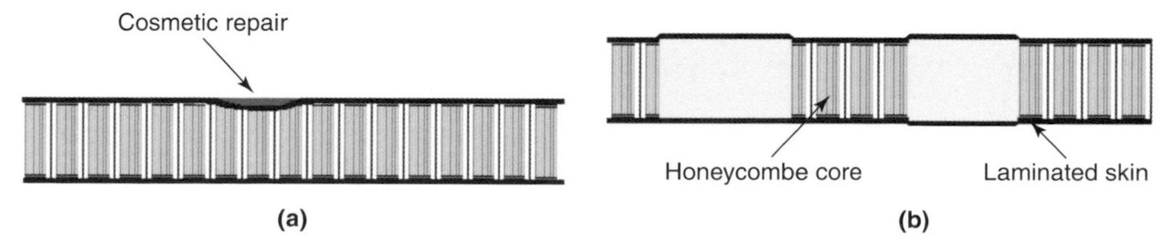

Figure 5.47 Repairs to a delaminated honeycomb composite: (a) cosmetic; (b) semistructural plug/patch repair.

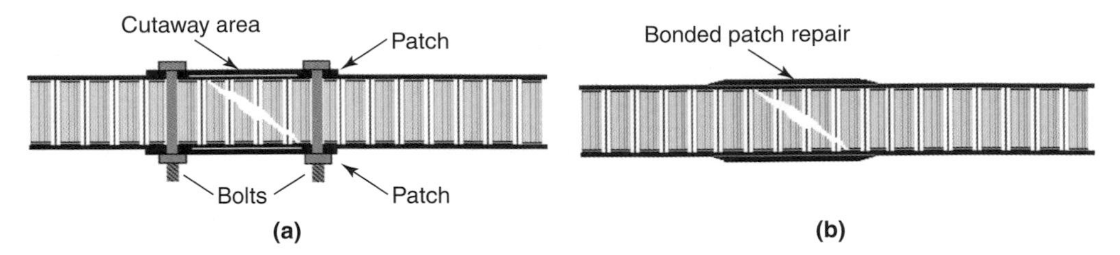

Figure 5.48 Patches: (a) bolted patch with doublers; (b) bonder doubler.

Structural bonded external doubler

Bonded external doublers (see Figure 5.48(b)) are often used to repair lightly loaded, thin laminate structures. This type of repair frequently uses wet lay-up materials. They may be cured at room temperature or high temperatures, depending on the matrix resin system used. These repairs can regain a significant portion of the original strength of the structure or even full strength, although with a significant stiffness and weight penalty in many cases. This type of repair is generally easy, relatively quick and does not require highly developed skills to carry it out.

Structural flush repair

This repair restores full structural properties by forming a joint between the prepared repair area and the repair patch. The repair patch is made by replacing each ply of the composite laminate that has been removed from the damage area. The size of the repair patch should fit exactly the area prepared for repair, except for a final cosmetic or sanding layer, which is often slightly larger to allow for sanding down to achieve a smooth and/or cosmetic surface.

5.11.3 Damage assessment

A damaged area in a composite structure may be difficult to locate, having few external signs. Internal damage of plies and delamination may result in no external marks of damage (see Table 5.4). The visibility of the damage can be influenced by the energy involved in the impact. High and medium

Table 5.4. Summary of strengths of NDT methods for composite damage detection

Type of defect	NDT method							
	Visual	Tap test	A-scan	C-scan	X-ray	Thermal	Dye penetration	Dielectric methods
Surface delamination	2	1	2	1	2	1	N/A	2
Deep delamination	N/A	3	1	1	2	2	N/A	1
Full disband	2	2	1	1	2	1	N/A	1
Kissing bond	N/A	3	3	3	N/A	N/A	N/A	3
Core damage	2	2	3	1	1	3	N/A	1
Inclusions	2	2	1	1	1	1	N/A	2
Porosity	2	N/A	2	1	N/A	N/A	2	1
Voids	2	2	2	2	2	2	2	1
Backing film	N/A	2	2	2	2	2	N/A	3
Edge damage	1	2	2	1	1	2	1	N/A
Heat damage	2	2	2	2	N/A	3	N/A	3
Severe impact	1	1	1	1	3	1	1	1
Medium impact	1	1	1	1	N/A	3	3	2
Minor impact	3	3	3	3	N/A	3	N/A	3
Uneven bond line	3	N/A	3	3	3	3	N/A	3
Weak bond	N/A	N/A	N/A	N/A	N/A	N/A	N/A	3
Water in core	N/A	2	3	1	2	1	N/A	1

N/A, not available.

energy impacts, while severe, are easy to detect. Low energy impacts can easily cause 'hidden' damage. There are a variety of nondestructive testing (NDT) techniques available to help determine the extent and degree of damage (see Table 5.4). Each NDT method has its own strengths and weaknesses, and more than one method may be needed to produce the exact damage assessment required. Table 5.4 provides a basic comparison between the NDT techniques. Each method is scored between 1 to 3, 1 indicating that the method is good at detecting the defects, 3 indicating that it is poor.

Table 5.4 indicates that there is no one method which is capable of detecting all types of defects. Some methods are better than others and some are more applicable than others for use in the field. Careful inspection of the surface coupled with tapping of the structure can often reveal changes in the integrity of the substructure and more detailed ultrasonic or X-ray measurements can be used to attempt to identify the nature and the location of the defects. Thermal imaging is useful for the study of aircraft and can be used on aircraft that have just returned from a long flight where the substructure will have been thoroughly cooled. Dielectric methods are relatively novel and whilst they have considerable potential they are not as yet widely used.

5.12 General physical characteristics of composites

Composite materials exhibit a broad range of attractive physical characteristics:

- *Toughness of the resin composite systems*: Toughness is a measure of the resistance to crack propagation, but this can be hard to measure accurately in a composite. The toughness of the composite is influenced not only by the resin but also by the strength of the interaction between the resin and the glass. Stress–strain curves provide an indication of the toughness: generally the greater the level of deformation a composite will accept before failure, the tougher and more crack-resistant the material will be. Conversely, a resin system with a low strain to failure will tend to create a brittle composite. It is generally accepted that to achieve optimum mechanical properties it is desirable to match the expansion coefficient of the resin to that of the reinforcing fibres.
- *Environmental properties of the resin system*: Good resistance to the environment, water and other aggressive substances, together with an ability to withstand constant stress cycling, are essential for any resin system. These properties are particularly important for use in a marine environment.
- *Core materials*: Theory indicates that the flexural stiffness of any panel is proportional to the cube of its thickness. The purpose of a core in a composite laminate is therefore to increase the laminate's stiffness by effectively 'thickening' it with a low-density core material. The resulting composite structure demonstrates a dramatic increase in stiffness for very little additional weight. The strengthening effect arises from the way in which the skins and the core interact. The sandwich laminate (see Figure 5.49) can be compared with an I beam with the core acting as the beam's shear web. In a three-point loading, the upper skin is put into compression, the lower skin into tension and the core into shear. Therefore the ability of the core to carry shear is a critical property of the sandwich composite. Connected with the ability of the core to sustain shear is also the adhesive bond between the skin and the core. A weak bond may be a major problem in the structure retaining its integrity.

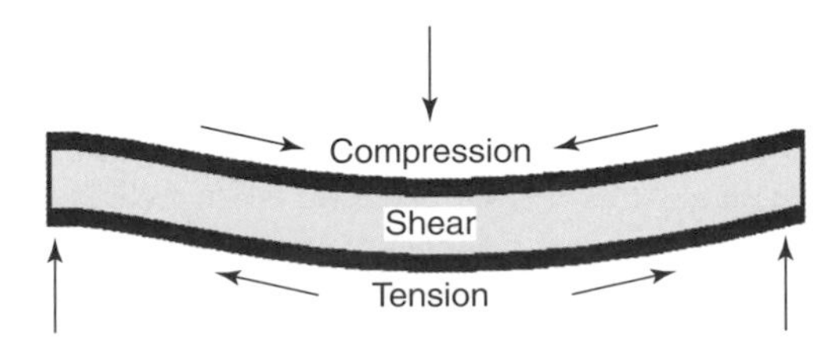

Figure 5.49 Force balance in a sandwich laminate structure.

In lightweight, thin laminate skins the core must be capable of taking a compressive loading without premature failure. This can help prevent thin-skinned structures from wrinkling and failing in a buckling mode.

Brief summary of chapter

- The processing of thermoplastic and thermoset materials has been reviewed. The strategy for the processing of a thermoplastic is very different form that required for a thermoset.
- Depending on the properties which are required in the final product, the method available for its fabrication may need to be varied.
- Most composites are produced from thermosets but the incorporation of fibres and fillers requires different processes to be used.

References and additional reading

Crawford R.J. *Plastics Engineering*, 2nd edn., Pergamon Press, Oxford, UK, 1987.

Griskey R.G. *Polymer Process Engineering*, Chapman and Hall, London, UK, 1995.

O'Brian, K.T. *Computer Modeling for Extrusion and Other Continuous Polymer Processes*, Hanser, Munich, Germany, 1992.

Pethrick R.A. and Hudson N.E. Rotational moulding a simplified theory. *Journal of Materials: Design and Applications*, 2008, **222**(3), 151–158.

Saechtling H. *International Plastics Handbook*, Hanser, Munich, Germany, 1984.

Tucker C.L. *Computer Modeling for Polymer Processing*, Hanser, Munich, Germany, 1989.

6

Composites

6.1 Introduction

Composites have been created to overcome some of the intrinsic deficiencies of polymer materials and create materials which are fit for purpose. The typical composite is created by incorporating stiffer fibrous or particulate material into a softer polymer material. The important factors that influence the physical characteristics of a composite are:

- the physical properties of the filler/fibrous phase
- the nature of the interactions between the resin and the filler/fibres
- the aspect ratio of the filler particles: rods, needles, plates, discs, spheres, etc.
- the nature of the surface chemistry of the filler/fibres
- the loading level and orientation of the filler/fibres

Each of these factors will influence the properties of the final composite material.

6.2 Classification of composites

Composites can be classified according to the nature of the matrix and reinforcement. Most composites can be created as either thermoplastic or thermoset and are usually differentiated by the nature of the reinforcing material.

- *Fibre-reinforced composites*: The fibres can be glass, carbon, ceramic, plastic or natural fibres. The fibres may be continuous, chopped or short.
- *Particulate composites*: This term covers materials in which the reinforcing phase has no particular shape.

A wide variety of materials can be used as fillers/fibres, including: glass, carbon fibres, polyethylene fibres, Kevlar fibres, clay, talc, quartz, sand, fly ash, carbon black and natural fibres (wood chips, etc.). For specialist applications metal, boron or ceramic fibres can be added. Each material has its particular advantages or disadvantages. Most composites are composed of two phases: the matrix, which is continuous and surrounds the other phase, known as the *dispersed phase*. The phase-separated styrene–butadiene–styrene triblock copolymer was discussed in Section 3.10.4. In that case the matrix was the butadiene phase and styrene formed spherical, cylindrical or lamellae phases which acted to reinforce the matrix and impart anisotropic characteristics to the final, solid, molecular composite material. If the material contains filler particles that have one of their dimensions in the nanoscale (10^{-9} m) then the material can be called a *nanocomposite*. Fillers can be classified according to their shape and size (see Figure 6.1). In the context of polymer composites, the working limit is determined by the organic phase that often loses mechanical stability at temperatures in excess of 300°C.

6.2.1 Why do we need composite materials?

When designing a particular article, the matrix polymer may meet the requirements of low weight and impact strength but may not have sufficient stiffness. Addition of reinforcing filler

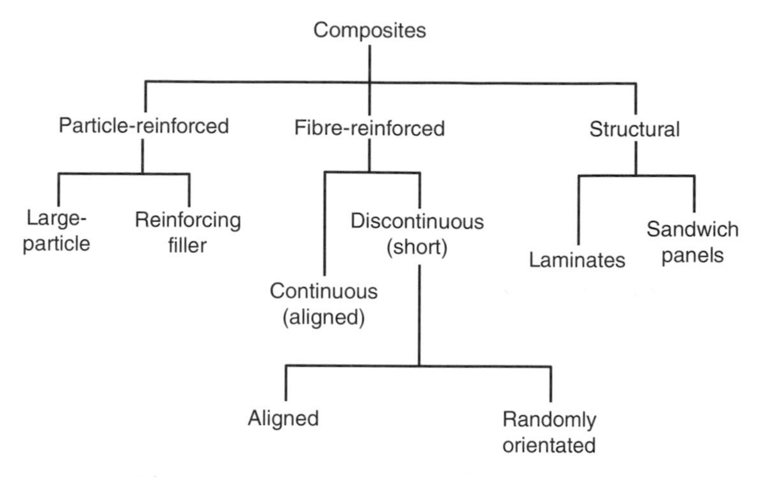

Figure 6.1 Classification of composite materials.

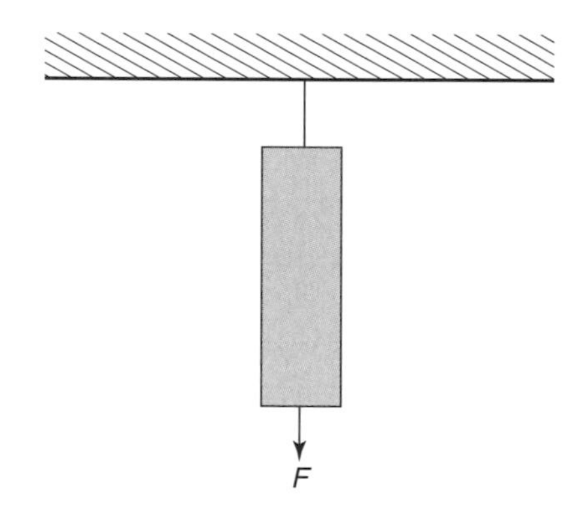

Figure 6.2 Tensile loading of a beam.

can often provide the necessary enhancement to its properties. Consider the tensile load (see Figure 6.2). The tensile load is defined by:

$$F = (ES / l)\Delta l \tag{6.1}$$

where E is Young's modulus, S is the cross-sectional area of the beam, l is the length and Δl the elongation of the beam. The beam stiffness K is defined by:

$$K = ES / l \tag{6.2}$$

which defines the characteristics of the beam in the elastic domain. In order to compare materials it is usual to look at the ratio of the beam stiffness. For two materials 1 and 2, we have:

$$K_1 / K_2 = \left(E_1 S_1 l_2\right) / \left(E_2 S_2 l_1\right) \tag{6.3}$$

When judging the advantages of a particular material it is appropriate to consider the ratio of the weights:

$$m_1 / m_2 = \left(S_1 l_1 \rho_1\right) / \left(S_2 l_2 \rho_2\right) \tag{6.4}$$

In judging composites the concept of a specific property is often used:

$$K_1 / K_2 = \left(E_1 \rho_2 m_1\right) / \left(E_2 \rho_1 m_2\right) \left(l_2 / l_1\right)^2 \tag{6.5}$$

If the two materials have the same length then Equation (6.5) becomes:

$$K_1 / K_2 = (E_1 \rho_2 m_1)/(E_2 \rho_1 m_2) \tag{6.6}$$

If the masses are identical then:

$$K_1 / K_2 = (E_1 \rho_2)/(E_2 \rho_1) \tag{6.7}$$

The composites are constructed from a variety of different fillers.

6.3 Particle-reinforced composites

This classification covers virtually all filled polymer materials, with the exception of fibre-reinforced materials which are a special case. Particle-reinforced composites can be divided into two groups:

- *Large particles*: A particle usually has a diameter of the order of 30 µm or greater. To put this in context a grain of sand will be typically 1000 µm and a carbon black particle, which is actually an aggregate of smaller particles, will be ~30 µm or smaller. Particles are termed large to indicate that they have a limited surface area for contact with the resin matrix. Once the scales are reduced to the micrometre scale or smaller, the contact area increases substantially and so it becomes possible to reinforce the physical properties of the material. The particulate phase is usually harder and stiffer than the matrix and increases the hardness, stiffness and modulus of the resultant material. If there is an interaction between the resin and particles then the movement of the matrix in the vicinity of the particles may be restricted.
- *Reinforcing fillers*: Many small particles and fillers which have been subjected to surface treatment (see Section 4.7) can act as reinforcing fillers. The filler is able to enhance some of the physical properties. The action of such fillers is not simply an increase in all physical properties. Very often enhancement of one may lead to a reduction in another. For instance, if an amorphous resin interacts with the filler then an increase in the glass transition temperature (T_g) may result from close packing of the chains. A closer packed matrix may have a lower coefficient of expansion, elongation to break and poorer impact resistance. Reinforcing fillers will either have a chemically modified, highly porous or large surface area or some combination of these properties. The addition of fumed silica to polydimethylsiloxane rubbers is essential to achieve acceptable tear properties and reflects the way in which all the factors can be used to enhance the physical properties of a material.
- *Nanofillers*: Recently nanoscale fillers have been used to create composites. They have large surface areas and the potential for strong interactions with the resin matrix. Silica beads with dimensions of the order of 30–60 nm can be formed by a sol-gel process. These very small particles can significantly increase the impact resistance of the resin. Remarkable enhancement of the tensile modulus can be achieved when carbon nanotubes are used to reinforce composites. Nanotubes may be several micrometres in length and allow stress transfer over large distances (see Figure 6.3(b)). Carbon nanotubes occur naturally as either single-walled structures (see Figure 6.3(a)) or as multiwalled structures (see Figure 6.3(c)) The multiwalled carbon nanotubes are often obtained when carbon black is produced in an electric arc.

6.3.1 Fibre-reinforced fillers

The effect of reinforcement depends critically upon the length–diameter ratio of the fibre. In the extreme situation, the fibre is a continuous fibre and the enhancement which is achieved is strongly influenced by the properties of the fibre, its orientation and the strength of the bond between the fibre and resin. The fibre usually has its surface treated to enhance the nature of the fibre–resin interaction. This topic has been discussed in detail in Section 4.7.

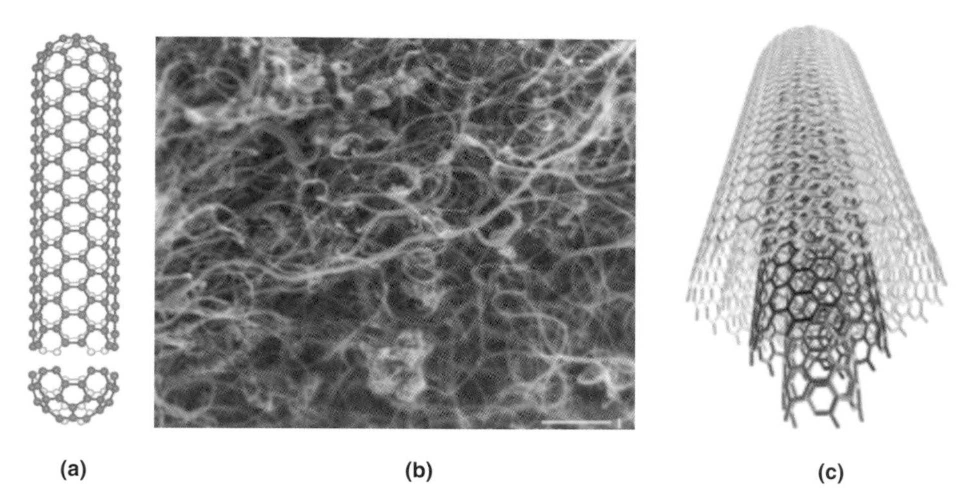

Figure 6.3 Pictorial representation of a nanotube and electron micrographs of single-walled nanotubes and a multiwalled nanotube.

- *Continuous fibres*: Fibres used in the formation of pipes and wound composites (see Section 5.9.2) will usually be continuous and it is possible to achieve very high hoop strength but this may be at the expense of the longitudinal strength of the pipe. To increase the latter, it is often good practice to wind the fibre at an angle in one direction and then reverse the angle on the return winding. The result is that the fibres can reinforce both the longitudinal and transverse strength of the item being manufactured.
- *Discontinuous short fibres*: For convenience of handling, it is common practice to use chopped fibres for many applications. The properties which are achieved with these materials depend critically upon the length–diameter ratio of the fibre, the volume fraction of the fibre and whether or not some preferred orientation has been achieved. Different properties are achieved for *random* and *aligned* orientation of short fibres.

6.3.2 Structures

To achieve mechanical property enhancement, it is often possible to assist the composite materials by the use of skilled design strategies. We can divide the strategies into two types: *laminates* and *sandwich* structures.

Laminates: The term laminate covers the use of prepreg construction and various forms of designed layered structures. It is usual when using carbon fibre to orientate the fibres in different directions so that the reinforcement is achieved in all directions rather than just one direction (see Figure 6.4). A new material, GLARE, which is a glass-reinforced fibre metal laminate, has been used in aerospace applications. It is composed of several very thin layers of metal (usually aluminium) interspersed with layers of glass fibre 'prepreg', bonded together with a matrix such as epoxy. The unidirectional prepreg layers may be aligned in different directions to suit the predicted stress conditions. Although GLARE is a composite material, as its mechanical properties are similar to those of aluminum, it has some major advantages:

- better 'damage tolerance' behaviour, especially with respect to impact properties
- better corrosion resistance
- better fire resistance
- lower specific weight
- high electrical conductivity, thus protecting an aircraft from lightning strikes

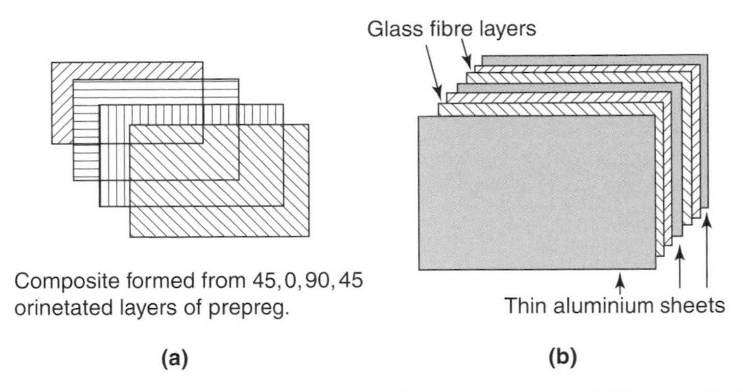

Figure 6.4 Layered structures. (a) in a typical carbon fibre laminate; (b) in a GLARE material.

A structure designed using GLARE will be significantly lighter and less complex than an equivalent metal structure. Carbon fibre laminates are usually constructed by placing one prepreg on top of another, with various orientations of the fibres. It is usual to change the direction of the reinforcement to produce an averaged property in all directions. However, in certain applications, it may be desirable to enhance the properties in a specific direction (see Section 6.5).

6.3.3 Sandwich structures

In many applications, it is desirable to reduce the weight of a structure without loss of load bearing characteristics. This can be achieved by using either a core shell or a honeycomb structure (see Figure 6.5). If correctly designed, the honeycomb structure can have a high compression strength and very high strength–weight ratio.

Core shell or *low density* structures are often used to achieve insulation in buildings (see Figure 6.5(a)). The core may be constructed from a foamed polyurethane, polyurea, melamine or urea formaldehyde. The foam may be created using a natural blowing agent which is associated with the chemical reactions that produce the foam or may have incorporated a blowing agent which has been selected to increase the insulation characteristics of the foam. The foam which is produced has to be carefully controlled to be a closed cell foam, so that gas cannot easily escape or pass through the material. If it is required that the foam should be compressible, it is necessary for the foam to be open celled, allowing gas to move through the foam structure. In insulating panels the outer coatings may be either melamine formaldehyde panels or may be aluminium sheets. This type of structure may be used in aerospace applications where sound or thermal insulation is desirable.

Honeycomb structures have been used in lightweight structures, such as aerofoils and similar components, where strength is required together with rigidity. Structures similar to those shown in Figure 6.5(b) were widely used in the construction of Concorde, a turbojet-powered supersonic passenger airliner which operated from 1969 to 2003 and illustrates the durability of such

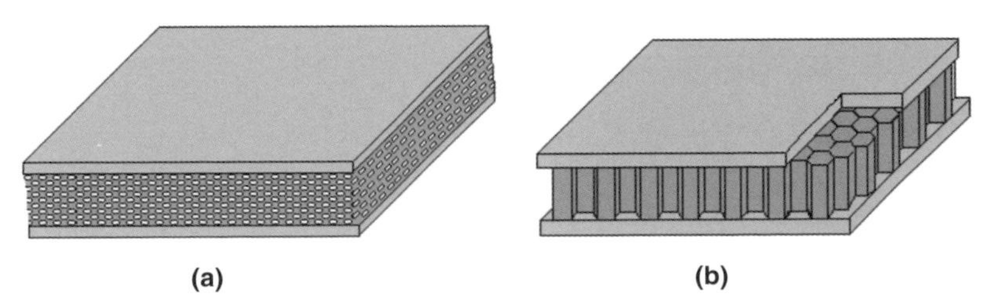

Figure 6.5 Sandwich structures: (a) using a low density core; (b) using a honeycomb structure.

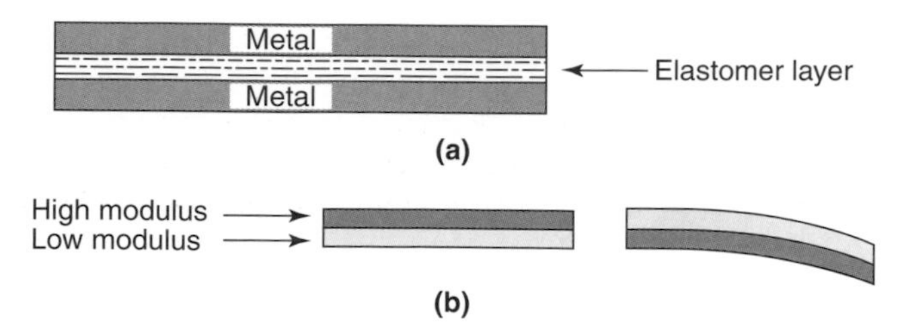

Figure 6.6 Sandwiched metal structure: (a) with elastomeric damping layer; (b) behaviour of layered structures which deform under their own weight.

structures, even in difficult conditions. In the case of Concorde, the honeycomb was constructed from Nomex, a thermoset rigid material produced using phenol formaldehyde-based resins. For the structure to be successful it is important that the honeycomb is securely bonded to the outer panels. In many applications the panels may be light aluminium or can be composite laminates from glass or carbon fibre.

Layered laminated structures

It is possible to achieve desirable properties by combining structural materials with different characteristics in layered structures (see Figure 6.6). In automobile applications, it is desirable to reduce vibration and damp noise. For instance, crankshaft cover housings have been constructed from mouldings created using two thin metal sheets bonded together with an elastic polymeric phase. The polymer may be either a cross-linked rubber or an elastomeric polyurethane. The vibrations in the metal are damped by the coupling material and engine noise is significantly reduced. It is possible to produce layer materials in which the moduli of the two layers differ significantly. Such materials will bend easily in one direction but be rigid in the other direction (see Figure 6.6). When the top layer is elastomeric and has a low modulus, it can bend, when it has a higher modulus, it cannot bend.

6.4 Prediction of characteristics of filled composite materials

The degree of reinforcement or improvement of mechanical behaviour depends on the strength of the bonding between the matrix and the particle. Because the strength of this bond is often not well-defined, there is a consequent uncertainty in the prediction of the resulting physical properties when the two phases are combined. Using the spring and dashpot analogy discussed in Section 2.6, it is possible to define two limiting cases corresponding to the series and parallel combination of the physical properties.

These *rule of mixtures* equations, which will be explored more fully in Section 6.5.2, predict that the elastic modulus should lie between an upper bound defined by:

$$E_c = E_m V_m + E_p V_p \tag{6.8}$$

and a lower bound limit:

$$E_c = \left(E_m V_p\right) / \left(V_m E_p + V_p E_m\right) \tag{6.9}$$

where the subscripts m and p refer, respectively, to the matrix and particulate phases. The above equations consider the modulus to be either combined in series or parallel.

6.4.1 Volume fractions

In many applications, we are concerned with the relative contribution from the fibre and the matrix. It is appropriate to define the volume fraction in the following manner:

$V_f = v_f / v_0$ where v_f and v_0 are, respectively, the volume of fibre and the total volume of material. The volume fraction of the matrix will be $V_m = v_m / v_0$ where v_m is the volume of the matrix. Since $v_0 = v_m + v_f$ then $V_m = 1 - V_f$. The composite is strengthened and hardened by the uniform dispersion of several volume percent of fine particles of very hard and inert material. It is often appropriate to convert the fibre weight fraction (*FWF*) to the fibre volume fraction (*FVF*):

$$FVF = 1 \big/ \left[1 + (\rho_F / \rho_R)((1 / FWF) - 1) \right] \tag{6.10}$$

and

$$FWF = (\rho_F FVF) \big/ \left[\rho_R + ((\rho_F - \rho_R) FVF) \right] \tag{6.11}$$

where ρ_R and ρ_F are, respectively, the densities of the resin and the fibre.

6.5 Fibre-reinforced composites

Technologically, the most important composites are those in which the dispersed phase is in the form of a fibre. The fibre adds high strength and/or stiffness to the structure for little addition in weight. The characteristics are expressed in terms of *specific strength* and *specific modulus*. The *specific strength* is the ratio of the tensile strength to the specific gravity. The *specific modulus* is the ratio of the modulus of elasticity to the specific gravity. Some examples of the specific strength plotted against specific stiffness for examples of glass fibre-reinforced plastic (GFRP) and carbon fibre-reinforced plastic (CFRP) compared with a metal (aluminium) are given in Figure 6.7. The specific strength and stiffness increase with the volume fraction of glass or carbon fibre in the composite. Similarly the values obtained with the unidirectional are significantly higher than the quasi-isotropic values. The specific strength and stiffness of both glass and carbon fibre are higher than those of a metal-loaded composite. The specific properties of the composite depend on the properties of the fibre: the values for the materials generated using T300 are inferior to those obtained using T800. The properties of carbon fibres have been discussed in Section 4.7.1.

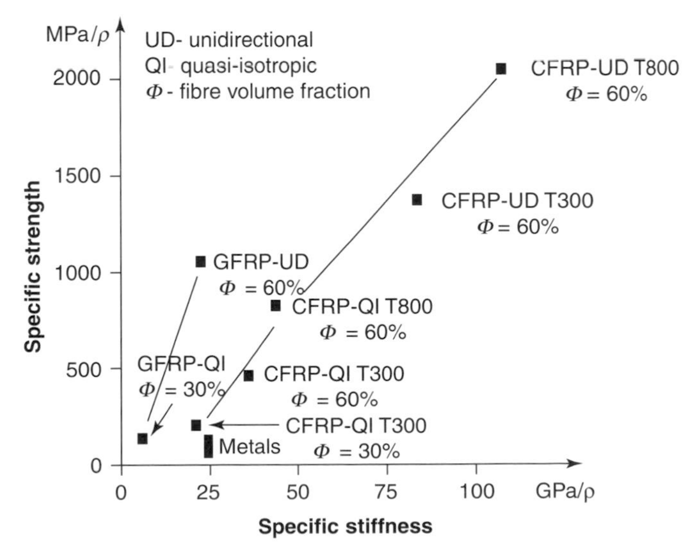

Figure 6.7 Specific strength–specific stiffness for some reinforced plastics.

6.5.1 Fibre performance

The performance of a material can be compared on the basis of its *specific* parameters. Ideally, the best composite material would be one in which the filler has a density which is close to that of the resin but it has significantly better physical properties. To illustrate the relative merits of various materials, a list of the specific tensile strength and the specific modulus are listed in Table 6.1. The reinforcement which is achieved will depend on the strength of the bond between the resin and the fibre and the shape and dimensions of the fibre. The data presented is typical of that found for epoxy resin matrix composites.

A comparison of the specific tensile strength leads to a different order compared with the simple ranking of the fibre according to its strength. Although graphite and silicon carbide whiskers have apparently the same tensile strength, when divided by the density, the specific tensile strength indicates that the graphite whiskers perform better than the denser silicon carbide. Aluminium oxide whiskers at the top end of their performance have a tensile strength of 298 MPa, and outperform graphite, but when the higher density at 3.9 $g\,cm^{-3}$ is taken into account the specific tensile strength at 7.1 MPa is significantly lower than the value for graphite at 9 MPa. Similar effects can be seen with the fibres, the higher density of glass compared with carbon fibre leading to lower values than those for carbon.

6.5.2 Influence of fibre length

The mechanical characteristics of a fibre composite depend not only on the properties of the fibre, but also on the degree to which an applied load is transmitted to the fibre by the matrix phase. Ideally, the stress should be distributed evenly along the fibre so that the interaction area is maximised and the stress minimised. The stress will be concentrated at the end of the fibres and the extent to which this occurs will depend on the length–diameter ratio. If the fibre is very short, it will effectively appear as a point and hence the stress will be concentrated at a point in space. Regions of stress concentration will usually not enhance the physical properties as they represent points of potential failure. There is a critical fibre length necessary for effective strengthening and stiffening of the composite. This critical length, l_c, is dependent on the fibre diameter, d, and its ultimate (or tensile) strength, σ_f, and on the fibre matrix bond strength (or shear yield strength of the matrix) τ_c, according to:

$$l_c = (\sigma_f d)/\tau_c \tag{6.12}$$

Table 6.1 Specific strength and specific modulus of a series of composites

Material	ρ ($g\,cm^{-3}$)	Tensile strength (MPa)	Specific tensile strength (TS/ρ)	Modulus (MPa)	Specific modulus (MPa ρ^{-1})
Whiskers					
Graphite	2.2	20	9	690	313
Silicon carbide	3.2	20	6.25	480	150
Silicon nitride	3.2	14	4.3	380	118
Al_2O_3	3.9	14–28	3.5–7.1	415–550	106–141
Fibre					
Aramid (Kevlar 49)	1.4	3.5	2.5	124	88
E-glass	2.5	3.5	1.4	72	29
Carbon	1.8	1.5–5.5	0.83–3.0	150–500	83–278
Al_2O_3	3.2	2.1	0.65	170	53
Silicon carbide	3.0	3.9	1.3	425	141

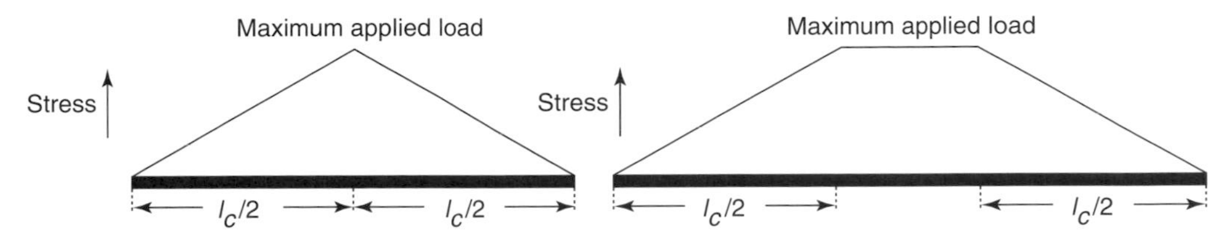

Figure 6.8 Stress–strain profiles for short and long fibres.

For a number of glass and carbon fibre matrix combinations this critical length is of the order of 1 mm, which ranges between 20 and 150 times the fibre diameter. When a stress equal to σ_f is applied to a fibre having just this critical length, the stress profile will be as shown in Figure 6.8. Once the fibre length is greater than this critical value, then a region of constant stress will occur along the fibre and the effective maximum load which can be carried by the fibre will be increased. It is common to refer to fibres that have a length shorter than l_c as being *discontinuous or short fibres* and those with a length in which $l >> l_c$ as being *continuous*. Ideally, the fibre length should be such that the maximum stress level which occurs along the fibre is only a fraction of the failure strain for the composite.

6.5.3 Influence of fibre orientation and concentration

The most important factor when considering the mechanical properties of a composite is the alignment of the fibre relative to the applied stress. The types of situation found in composites are (see Figure 6.9):

- continuously aligned
- discontinuous and aligned
- discontinuous and randomly aligned

The separation into types recognises that there is a critical length for a particular fibre–resin combination and the dominant effects of fibre orientation.

Continuous and aligned composites

The properties of a composite having its fibres aligned are highly anisotropic, i.e. they depend on the direction in which they are measured. Assuming that the fibre–matrix bond is good and is

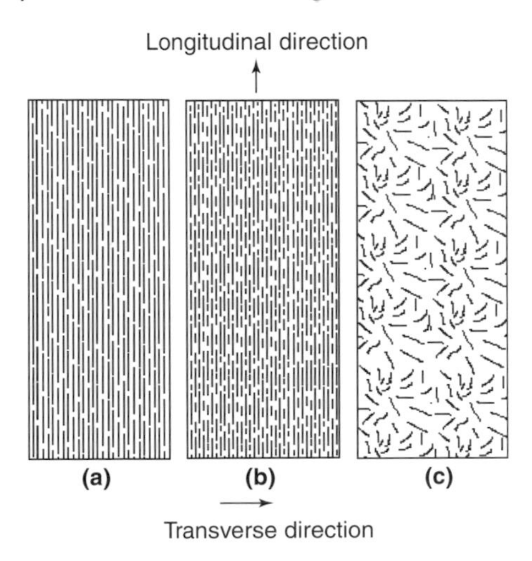

Figure 6.9 Various classes of composites: (a) continuously aligned; (b) discontinuous and aligned; (c) discontinuous and randomly aligned.

not a limiting factor, then the total load sustained by the composite F_c is equal to the sum of the loads carried by the matrix F_m and fibre F_f phases:

$$F_c = F_m + F_f \quad (6.13)$$

The stress can be defined as $F = \sigma A$ and the above equation becomes:

$$\sigma_c A_c = \sigma_m A_m + \sigma_f A_f \quad (6.14)$$

Dividing through by the total cross-section area of the composite A_c, we then have:

$$\sigma_c = \sigma_m \left(A_m / A_c\right) + \sigma_f \left(A_f / A_c\right) \quad (6.15)$$

where A_m/A_c and A_f/A_c are the fractions of the matrix and fibre phases, respectively. It is then possible to approximate these fractions to the volume fractions V_f and V_m:

$$\sigma_c = \sigma_m V_m + \sigma_f V_f \quad (6.16)$$

The previous assumption of an isostrain state means that:

$$\varepsilon_c = \varepsilon_m = \varepsilon_f \quad (6.17)$$

and when each term in Equation (6.16) is divided by its respective strain:

$$\sigma_c / \varepsilon_c = \left(\sigma_m / \varepsilon_m\right) V_m + \left(\sigma_f / \varepsilon_f\right) V_f \quad (6.18)$$

If we assume that the matrix and fibre both behave as perfect elastic media $\sigma_i / \varepsilon_I = E_I$ then the moduli of elasticity can be expressed as:

$$E_c = E_m V_m + E_f V_f \quad (6.19)$$

but since $V_f + V_m = 1$, then:

$$E_c = E_m \left(1 - V_f\right) + E_f V_f \quad (6.20)$$

The above equation is the upper bound of the fibre equation and is obtained from the rule of mixtures. The ratio of the load carried by the fibres to that carried by the matrix is:

$$F_f / F_m = \left(E_f V_f\right) / \left(E_m V_m\right) \quad (6.21)$$

Discontinuous and aligned fibre composites

Chopped fibres are used extensively and produce composites which have moduli of elasticity and tensile strengths that approach 90% and 50%, respectively, of those of continuous fibre values. For a discontinuous and aligned fibre composite having a uniform distribution of fibres and $1 > l_c$ the longitudinal strength $(TS)_c$ is given by the relationship:

$$(TS)_c = (TS)_f V_f \left(1 - \left(l_c / 2l\right)\right) + (TS)_m \left(1 - V_f\right) \quad (6.22)$$

where $(TS)_i$ are the fracture strengths of the fibre and the stress in the matrix when the composite fails. If the fibre length is less than the critical length $1 < l_c$, then the longitudinal strength is given by:

$$(TS)_c = \left(l\tau_c / d\right) V_f + (TS)_m \left(1 - V_f\right) \quad (6.23)$$

where d is the diameter of the fibre.

Discontinuous and randomly orientated fibre composites

Under these circumstances, a 'rule of mixtures' operates for the elastic modulus as follows:

$$E_c = K E_f V_f + E_m V_m \quad (6.24)$$

Table 6.2 Summary of reinforcement factors for various orientations

Fibre orientation	Stress direction	Reinforcement efficiency
Fibres parallel	Parallel to fibres	1
	Perpendicular to fibres	0
Fibres randomly within a specific plane	Any direction in plane	3/8
Fibres randomly within three dimensions	Any direction	1/5

Table 6.3 Properties of polycarbonate and composites reinforced with orientated glass fibres

Property	Pure polymer	Volume of glass (%)		
	0	20	30	40
Specific gravity	1.19–1.22	1.35	1.43	1.52
Tensile strength (MPa)	59–62	110	131	159
Modulus	2,240–2,345	5,930	8,620	11,590
Elongation (%)	90–115	4–6	3–5	3–5
Impact strength (J)	1.36–1.80	0.23	0.23	0.28

where K is a fibre efficiency parameter, which depends on V_f and the ratio E_f/E_m and has a value less than unity, typically in the range 0.11–0.6 (see Table 6.2). The modulus increases in proportion to the load taken by the fibres. The fibres can be aligned in a plane when the processing induces a high shear flow on the resin mixture prior to being cured.

To illustrate the effect of the enhancement consider the mechanical properties of a series of filled polycarbonate composites, reinforced with glass fibres (see Table 6.3). In the case of polycarbonate, the reinforcement increases the stiffness but there is a loss of elongation and impact strength. Filling a polymer does not always lead to improvement in the physical properties, a weak interface may lead to a loss of physical properties!

6.5.4 Fibre phase

Glass fibre, carbon fibre, Kevlar and polyethylene fibres are used as reinforcing fillers for plastics. Incorporation of fibre will lead to an increase in the modulus and stiffness of the material allowing creation of light load-bearing structures.

Matrix materials

The matrix materials which are typically used are:

- epoxy resins, which are usually high temperature resins
- vinyl ester resins, which are cheap and stable to low temperatures
- thermoplastic resins, e.g. polyethersulfone
- elastomeric materials, such as silicone, polyurethane or rubber-based systems

In general, the matrix has the role of transmitting and distributing load as well as protecting the individual fibres from surface damage as a result of mechanical abrasion or chemical reactions with the environment. Surface flaws will lead to cracks and failure at low levels of tensile stress. The role of the matrix is to separate the fibres and prevent the propagation of brittle cracks from fibre to fibre resulting in catastrophic failure.

Relative merits of various fibres

Glass is popular as a reinforcement for various reasons:

- It is easily drawn into high strength fibres from the molten state.
- It is readily available and may be used economically in the creation of a wide variety of structures.

- It is relatively strong and when embedded in a plastic it produces a composite which has a very high specific strength.
- When coupled with various plastics it is chemically inert making the composite useful in a variety of corrosive environments. It is, however, sensitive to alkaline environments.

Carbon fibres are more attractive but also more expensive and are used when weight or performance are critical. Hybrid composites can allow the advantages to be balanced. Polyethylene fibre can be used in aggressive alkaline environments, aramid and related fibres are used when high strength is required.

6.6 Fabrication

The cure of the composite to produce a homogeneous structure requires a number of factors:

- The resin must be sufficiently viscous to be retained between the fibres of the *prepreg* before it is laid up for cure. This will usually mean that the resin is used in a partially cured state, known as a *Class 1* resin.
- During cure the resin must have a sufficiently low viscosity to flow between the laminates but must not be so low as to drain through the laminate structure.
- The air between the laminates must be excluded and this is usually achieved by applying a vacuum.
- The final material must be consolidated by the application of pressure to produce a uniform structure.

The curing process uses a combination of heating and pressure to achieve these objectives. Incorrect cure conditions will lead to voids being left between laminates due to poor flow, resin-rich areas as a consequence of draining phenomena, fibre twisting and bending due to loss of resin and high pressures (see Section 5.9).

A variety of methods are used to achieve a composite structure. The most common are:

- *Pultrusion*: Resin-coated fibres are pulled through a heated mould and form continuous composite structures (see Section 5.9.3).
- *Laminate lay-ups*: The laminates created by doctor blade coating of the fibres can be cured in autoclaves (see Section 5.8.1).
- *Filament winding*: As the name implies, the coated filaments are wound around a former and the structure created is cured either as it is created or after the winding process has been completed (see Section 5.9.2).

6.7 Failure

The failure of composites is controlled by the quality of the interaction between the fibre and the resin. In general, the fibres are hydrophobic and are not wetted out by the resin. It is usually necessary to modify the fibre surface to achieve greater stability and shear strength. To improve the nature of the interaction the fibres may be treated with compounds which will increase the interaction between the fibre and the matrix.

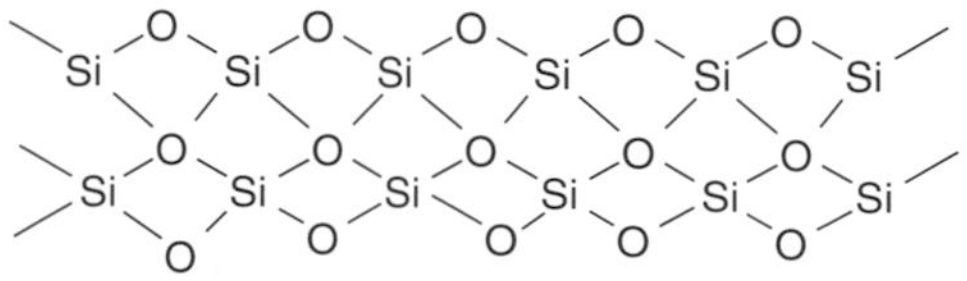

Hydrophobic glass surface

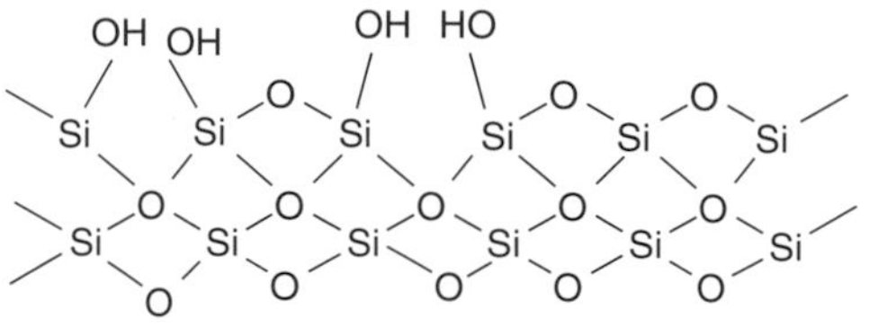

Hydrophilic glass surface

Glass fibres: The glass fibres can be treated with silane coupling agents which will increase the binding of the polymer to the fibre. The reactive hydroxyl groups generated at the surface may then be reacted with silane coupling agents, which are usually amine, epoxy or vinyl functionalised depending on the chemistry of the polymer matrix being formed.

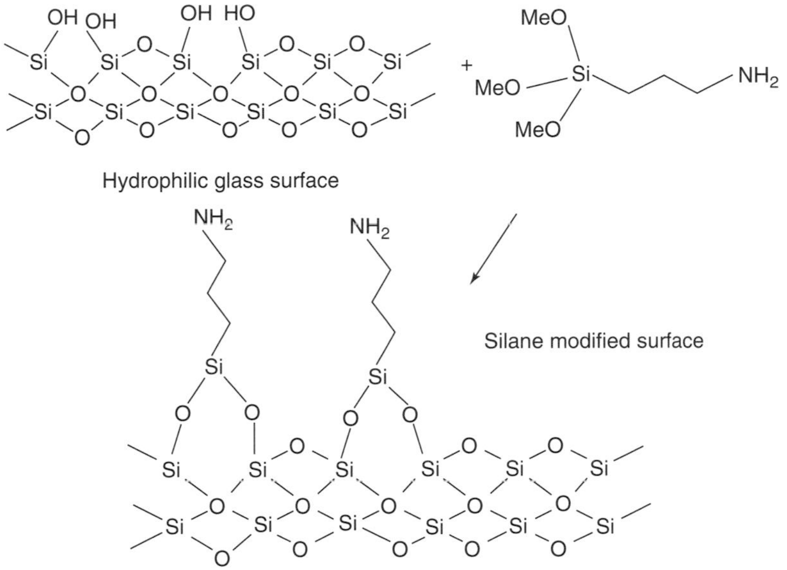

The amine coupling agent can react with epoxy to form a chemical bond with the surface.

Carbon fibres: In general, the carbon fibres will not have functional groups which can be used to attach reactive groups so the shear strength has to be enhanced via a physical mechanism. The surface of the fibre is slightly porous and hence a low viscosity resin can flow into the fibre. The carbon fibre is pretreated with low viscosity resin and then incorporated into the Class 1 resin material. The *lock and key* mechanism improves the stability of the interface and hence the durability of the composite.

The wetting of the fibre by the resin is critical to obtaining a durable composite and will help with the elongation strength of the material, which in this case is dominated by the properties of the fibre. The toughness of the composite will be influenced by the strength of the interaction between fibre and matrix. However, if the level of cross-linking agent in the surface is too high, a rigid structure will be obtained which may be brittle.

6.8 Factors influencing the performance of composites

6.8.1 Adhesive properties

The strength of the interface between resin and fibre is critical to achieving the theoretical enhancement of the physical properties. Polyester resins generally have the lowest adhesive properties, whereas vinyl ester resin shows improved adhesive properties, but epoxy systems are the best. Adhesive properties are primarily due to chemical interactions between the polymer and the fibre surface. Epoxies cure with low shrinkage which means that the surface contacts generated during the cure process are not disturbed. Epoxy resins are useful in the construction of honeycomb-cored laminates where the small bonding surface area requires that maximum adhesion is achieved.

6.8.2 Mechanical properties

Shrinkage is a very important factor to be considered when designing composite structures. A resin shrinks due to the molecules rearranging and re-orienting themselves in the semigelled phase. Polyester and vinyl esters require considerable molecular rearrangement to reach a fully cured state and shrink by up to 8%. In contrast, epoxy resins usually require little rearrangement and shrinkage can be around 2%. The absence of shrinkage and associated build up of stresses is, in part, responsible for the improved mechanical properties of epoxies over polyester. Shrinkage through the thickness of a laminate can lead to 'print-through' of the pattern of the reinforcing fibres, a cosmetic defect that is difficult to eliminate.

6.8.3 Microcracking

The strength of a laminate is reflected in the load it can withstand before it suffers complete failure. This ultimate or breaking strength is the point at which the resin exhibits catastrophic breakdown. However, before catastrophic failure occurs, the laminate will reach a stress level where, for fibres not aligned with the load, the resin begins to crack away from fibre and these cracks spread through the resin matrix. This process is known as *transverse microcracking* and is the point at which breakdown has commenced. Engineers who want a long-lasting structure must ensure that composites do not exceed this point under regular service loads (see Figure 6.10).

The strain that a laminate can reach before microcracking depends strongly on the toughness and adhesive properties of the resin system. For brittle resin systems, such as most polyesters, this point occurs a long way before laminate failure. In a polyester/glass woven roving laminate, microcracking typically occurs at about 0.2% strain with ultimate failure not occurring until 2.0% strain, which is only 10% of the ultimate strength. As the ultimate strength of a laminate in tension is governed by the strength of the fibres, these resin microcracks do not immediately reduce the ultimate properties of the laminate. However, in an environment such as water or moist air, the microcracked laminate will absorb considerably more water than an uncracked laminate, leading to an increase in weight, moisture attack on the resin and fibre sizing agents, loss of stiffness and ultimate properties.

6.8.4 Fatigue resistance

Generally composites show excellent fatigue resistance when compared with most metals. However, since fatigue failure results from the gradual accumulation of small amounts of damage, the fatigue behaviour of any composite will be influenced by the toughness of the resin, its resistance to microcracking and the quantity of voids and other defects which occur during manufacture. Epoxy-based laminates tend to show very good fatigue resistance when compared with both polyester and vinyl ester and are used in aircraft structures.

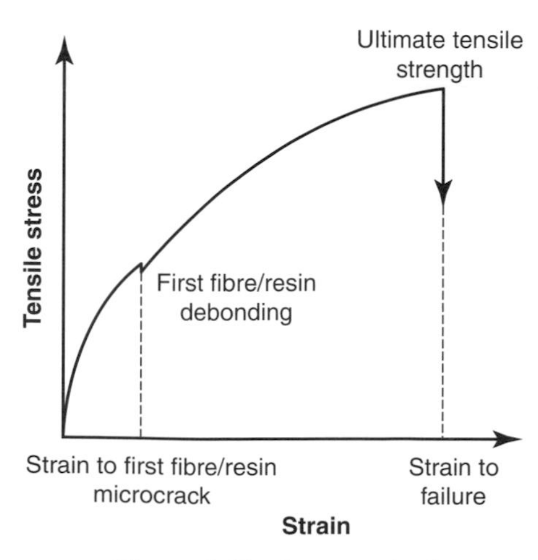

Figure 6.10 Stress–strain plots.

6.8.5 Water ingress

An important property of any composite to be used in a marine environment is its ability to withstand degradation as a result of water ingress. All resins absorb moisture to some extent. However, the real issue is the way the absorbed water affects the resin and the resin–fibre bond in a laminate. Both polyester and vinyl ester resins are prone to water degradation due to the presence of hydrolysable ester groups in their molecular structures. A thin polyester laminate can be expected to retain only 65% of its interlaminar shear strength (ILSS) after immersion in water for a period of one year, whereas an epoxy laminate immersed for the same period will retain around 90% of its ILSS. Figure 6.11 demonstrates the effects of water on epoxy- and polyester-woven glass laminates which have been subjected to water at 100°C. Use of elevated temperatures accelerates the degradation of the laminate.

6.8.6 Osmosis

All laminates in a marine environment will transmit low quantities of water in vapour form. The water passes through the microscopic cells of the laminated form, which can concentrate mobile ions and degradation products. During the osmotic cycle, more water is drawn through the semi-permeable membrane of the laminate in an attempt to dilute this solution. This additional water increases the fluid pressure by as much as 700 psi. Eventually the pressure distorts or bursts the laminate or gel coat, and can lead to a characteristic 'chicken-pox' surface (see Figure 5.46(d)). Hydrolysable components in a laminate can be residues left after synthesis and degradation products such as acids produced from breaking the ester linkages in a polyester or to a lesser extent a vinyl ester. For polymer resins, it is essential to use resin layers next to the gel coat in order to minimise this type of degradation. To prevent the onset of osmosis effects, it is necessary to use a resin which has both a low water transmission rate and a high resistance to attack by water. An epoxy backbone is significantly better than many other resins at resisting the effects of water.

For the full mechanical properties of the composite to be achieved in tension, the resin must be able to deform to at least the same extent as the fibre. Figure 6.12 gives the strain to failure for E-glass, S-glass, aramid and high-strength grade carbon fibres on their own (i.e. not in a composite form). The S-glass fibre, with an elongation to break of 5.3%, will require a resin with an elongation to break of at least this value to achieve maximum tensile properties.

6.8.7 Adhesive properties of the resin system

High adhesion between resin and reinforcement fibres is necessary for any resin system. This will ensure that the loads are transferred efficiently and will prevent cracking or fibre–resin debonding when stressed.

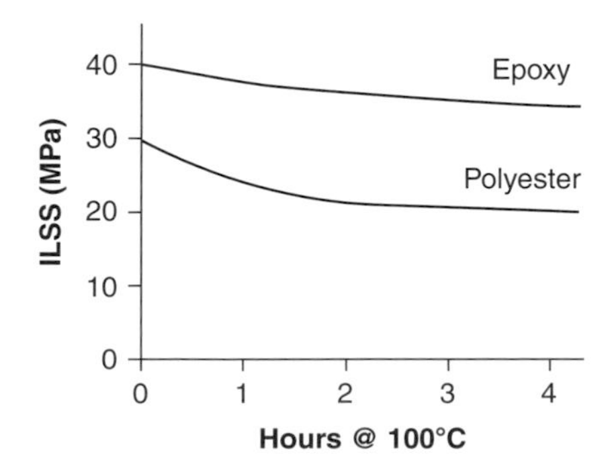

Figure 6.11 ILSS versus exposure time at 100°C.

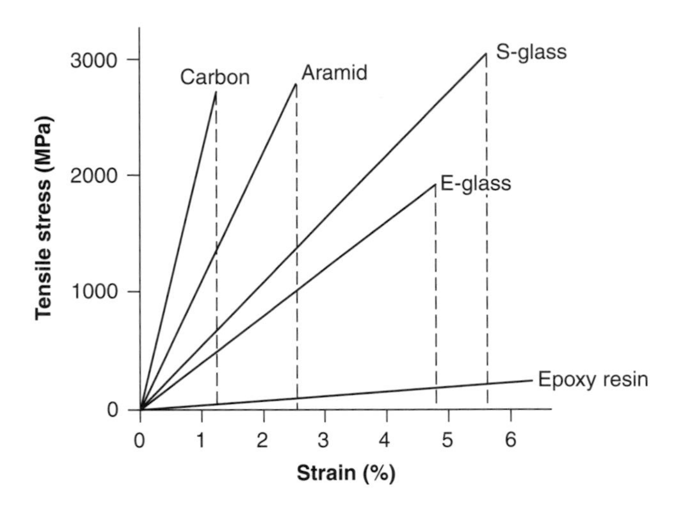

Figure 6.12 Tensile stress versus strain for some common composites.

6.9 Uses of plastic composites

The range of uses is very wide: water pipes, water tanks, electrical insulation, window frames, seals and gaskets in engines, fuel pipes, water pumps, pressure vessels, bridge sections, boats, tail sections of aircraft, wings, rubbers, dashboards in motor cars, bonnets, bumpers, computer cabinets, encapsulations for semiconductors, etc.

6.10 Elastic behaviour of composite materials

Many composite materials will be *anisotropic* and therefore it is necessary to consider the constituent elastic constants. It is usual to assume that the elastic behaviour is independent of temperature, which is not strictly correct but is a good first approximation. The linear elasticity relation can be written in the following matrix form:

$$\begin{bmatrix} \sigma_1 \\ \sigma_2 \\ \sigma_3 \\ \sigma_4 \\ \sigma_5 \\ \sigma_6 \end{bmatrix} = \begin{bmatrix} C_{11} & C_{12} & C_{13} & C_{14} & C_{15} & C_{16} \\ C_{12} & C_{22} & C_{23} & C_{24} & C_{25} & C_{26} \\ C_{13} & C_{23} & C_{33} & C_{34} & C_{35} & C_{36} \\ C_{14} & C_{24} & C_{34} & C_{44} & C_{45} & C_{46} \\ C_{15} & C_{25} & C_{35} & C_{45} & C_{55} & C_{56} \\ C_{16} & C_{26} & C_{36} & C_{46} & C_{56} & C_{66} \end{bmatrix} \begin{bmatrix} \varepsilon_1 \\ \varepsilon_2 \\ \varepsilon_3 \\ \varepsilon_4 \\ \varepsilon_5 \\ \varepsilon_6 \end{bmatrix} \quad (6.25)$$

or in the condensed form:

$$\sigma = \mathbf{C}\varepsilon \quad (6.26)$$

This is the so-called *generalised Hooke's law*, which introduces the symmetric stiffness matrix **C**. So the linear behaviour of a material is described in the general case by 21 independent coefficients, here the 21 stiffness constants C_{ij}.

Compliance matrix

The elasticity relation, Equation (6.26), can be written in the inverse form as:

$$\varepsilon = \mathbf{S}\sigma \quad (6.27)$$

by introducing the inverse of the stiffness matrix, **S**, called the *compliance matrix* and in the general case written as:

$$\mathbf{S} = \begin{bmatrix} S_{11} & S_{21} & S_{31} & S_{41} & S_{51} & S_{61} \\ S_{12} & S_{22} & S_{23} & S_{24} & S_{25} & S_{26} \\ S_{13} & S_{23} & S_{33} & S_{34} & S_{35} & S_{36} \\ S_{14} & S_{24} & S_{34} & S_{44} & S_{45} & S_{46} \\ S_{15} & S_{25} & S_{35} & S_{45} & S_{55} & S_{56} \\ S_{16} & S_{26} & S_{36} & S_{46} & S_{56} & S_{66} \end{bmatrix} \quad (6.28)$$

where

$$S = C^{-1} \quad (6.29)$$

The coefficients S_{ij} are known as the *compliance constants*.

6.10.1 Different types of anisotropic materials

In the most general case, the stiffness matrix and compliance matrix are each determined by 21 independent constants. This case corresponds to a material which does not have symmetry properties. Such a material is called a *triclinic* material. Most anisotropic materials have a structure which exhibits one or more elements of symmetry. For example, in the case of monocrystals, fibrous structures, fibre or cloth composite materials, the properties of the geometric symmetries reduce the number of independent constants needed to describe the elastic behaviour of the material. This reduction is a function of the symmetries exhibited by the material considered.

6.10.2 Monoclinic materials

A monoclinic material has a plane of symmetry and when the plane of symmetry is in the (1,2) plane, the stiffness matrix has the form:

$$\begin{bmatrix} C_{11} & C_{12} & C_{13} & 0 & 0 & C_{16} \\ C_{12} & C_{22} & C_{23} & 0 & 0 & C_{26} \\ C_{13} & C_{23} & C_{33} & 0 & 0 & C_{36} \\ 0 & 0 & 0 & C_{44} & C_{45} & 0 \\ 0 & 0 & 0 & C_{45} & C_{55} & 0 \\ C_{16} & C_{26} & C_{36} & 0 & 0 & C_{66} \end{bmatrix} \quad (6.30)$$

The compliance matrix has the same form. Thus, the number of independent elasticity constants is reduced to 13.

6.10.3 Orthotropic material

An orthotropic material has three symmetry planes that are mutually orthogonal. The existence of two orthogonal symmetry planes implies the existence of the third. The invariance of the stiffness matrix leads to a stiffness matrix of the form:

$$\begin{bmatrix} C_{11} & C_{12} & C_{13} & 0 & 0 & 0 \\ C_{12} & C_{22} & C_{23} & 0 & 0 & 0 \\ C_{13} & C_{23} & C_{33} & 0 & 0 & 0 \\ 0 & 0 & 0 & C_{44} & 0 & 0 \\ 0 & 0 & 0 & 0 & C_{55} & 0 \\ 0 & 0 & 0 & 0 & 0 & C_{66} \end{bmatrix} \quad (6.31)$$

The compliance matrix has the same form; the number of independent elasticity constants is reduced to 9.

6.10.4 Unidirectional material

The prepreg is often a unidirectional material and if the fibres are orientated in one direction then a unidirectional material will be obtained. The elementary cell of a unidirectional composite material can be considered to be composed of fibre embedded in a cylinder of matrix material (see Figure 6.13).

The material thus behaves as an orthotropic material having one axis of revolution. The material is called a *revolution orthotropic* or *transverse isotropic* material. A change of the reference system obtained by an arbitrary rotation around the revolution axis must leave the stiffness (or compliance) matrix unchanged. The application of this transform leads to:

$$C_{13} = C_{12} \quad C_{33} = C_{22}$$

$$C_{55} = C_{66} \quad C_{44} = 1/2(C_{22} - C_{23})$$

and

$$S_{13} = S_{12} \quad S_{33} = S_{22}$$

$$S_{55} = S_{66} \quad S_{44} = 2(S_{22} - S_{23})$$

The stiffness matrix is therefore written as:

$$\begin{bmatrix} C_{11} & C_{12} & C_{12} & 0 & 0 & 0 \\ C_{12} & C_{22} & C_{23} & 0 & 0 & 0 \\ C_{12} & C_{23} & C_{22} & 0 & 0 & 0 \\ 0 & 0 & 0 & 1/2\left(C_{22} - C_{23}\right) & 0 & 0 \\ 0 & 0 & 0 & 0 & C_{66} & 0 \\ 0 & 0 & 0 & 0 & 0 & C_{66} \end{bmatrix} \tag{6.32}$$

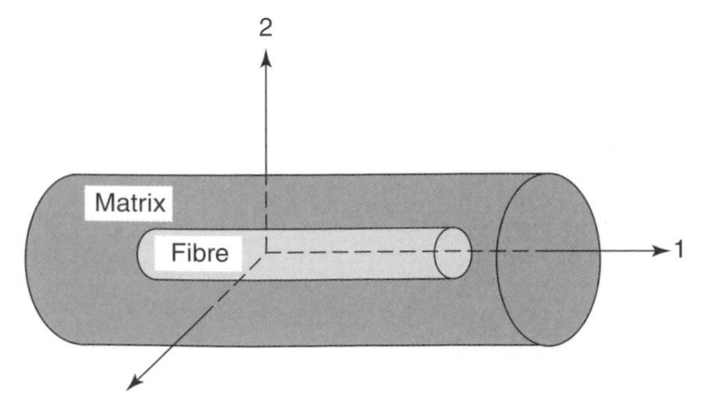

Figure 6.13 Unidirectional composite material.

and the compliance matrix is:

$$\begin{bmatrix} S_{11} & S_{12} & S_{12} & 0 & 0 & 0 \\ S_{12} & S_{22} & S_{23} & 0 & 0 & 0 \\ S_{12} & S_{23} & S_{22} & 0 & 0 & 0 \\ 0 & 0 & 0 & 2(S_{22}-S_{23}) & 0 & 0 \\ 0 & 0 & 0 & 0 & S_{66} & 0 \\ 0 & 0 & 0 & 0 & 0 & S_{66} \end{bmatrix} \qquad (6.33)$$

The elasticity properties of a unidirectional material are described by five independent elasticity constants.

6.10.5 Isotropic materials

A material is isotropic if its properties are independent of the choice of the reference system. Many materials, with the exception of wood or other anisotropic materials, satisfy this description at a macroscopic scale and there is no preferred direction. The application of this limitation to a unidirectional material leads to the relation:

$$C_{22} = C_{11}, \quad C_{23} = C_{12}, \quad C_{66} = 1/2(C_{11} - C_{22}) \qquad (6.34)$$

The number of independent elasticity constants is thus reduced to two and the stiffness matrix is written as:

$$\begin{bmatrix} C_{11} & C_{12} & C_{12} & 0 & 0 & 0 \\ C_{12} & C_{11} & C_{12} & 0 & 0 & 0 \\ C_{12} & C_{12} & C_{11} & 0 & 0 & 0 \\ 0 & 0 & 0 & 1/2(C_{11}-C_{12}) & 0 & 0 \\ 0 & 0 & 0 & 0 & 1/2(C_{11}-C_{12}) & 0 \\ 0 & 0 & 0 & 0 & 0 & 1/2(C_{11}-C_{12}) \end{bmatrix} \qquad (6.35)$$

Usually the stiffness constants are expressed by introduction of the Lame coefficients, λ and μ:

$$C_{12} = \lambda, \quad 1/2(C_{11} - C_{12}) = \mu \qquad (6.36)$$

Hence:

$$C_{11} = \lambda + 2\mu \qquad (6.37)$$

The elasticity relation can be written as:

$$\sigma_{ij} = \lambda\delta_{ij}\varepsilon_{kk} + 2\mu\varepsilon_{ij} \quad \text{or} \quad \sigma_{ij} = \lambda\delta_{ij}tr\varepsilon + 2\mu\varepsilon_{ii} \qquad (6.38)$$

where $tr\varepsilon = \varepsilon_{kk} = \varepsilon_{11} + \varepsilon_{22} + \varepsilon_{33}$ is the volumetric strain of the isotropic material. The normal stresses ($j = i$) are therefore written as:

$$\sigma_{ii} = \lambda tr\varepsilon + 2\mu\varepsilon_{ii} \qquad (6.39)$$

and the tangential stresses are $(j \neq i)$:

$$\sigma_{ij} = 2\mu\varepsilon_{ij} \qquad (6.40)$$

From Equation (6.39) we deduce that:

$$\sigma_{11} + \sigma_{22} + \sigma_{33} = (3\lambda + 2\mu)(\varepsilon_{11} + \varepsilon_{22} + \varepsilon_{33}) \quad (6.41)$$

or

$$tr\sigma = (3\lambda + 2\mu) tr\varepsilon \quad (6.42)$$

The strains as a function of stresses are derived from Equations (6.41) and (6.42). The normal strains are written as:

$$\varepsilon_{ii} = -\left(\lambda / \left[2\mu(3\lambda + 2\mu)\right]\right) tr\sigma + (1/2\mu)\sigma_{ii} \quad (6.43)$$

and the tangential strains are:

$$\varepsilon_{ij} = (1/2\mu)\sigma_{ij} \quad (6.44)$$

6.10.6 Moduli of elasticity

For an isotropic material, the elasticity is expressed in terms of engineering constants: Young's modulus, the Poisson ratio and the shear modulus.

6.10.7 Uniaxial tension or compression

In the case of uniaxial tension or compression in the 1-direction, the stress matrix is:

$$\begin{bmatrix} \sigma_1 & 0 & 0 \\ 0 & 0 & 0 \\ 0 & 0 & 0 \end{bmatrix} \quad (6.45)$$

The tangential strains (ε_{ij}, with $j \neq i$) are zero. The normal strain in the test direction is:

$$\varepsilon_{11} = -\left[\lambda / (2\mu(3\lambda + 2\mu))\right]\sigma_1 + (1/2\mu)\sigma_1 \quad (6.46)$$

Thus using $\sigma_1 = E\varepsilon_1$, where E is Young's modulus:

$$E = \left[\mu(3\lambda + 2\mu)/(\lambda + \mu)\right] \quad (6.47)$$

The transverse strains are similarly:

$$\varepsilon_2 = \varepsilon_3 = -(\lambda / \left[2\mu(3\lambda + 2\mu)\right]\sigma_1 = -\left[\lambda / (2(\lambda + \mu))\right]\varepsilon_1 \quad (6.48)$$

or using $\nu = \lambda / \left[2(\lambda + \mu)\right]$, the *Poisson ratio* of the material:

$$\varepsilon_2 = \varepsilon_3 = -(\nu / E)\sigma_1 = -\nu\varepsilon_1 \quad (6.49)$$

6.10.8 Shear modulus

In the shear experiment, the stress matrix refers to the directions (1,2) and thus:

$$\sigma_{11} = \sigma_{22} = \sigma_{33} = 0, \quad \sigma_{13} = \sigma_{23} = 0, \quad \sigma_{12} = \tau \quad (6.50)$$

The stress–strain relationships are then reduced to:

$$\sigma_{12} = 2\mu\varepsilon_{12} = \mu\gamma_{12} \quad (6.51)$$

where γ_{12} is the shear strain between the two coefficients orthogonal to directions 1 and 2. The coefficient μ is thus the shear modulus, which is usually denoted by G, the other strains being zero.

6.10.9 Spherical compression or tension

In these tests the stresses are applied in all direction simultaneously. If p is the hydrostatic compression $p < 0$ then:

$$\sigma_{ij} = p\delta_{ij} \quad \text{and} \quad \sigma_{11} = \sigma_{22} = \sigma_{33} = 1/3tr\varepsilon \tag{6.52}$$

and the tangential stresses are zero: $\sigma_{12} = \sigma_{13} = \sigma_{23} = 0$. Then the strains are also spherical:

$$\varepsilon_{11} = \varepsilon_{22} = \varepsilon_{33} = 1/3tr\varepsilon \tag{6.53}$$

with:

$$p = (\lambda + 2/3\mu)tr\varepsilon \tag{6.54}$$

The coefficient k the *bulk* modulus is given by:

$$k = \lambda + 2/3\mu \quad \text{and} \quad p = ktr\varepsilon \tag{6.55}$$

6.11 Elastic behaviour of composite materials

The elastic behaviour is often influenced by molecular interactions which have to be scaled to macroscale to describe engineering properties. The above equations are describing the macroscale observations and do not necessarily scale to the molecular interactions. Hence the modulus measured represents an average of all the molecular interactions and can be used in engineering calculations. Thus for a unidirectional composite material (see Figure 6.14), the stiffness and compliance matrices have the form:

$$\begin{bmatrix} \sigma_1 \\ \sigma_2 \\ \sigma_3 \\ \sigma_4 \\ \sigma_5 \\ \sigma_6 \end{bmatrix} = \begin{bmatrix} C_{11} & C_{12} & C_{12} & 0 & 0 & 0 \\ C_{12} & C_{22} & C_{23} & 0 & 0 & 0 \\ C_{13} & C_{23} & C_{22} & 0 & 0 & 0 \\ 0 & 0 & 0 & 1/2(C_{22} - C_{23}) & 0 & 0 \\ 0 & 0 & 0 & 0 & C_{66} & 0 \\ 0 & 0 & 0 & 0 & 0 & C_{66} \end{bmatrix} \begin{bmatrix} \varepsilon_1 \\ \varepsilon_2 \\ \varepsilon_3 \\ \varepsilon_4 \\ \varepsilon_5 \\ \varepsilon_6 \end{bmatrix} \tag{6.56}$$

or the inverse form:

$$\begin{bmatrix} \varepsilon_1 \\ \varepsilon_2 \\ \varepsilon_3 \\ \varepsilon_4 \\ \varepsilon_5 \\ \varepsilon_6 \end{bmatrix} = \begin{bmatrix} S_{11} & S_{12} & S_{12} & 0 & 0 & 0 \\ S_{12} & S_{22} & S_{23} & 0 & 0 & 0 \\ S_{12} & S_{23} & S_{22} & 0 & 0 & 0 \\ 0 & 0 & 0 & 2(S_{22} - S_{23}) & 0 & 0 \\ 0 & 0 & 0 & 0 & S_{66} & 0 \\ 0 & 0 & 0 & 0 & 0 & S_{66} \end{bmatrix} \begin{bmatrix} \sigma_1 \\ \sigma_2 \\ \sigma_3 \\ \sigma_4 \\ \sigma_5 \\ \sigma_6 \end{bmatrix} \tag{6.57}$$

The appropriate constants are therefore averaged or composite values of the matrix and the resin as indicated in Equations (6.13)–(6.20). As indicated above there are bounds on the values of the elasticity moduli which reflect the way in which each element interacts with the other. If we adopt the simplifying hypothesis that the element of the composite shown in Figure 6.15 is

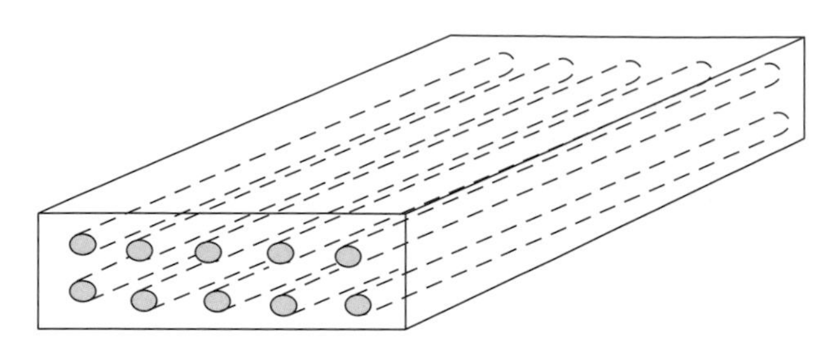

Figure 6.14 Unidirectional composite.

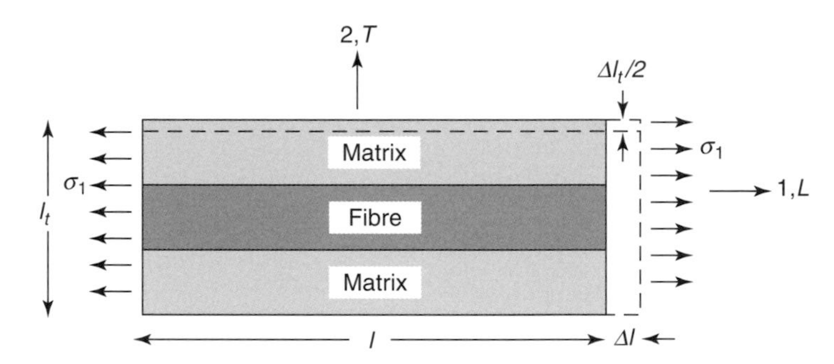

Figure 6.15 Composite element subjected to longitudinal tension.

elongated such that the change in length of both the fibre and resin are identical then if Δl is the extension, the longitudinal strain imposed on the cell is:

$$\varepsilon_1 = \Delta l / l \tag{6.58}$$

where l is the length of the element under consideration. The constraint that the two elements have identical strains imposes:

$$\varepsilon_f = \varepsilon_m = \varepsilon_1 \tag{6.59}$$

where the subscripts f and m refer to the fibre and matrix, respectively. Assuming that both elements are within their linear elastic limits then:

$$\sigma_f = E_f \varepsilon_1 \quad \text{and} \quad \sigma_m = E_m \varepsilon_1 \tag{6.60}$$

The resultant load applied to the composite is:

$$F_1 = \sigma_f S_f + \sigma_m S_m \tag{6.61}$$

where S_f and S_m are, respectively, the cross-sectional areas of the fibre and matrix and F is the load. If S is the cross-sectional area of the average cell, the average stress is:

$$\sigma_1 = F_1 / S \quad \text{and} \quad \sigma_1 = \sigma_f V_f + \sigma_m \left(1 - V_f\right) \tag{6.62}$$

where V_f is the volume fraction of fibre in the composite. This average stress is related to the strain of the element by the longitudinal Young's modulus:

$$\sigma_1 = E_L \varepsilon_1 \tag{6.63}$$

Combining Equations (6.60) with Equation (6.63) yields:

$$E_L = E_f V_F + E_m \left(1 - V_f\right) \tag{6.64}$$

This expression is the *law of mixing* which is often assumed to be valid for the simple composite elements and was introduced in Section 6.5.3. Essentially the law assumes that the modulus varies linearly between the two limits of the resin and fibre.

6.11.1 Transverse Young's modulus

The transverse Young's modulus is determined by loading the sample in the direction normal to the fibres. Such a sample (see Figure 6.16) will be assumed to have a neutral plane through the centre of the slab.

The load F_2 is assumed to impose equal stresses on the fibre and the matrix. The height of the layers which constitute the composite are related to the volume fraction by:

$$V_f = h_f / \left(h_f + h_m\right) \quad \text{and} \quad 1 - V_f = h_m / \left(h_f + h_m\right) \tag{6.65}$$

where h_m and h_f are, respectively, the heights of the matrix and fibre layers. The load F_2 is applied in the transverse direction and will be transmitted to both fibres and matrix:

$$\sigma_m = \sigma_f = \sigma_2 \tag{6.66}$$

Thus the respective strains in fibres and matrix in the transverse direction will be:

$$\varepsilon_f = \sigma_2 / E_f \quad \text{and} \quad \varepsilon_m = \sigma_2 / E_m \tag{6.67}$$

The transverse extension of the element (see Figure 6.16) results from the cumulative elongations in fibre and matrix. Thus:

$$\Delta l_2 = \varepsilon_f h_f + \varepsilon_m h_m \tag{6.68}$$

The transverse strain can be written as:

$$\varepsilon_2 = \Delta l_2 / \left(h_f + h_m\right) = \varepsilon_f h_f / \left(h_f + h_m\right) + \varepsilon_m h_m / \left(h_f + h_m\right) \tag{6.69}$$

which reduces to:

$$\varepsilon_2 = e_f V_f + \varepsilon_m \left(1 - V_f\right) \tag{6.70}$$

This strain is related to the transverse stress applied to the element by the transverse modulus:

$$\sigma_2 = E_T \varepsilon_2 \tag{6.71}$$

Combining Equations (6.67) and (6.71) leads to an expression for the transverse modulus:

$$1 / E_T = V_f / E_f + \left(1 - V_f / E_m\right) \tag{6.72}$$

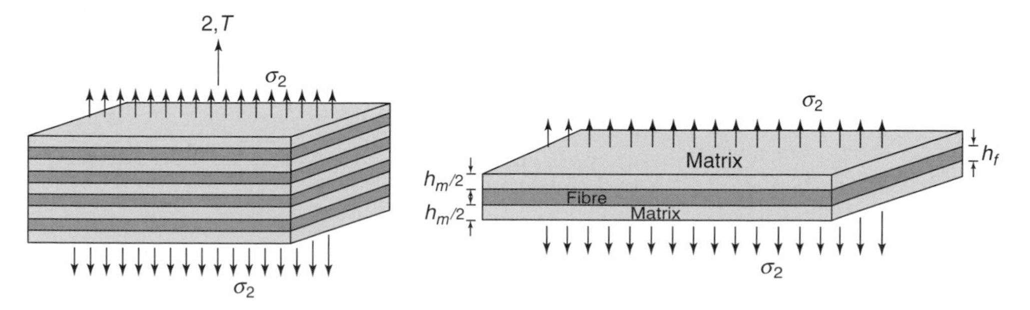

Figure 6.16 Model of layers of a unidirectional composite.

This expression is know as the *inverse law of mixing* and can be written as a dimensionless expression in the form:

$$E_T / E_m = \left(1 + \left(E_m / E_f - 1\right) V_f\right)^{-1} \tag{6.73}$$

Taking reasonable values for the E_f/E_m ratio ~100, it will be seen that for Young's modulus to be twice that of the pure matrix, the volume fraction of fibre has to be 50%. In practice, the fibres will only weakly contribute to the transverse Young's modulus.

6.11.2 Longitudinal Poisson ratio

The longitudinal Poisson ratio, ν_{LT}, determined by a longitudinal tensile measurement, can be obtained using the element described in Figure 6.17. The fibre and matrix are subjected to identical longitudinal deformations such that:

$$\varepsilon_{2m} = -\nu_m \varepsilon_1 \quad \text{and} \quad \varepsilon_{2f} = -\nu_f \varepsilon_1 \tag{6.74}$$

The transverse extension of the elementary cell is:

$$\Delta l_t = -\nu_m \varepsilon_1 h_m - \nu_f \varepsilon_1 h_f \tag{6.75}$$

and the transverse strain is given by:

$$\varepsilon_2 = \Delta l_t / \left(h_f + h_m\right) = -\left[\nu_m \left(1 - V_f\right) + \nu_f V_f\right] \varepsilon_1 \tag{6.76}$$

Thus the expression for the Poisson ratio is:

$$\nu_{LT} = \nu_f V_f + \nu_m \left(1 - V_f\right) \tag{6.77}$$

This is the *law of mixing* for the longitudinal Poisson ratio.

6.11.3 Longitudinal shear modulus

The longitudinal shear modulus is determined by measurement in which the element is subjected to both longitudinal and transverse forces as shown in Figure 6.18. The shear strains are as follows:

$$\gamma_f = \tau / G_f \quad \text{and} \quad \gamma_m = \tau / G_m \tag{6.78}$$

The shear deformations induced in the fibre and the matrix are:

$$\delta_f = h_f \gamma_f \quad \text{and} \quad \delta_m = h_m \gamma_m \tag{6.79}$$

The total deformation of the cell is given by:

$$\delta = \delta_f + \delta_m = h_f \gamma_f + h_m \gamma_m \tag{6.80}$$

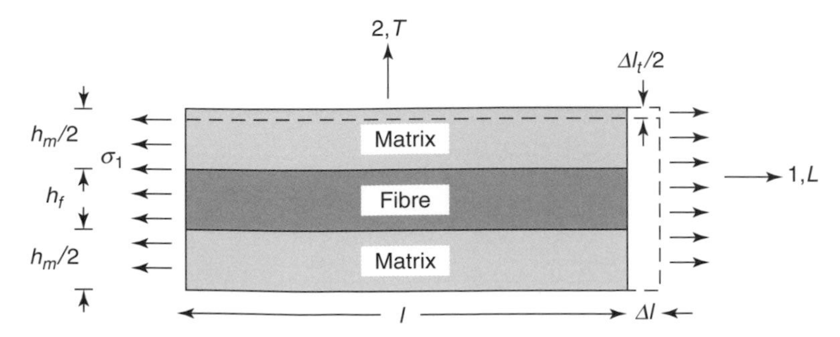

Figure 6.17 Model of layers for longitudinal tensile measurements.

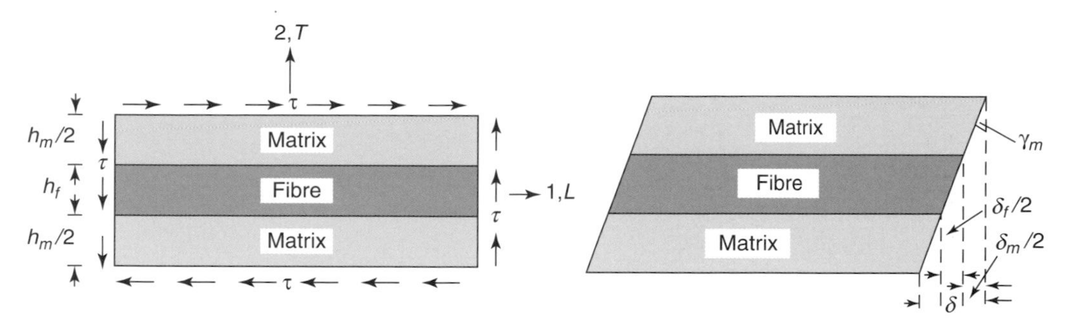

Figure 6.18 Model of layers for longitudinal shear.

and the shear angle of the element is determined by the expression:

$$\gamma = \delta / \left(h_f + h_m\right) = \gamma_f V_f + \gamma_m \left(1 - V_f\right) \tag{6.81}$$

This shear angle is related to the shear stress by the longitudinal shear modulus G_{LT} by:

$$\gamma = \tau / G_{LT} \tag{6.82}$$

Combining Equations (6.80) and (6.82) yields:

$$1 / G_{LT} = V_f / G_f + \left(1 - V_f / G_m\right) \tag{6.83}$$

6.11.4 Halpin–Tsai equations

These are general equations that describe the moduli of a unidirectional composite and are expressed by the law of mixtures for the modulus E_L and the coefficient ν_{LT}:

$$E_L = E_f V_f + E_m \left(1 - V_f\right) \quad \nu_{LT} = \nu_f V_f + \nu_m \left(1 - V_f\right) \tag{6.84}$$

the general expression for the other moduli is:

$$M / M_m = \left(1 + \xi \eta V_f\right) / \left(1 - \eta V_f\right) \tag{6.85}$$

an expression in which the coefficient η is given by:

$$\eta = \left[\left(M_f / M_m\right) - 1\right] / \left[\left(M_f / M_m\right) + \xi\right] \tag{6.86}$$

where M is the modulus under consideration: E_T, G_{LT} or ν_{TT} and M_f are the corresponding modulus of fibres: E_f, G_f or ν_f, and M_m is the modulus of the matrix: E_m, G_m or ν_m. The factor ξ is a measure of the fibre reinforcement and depends on the geometry of the fibres and the type of measurement being performed.

The equations rely on the correct choice of ξ, which is really a scaling factor to achieve a fit between theory and experiment. For instance, in the case of cylindrical fibres distributed in a square arrangement and for a fibre volume fraction of 0.55 $\xi = 2$ for determination of E_T and $\xi = 1$ for the determination of the modulus G_{LT}.

6.12 Orthotropic composites

Laminates consist of layers of unidirectional composite materials or composites reinforced with woven fabric. Usually the woven fabrics are made of unidirectional filaments interlaced at 90°, one in the warp direction the other in the weft direction. These layers have three mutually orthogonal symmetry planes and from the elastic viewpoint they behave like an orthotropic

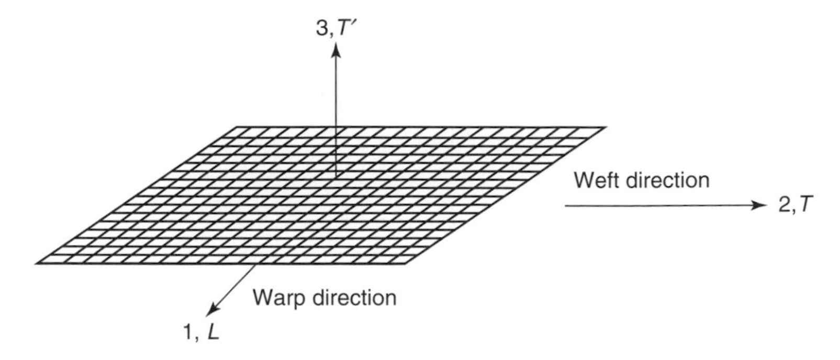

Figure 6.19 Layer of orthotropic composite woven material.

material. The material directions (1,2) will be taken in the warp and weft directions, respectively. These directions will be denoted L and T (see Figure 6.19).

The direction 3 orthogonal to the plane of the layer will be denoted T'. The calculation of the elastic behaviour (stiffness and compliance) uses the matrices summarised in Equations (6.31) etc. In this analysis, the directionality of the reinforcing action of the fibres does not obviously come out of the analysis. It is obvious that in the case of a unidirectional composite, the mechanical properties will be enhanced in the direction of the fibre orientation and would be reduced orthogonally to that direction. In many practical situations, it would be useful to understand the way in which the various elastic constants vary as the direction in which the force is applied varies relative to the orientation of the fibres.

6.12.1 Elasticity relations for an off-axis orientation

The analysis of the off-axis orientation is relatively straightforward and involves the transformation of the stiffness and compliance matrices for a reference state where the 1 direction is parallel to the fibre direction to a new set of axes where 1′ is the result of rotation through some angle θ (see Figure 6.20). The transformation of the stress and strain T matrices involves the relation:

$$\mathbf{C}' = \mathbf{T}_\sigma^{-1}\mathbf{C}\mathbf{T}_\varepsilon \quad \text{and} \quad \mathbf{S}' = \mathbf{T}_\varepsilon^{-1}\mathbf{S}\mathbf{T}_\sigma \tag{6.87}$$

where the transformation involves change of the axes from the system (1,2,3) → (1′,2′,3′) by performing a rotation of angle θ.

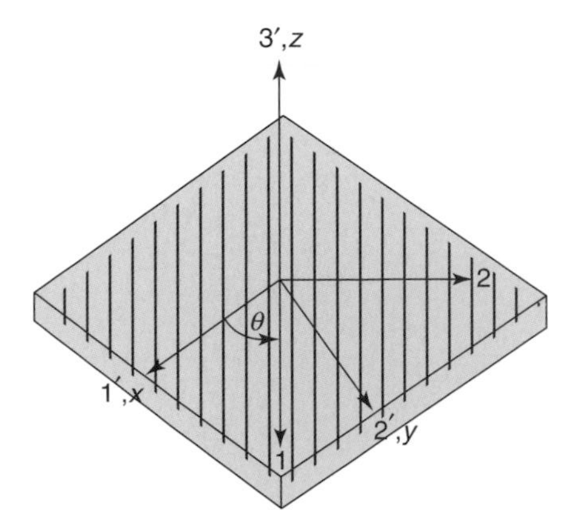

Figure 6.20 Materials directions are selected to be in fibre direction (1,2,3) of a laminate layer and reference system is (1′,2′,3′) = (x,y,z) of laminate.

The stiffness and compliance matrices can be written in the form:

$$\begin{bmatrix} A_{11} & A_{12} & A_{13} & 0 & 0 & A_{16} \\ A_{12} & A_{22} & A_{23} & 0 & 0 & A_{26} \\ A_{13} & A_{23} & A_{33} & 0 & 0 & A_{36} \\ 0 & 0 & 0 & A_{44} & A_{45} & 0 \\ 0 & 0 & 0 & A_{45} & A_{55} & 0 \\ A_{16} & A_{26} & A_{36} & 0 & 0 & A_{66} \end{bmatrix} \quad (6.88)$$

with $A_{ij}=C'_j$ or S'_{ij}. The stiffness constants which are obtained from the application of the transform to the above matrix are summarised in Table 6.4. Comparison of the expressions in Table 6.4 with those for a unidirectional material and to an orthotropic material indicated that these are identical to the terms C'_{ij} or S'_{ij} with I,j = 1,2,6. For example:

Identical relations   Different relations

$$\left[\begin{array}{ccc:cccc} C'_{11} & C'_{12} & C'_{16} & C'_{13} & 0 & 0 \\ C'_{12} & C'_{22} & C'_{26} & C'_{23} & 0 & 0 \\ C'_{16} & C'_{26} & C'_{66} & C'_{36} & 0 & 0 \\ \hdashline C'_{13} & C'_{23} & C'_{36} & C'_{33} & 0 & 0 \\ 0 & 0 & 0 & 0 & C'_{44} & C'_{45} \\ 0 & 0 & 0 & 0 & C'_{45} & C'_{55} \end{array}\right]$$

Table 6.4 Stiffness constants for a unidirectional composite, fibre direction making an angle of θ with reference direction

$C'_{11} = C_{11}\cos^4\theta + C_{22}\sin^4\theta + 2(C_{12}+2C_{66})\sin^2\theta\cos^2\theta$	
$C'_{12} = (C_{11}+C_{22}-4C_{66})\sin^2\theta\cos^2\theta + C_{12}(\sin^4\theta+\cos^4\theta)$	$C'_{15}=0$
$C'_{13} = C_{12}\cos^2\theta + C_{23}\sin^2\theta$	$C'_{14}=0$
$C'_{16} = (C_{11}-C_{12}-2C_{66})\sin\theta\cos^3\theta + (C_{12}-C_{22}+2C_{66})\sin^3\theta\cos\theta$	
$C'_{22} = C_{11}\sin^4\theta + C_{22}\cos^4\theta + 2(C_{12}+2C_{66})\sin^2\theta\cos^2\theta$	$C'_{25}=0$
$C'_{23} = C_{12}\sin^2\theta + C_{23}\cos^2\theta$	$C'_{33}=C_{22}$
$C'_{26} = (C_{11}-C_{12}-2C_{66})\sin^3\theta\cos\theta + (C_{12}+C_{22}+2C_{66})\sin\theta\cos^3\theta$	$C'_{35}=0$
$C'_{36} = (C_{12}-C_{23})\sin\theta\cos\theta$	$C'_{24}=0$
$C'_{44} = 1/2(C_{22}-C_{23})\cos^2\theta + C_{66}\sin^2\theta$	$C'_{46}=0$
$C'_{45} = \left[C_{66} - 1/2(C_{22}-C_{23})\right]\sin\theta\cos\theta$	$C'_{56}=0$
$C'_{55} = 1/2(C_{22}-C_{23})\sin^2\theta + C_{66}\cos^2\theta$	$C'_{34}=0$
$C'_{66} = \left[C_{11}+C_{12}-2(C_{12}+C_{66})\right]\sin^2\theta\cos^2\theta + C_{66}(\sin^4\theta+\cos^4\theta)$	

6.12.2 Off-axis tensile testing

In the case of tensile testing in the x-direction all stresses are zero, except for the stress σ_{xx}. Introducing the compliance constants, the elasticity relations are written as:

$$\varepsilon_{xx} = S'_{11}\sigma_{xx} \quad \varepsilon_{yy} = S'_{12}\sigma_{xx} \quad \varepsilon_{zz} = S'_{13}\sigma_{xx} \quad \gamma_{yz} = \gamma_{xz} = 0 \quad \gamma_{xy} = S'_{16}\sigma_{xx} \tag{6.89}$$

Young's modulus E_x in the x-direction is defined by:

$$E_x = \sigma_{xx} / \varepsilon_{xx} = 1 / S'_{11} \tag{6.90}$$

From the analysis of the off-axial problem:

$$1 / E_x = S_{11}\cos^4\theta + S_{22}\sin^4\theta + \left(2S_{12} + S_{66}\right)\sin^2\theta\cos^2\theta \tag{6.91}$$

This expression can be written by introducing the engineering constants of the unidirectional or orthotropic composite, measured in its material directions:

$$E_L = 1 / S_{11}, \quad \upsilon_{LT} = -S_{12} / S_{11}, \quad E_T = 1 / S_{12}, \quad G_{LT} = 1 / S_{66} \tag{6.92}$$

whence:

$$1 / E_x = 1 / E_L \cos^4\theta + 1 / E_T \sin^4\theta + \left(1 / G_{LT} - 2\nu_{LT} / E_L\right)\sin^2\theta\cos^2\theta \tag{6.93}$$

The normal strains ε_{yy} and ε_{zz} in the transverse directions are related to the normal strain ε_{xx} in the x-direction by the expressions in Equation (6.89) which lead to:

$$\varepsilon_{yy} = S'_{11} / S'_{11}\varepsilon_{xx} \quad \varepsilon_{zz} = S'_{13} / S'_{11}\varepsilon_{xx} \tag{6.94}$$

These relations allow the derivation of the Poisson ratios ν_{xy} and ν_{xz} defined by:

$$\varepsilon_{yy} = -\nu_{xy}\varepsilon_{xx} \quad \varepsilon_{zz} = -\upsilon_{xz}\varepsilon_{xx} \tag{6.95}$$

whence the expression for the Poisson ratios:

$$\nu_{xy} = E_x\left[\nu_{LT} / E_L\left(\cos^4\theta + \sin^4\theta\right) - \left(1 / E_L + 1 / E_T - 1 / G_{LT}\right)\sin^2\theta\cos^2\theta\right] \tag{6.96}$$

$$\nu_{xz} = E_x\left(\nu'_{LT} / E_T\cos^2\theta + \nu'_{TT} / E_T\sin^2\theta\right) \tag{6.97}$$

with $\nu'_{LT} = \nu_{LT}$ for a unidirectional composite. Lastly, Equation (6.89) shows that the off-tension induces an in-phase shear strain γ_{xy}. We can define a *coupling coefficient* $\eta_{xy,x}$, analogous to a Poisson ratio, which relates the in-phase shear strain to the normal strain ε_{xx} in the x-direction by the relation:

$$\gamma_{xy} = -\eta_{xy,x}\varepsilon_{xx} \quad \eta_{xy,x} = -S'_{16}E_x \tag{6.98}$$

The coupling coefficient is finally expressed as follows:

$$\eta_{xy,x} = E_x\left[\left(1 / G_{LT} - 2\nu_{LT} / E_L - 2 / E_L\right)\cos^3\theta\sin\theta + \left(2 / E_T + 2\nu_{LT} / E_L - 1 / G_{LT}\right)\sin^3\theta\cos\theta\right] \tag{6.99}$$

To appreciate the variations of the elasticity moduli, we can plot the graphs of the various functions with θ as a variable (see Figure 6.21) for carbon fibre epoxide composites, with the following constants:

$$E_L = 159\ \text{GPa} \quad E_T = 14\ \text{GPa} \quad \nu_{LT} = 0.32 \quad G_{LT} = 4.8\ \text{GPa} \quad G_{TT'} = 4.3\ \text{GPa}$$

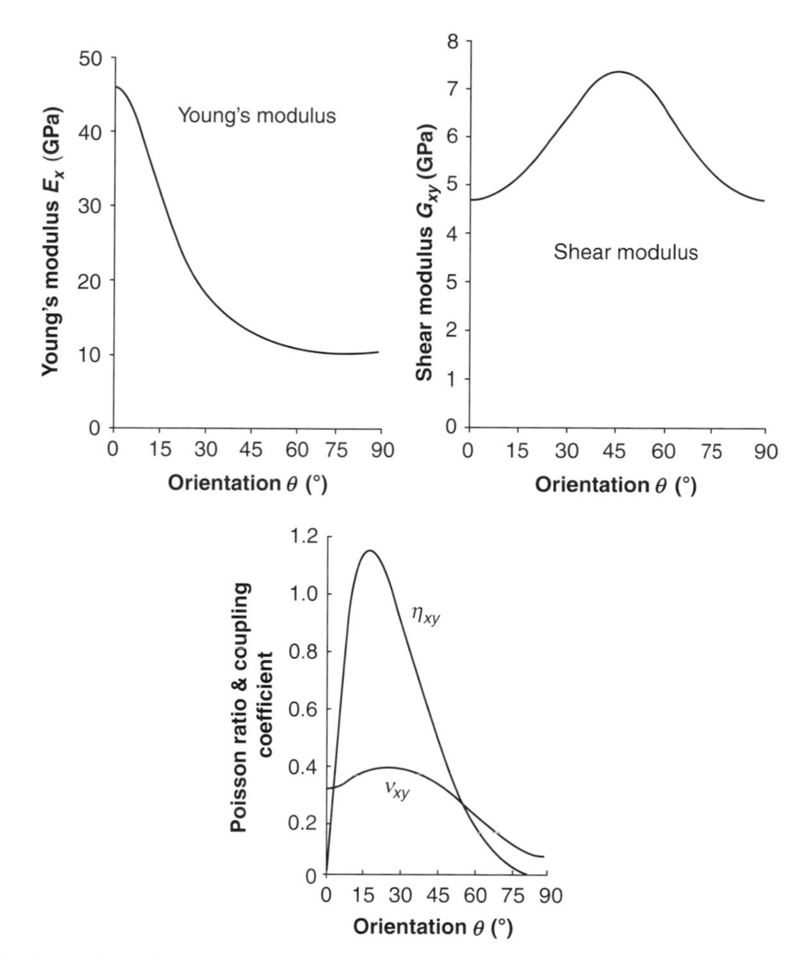

Figure 6.21 Young's modulus, shear modulus, Poisson ratio and coupling coefficient variation with orientation of fibre direction.

Young's modulus in the x-direction, E_x, decreases monotonically from the value E_L for $\theta = 0°$ to the value E_T for $\theta = 90°$. The shear modulus G_{xy} passes through a maximum for $\theta = 45°$, and its variation is symmetric about either side of this value. The Poisson ratio ν_{xy} passes through a maximum for a value of the angle that depends on the composite. The coupling coefficient $\eta_{xy,x}$ is zero for $\theta = 0°$ and $\theta = 90°$, and reaches high values for intermediate angles. The curves show that the extreme values of G_{xy}, ν_{xy}, $\eta_{xy,x}$ are reached for fibre orientations that differ from the directions of the materials. If one searches for extreme values, one finds that E_x passes through a maximum greater than E_L for a value of θ different from 0°. If:

$$G_{LT} > E_L / \left[2\left(1 + \nu_{LT}\right)\right] \tag{6.100}$$

and the modulus E_x passes through a minimum lower than E_T for a value of θ different from 90° then:

$$G_{LT} < E_L / \left[2\left((E_L / E_T) + \nu_{LT}\right)\right] \tag{6.101}$$

6.13 Fracture mechanisms induced in composite materials

The main cause of fracture within a composite is the creation of a 'discontinuity' within the material and more specifically the formation of a 'crack'. The initiation of fracture can be considered as the nucleation of *microcracks* at the microscopic level. The *microcracking* process is the result of the creation of new fracture surfaces at the macroscopic level which are the result of cracks formed at a microscopic level. The final rupture of a unidirectional composite is the result of the accumulation of different elementary mechanisms:

- *Fibre fracture*: Fibres are often formed from high modulus, but brittle, materials which are sensitive to impact damage.
- *Transverse fracture of the matrix*: As indicated above the reinforcement is not necessarily very effective at improving the transverse properties and hence the susceptibility to impact damage.
- *Longitudinal fracture of matrix*: Although the fibres will reinforce in the longitudinal direction, damage can leave the load on the matrix which will often be less able than the fibres to carry load especially when the fibre has been damaged.
- *Loss of integrity of the fibre–resin interface*: If the wetting of the fibre has been inefficient then a weak interface may be created. Under load this may fail and so the load transfer between matrix and fibre will be very inefficient and failure will occur.

Failure is rarely a result of one mechanism but is a consequence of a number of processes acting together. Some of the possible modes of fracture are illustrated in Figure 6.22.

The composite (see Figure 6.22(a)) is subjected to uniaxial tension. Fibres may crack (see Figure 6.22(c)). The fracture of the matrix can produce *transverse cracks* (see Figure 6.22b) when the tension stress σ_m reaches the ultimate stress of the matrix σ_{mu} or by *longitudinal cracking* (see Figure 6.22(d)) when the shear stress τ_m in the matrix reaches the ultimate shear stress, τ_{mu}, generally close to the fibre. This latter mode, called *splitting*, is produced when the debonding stress is greater than the ultimate shear stress of the matrix: $\tau_d < \tau_{mu}$, *debonding fracture* is induced at the

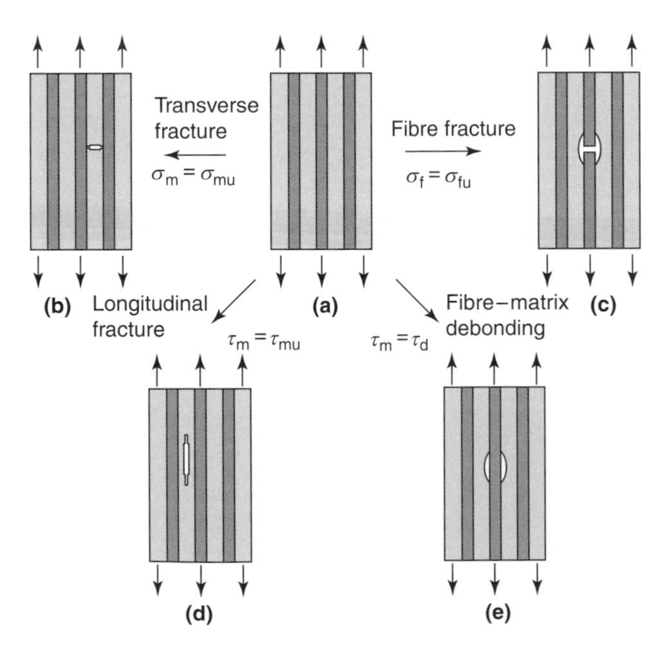

Figure 6.22 Various forms of composite failure.

fibre–matrix interface (see Figure 6.22(e)). The final rupture of a unidirectional fibre composite is the result of the accumulation of these different elementary processes. The initiation and propagation of the fracture depend upon the properties of the fibres and matrix and of the fibre–matrix interface, the volume fraction of fibre and the conditions used to apply the mechanical load.

6.13.1 Unidirectional composite subjected to longitudinal tension

For a unidirectional composite the most common failures are associated with fibre fracture when the fracture strain is less than that of the matrix, $\varepsilon_{fu} < \varepsilon_{mu}$, or the transverse fracture of the matrix,which is the opposite case. For the case where $\varepsilon_{fu} < \varepsilon_{mu}$, the stress–strain curve is shown in Figure 6.23.

Assuming the strains in the fibre and matrix are equal then:

$$\sigma_{cu} = \sigma_{fu} V_f + (\sigma_m)_{\varepsilon_{fu}} (1 - V_f) \tag{6.102}$$

where σ_{cu} is the ultimate stress of the composite, σ_{fu} is the ultimate stress of the fibres and $(\sigma_m)_{\varepsilon_{fu}}$ is the stress in the matrix for a strain equal to the ultimate strain ε_{fu} of the fibres. The stress $(\sigma_m)_{\varepsilon_{fu}}$ is less than the stress σ_{mu} at the matrix facture. Hence:

$$\sigma_{cu} \leq \sigma_{fu} + \sigma_{mu} (1 - V_f) \tag{6.103}$$

Usually, the expression used for the ultimate stress of the composite material is the law of mixing values:

$$\sigma_{cu} = \sigma_{fu} V_f + \sigma_{mu} (1 - V_f) \tag{6.104}$$

where:

$$\sigma_{cu} \approx \sigma_{fu} V_f \tag{6.105}$$

When the strain in the matrix fracture is less than that of the fibres, the ultimate stress of the composite is given by:

$$\sigma_{cu} = (\sigma_f)_{\varepsilon_{mu}} V_f + \sigma_{mu} (1 - V_f) \tag{6.106}$$

where $(\varepsilon_f)_{\varepsilon_{mu}}$ is the stress in the fibre at the instant the matrix fractures. For most matrices, the ultimate strain is of the order of 2–5%; for rigid polyesters it is ~2–5%; for phenolic resins it is ~2–3%;

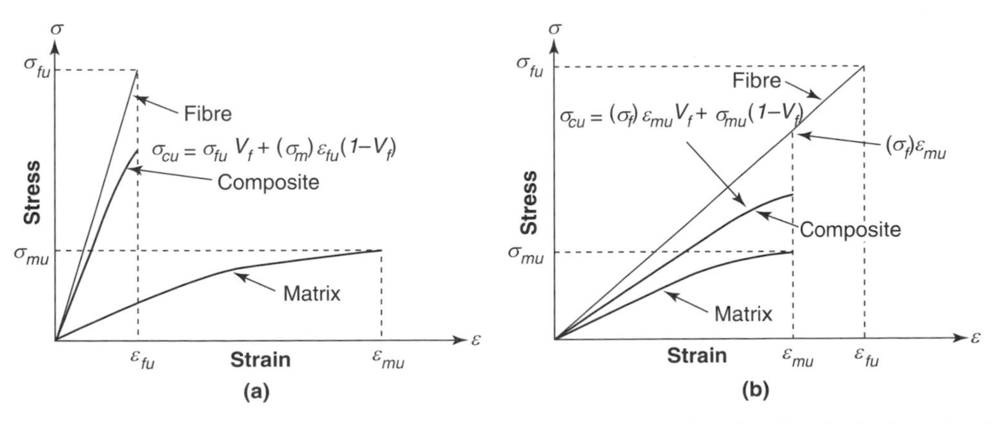

Figure 6.23 Stress–strain plots for a unidirectional fibre composite subjected to longitudinal tension for: (a) $\varepsilon_{fu} < \varepsilon_{mu}$; (b) $\varepsilon_{fu} > \varepsilon_{mu}$.

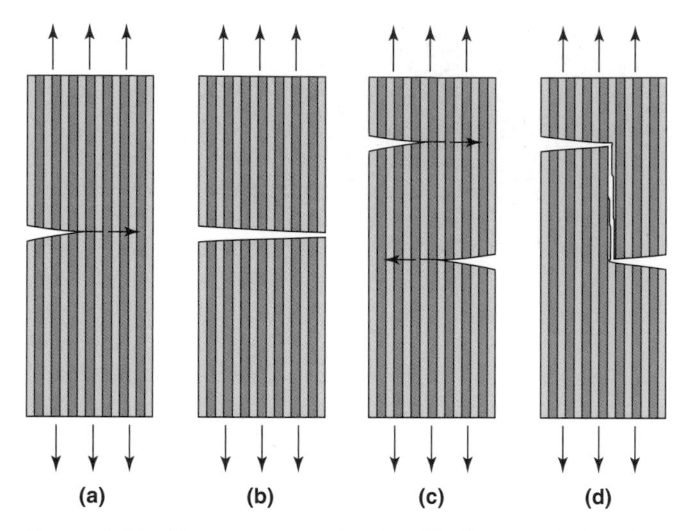

Figure 6.24 Fracture propagation for high fibre–matrix bonding.

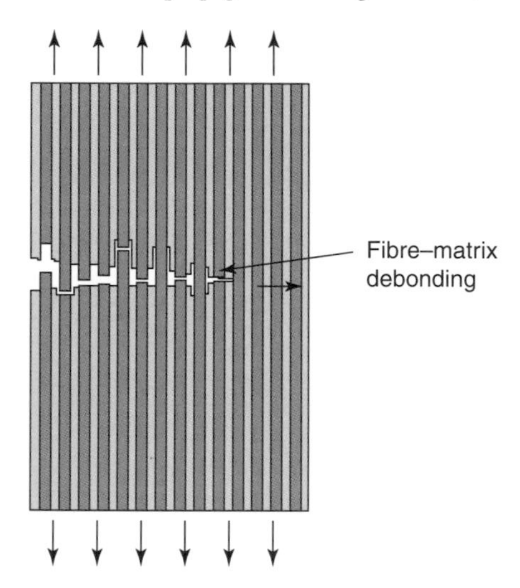

Figure 6.25 Fracture propagation for poor fibre–matrix bonding.

and for epoxide resins it is 2–5%. The initial fracture will propagate according to the nature of the fibre–matrix interface. In the case of high fibre–matrix bonding, the fracture initiated either by fibre fracture or by matrix fracture induces a high stress concentration near the crack tip and this leads to successive propagation of the fracture in the fibres and in the matrix (see Figure 6.24).

The fracture observed is usually brittle (see Figure 6.24(b)). When there is more than one propagating crack, it is possible to observe bridging of the cracks (see Figure 6.24(d)). A characteristic of this type of failure is that the cracks are fairly clean. However, in the case of poor fibre–matrix bonding, the cracks tend to reflect a high degree of fibre pull-out occurring in the crack interface (see Figure 6.25).

6.13.2 Fracture mechanisms induced in composite materials

Figure 6.26 illustrates some of the common forms of failure found in carbon fibre laminates. Under the influence of a tensile stress, failure can occur as a consequence of fibre pull-out

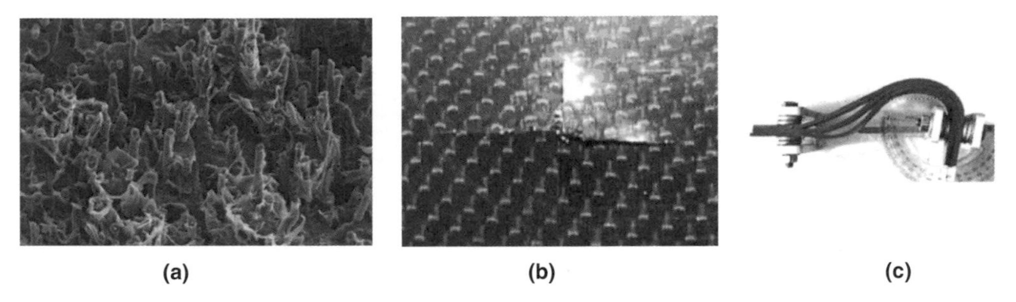

Figure 6.26 Failure in carbon fibre laminates: (a) failure surface of a carbon fibre; (b) failure in woven composite structure; (c) delamination of a laminate structure under buckling load.

and disbonding of fibres from the composite matrix. In a woven fabric (see Figure 6.26(b) a diamond-shaped failure can occur as a consequence of the interaction of the fibre alignment with the applied load. Under the influence of a buckling load delamination of a laminate structure can occur due to interfacial failure between the inert laminar layers (see Figure 6.26(c)).

Note the diamond shape of the failure as a consequence of the interaction of the fibre alignment with the applied load. If a laminate structure delaminates under a buckling load then the failure follows the interplay layer.

6.13.3 Practical composite structures

A simple, unidirectional laminate is rarely used in practice unless strengthening is specifically required in one direction and not in another. In many applications, it is desirable to achieve an isotropic distribution of the moduli. This can be achieved by using layered structures in which the directions of the layers are changed with the composite build up. Some typical examples are shown in Figure 6.27.

6.13.4 Fracture in laminate structures

In the case of laminates, besides the above failure modes there are additional factors to be considered (see Figure 6.28). The fracture processes induced depend on the nature of the constituents,

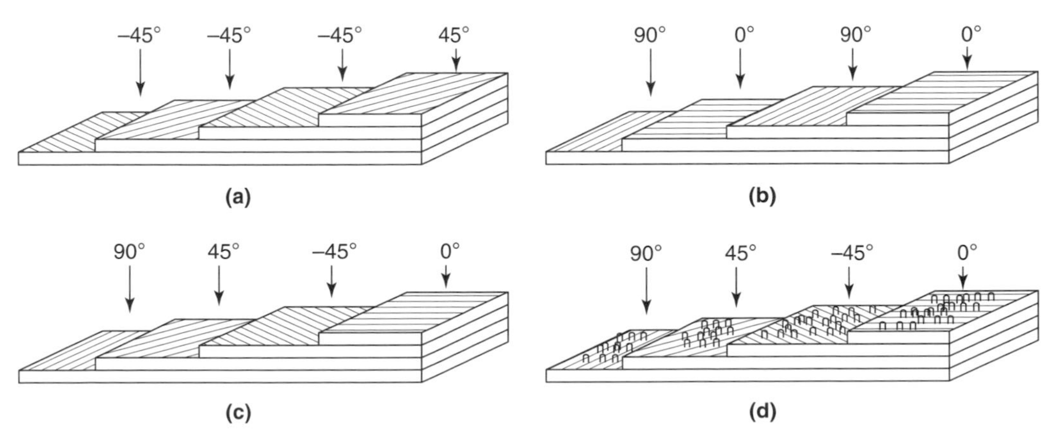

Figure 6.27 Some laminate structures: (a) [45°,−45°]; (b) [0°,90°]; (c) [0°,−450,45°,90°]; (d) [0°, −45°,45°,90°] (knitted).

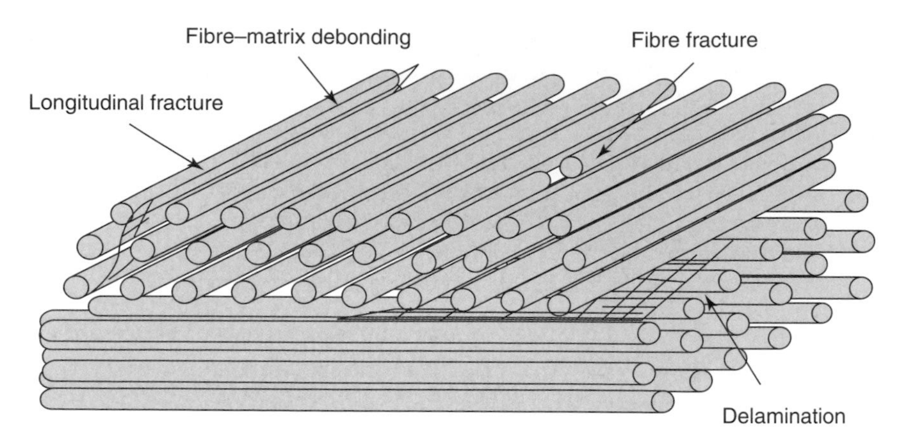

Figure 6.28 Fracture mechanisms in a laminate structure.

the architecture of the laminates and type of mechanical loading to which it is subjected. Some examples of the types of fracture that are possible are considered below.

Cross-ply laminate subjected to tensile loading in the 0° direction

The first process of fracture generally observed is the fracture of the layers with a 90° orientation. Fracture is induced by longitudinal cracking of the matrix or by fracture of the fibre–matrix interface in the 90° layers. This fracture process leads to the development of cracks which are transverse to the direction of the load. As indicated above, this initial fracture in the 90° direction corresponds to transverse cracking of the cross-ply laminate. When the mechanical loading is increased, the number of crack increases up to some saturation level. The transverse cracks induced at the crack tips, between the 90° and 0° layers, induce stress concentrations which ultimately lead to the initiation and then propagation of delamination at the interface between the 90° and 0° layers. Thus the delamination process continues until failure occurs. Figure 6.29 shows a typical failure of a composite laminate.

A ±45° angle-ply laminate subjected to longitudinal tension in the 0° direction

In this situation, longitudinal fracture of the ±45° layers is followed by the delamination of the interface between the layers.

A plate made from [0°, ±45°, 90°] laminate with a hole at its centre and subject to tensile loading in a 0° direction

There are several stages in the cracking process. The first stage is usually longitudinal cracking in the matrix induced in the 90° layers. In the second stage, matrix cracking is initiated in

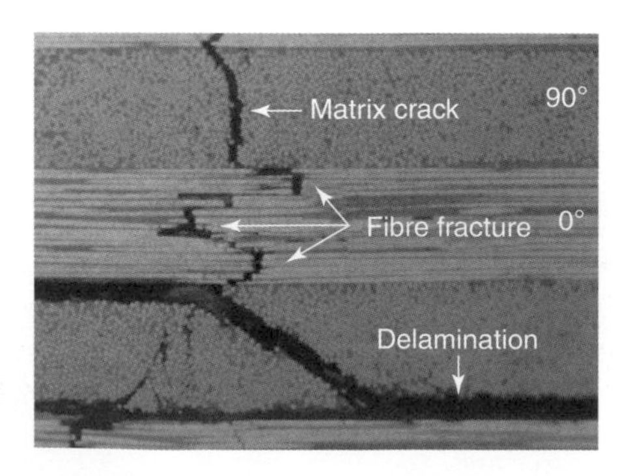

Figure 6.29 Typical cracks in a [0°,90°] laminate layered structure.

the ±45° layers and cracks are observed to propagate in the 90° layers with a limited propagation in the ±45° layers. The third stage is characterised by the initiation of longitudinal matrix cracks emanating from the hole, which propagate in the 0° layers. These cracks generate secondary cracks in the ±45° layers. In the final stage, the longitudinal cracks in the 0° layers induce a delamination of the layers, followed by the fracture of the 90° layers, then of the ±45° layers followed by the fracture of the fibres in the 0° layers, leading to the final fracture of the plate.

Knitted composite structures

The failure in these structures is initially similar to that considered by Hill (1965) except that the delamination process is suppressed by the fibres which are oriented transverse to the layer direction. The propagation and delamination are now controlled by the strength of these knitted fibres. Kinked fibres tend to have a lower strength that axially oriented fibres and hence the enhancement is limited by the nature of the knitting process. The failure of a knitted composite is illustrated in Figure 6.30. The original composite is shown in Figure 6.30(a) and an image of the failed structure is shown in Figure 6.30(b)).

6.13.5 Failure criteria

Whilst it is clearly difficult to predict the failure characteristics of a particular laminate structure, it is appropriate to attempt to develop an approach to estimating the failure behaviour. The failure of a material is a consequence of the irreversible degradation of that material. In general, materials are able to undergo deformation up to their elastic limit, but beyond this stress level, irreversible changes occur which are often associated with the failure mechanism discussed above. For simplicity we will consider the case of a single laminate. For this case there will be a maximum stress which can be accommodated, a maximum strain and an interactive criterion, usually known as the *energy criterion*.

6.13.6 Maximum stress criterion

The maximum tensile and compressive strengths in the longitudinal directions will be designated X_t and X_c, respectively (see Figure 6.31). Likewise the tensile and compressive strengths in the transverse directions will be designated Y_t and Y_c, respectively. The other parameter which needs to be defined is the in-plane shear strength which is designated S. The maximum stress criterion can be written in the form:

$$-X_c < \sigma_L < X_t \quad \text{and} \quad -Y_c < \sigma_T Y_t \quad \text{and} \quad -S < \sigma_{LT} < S \tag{6.107}$$

If the six inequalities are satisfied then the layer does not fail. If any one of the inequalities is not satisfied then the layer will fail, this will initiate the propagation of the crack, and ultimately failure will result.

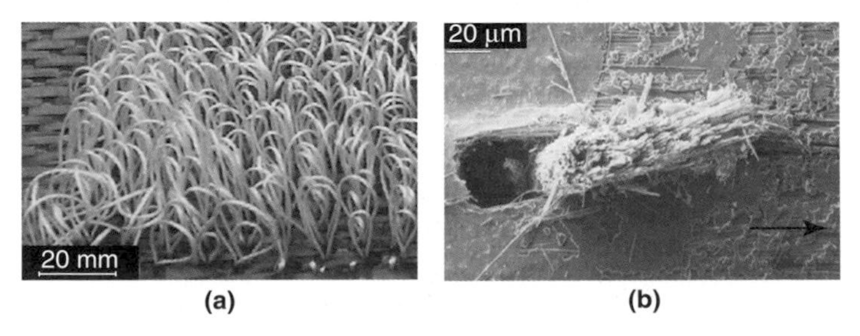

Figure 6.30 Optical images (a) knitted composite; (b) fracture surface showing broken fibre tie strands.

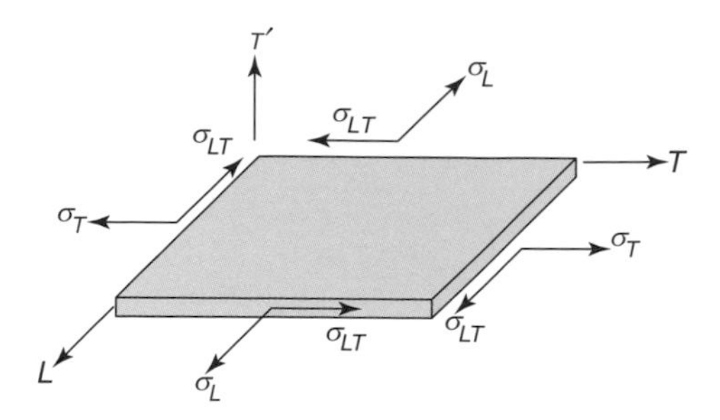

Figure 6.31 Stresses in laminate material in various directions.

6.13.7 Off-axis failure criterion

When a laminate is loaded, it is rare that the tension will coincide with the direction of the fibre. Hence, it is appropriate to consider how the criteria change when the tension is applied at an angle θ. Using the transformations described above, the criteria become:

$$
\begin{aligned}
&-X_c < \sigma_{xx}\cos^2\theta + \sigma_{xy}\sin^2\theta + 2\sigma_{xy}\sin\theta\cos\theta < X_t \\
&-Y_c < \sigma_{xx}\sin^2\theta + \sigma_{yy}\cos^2\theta - 2\sigma_{xy}\sin\theta\cos\theta < Y_t \qquad (6.108)\\
&-S < \left(\sigma_{yy} - \sigma_{xx}\right)\sin\theta\cos\theta + \sigma_{xy}\left(\cos^2\theta - \sin^2\theta\right) < S
\end{aligned}
$$

The criteria can be represented graphically by plotting the maximum value of σ_{xu} of the tensile or compressive stress σ_{xx} for which one of the criteria is reached as a function of the angle θ between the loading direction and the longitudinal, material direction (see Figure 6.32). The ultimate

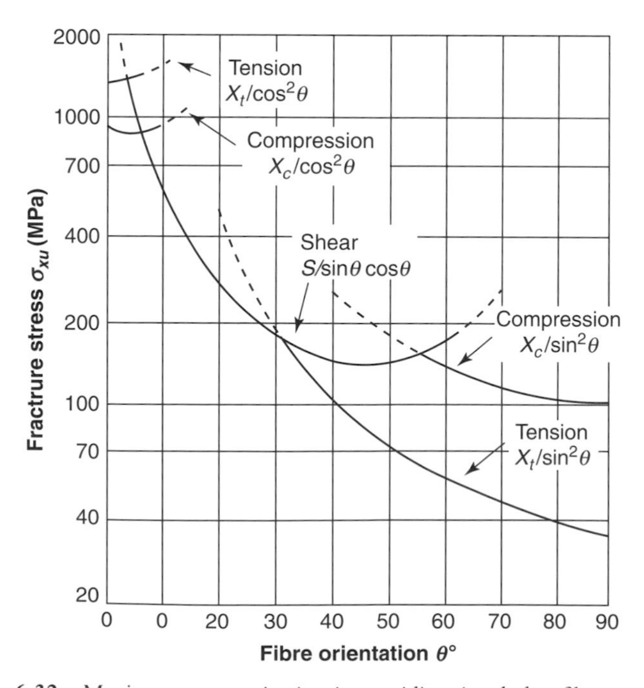

Figure 6.32 Maximum stress criterion in a unidirectional glass fibre composite.

values of the tensile stress σ_{xu} and compressive stress σ_{xu} at failure are correspondingly the smallest values of the following equalities:

$$\sigma_{xu} = X_t / \cos^2\theta, \quad \sigma_{xu} = Y_t / \sin^2\theta, \quad \sigma_{xu} = S / \sin\theta\cos\theta \tag{6.109}$$

$$\sigma_{xu} = X_c / \cos^2\theta, \quad \sigma_{xu} = Y_c / \sin^2\theta, \quad \sigma_{xu} = S / \sin\theta\cos\theta \tag{6.110}$$

The sticking feature which emerges from Figure 6.32 is that whilst the ultimate stress in the materials direction is 1400 MPa, just a misalignment of 25° lowers the ultimate stress to 200 MPa: as the angle is increased, so the dominant mechanism changes.

6.13.8 Interactive criteria

Comparison between experiment and prediction indicates that sometimes there are differences which are not accounted for by the theory. The above criteria do not consider interactions between the modes of fracture: longitudinal, transverse and shear fracture, which are assumed to occur independently. The interactive criteria can be introduced in order to allow for possible interaction between the various modes of fracture. The Von Mises' criterion can be applied to orthotropic materials; it is related to the deformation energy per stored unit volume of strained material and the interactive criteria are sometimes known as *energy criteria*.

6.13.9 Hill's criterion

This criterion (Hill, 1965) can be formulated by stating that the fracture of an anisotropic material does not occur as long as the following inequality is satisfied:

$$F(\sigma_T - \sigma_{T'})^2 + G(\sigma_{T'} - \sigma_L)^2 + H(\sigma_L - \sigma_T)^2 + 2L\sigma_{TT'}^2 + 2M\sigma_{LT}^2 + 2N\sigma_{LT}^2 < 1 \tag{6.111}$$

The fracture of the material will happen when the equality is satisfied:

$$F(\sigma_T - \sigma_{T'})^2 + G(\sigma_{T'} - \sigma_L)^2 + H(\sigma_L - \sigma_T)^2 + 2L\sigma_{TT'}^2 + 2M\sigma_{LT}^2 + 2N\sigma_{LT}^2 = 1 \tag{6.112}$$

and refers to the materials directions (L, T, T'). Equation (6.110) can be rewritten as:

$$\begin{aligned}(G+H)\sigma_L^2 + (F+H)\sigma_T^2 + (F+C)\sigma_{T'}^2 \quad 2H\sigma_L\sigma_T - 2G\sigma_L\sigma_{T'} \\ - 2F\sigma_T\sigma_{T'} + 2M\sigma_{LT'}^2 + 2N\sigma_{LT}^2 - 1\end{aligned} \tag{6.113}$$

The quantities F, G, H, L, M and N are characteristic parameters of the material under consideration and are related to the fracture stresses X, Y and S of the material according to the relations:

$$G + K = 1/X^2 \quad F + H = 1/Y^2 \quad F + G = 1/Z^2 \tag{6.114}$$

and for shear:

$$2N = 1/S_{LT}^2 \quad 2M = 1/S_{LT'}^2 \quad 2L = 1/S_{TT'}^2 \tag{6.115}$$

where S_{LT} and $S_{TT'}$ are the shear strengths in the respective planes (L, T') and (T, T').

Combining the above equations we obtain:

$$\begin{aligned}&(\sigma_L / X)^2 + (\sigma_T / Y)^2 + (\sigma_{T'} / Z)^2 - (1/X^2 + 1/Y^2 + 1/Z^2)\sigma_L\sigma_T \\ &- (1/X^2 + 1/Z^2 - 1/Y^2)\sigma_T\sigma_{T'} + (\sigma_{LT} / S_{LT})^2 + (\sigma_{LT'} / S_{LT'})^2 + (\sigma_{TT'} / S_{TT'})^2 = 1\end{aligned} \tag{6.116}$$

Note that the Hill criterion does not take into account the difference between the behaviour of the material under tension and under compression. In the case of a plane stress in the plane (L, T) of the laminate, we have $\sigma_{T'} = \sigma_{LT'} = \sigma_{TT'} = 0$ and the Hill criterion simplifies to:

$$(\sigma_L / X)^2 + (\sigma_T / Y)^2 - (1/X^2 + 1/Y^2 + 1/Z^2)\sigma_L \sigma_T + (\sigma_{LT} / S_{LT})^2 = 1 \qquad (6.117)$$

Other modifications of these approaches have been produced and lead to the so-called Tsui–Hill criterion and the Hoffman criterion. All these formulations follow the same general approach and attempt to achieve a better fit between theory and experiment. In summary: for typical composite structures the failure is dictated by the way in which the load reaches a critical limit within the particular structure. In the majority of failures the matrix is a determining factor and it is appropriate to attempt to reduce its susceptibility to damage. To illustrate the use of the theory, the calculation of the compliance for a unidirectional off its material direction and the stiffness constant of an orthotropic composite with the material direction making and angle θ with the reference x-direction are presented in the Appendix at the end of this chapter.

Brief summary of chapter

- The addition of fibres and particulate material to a polymer resin will create a material with enhanced mechanical properties.
- The properties of these composite materials can be predicted using simple additive relationships and the anisotropy produced by aligning the fibres can be estimated using simple theory.
- The physical properties of a composite are intimately connected with the stability of the polymer–filler interface.
- Failure in composites can be attributed to stresses exceeding a critical value which is reflected in the strength of the interface.

References and additional reading

Berthelot J.-M. *Composite Materials, Mechanical Behaviour and Structural Analysis*, Springer, New York, 1999.

Christensen R.M. *Mechanics of Composite Materials*, Wiley, New York, 1979.

Hill R. Theory of mechanical properties of fibre strengthened materials I Elastic behaviour. *Journal of the Mechanics and Physics of Solids* 1965, **13** 119.

Appendix

Tables 6.5–6.7 are intended to complement Table 6.4

Table 6.5 Compliance constants for a unidirectional composite off its material direction

$S'_{11} = S_{11}\cos^4\theta + S_{22}\sin^4\theta + 2(S_{12} + 2S_{66})\sin^2\theta\cos^2\theta$	
$S_{12} = (S_{11} + S_{22} - S_{66})\sin^2\theta\cos^2\theta + S_{12}(\sin^4\theta + \cos^4\theta)$	$S'_{15} = 0$
$S'_{13} = S_{12}\cos^2\theta + S_{23}\sin^2\theta$	$S'_{14} = 0$
$S'_{16} = (2(S_{11} - S_{12}) - S_{66})\sin\theta\cos^3\theta + (2(S_{12} - S_{22}) + S_{66})\sin^3\theta\cos\theta$	
$S'_{22} = S_{11}\sin^4\theta + S_{22}\cos^4\theta + (2S_{12} + S_{66})\sin^2\theta\cos^2\theta$	$S'_{25} = 0$
$S'_{23} = S_{12}\sin^2\theta + S_{23}\cos^2\theta$	$S'_{33} = S_{22}$
$S'_{26} = [2(S_{11} - S_{12}) - S_{66}]\sin^3\theta\cos\theta + [2(S_{12} - S_{22}) + S_{66}]\sin\theta\cos^3\theta$	$S'_{35} = 0$
$C'_{36} = (C_{12} - C_{23})\sin\theta\cos\theta$	$S'_{24} = 0$
$S'_{44} = 2(S_{22} - S_{23})\cos^2\theta + S_{66}\sin^2\theta$	$S'_{46} = 0$
$S'_{45} = [S_{66} - 2(S_{22} - S_{23})]\sin\theta\cos\theta$	$S'_{56} = 0$
$S'_{55} = 2(S_{22} - S_{23})\sin^2\theta + S_{66}\cos^2\theta$	$S'_{34} = 0$
$S'_{66} = 2[2(S_{11} + S_{12} - S_{12}) - S_{66}]\sin^2\theta\cos^2\theta + S_{66}(\sin^4\theta + \cos^4\theta)$	

Table 6.6 Stiffness constant of orthotropic composite, material direction 1 making and angle θ with reference x-direction

$C'_{11} = C_{11}\cos^4\theta + C_{22}\sin^4\theta + 2(C_{12} + 2C_{66})\sin^2\theta\cos^2\theta$	
$C_{12} = (C_{11} + C_{22} - 4C_{66})\sin^2\theta\cos^2\theta + C_{12}(\sin^4\theta + \cos^4\theta)$	$C'_{15} = 0$
$C'_{13} = C_{13}\cos^2\theta + C_{23}\sin^2\theta$	$C'_{14} = 0$
$C'_{16} = (C_{11} - C_{12} - 2C_{66})\sin\theta\cos^3\theta + (C_{12} - C_{22} + 2C_{66})\sin^3\theta\cos\theta$	
$C'_{22} = C_{11}\sin^4\theta + C_{22}\cos^4\theta + 2(C_{12} + 2C_{66})\sin^2\theta\cos^2\theta$	$C'_{25} = 0$
$C'_{23} = C_{12}\sin^2\theta + C_{23}\cos^2\theta$	$C'_{33} = C_{22}$
$C'_{26} = (C_{11} - C_{12} - 2C_{66})\sin^3\theta\cos\theta + (C_{12} - C_{22} + 2C_{66})\sin\theta\cos^3\theta$	$C'_{35} = 0$
$C'_{36} = (C_{13} - C_{23})\sin\theta\cos\theta$	$C'_{24} = 0$
$C'_{44} = C_{44}\cos^2\theta + C_{55}\sin^2\theta$	$C'_{46} = 0$
$C'_{45} = (C_{55} - C_{44})\sin\theta\cos\theta$	$C'_{56} = 0$
$C'_{55} = C_{44}\sin^2\theta + C_{66}\cos^2\theta$	$C'_{34} = 0$
$C'_{66} = [C_{11} + C_{12} - 2(C_{12} + C_{66})]\sin^2\theta\cos^2\theta + C_{66}(\sin^4\theta + \cos^4\theta)$	

Table 6.7 Compliance constant of an orthotropic composite, material direction 1 making and angle θ with reference x-direction

$S'_{11} = S_{11}\cos^4\theta + S_{22}\sin^4\theta + (2S_{12} + S_{66})\sin^2\theta\cos^2\theta$	
$S_{12} = (S_{11} + S_{22} - S_{66})\sin^2\theta\cos^2\theta + S_{12}(\sin^4\theta + \cos^4\theta)$	$S'_{15} = 0$
$S'_{13} = S_{12}\cos^2\theta + S_{23}\sin^2\theta$	$S'_4 = 0$
$S'_{16} = (2(S_{11} - S_{12}) - S_{66})\sin\theta\cos^3\theta + (2(S_{12} - S_{22}) + S_{66})\sin^3\theta\cos\theta$	
$S'_{22} = S_{11}\sin^4\theta + S_{22}\cos^4\theta + (2S_{12} + S_{66})\sin^2\theta\cos^2\theta$	$S'_{25} = 0$
$S'_{23} = S_{12}\sin^2\theta + S_{23}\cos^2\theta$	$S'_{33} = S_{22}$
$S'_{26} = [2(S_{11} - S_{12}) - S_{66}]\sin^3\theta\cos\theta + [2(S_{12} - S_{22}) + S_{66}]\sin\theta\cos^3\theta$	$S'_{35} = 0$
$C'_{36} = (C_{12} - C_{23})\sin\theta\cos\theta$	$S'_{24} = 0$
$S'_{44} = S_{44}\cos^2\theta + S_{55}\sin^2\theta$	$S'_{46} = 0$
$S'_{45} = (S_{55} - S_{44})\sin\theta\cos\theta$	$S'_{56} = 0$
$S'_{55} = 2(S_{22} - S_{23})\sin^2\theta + S_{66}\cos^2\theta$	$S'_{34} = 0$
$S'_{66} = 2[2(S_{11} + S_{12} - S_{12}) - S_{66}]\sin^2\theta\cos^2\theta + S_{66}(\sin^4\theta + \cos^4\theta)$	

7

Case studies

7.1 Introduction

In this chapter a series of topics of importance to engineers will be considered as case studies. The topics to be considered are:

- environmental stress cracking
- energy absorption and vibration damping
- adhesion and adhesives
- polymers in corrosion protection
- gas diffusion through polymer matrices
- selection of polymeric materials for particular applications

7.2 Environmental stress cracking: some case studies

Environmental stress cracking (ESC) is a major cause of failure in plastics (the term 'crazing' can be used interchangeably with the term 'cracking'). The website of the Rubber and Plastics Research Association (RAPRA) discusses ESC in great detail and has been used as a source of some of the data presented below (see http://www.rapra.net/search.asp?searchterm=environmentla+stress+cracking).

7.2.1 Failure in 'sight glasses'

Sight glasses are the transparent tubes used to observe the flow of fluids. Traditionally these tubes have been made from glass; however, the better impact strength and ease of fabrication have made plastics an attractive alternative, with polymethylmethacrylate (PMMA), styrene acrylonitrile (SAN) or polycarbonate being widely used. Polar polymers have the ability to absorb small amounts of water, leading to plasticisation. A sight glass constructed from PMMA and subjected to water at a temperature of 50°C and a maximum stress of 5 MPa failed after eight years of service, whereas it might have been expected to have had a longer service life. Failure of the sight glass is usually accompanied by a loss of optical clarity as a consequence of crazing being generated on the surface exposed to the water.

Crazing in PMMA can occur as a result of sustained stress, however, the presence of certain fluids hastens the arrival of ESC. The temperature of the fluid is the critical factor in determining when failure occurs. In the case study quoted by RAPRA, the water in a boiler was replaced by a corrosion-inhibiting water mixture. The inhibitor was able to interact aggressively with the PMMA and subsequent tests showed that it significantly shortened the life of the sight glass (see Table 7.1).

The data obtained at 50°C and a constant stress rate of 4 MPa h^{-1} indicate a significant reduction in the critical time to failure. The sight glasses were not used very often and a small change in opacity had gone unnoticed, leading to a dramatic failure when the glass broke. The water–inhibitor mixture is a weak electrolyte and so can swell the PMMA more effectively than water can. It should have been many years before the sight glass would have cracked under a stress of 5 MPa at 50°C in water.

Table 7.1 Critical conditions for craze initiation in PMMA at 50°C and 4 MPa h^{-1} demonstrating influence of corrosion inhibitor

Exposure fluid	Critical strain (%)	Critical time (s)	Critical stress (MPa)
Water	0.85	18,180	20.2
Water/inhibitor	0.62	11,970	13.3

7.2.2 High density polyethylene blow moulded containers

Polyethylene (PE) is a paracrystalline material, the crystals being held together by small numbers of tie molecules (see Section 3.4). The ability of 'solvents' to swell the solid will depend on the number of tie molecules between the crystallites.

7.2.2.1 *Agrochemical containers*

HDPE blow-moulded containers are often used for the short-term storage of agrochemicals. Containers are usually stacked in pallets for several months in warehouses prior to distribution. After a couple of weeks a number of containers were found to have leaked. The failures appeared as microcracking and resulted in the containers buckling and ultimately leaking (see Figure 7.1). The 'linear and stepped' appearance of the microcracks is strongly indicative of ESC.

Microscopic examination of the inside of the container revealed the presence of extensive microcracking. PE had been used to manufacture these containers for many years and the failure was unexpected. The ESC resistance measurements showed that the material that had been used was prone to failure. Consultation with the container supplier indicated that the blow moulder had changed the grade of PE that they had used to produce these containers. The grade used was easier to process (see Section 5.4.19), having a narrow distribution lower molar mass than that usually used, which has a broader molar mass distribution and contains some higher molar mass polymer chains. The lack of the longer chain polymers in this new material resulted in fewer tie molecules between crystallites and a greater susceptibility to ESC by the agrochemicals.

7.2.2.2 *Blow-moulded containers for alcohol*

Blow-moulded HDPE containers are widely used for the storage and transport of industrial alcohols. Highly polar alcohols are not expected to have any significant interaction with nonpolar polymers, however, a batch of PE containers exhibited ESC after three months (see Figure 7.2).

The containers were usually stored in an upright position and cracks in the screw-caps went undetected until an accident caused a pallet to fall on its side and alcohol was released.

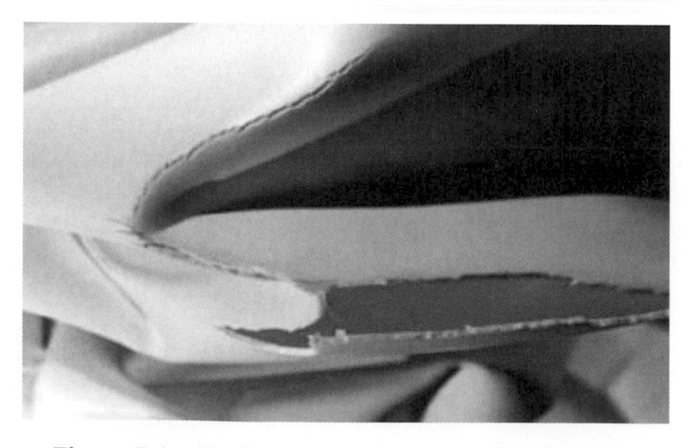

Figure 7.1 Cracking on tensile surface of buckled fold.

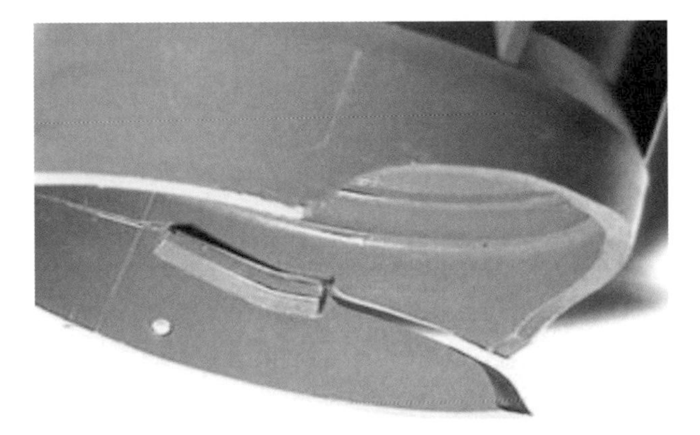

Figure 7.2 Cracked HDPE screw-cap.

Approximately 60% of the screw-caps had failed (see Figure 7.2). All the caps showed failures initiating close to one or both of the tightening lugs. The screw-cap was sealed to the container using an O-ring coated with silicone grease. The O-ring maintained its seal under compression and stopped the escape of alcohol vapour. Silicone grease is a mild ESC agent for HDPE, causing slow crack growth at stresses that approach the yield stress of the material. Comparing the condition of fully torqued screw-caps after three months in the presence and absence of the grease indicated that this was the cause of the failure.

7.2.3 Cracks in communication wiring

Low density cellular polyethylene with a dielectric constant of ~1.4 is used to insulate high-speed communication wire and cable. Although it had performed perfectly for two years, a problem was found with the intranet system that had been installed in a large office complex. The bundles of internet wires exhibited cracking of the insulation (see Figure 7.3). Inspection of the inside surfaces of the conduit and the outer wires of the bundles indicated a liberal coating of silicone oil. The oil had been used to ease the threading of the wire bundles into difficult-to-access parts of the conduit system.

7.2.4 Failure of polycarbonate electronic housing

An instrument used for remote monitoring of telecommunications systems was fabricated from UV stabilised polycarbonate. The instrument was designed to be a leak-tight container which

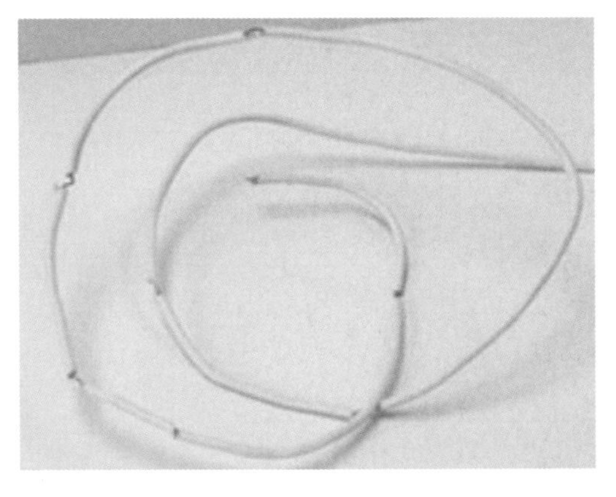

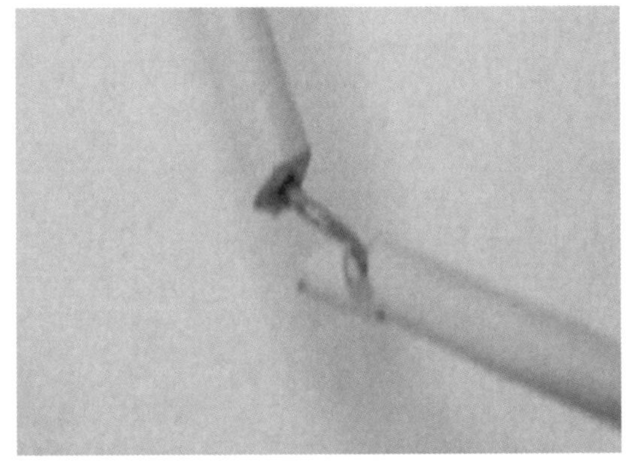

Figure 7.3 Cracking of PE wire insulation.

would withstand rainwater ingress and a wide range of climatic conditions. The first failure occurred in an arid tropical environment following a rare rainstorm. Although the instrument had been in service for about two years, records revealed that it had only been exposed to direct sunlight for about 10 days, allowing UV degradation to be eliminated as a contributory factor to the failure. The container had failed by brittle fracture, adjacent to the location of a 'company logo' which had been film bonded to the surface. Investigation revealed that the plastic had extensive microcracking beneath the film and that PVC plasticised with dioctylphthalate (DOP), which is a powerful ESC agent for polycarbonate, had been used as the adhesive film. The linear and stepped nature of the microcracks, without discolouration or other symptoms of material degradation, is characteristic of ESC. Moulded-in stress in the instrument case provided the driving force and the DOP was sufficiently mobile to migrate from the film to the surface of the polycarbonate.

7.2.5 Environmental stress cracking in polyethylene

Polyethylene is considered to be inert to all liquids, however, it is sensitive to ESC when subjected to a number of liquids. The effect of melt index, molecular-weight distribution, crystallinity and orientation all can have an effect on ESC.

7.2.6 Model for environmental stress cracking failure

In order to understand ESC at a molecular level, one must visualise the structural variables that directly influence cracking. In the case of polyethylene the semicrystalline nature of the solid and distribution of the tie molecules are the controlling factors in determining the ability of a small molecule to swell the solid. In the case of amorphous polymers the ability to swell the matrix will depend on the extent to which the polymer chains are physically entangled. When solvent molecules enter the polymer, swelling will occur but it is controlled by the morphology. In semicrystalline polymers, the lamellae are held together by tie molecules which are a result of entanglements in the melt phase resisting the incorporation of individual polymer chains into the same lamellae. Solvent entering between lamellae is restricted by the tie molecules and their integrity is critical in order to avoid creep fracture.

7.2.6.1 *Ductile failure*

At high stresses, the chains in semicrystalline polymers can be pulled out of the lamellae and so reduce the 'local' stresses on individual polymer chains. The matrix will undergo deformation which is aided by the presence of the plasticising liquid. The plasticisation will aid relaxation of the stresses in amorphous polymers. The result is a slow elongation of the sample but without failure. However, a point will be reached where yielding will occur and the sample will fail. These processes are illustrated in Figure 7.4.

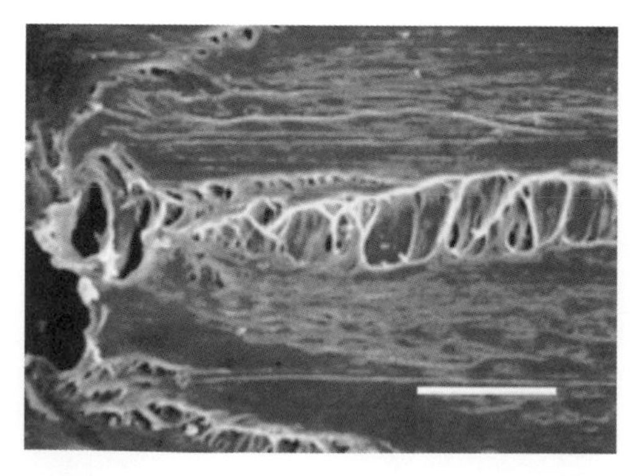

Figure 7.4 Structure of polymer around crack tip for ductile failure showing elongation of fibre structure.

7.2.6.2 *Brittle failure*

Brittle-type creep fracture is a consequence of slow crack growth taking place over longer periods of time at lower stress levels than are required for ductile deformation. The force applied to the material is insufficient to achieve large-scale fibre pullout from the lamellae. However, after a long time at a low-level stress, tie molecules can begin to untangle and relax with the result that only a small number of tie molecules are supporting the load. At a certain point, this number of molecules becomes insufficient and brittle failure is observed. Immersion in a surfactant accelerates the process of brittle failure as it helps the ingress of the fluid into the polymer and aids the swelling of the matrix. Interlamellar failure is a rate-dependent process and is influenced by temperature. The higher the temperature, the faster the polymer chains can relax, and the shorter the time to creep failure.

7.2.6.3 *Structure–property relationships*

Polyethylene materials containing relatively few tie molecules are more susceptible to the various modes of brittle failure. However, if the proportion of tie molecules to crystalline molecules is too high, the materials will display high ductility, but also very low stiffness as the crystallite growth has been inhibited by the tie chains. There are several important molecular parameters for optimising the environmental stress cracking resistance (ESCR) in polymers. These parameters include:

- *Molecular weight*: The higher the molecular weight, the longer the polymer chains, the greater the number of tie molecules. Because polymers are polydisperse, the entire molecular weight distribution of the material is a critical factor.
- *Comonomer content*: In the case of semicrystalline polymers a small amount of a comonomer, such as 1-butene or 1-hexene, in the case of MDPE and LDPE tends to inhibit crystallinity. A higher comonomer concentration will result in better resistance to brittle fracture, because the portions of polymer chains with the longer branches (that is 1-hexene or longer) do not enter the tightly packed lamellar lattice. Thus, the chains with these branches increase the number of intercrystalline tie molecules.
- *Density/degree of crystallinity*: The more crystalline the material, the fewer amorphous intercrystalline tie molecules that hold it together and hence the poorer the ESCR.
- *Lamellar orientation*: If the lamellae are predominantly oriented perpendicular to the tensile stress direction, there is a greater susceptible to interlamellar failure than if they were parallel to the stress. This effect would be minimised in the case of a spherulitic polyethylene, since the lamellae in spherulites are oriented radially. Some of the metallocene polyethylene polymers favour lamellar growth and exhibit a greater susceptibility to ESC.

7.2.7 How is resistance to environmental stress cracking assessed?

The simplest test method is based on ASTM Standard test D1693, which characterises ESC by the time to failure of 50% of the specimens tested, the sample being held under constant stress. ESCR testing is performed by slowly bending the test specimens and placing them in a holding clamp (see Figure 7.5). The clamp and specimens are then placed in a test tube and immersed in

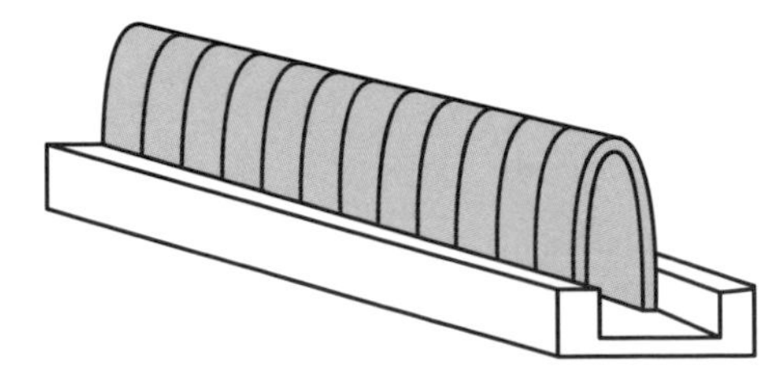

Figure 7.5 Clamp loaded with samples for ESC testing.

a specified reagent. The test tube is sealed and placed in a constant-temperature bath. Multiple test specimens are tested at one time. Specimens are inspected periodically for failure. Cracks generally develop at the notch, perpendicular to the notch, and run to the edge of the specimen. Any cracks constitute failure, not just cracks that reach the edge of the specimen. Cracks sometimes appear beneath the surface and are visible as surface depressions. If a depression develops into a surface crack, the time at which the depression is noted is taken as the time of failure.

A limitation of ESCR testing is the inability to isolate the yield stress, independent of the other mechanical properties of the polymer. In the test described in Figure 7.5, the specimen is under a constant strain. However, the stiffness will vary from sample to sample and the question arises as to whether the differences observed are real or reflect the higher stress levels in the stiffer specimens. The question is whether a material fails quickly as a result of its low ESCR, or because it has low stiffness making it more deformable under constant load. In the case of HDPE and LDPE, the stiffer material fails faster than more flexible material. On the other hand, the same samples exhibit the opposite effects in a constant-tensile-load ESCR test, in which notched strips are subjected to a constant tensile load. The reason for the difference in failure times between the two tests becomes clearer when one considers the influence of mechanical properties on a material's response to a load. Because of its relative stiffness, HDPE is stressed close to or beyond the yield point in a constant-strain test; cracking takes place in that portion of the bend in which the material is just below the yield strain. In contrast, LDPE does not come close to its yield point under the same test conditions. However, the opposite occurs with the constant-tensile-load test, in which LDPE reaches its yield point more readily than HDPE, and is thus more susceptible to failure.

Neither a constant-stress nor a constant-strain test provides a good criterion for discerning ESCR. Rather, the parameter that should be examined as the ordinate of an environmental stress cracking plot in a constant-load situation is the percentage of yield stress or reduced stress. For a more realistic comparison of polyethylene, this percentage should be kept constant, although the actual stress may differ widely among specimens. The practical test involves studying the creep rupture for samples subjected to a constant load and stored in an environment for a particular time (see Figure 7.6).

In this experiment dumbbells cut from the material are subjected to a constant load and the time at which failure occurs is recorded. A set of dumbbells after testing are shown schematically in Figure 7.7. This experiment is performed at constant temperature and the samples are immersed in the fluid of

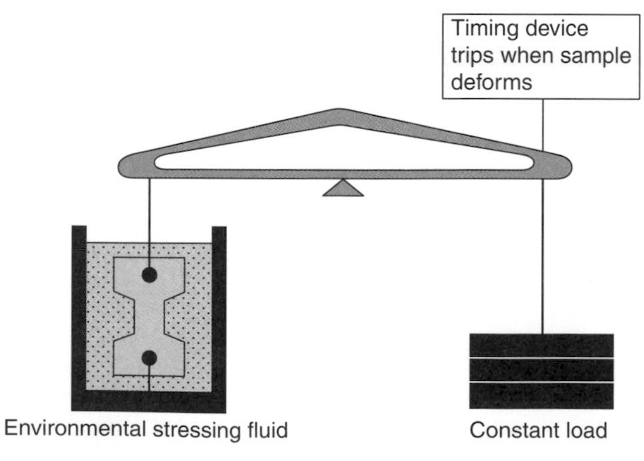

Figure 7.6 ESC test apparatus for studies at constant load. Dumbbell is immersed in fluid and kept at constant temperature.

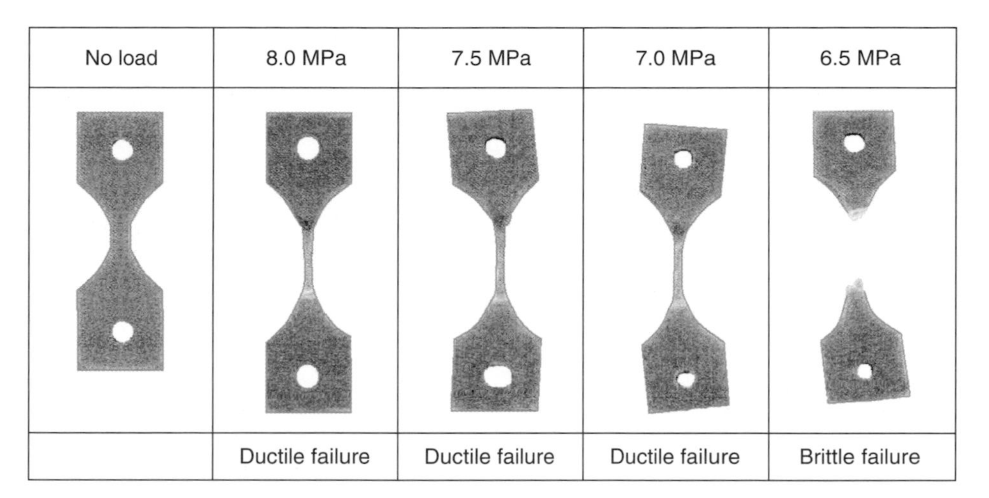

Figure 7.7 Original dumbbell and failed samples after different exposure times.

interest. Elevated temperatures are often used to accelerate ageing. At higher temperatures, the fluid will diffuse into the polymer more quickly and the creep rupture will occur faster. The time dependence of the transition from ductile to brittle failure can often be observed to follow an approximately Arrhenius-type behaviour. A set of dumbbells after testing is shown in Figure 7.7. The dumbbells are shaped to give a defined cross-section and ensure that the failure will occur in the tapered section.

The samples fail initially by ductile failure, the polymers undergoing significant elongation before they break. The pictures indicate that at the high loads the dumbbells are failing by ductile fracture. This type of failure is observed at 8.0 MPa and down to 7.0 MPa. At a load of approximately 6.5 MPa the mechanism of failure changes to brittle fracture and stays that way for all lower loads. Although the failure is brittle, the time to failure is significantly longer than for the ductile failure observed at shorter times and higher loads. The key parameter in constant-tensile-load testing is the so-called 'ductile brittle transition', corresponding to that area of the stress–time plot in which a downward inflection, or 'knee', becomes evident. This inflection point represents the region of the curve in which ductile-creep-type deformation ends and brittle-stress-cracking behaviour begins. The later this transition occurs, the better the resistance of the material to ESC. Data for some typical polyethylene materials is given in Figure 7.8.

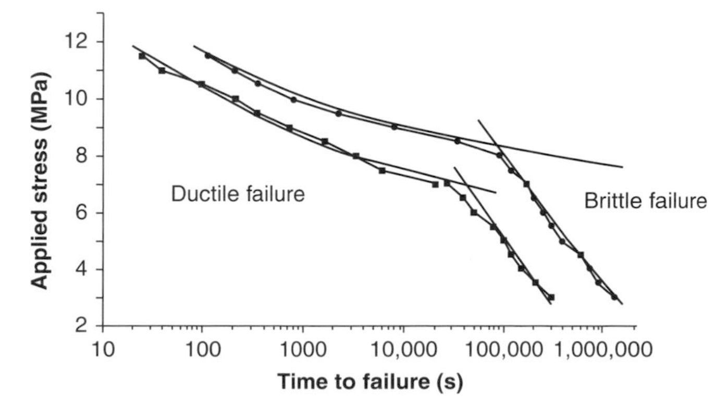

Figure 7.8 Typical plot for ESC data on PE aged in Igepal (surfactant).

When polyethylene is tested under constant load in the presence of a surfactant, it can be shown that the time to failure represents an acceleration of the intrinsic brittle-failure process that would take place in air at a later time. In general, the better a polymer is resistant to ESCR, the better its resistance to slow crack growth, in the absence of any obvious accelerating factors. The fracture surfaces of stress-crack failures in a surfactant and in the absence of any accelerating environment are very similar. ESCR is extremely sensitive to temperature. Crack-growth data generated on polyethylene suggest that for every 7°C increase in temperature, the crack-growth rate is doubled. Increased temperature can therefore be regarded as a type of crack-growth accelerator.

The problem of ESCR is commonly encountered when the wrong grade of polymer is used for an application. In general, if ESCR is likely to be an issue and the material is paracrystalline then it is important to have sufficient tie molecules between the crystallites to counter the swelling of the polymer. In the case of polyethylene, material to be used for gas pipeline distribution applications has to have a certain proportion of a higher molar mass component or else the pipes are prone to exhibit ESCR, which can be explosively dangerous!

7.3 Energy absorption and vibration damping

Polymers are increasingly used in a variety of applications in order to reduce vibration in automobiles, ships, and other situations where machinery produces vibration or there are natural causes of vibration which should be eliminated from the environment in question. Acoustic engineers have anechoic chambers to allow accurate measurement of sound energy etc. Energy can be dissipated or absorbed through a variety of mechanisms. In an anechoic chamber, the walls are covered with shaped hollow wedges and sound energy is dissipated by scattering and absorption. In other situations, it is desirable to be able to absorb the energy by decoupling the vibrating element from the structure. In the simplest solution, the damping is achieved by a spring–dashpot combination, rather like that discussed in relation to viscoelasticity (see Section 2.6). Detailed examination of the way in which such structures/materials work indicates that they are most efficient when operated

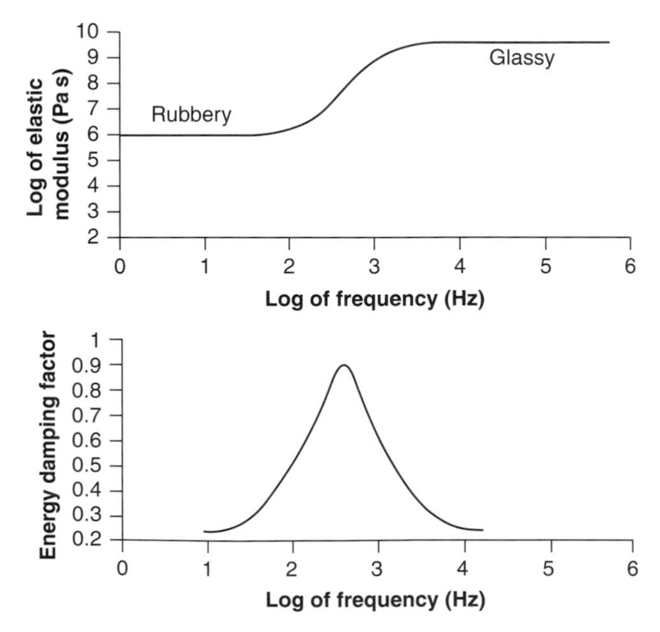

Figure 7.9 Idealised plot for frequency response of a typical elastomer.

at resonance. If we explore how a viscoelastic material will respond to increasing frequency, we find the following type of response if the material is above its glass–rubber transition (see Figure 7.9).

At low frequency, there is sufficient time for the material to respond to the vibration and the effective modulus will have a value which is equal to that of a rubber (~10^6 Pa s). As the frequency is increased, the ability of the material to respond will be reduced until at very high frequency the chains can no longer respond and the material behaves like a glass and has an effective modulus of the order of 10^{10} Pa s. Stretching and compressing the chains in the region where the time scale of the vibration matches that of the polymer motion means that part of the elastic response is not recovered and energy is absorbed and given out as heat. If a rubber band is quickly stretched, an increase in its temperature can be felt. This energy absorption will be a maximum at the peak of the loss process and will fall off either side. In a practical application, the temperature at which the 'damper' may be required to work may change. A rubber mount for an engine may have to operate between −20°C and 40°C. The curve shown in Figure 7.9 is a function of temperature, the location of the peak reflecting the Williams–Landel–Ferry nature of the process producing the change in energy. As a consequence, it is possible to find that a material which is a good energy damper at one temperature may be less good or even ineffective at another. Ideally, one often requires a material which will absorb energy over a broad temperature frequency range and a broad loss curve is desirable.

7.3.1 Which materials are useful for energy damping?

In general, elastomers are ideal for damping applications. Rubbery behaviour can be observed in a range of materials and especially in block copolymers of styrene with butadiene or isoprene and polyurethanes. In Section 3.12, it was pointed out that both the latter systems exhibit significant ranges of temperature in which the modulus and energy dissipation factor are essentially independent of temperature. With a judicious selection of materials, it is possible to fit the required characteristics.

7.3.2 Rubber balls and tyres

Tyres need to be able to absorb energy and carry a load, thus the selection of materials to form an automobile tyre or conveyor belt is an interesting challenge. In the case of a rubber, the energy dissipation is related to the work which is required to rotate the segments of the polymer chain. As a rule of thumb, we can say that rubbers with large energy barriers will require more work and absorb more energy than chains with low energy barriers. Three commonly used rubbers are: polybutadiene, natural rubber and butyl rubber (see Figure 7.10).

The unhindered nature of the structure of polybutadiene allows it to retain its ability to undergo backbone rotation to lower temperatures compared with cis-1,4 polyisoprene, which in turn is more flexible than polyisoprene-co-isobutene. The change in the stereochemistry has an effect on the barrier to internal rotation and on the energy difference between the states (see Figure 7.11).

The polybutadiene has the lowest barrier to rotation and the lowest energy difference between states, resulting in a low capacity to absorb energy. Polyisoprene-co-butene has the highest barrier and the largest energy difference, which allows it to operate at high temperatures where butadiene is becoming ineffective. A low loss rubber, like polybutadiene, reacts to deformation by recovering rapidly and returning most of the energy applied by the original deforming stress. This property is termed *resilience*. Objects with high resilience fly further when hit (as with golf balls) or bounce higher when dropped. A high loss rubber, like butyl rubber, absorbs much of the energy when deformed. Incorporation of butyl rubber into a vibration damper will produce a more effective system than if butadiene is used.

7.3.3 Tyre technology

The energy absorbing properties of a rubber are of tremendous importance in vehicle tyre technology (Pethrick and North, 2008). The different parts of a tyre are constructed of rubbers. The cross-section of an unloaded tyre is shown in Figure 7.12.

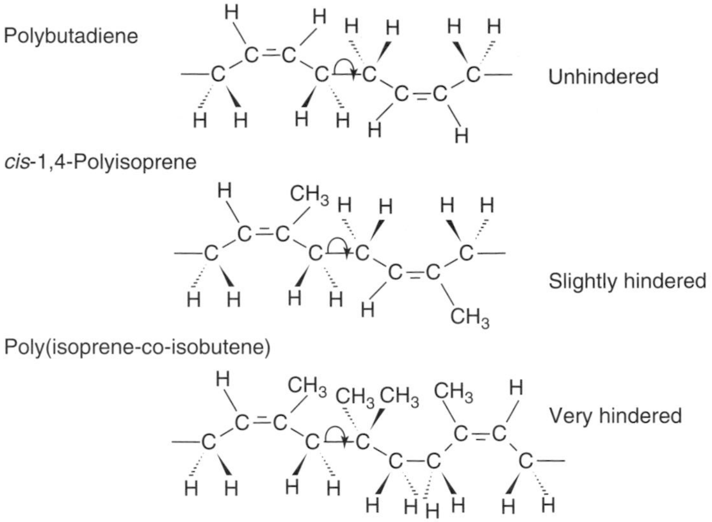

Polybutadiene — Unhindered

cis-1,4-Polyisoprene — Slightly hindered

Poly(isoprene-co-isobutene) — Very hindered

Figure 7.10 Chemical structures of some typical rubbers showing varying degrees of hindrance to internal rotational motion.

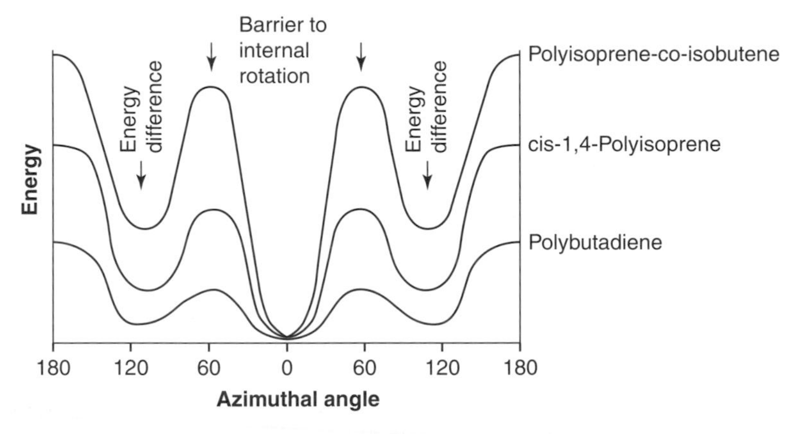

Figure 7.11 Potential energy diagram for rotational isomerism in some common elastomers.

The area in contact with the road will be under compression and as a consequence the walls of the tyre will be deformed. To retain the overall profile of the tyre, it is desirable to grade the modulus of the material used to construct its various parts. In practice, up to seven or eight different materials may be used, and cord is also used to enhance the radial modulus. Typically, a tyre may be constructed from a poly(isoprene-co-butene) rubber which has a high modulus and low resilience. The bulk of the carcass may be constructed from polyisoprene, and the walls from polybutadiene, which has high resilience and can accommodate the rapid cyclic deformations which will occur as the tyre rotates. The rate at which the tyre deforms will influence the ability of the various rubber components to store or dissipate energy. In practice, the sidewall of the tyre is alternately deformed and relaxed, a movement that creates an alternating stress/strain on the rubber. At each cycle, energy is absorbed, and the temperature of the tyre rises until the rate of heating is balanced by the rate of cooling. If the deformation is excessive (due to overloading

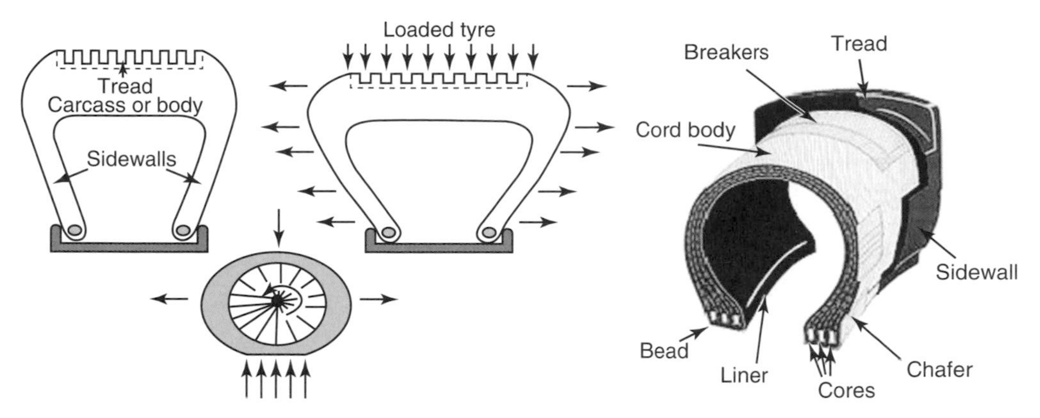

Figure 7.12 A car tyre as it rotates will be subjected to compression and stretching due to changing forces which are applied to it. Forces and deformations are indicated by arrows.

of the vehicle), or if the frequency of deformation is high (due to speeding), the rate of heating exceeds that of cooling and the temperature rises to the point where disintegration may occur: the rubber weakens, or even melts, the tyre sheds its tread, explodes, and the vehicle may crash.

To avoid high energy dissipation the sidewalls of a tyre are constructed of rubber with the least energy loss. When cost is not a factor, this means the use of a special low loss rubber such as polybutadiene. However, cost is often a factor and less expensive tyres have sidewalls constructed of cheaper natural rubber, with consequent higher heat generation and more severe restrictions on vehicle speed. Low loss rubbers like polybutadiene are extremely *resilient*. If the tyres are too bouncy then the ride may be uncomfortable. Reinforcement with cord helps to reduce this effect. Often the tyre in contact with the road requires an element of deformability to achieve road hold and butyl rubbers give better *road adhesion*, particularly in the wet. Cheaper tyres, with a tread constructed of natural rubber, are more prone to skid on wet roads than are those constructed with a tread of special high loss rubber. The tyre pattern influences the ability of the tyre to remove water from the contact area between the tyre and the road.

The carcass (or body) of the tyre, which is neither deformed nor in contact with the road, can be constructed of the cheapest rubber. It will usually be a reinforced structure and heat dissipation between the rubber and cord impregnation will dictate the extent to which heating becomes an issue. In high-performance car tyres, several different rubbers will be used. The sidewall will be produced from a blend containing polybutadiene, the carcass from natural rubber (polyisoprene) and the tread from a copolymer which contains isoprene and isobutene. For a less expensive tyre, constructed totally of natural rubber, a similar, though lesser effect, can be achieved by using more flexible, lightly vulcanised, rubber in the sidewall and higher modulus, more extensively vulcanised and so more tightly cross-linked, rubber in the tread. The modern tyre is a mixture of steel belts, advanced fabrics and rubber compounds, which are made of around 40% natural rubber and 60% synthetic rubber. Most manufacturers now include, amongst many others, silica as a filler and allow the modulus of the material to be varied. These modifications, along with tread design, help to optimise the performance of the tyre in wet weather conditions.

7.3.4 Effect of cross-linking on rubber characteristics

The manufacture of a tyre will involve cross-linking or vulcanisation of the rubber. The vulcanisation process has been traditionally carried out using sulfur but this leads to the creation of high levels of mercaptans and is a very smelly process. Alternative cross-linking chemistry is now available, which uses chemicals containing organic sulfur and sometimes peroxides. A tyre contains fillers,

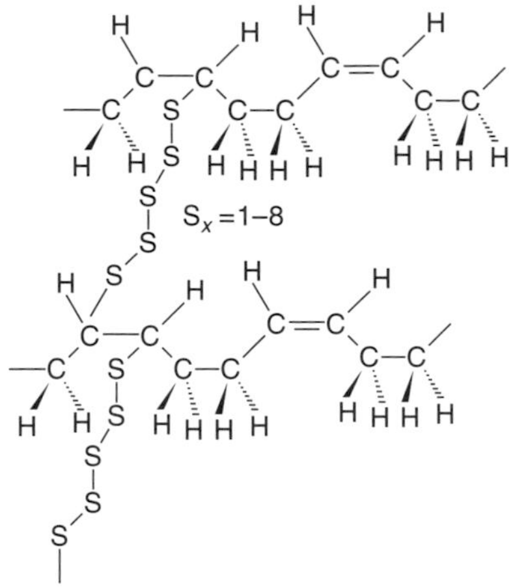

Figure 7.13 Vulcanisation of polyisoprene (natural rubber).

usually carbon black and silica, but may be complemented in a very high-performance tyre by the use of proprietary nanofillers. The net effect of the vulcanisation process is the creation of a thermoset structure and formation of a three-dimensional network (see Figure 7.13).

The cross-linking prevents creep and flow at high temperatures. Since flow is necessary to process articles into their final shape, the cross-linking reaction is usually carried out using the elevated temperatures of the shaping process (either extrusion or moulding). For natural rubber, the most important industrial cross-linking process is vulcanisation which is achieved using sulfur and a catalyst. Short polysulfide chains are added to the double bonds of the polyisoprene chain (see Figure 7.13). When a rubber is very lightly cross-linked, the modulus is barely affected, but the rubber region extends to higher temperatures and the creep/flow region is eliminated. However, as the cross-linking becomes tighter and tighter, shape change in the chains between the cross-links becomes more and more difficult, so that the modulus, the restoring force and the loss all rise.

7.4 Adhesion and adhesives

Polymers are extensively used as adhesives to bond materials together. The important characteristics of an adhesive are that it is able to form a strong bond to the substrate and that it is able to effectively transfer the stress from one element to the other.

7.4.1 Polymers as adhesives

Many engineers will encounter polymers as adhesives. Adhesive bonding has grown in popularity in recent years as it has many advantages over conventional methods of joining materials. This point can be simply illustrated by considering the case of joining two sheets of aluminium of the type which might be found in the manufacture of aircraft (see Figure 7.14).

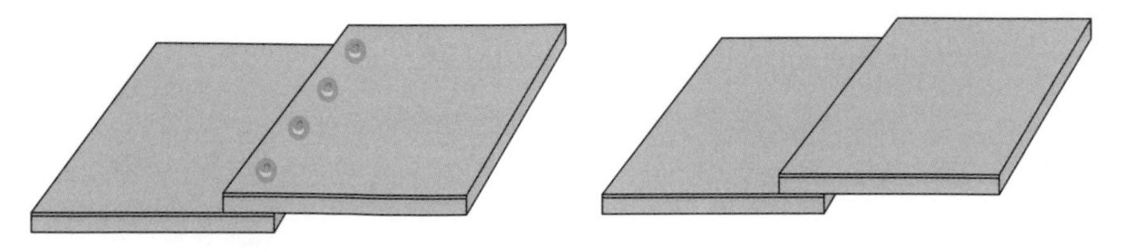

Figure 7.14 Structural joints: (a) riveted; (b) bonded.

Aluminium is very difficult to weld and the alternative method of joining is to rivet the panels together. However, the load is carried by the rivets and so the joining needs to be carried out at a relatively low pitch. Each rivet adds additional weight to the structure and becomes a point at which stress can be focused. If, instead of a rivet, the joint is adhesively bonded then the stress is distributed over the whole length of the joint and the weight of the lower density thin adhesive film can be significantly less than the equivalent weight of the rivet. In a riveted structure, the area forming the joint will usually be thickened and this adds weight to the structure. Adhesive bonding is therefore a very effective alternative for the bonding of a structure where weight is a critical consideration.

The advantages of adhesive bonding are:

- It is possible to bond dissimilar materials: metal to plastics, ceramics to metals, etc.
- It allows relatively thin sheets of material to be jointed together.
- It improves the stress distribution in the structure.
- Bonding is a relatively easily automated process, which is usually relatively cheap.
- It allows flexibility of design, allowing complex shapes to be assembled from simple components.
- It produces a smooth profile and avoids the inclusion of disruptions in the surface profile without having to counter sink fastenings.

However, there are disadvantages:

- Adhesives have poorer mechanical properties than metals, this can be a limiting factor in a design.
- Adhesives tend to suffer limitations as a consequence of their strength and toughness.
- The properties of adhesives will tend to be degraded as a result of exposure to various environments.

Despite these limitations there is an increasing growth in the use of adhesives in the manufacture of aircraft and automobiles and in particular where composites are being joined to metals.

7.4.2 Adhesion mechanisms

There are a number of factors which contribute to the strength of an adhesive bond. Adhesives are usually applied as liquids and therefore a critical factor in obtaining a good bond is the ability of the liquid to wet the substrate. The interaction of the adhesive and the substrate may involve several effects, which we now consider.

7.4.2.1 *Van der Waals interactions*

These are the normal interactions between solids and will usually be weak but attractive. A general rule is that like interacts with like. Therefore bonding to a substrate such as a nonpolar polymer (polyethylene) requires an adhesive which is similarly nonpolar, whereas bonding to a more polar substrate would be best achieved using a more polar adhesive. The process of adhesion at a molecular level is illustrated in Figure 7.15.

7.4.2.2 *Electrostatic interactions*

Charge interactions act over long distances and one theory of adhesion uses the concept of charges in the substrate interacting with those in the adhesive to form part of the bond strength (see Figure 7.16). Although the separation of an adhesive and a substrate can create charges, it is more contentious to state that these charges normally exist in either component in a normal bonding situation. The adhesion is associated with the creation of a double layer due to the inability of charge to migrate across an interface.

Whilst it is possible that charged entities such as carboxylic acid and quaternary ammonium ions can be incorporated into the adhesive to improve the adhesion, electrostatic interactions are not generally used in adhesive formulations. The charges tend to pick up moisture and do

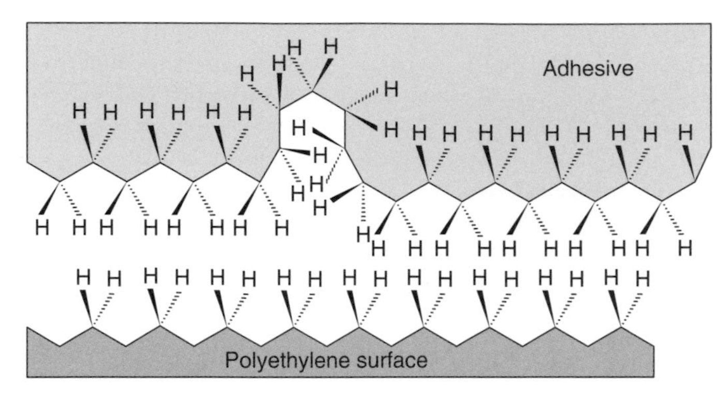

Figure 7.15 Schematic of interaction of a hydrocarbon-based adhesive with a PE surface.

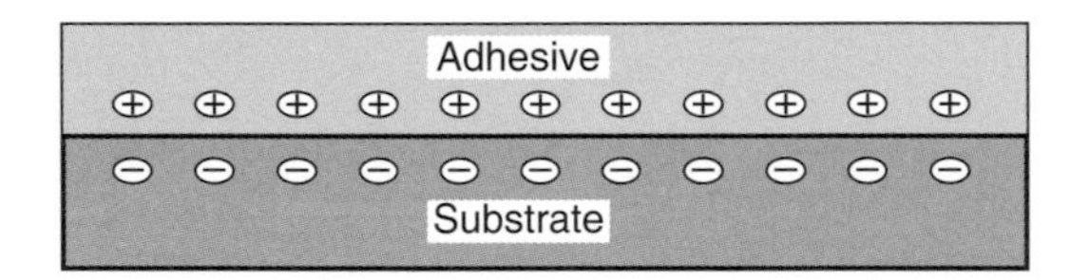

Figure 7.16 Schematic of a double-layer adhesive system.

not provide a stable performance. In both of the above situations, the interfacial strength arises from there being good contact between the substrate and adhesive. One way in which the bond strength can be enhanced is to increase the area of contact.

7.4.2.3 *Mechanical interlocking and surface roughness*

In many practical situations the adhesive strength is substantially increased if the surface of the substrate is modified so as to increase the surface area:

- *Profiling of the surface*: The adhesive strength of a bond can be substantially increased by creating patterns such as grooves, troughs or channels in an otherwise planar surface. The enhancement which is achieved depends on the profile but can equate to a doubling of the bond strength in certain situations.
- *Chemical modification*: There are two types of chemical modification. First, etching, in which the surface is roughened by etching away material form the surface. This process can be used for alloys and for polymer substrates. The second approach is to use chemical functions in the surface to graft molecules which have a greater ability to interact with the adhesive. The additional interactions, which may be hydrogen or covalent bonding, improve the bond strength.
- *Growth of a surface layer*: In the case of metal substrates, it is common to carry out a controlled oxidation of the metal. The oxide created will often have a specific structure which will substantially increase the area of interaction. The nature of the oxide created depends on the metal and the method used for oxide growth.
- *Grit blasting and mechanical abrasion*: One of the simplest and most effective methods of increasing the adhesive strength is to abrade the surface. This process of roughening can be used on plastics and metals and uses compressed air to blast small grit particles at the surface. The result of the impact is to rip bits of material from the surface, exposing a pristine substrate and increasing the roughness. In the case of metals, this process will remove loose oxide layers which have been contaminating the surface and will create a stable surface for bonding.
- *Plasma treatment and chemical modification*: Plastics are often difficult to bond together and may require special treatment. Plastic sheet is often subjected to corona discharge in order to create new chemical entities, which are usually associated with the interaction of oxygen (see Figure 7.17).

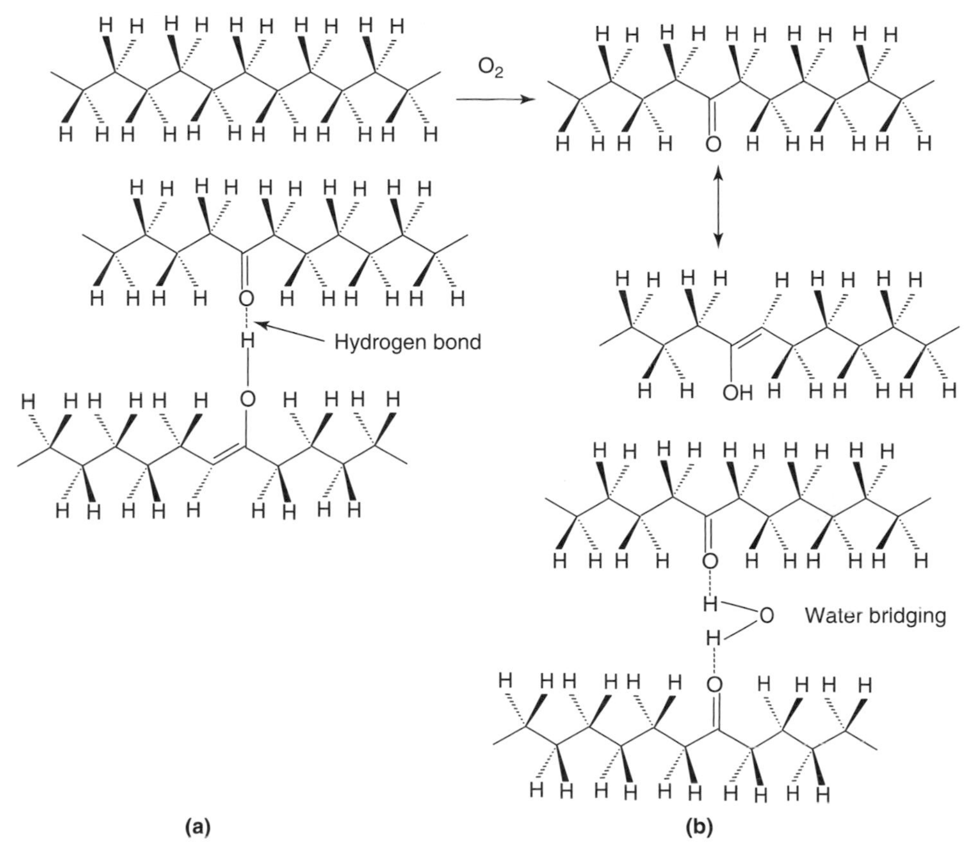

Figure 7.17 Modification of PE by oxidation: (a) tautomerism and interaction between hydroxyl and carbonyl groups; (b) bridging when water is present.

Similar modification of the surface can be achieved by the use of oxidising acids. In the case of semiconductor applications, various gas plasmas may be used, oxygen usually being a significant component of the gas mixture. Treatment of polyethylene will create a carbonyl function which can tautomerise into hydroxyl groups and the creation of a double bond. The hydroxyl group can interact with the carbonyl and create a hydrogen-bonded interaction which will substantially enhance the bond strength.

7.4.2.4 *Diffusion across the interface*

This mechanism is only really appropriate if the adhesive and the substrate are physically able to allow diffusion to occur. For instance, if a solid such as PMMA is treated with a solvent such as MEK, the surface will become soft and tacky as a result of the solvent lowering the glass transition point to below ambient temperature. If an adhesive, which is typically a solvent-based product, is applied then some interdiffusion of the polymer chains at the surface may occur. The adhesive bond strength will be enhanced by the polymer entanglement. This approach is used when 'permanent' labelling of goods is required.

7.4.2.5 *Adsorption and chemical bond formation*

The main contribution to the bond energy in most practical situations arises from molecular interactions of the type illustrated in Figure 7.15. Whilst for Van der Waals (VdW) interactions

Table 7.2 Bond type and typical bond energies

Bond type	Bond energy (kJ mol^{-1})
Van der Waals interactions	0.08–4
Dipole-induced dipole	Less than 2
Dipole–dipole interactions	4–20
Hydrogen bonds excluding fluorine	10–25
Hydrogen bonds including fluorine	Up to 40
Donor–acceptor bonds	
Bronstead acid–base interactions	Up to 1000
Lewis acid–base interactions	Up to 80
Covalent bonds	110–350
Ionic bonds	600–1100
Metallic	600–1100

these are weak, the addition of dipole, quadrupole and more specific interactions can create significant bond strengths (see Table 7.2).

The fluorine atom can introduce quadrupole interactions which are significantly stronger than normal dipole interactions. In general, covalent bonds are weaker than ionic or metallic bonds. Thus, provided that the substrate, which will usually be a metal or glass, does not contain a significant level of defects, failure will usually occur in the interface of the adhesive. If the failure occurs in the adhesive–substrate interface, it is known as *cohesive* failure, whereas if failure is in the adhesive, it is known as *adhesive* failure. For a good bond, the failure should be *adhesive* but may become *cohesive* if the interface strength is in some way reduced through ageing.

7.4.3 Examples of specific interactions which can occur in surfaces

7.4.3.1 *Adsorption of PMMA onto a silica surface*

If PMMA is deposited on silica from solutions which have had their pH adjusted, the level of adsorption varies with the effective interaction energy measured as the ΔH_{ab} of acid–base interactions relative to that of the solvent butyl alcohol or ethyl acetate. The variation of the adsorption is shown in Figure 7.18. The variations that are observed reflect the effects which pH have on the silica and the creation of different types of polar and electrostatic interactions.

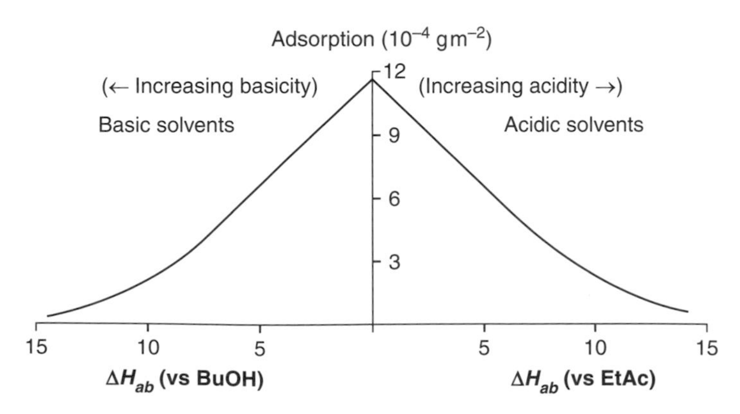

Figure 7.18 Adsorption of PMMA, a basic polymer, onto acidic surface of silica from basic, neutral and acidic solvents.

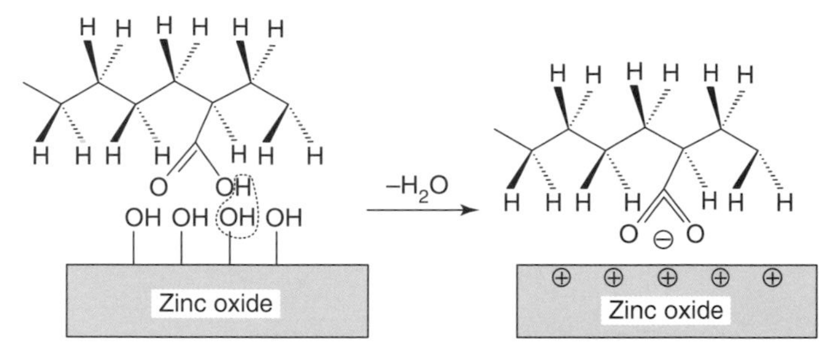

Figure 7.19 Possible interactions of methacrylic acid element with a zinc oxide surface.

7.4.3.2 *Electrostatic interactions between polymer and metal surfaces*

It is common for aliphatic polymers to be modified to improve their interaction with metal substrates. For instance, the incorporation of methacrylic acid into the backbone creates additional electrostatic interactions which can improve the adhesive energy (see Figure 7.19).

7.4.3.3 *Organic modification of reactive surfaces*

A very commonly used surface modifying agent is γ-aminopropyltriethoxysilane (APTES), which is able to react with surface hydroxyl functions on metals, silica and polymers which contain hydroxyl functions (see Figure 7.20). The reaction is the hydrolysis of the ethoxy group, which leads to a covalent-bonded surface structure which is capable of reacting with epoxy functions etc. In a humid atmosphere the APTES can react with moisture to produce a silanol function which will react with the surface hydroxyl group to form a covalent bond with the surface. The APTES can also interact with the surface through the amine tail group and other surface hydroxyl groups. When the adhesive contains epoxy groups, it is possible for reactions to occur with the amine and for a covalent bond to form, which will lead to significant adhesive bond strength. Another popular modifying agent is the vinyl analogue of APTES.

7.4.3.4 *Use of peel ply*

When fabricating composites, it is common to use peel ply to prepare a substrate for bonding. Peel ply is usually a fabric, often a polyester material, which is surface treated to have a weak

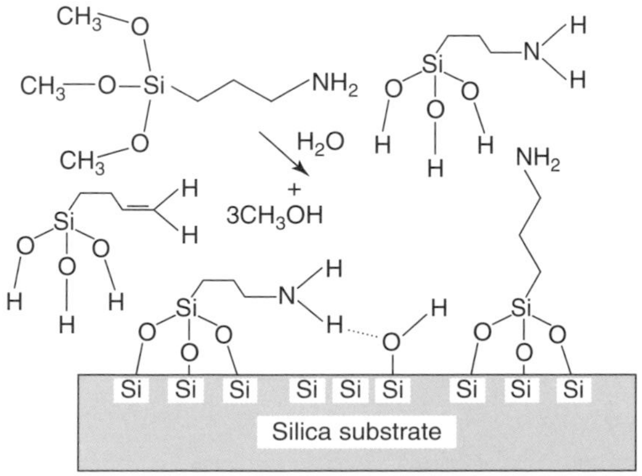

Figure 7.20 Interaction of γ-aminopropyltriethoxysilane with hydroxyl functions on a silica surface.

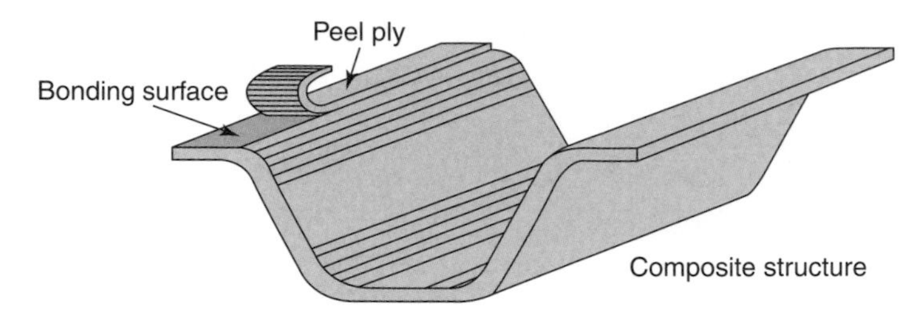

Figure 7.21 Peel ply used to create a good bonding surface for composites.

interface with the cured substructure (see Figure 7.21). The peel ply is stripped from the substructure and a roughened clean surface is created. However, it is important that all of the release agent should be removed as residues can significantly reduce the bond strength.

7.4.4 How does surface tension help to achieve a good bond?

To create a good adhesive bond it is essential that there is adequate contact between the adhesive and substrate: the adhesive should wet out the substrate. To understand the process of wetting we must understand *surface tension*. Surface tension is the result of an imbalance between the molecules on the two sides of the interface. The molecules in one phase are constrained within certain physical dimensions by the other molecules around them and the molecules of the second phase cannot mix with them because of the energy of the interface. When the energies of the two phases approach one another then mixing can occur. If the two phases are the same then diffusion will occur and a homogeneous mixture will be created. One liquid will mix with another liquid if their energies are similar and the surface tension is low.

7.4.4.1 *Surface tension: liquids*

Gibbs used the term 'surface tension' to define the imbalance between the two phases. The surface tension can easily be measured using the 'expanding film' method (see Figure 7.22).

Work is required to create a 'new' surface, Fdx, therefore $Fdx \propto dA$, where dA = change in area of the surface. So $Fdx = \gamma\, dA$, wher γ is the surface tension. The surface tension changes with the molecular structure of the liquid and reflects the intermolecular forces that exist within the liquid. Hydrogen-bonding liquids have the highest surface tensions and polar molecules have higher surface tensions than nonpolar molecules (see Table 7.3). Physically, mercury forms a drop-

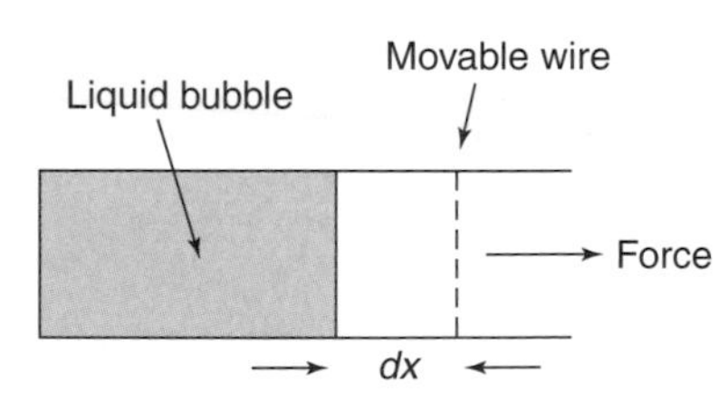

Figure 7.22 Schematic of expansion of a liquid bubble.

Table 7.3 Typical values of surface tension for some common liquids

Liquid	γ (mN m^{-1})	Force involved
Mercury	476	Metallic
Water	72.75	Hydrogen bonds
Octane	21.69	VdW forces
Benzene	28.88	VdW forces + π–π induced dipole interaction
Carbon tetrachloride	26.77	VdW forces + collisional dipole interaction

let on a table, water forms a drop that spreads and octane forms an oily film. The surface tension dictates whether or not the liquid spreads on a surface. To understand the spreading process we will consider Young's equation.

7.4.4.2 *Derivation of Young's equation and definition of contact angle*

Consider a droplet in contact with a solid. There are several interfaces: between the liquid and solid, the liquid and gas, and the solid and gas. The balance of forces at the point of contact of the edge of the liquid and the solid surface dictates where a liquid will spread and represents all three phases in balance (see Figure 7.23). To balance the forces γ^{α} is treated as a mechanical force and balanced in terms of its vertical and horizontal components.
When the droplet is at equilibrium, *Young's equation* applies:

$$\gamma^{\beta} = \gamma^{\alpha\beta} + \gamma^{\alpha} \cos\theta \tag{7.1}$$

where γ^{α} is the surface tension of the liquid–air interface, γ^{β} is the surface tension of the substrate–air interface and $\gamma^{\alpha\beta}$ is the liquid–substrate surface tension.

It is not practical to measure the force $\gamma^{\alpha\beta}$, thus an alternative approach is required. If you take a column of liquid and split it into two (see Figure 7.24) then the work done is:

$$\text{work} = \gamma^{\alpha} + \gamma^{\beta} - \gamma^{\alpha\beta} \tag{7.2}$$

The work of adhesion is defined by Equation (7.2). In this process two surfaces have been created. Taking Young's equation (Equation 7.1) and combing it with Equation (7.2) for the work of adhesion we obtain:

$$\gamma^{\beta} = \gamma\theta^{\alpha\beta} + \gamma^{\alpha} \cos\theta \quad \text{and} \quad W^{\alpha\beta} = \gamma^{\alpha} + \gamma^{\beta} - \gamma^{\alpha\beta} \tag{7.3}$$

The work of adhesion is the difference between the surface tension of the two phases in contact. So:

$$W^{\alpha\beta} = \gamma^{\alpha} + \gamma^{\alpha\beta} + \gamma^{\alpha} \cos\theta - \gamma^{\alpha\beta} \quad \text{and} \quad W^{\alpha\beta} = \gamma^{\alpha} + \gamma^{\alpha} \cos\theta \tag{7.4}$$

Therefore the strength of adhesion of a liquid to a surface:

$$W^{\alpha\beta} = \gamma^{\alpha}(1 + \cos\theta) \tag{7.5}$$

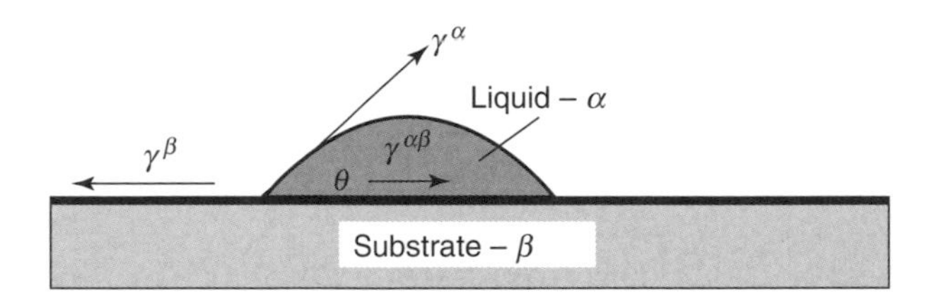

Figure 7.23 Schematic of forces balancing bubble on surface.

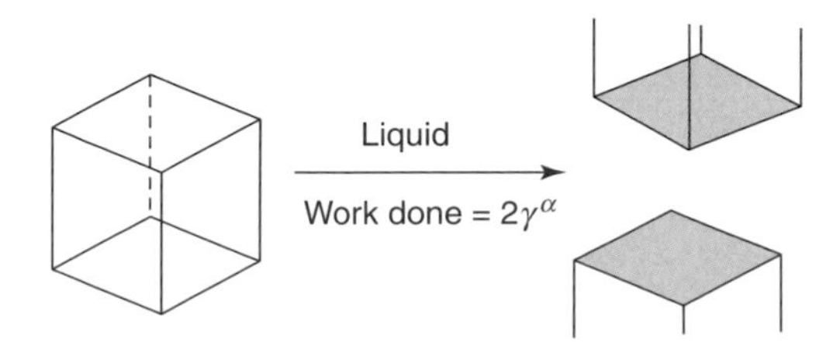

Figure 7.24 Energy change when a liquid column is split to create two surfaces.

where θ is the contact angle. Depending on the value of the contact angle then the liquid in contact with the solid will form either:

(a) spreading film $\theta = 0°, \cos\theta = 1$ (b) wetting droplet $\theta = 90°, \cos\theta = 0$ (c) nonwetting droplet $\theta = 180°, \cos\theta = -1$

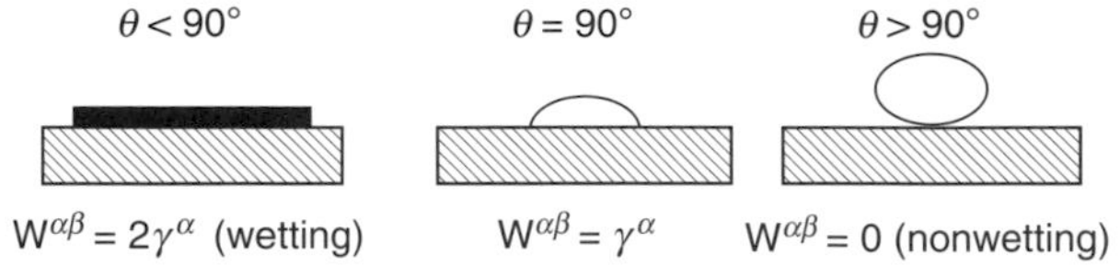

$\theta < 90°$ $\theta = 90°$ $\theta > 90°$

$W^{\alpha\beta} = 2\gamma^{\alpha}$ (wetting) $W^{\alpha\beta} = \gamma^{\alpha}$ $W^{\alpha\beta} = 0$ (nonwetting)

The work of cohesion = $2\gamma^{\alpha}$, the attraction of liquid to itself, i.e. when the contact angle = 0°

The work of adhesion is equal to the work of cohesion. When $\theta < 90°$, the liquid is *wetting* the solid, but when $\theta > 90°$, the liquid is *not wetting* the solid. For a good bond to be formed it is essential that the liquid adhesive wets out the substrate. It may be necessary to modify the substrate to improve the wetting characteristics. For spontaneous wetting to occur the condition:

$$\gamma_{sv} \geq \gamma_{sl} + \gamma_{lv} \tag{7.6}$$

must apply. If this condition is not fulfilled then a droplet will be formed and the liquid will not effectively wet the substrate. In the context of adhesion this would lead to a poor bond.

7.4.4.3 *Surface energy of some typical polymeric materials*

Changes in the chemical structure of the polymer can produce significant differences in the interactions between a polymer and another component, such as glue. Table 7.4 lists some typical values of the surface energy for some common polymeric materials. There is a significant spread in the values of the critical surface tension which reflects the ways in which the various components of the surface energy change with the chemical structure of the polymer. The implication is that as the critical surface tension changes so the nature of the glue will have to be adjusted to ensure that the wetting condition is retained and that a strong interface is developed.

The surface energy for various liquids can vary over a significant range and the values for the solids presented above are usually obtained by determining the nonwetting condition for the solid in contact with a liquid. Values for various liquids are presented in Table 7.5.

Table 7.4 Values for surface free energy for various plastics

Polymer	**Critical surface tension, γ_c (mN m^{-1})**
Polytetrafluoroethylene	18.5
Polyvinylidenefluoride	25.0
Polyvinylfluoride	28.0
Polychlorotrifluoroethylene	31
Polyethylene	31
Polypropylene	31
Polystyrene	32.5
Polyvinylchloride	39
Polymethylmethacrylate	39
Polyvinylidenechloride	40
Nylon 6,6	42.5
Polyethyleneterephthalate	43
Epoxy resin	~46
Phenol–resorcinol resin	52
Urea–formaldehyde	61

Table 7.5 Values for surface free energy for various common liquids

Liquid	Surface free energy (mJ m^{-2})
Water	72.2
Glycerol	64.0
Formamide	58.3
Diiodomethane	50.8
Ethane-1,2-diol	48.3
1-Bromonapthalene	44.6
Dimethylsulfoxide	43.6
Pyridine	38.0
Dimethylformamide	37.3
Polyglycol E-200	43.5
Polyglycol 15–200	36.6
2-Ethoxyethanol	28.6
Hexadecane	27.6
Tetradecane	26.7
Dodecane	25.4
Decane	23.9
Octane	21.8
Hexane	18.4

The precise value is the result of a balance between polar and dispersive forces and changes with the chemical structure of the liquid. As the liquid becomes less polar, so the value of the surface free energy is reduced and these liquids will wet better the lower surface energy plastics than those with higher values.

The general rule which is useful to remember is that 'like wets like'. Therefore a hydrocarbon-like liquid will better wet polyethylene than water. When selecting a glue for a plastic it is appropriate to select a material with a surface energy which is close to that of the solid to which it will be bonded. It is possible to predict the values of the surface energy theoretically by adding together the components of the forces which a molecule exhibits. The effective surface energy is an average of the various interactions.

7.4.4.4 *Kinetics of wetting*

As with many situations encountered in a molecular system, whilst the energy of the interaction determines the final equilibrium situation, there are often kinetic factors which will tell us how quickly this equilibrium will be achieved.

7.4.4.5 *Surface tension gradients*

Surface tension gradients arise from thermal gradients or, in the case of liquids or glues which contain a component which is volatile component, a difference in surface free energy which results in a concentration gradient (see Figure 7.25).

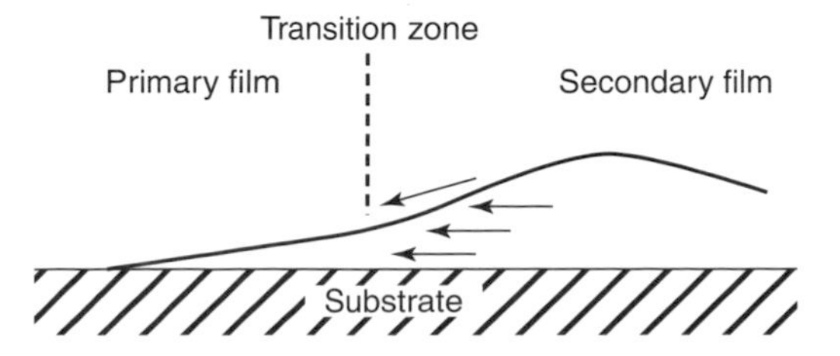

Figure 7.25 Effect of concentration–surface tension gradient on flow through transition region.

At the leading edge of the droplet there is a thinning of the liquid to what is virtually a monolayer film which is connected to the main droplet by a transition region in which thermal and concentration gradients create a surface flow that drags the liquid across the surface. The rate of flow in the transition region may exceed the rate of gravity flow in the much thicker secondary film so that a ridge of liquid develops just behind the advancing edge. The rate depends on the evaporation rate and will typically be relatively low for most systems, unless they contain a very volatile solvent. It must be remembered that the inclusion of surface active components in the liquid can lead to similar effects and does not require evaporation to occur.

7.4.4.6 *Dynamic contact angle*

When an adhesive is applied to a substrate, it is usually forcibly spread across the surface using a palette knife or similar object. If the viscosity of the liquid is greater than a few centipoises, then the liquid near the surface cannot keep up with the advancing front and this creates a dynamic contact angle which is different from the equilibrium value. For liquid in the range 10–1000 cp the dynamic contact angle, θ_d, is described by the Fritz equation:

$$\tan\left(\theta_d\right) = m\left(\eta v_s / \gamma_{lv}\right)^n \tag{7.7}$$

where η is the viscosity of the liquid, γ_{lv} is the surface energy of the liquid vapour interface, v_s is the spreading velocity and m and n are constants. In practice, the effect is that the extent of wetting is more limited, the dynamic contact angle being higher than the equilibrium value. The implication is that during the spreading/application process, if the glue is spread at high speed it may become more difficult to wet the substrate than if the process is carried out at lower speeds. This can be a significant issue when the application of the glue is automated.

7.4.4.7 *Influence of surface roughness*

In the above development of the theory, it was assumed that the substrate is a perfectly smooth surface. Plastics, polished metals and ceramics may have what appear to be smooth surfaces. In order to improve adhesion it is often good practice to roughen the surface, thus, in many situations the glue is applied to a surface which contains pits and crevices. Capillary action and the fact that the liquid will form a contact angle with the textured surface which is less than 90° will aid the spreading of the glue. These effects are important in achieving the development of a good interface but do not have a major effect in achieving the spread of the liquid.

7.4.4.8 *Temperature effects*

The contact angle is a function of the temperature and deceases with increasing temperature. Similarly the viscosity of the glue will also decrease as the temperature is increased. Adhesives are often applied at an elevated temperature, however, if the adhesive is a reactive mixture, as in the case of an epoxy system, the effects of increase in the viscosity as a consequence of the cure reaction can offset the advantages gained by application at increased temperature. In practice, a compromise is usually found which achieves the best bonding.

7.4.5 Issues which arise during the bonding process

Whilst the ideal bond may have significant advantages over other forms of jointing technology, it is critical that the joints are not compromised by poor bonding practice. For an ideal bond to be formed, it is crucial that the adherent is correctly prepared for bonding (abraded, primed, etc.) and that the adhesive is correctly applied at the correct thickness and correctly cured to give the maximum strength. Some of the most common problems can be easily recognised and corrected.

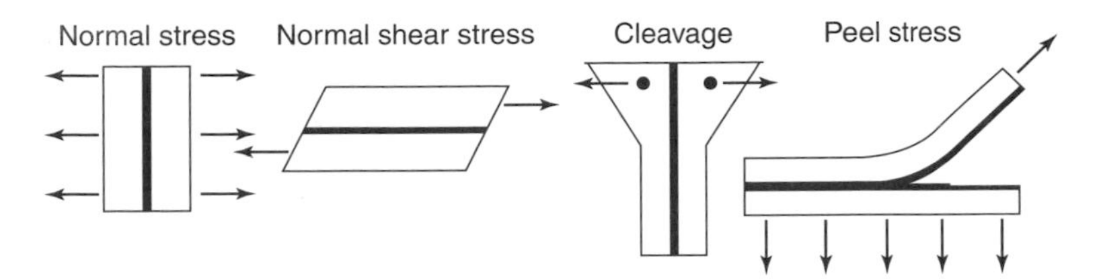

Figure 7.26 Common types of stresses applied to joint structures.

7.4.5.1 *Air entrapment*

The liquid glue has to form an intimate bond with the substrate and the process of application can be considered to be one in which air is being displaced from the surface. If the spreading and wetting are not efficient then the air is not effectively displaced and bubbles will be entrapped in the interface. This problem can be addressed by applying external pressure which will thin the bond line and can eliminate the trapped bubbles. With some resins, dissolved water and other gases can form bubbles when the resin is cured. It is good practice to degas the glue prior to application.

7.4.5.2 *Bonding environment*

Adhesives are of various types and the effect of the environment can depend on which type is being considered. However, there is a general issue which is common to all systems: the influence of the environment on the ability to wet the substrate. Adhesives are widely used to bond metals together. Metals are usually considered to be high energy surfaces having surface free energies typically greater than 500 $mJ\,m^{-2}$. Metals such as steel and aluminium will usually be covered by a thin layer of oxide which will change the surface energy, but the surface energy will still be sufficiently high that it can be predicted that wetting should not be a problem. However, the surface may become covered with adsorbed water and, particularly when the piece to be bonded has been machined, the surface can be covered with hydrocarbon cutting oils. This contamination may lower the surface energy to values as low as ~40 $mN\,m^{-1}$. The contaminated surface may therefore be difficult to wet and so it may be difficult to form a good adhesive bond. The quality of the bond formed depends critically on the removal of contamination, which is usually achieved by sand blasting, treatment with a volatile solvent, then etching. The process usually ends with baking in an oven to ensure that no solvent remains on the surface.

7.4.6 Adhesive bond design

As in the case of composites and metal bonds, the design of the structure can have a profound influence on the mechanical properties which are achieved. It is important to understand how the load acts on a joint and how the thickness of the adhesive layer influences the failure characteristics of the joint. Figure 7.26 illustrates the common types of stress profile which are likely to be encountered in joints.

There are a number of simple rules which aid the creation of a good joint. For the most cost-effective assembly it is desirable to apply the adhesive as either a stripe or a cross-pattern. The action of bringing two surfaces together to be bonded should compress the liquid layer and avoid the possibility of air entrapment. Air entrapment can be avoided by bringing two edges together and then compressing the wedge of liquid adhesive. It is better to form a joint by compression than by sliding surfaces over one another. In critical areas, rather than using a liquid adhesive, tape adhesive (adhesive containing a fabric mesh) or adhesive which contains small glass balls may be used. Examples of good and poor joint design are shown in Figure 7.27. Good design is characterised by minimising the shear forces, keeping an element of the adhesive under compression and maximising the area of the joint, thus minimising the shear stresses (see Table 7.6).

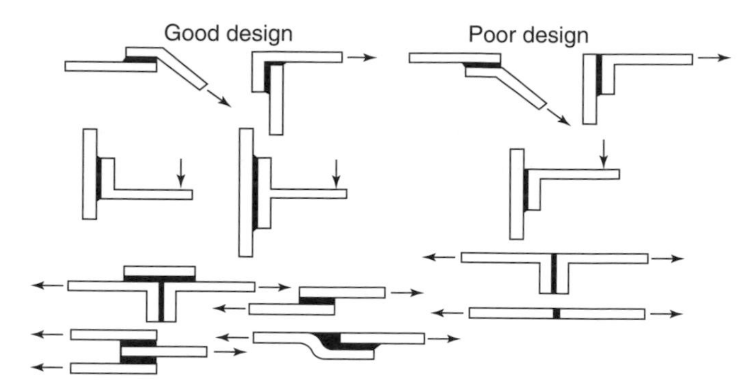

Figure 7.27 Examples of good and poor joint design.

Table 7.6 Summary of geometries and associated test standards

Test geometry	Standard	Comment
Axially loaded (tensile) butt joints	ASTM D897–78	For substances in 'block' form
	ASTM D 2094–69 and D 2095–72	Specifically for bar and rod shaped substances
	BS 5350 Part C3	UK version for bar and rod
	ASTM 816–82	Specifically for rubbery adhesives
	ASTM 1344–78	Cross-lap specimen specifically for glass substrates
Lap joints loaded in tension	ASTM 1002–72	Metal–metal single lap joint
	BS 5350 Part C5	UK version
	ASTM D 2295–72	Tests at elevated temperatures
	ASTM D 2557–72	Tests at low temperatures
	ASTM 3163–73	Using rigid plastic substrates
	ASTM D 3983–81	Uses thick substrates
Double lap joint	ASTM D 3528–76	Double lap joint
	BS 5350 Part C5	UK version
Modified double shear	ASTM D 3165–73	Metal–metal laminate test uses defined bond area
	ASTM D 906 -82	
	ASTM D 2339–82	Specifically for plywood laminates
Cleavage	ASTM D 1062–78	Metal–metal joints
	ASTM D 3807–79	Engineering plastics
	ASTM D 3433–75	Flat and contoured cantilever beam specimens for adhesive fracture energy, G_{Ic}
Peel joints	BS 5350 Part C14	90° peel test for flexible–rigid joints

Control of the bond line thickness is also an important factor in achieving a good bond. The testing of adhesives has led to the creation of a number of test standards. These tests are designed to measure a particular parameter for the adhesive, i.e. tensile, cleavage peel or shear strength. Some of the more common standards are listed in Table 7.6. It is also important to consider how easy it will be to apply the adhesive to the substrate. It is always important to ensure that the surface to be bonded has been correctly prepared so that all of it can easily be wetted. Thus, the bond where the metal is curved represents an example of good design but might in practice be difficult to bond because the S-shaped section may be difficult to clean. Plane surfaces are usually easier to prepare and bond than more complex surfaces. In the preparation of bonds for use in aircraft, the bonding surfaces are carefully etched and cleaned and a low molar mass adhesive is then applied to ensure that the surface is maintained. This allows the application of tape adhesive to create the final bond.

7.4.7 Stresses in adhesive joints

To deduce the nature and magnitude of the stresses in an adhesive joint it is necessary to know the basic mechanical properties of the adhesive: tensile strength, Young's modulus (E_a) and shear modulus (G_a), also the yield stresses, facture stresses and strains in uniaxial tension and pure shear.

The forces which are created in a joint depend on the interaction between the force applied to the substrate and the joint and are influenced by the thickness of the adhesive layer. Two approaches are used to determine the properties: the first is the direct measurement of the physical properties. However, this approach requires the use of thick sections of adhesive and it can be argued that the physical properties measured do not necessarily replicate those in the thin layer used to create the adhesive. The second approach is to measure the properties using specific joint geometries such as those shown in Table 7.6. In the latter approach it is important that the joint structure that is created produces simple stresses and that no stress concentrations are present. It is important that the surface treatment that is used ensures that the failure occurs in the adhesive (a cohesive failure).

The measurement of the tensile properties is relatively straightforward and uniaxially loaded tensile configurations are often used. Measurement of the shear properties is slightly more problematic. One of the most frequently used configurations is the 'napkin ring' test or the use of a 'thick adherend lap-shear' test (see Figure 7.28).

7.4.8 Axially loaded butt joints: tensile measurements

In principle, the application of a load to a butt joint (see Table 7.6) should allow direct measurement of the tensile stress. However, there are a number of issues which need to be considered. First, it is very difficult to set up the test so that the load which is applied is purely tensile and does not contain a shear element. A further complication is that the application of a tensile load to the butt joint will cause both the adhesive and the substrate to undergo lateral deformation and this will occur to different extents in the two materials creating a shear force within the adhesive layer. Detailed finite element analysis of the stress distribution indicates that stress will often be uniform at the centre of the butt joint but may deviate because of the shear effects as the edge is approached.

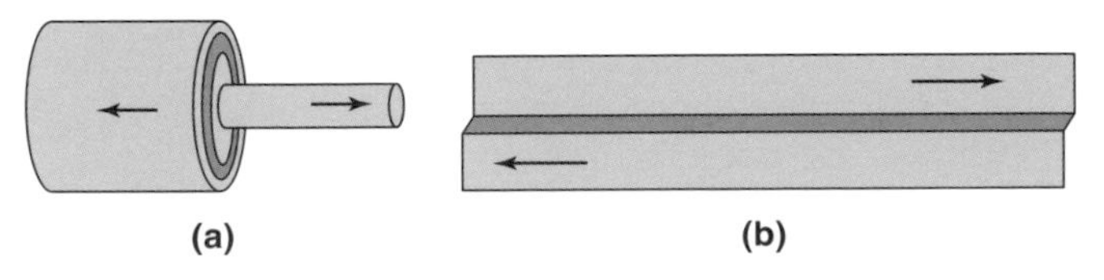

Figure 7.28 Configurations used to measure shear properties of adhesives: (a) napkin ring test; (b) thick adherend lap shear test.

In the analysis of the stresses for the butt joint, it is assumed that the tensile stresses are uniform and the shear stress, τ_{rz}, is zero at the centre, but will vary as the edge is approached. The normal applied stress, σ_0, is applied to the butt and induces a radial stress, σ_r, and a circumferential stress, σ_θ, which arises as a consequence of the different Poisson ratios, v, for the substrate and adhesive. If it is assumed that the radial and circumferential strains in the adhesive layer are equal to the Poisson ratio of the strains in the substrates, then:

$$1/E_a\left(\sigma_\theta - v_a\left[\sigma_z + \sigma_\theta\right]\right) = -\left(v_s / E_s\right)\sigma_z \tag{7.8}$$

Hence:

$$\sigma_\theta = \theta_r = \left[v_a - \left(E_a v_s / E_s\right)\right]\left[\sigma_z / \left(1 - v_a\right)\right] \tag{7.9}$$

where the subscripts *a* and *s* refer to the adhesive and substrate, respectively. The stresses will vary with the thickness of the adhesive layer and ideally the layer should be as thin as is practical whilst still achieving a uniform layer in the butt joint. It is important to ensure that the edge of the butt joint is free from 'sprew' which can transfer the load across the adhesive. If it is assumed that the adhesive is perfectly constrained by the substrates and if the substrate is assumed to be infinitely rigid then the radial and hoop strains in the adhesive and substrate are zero and the apparent Young's modulus, E'_a, of the adhesive is given by:

$$E'_a = \left[\left(1 - v_a\right) / \left[\left(1 - v_a\right)\left(1 - 2v_a\right)\right]\right]E_a \tag{7.10}$$

For a structural adhesive, such as an epoxy resin, Poisson's ratio will have a value of ~0.35 which implies that the measured modulus will be about 50% greater than the true value. For less rigid materials, Poisson's ratio can be higher, leading to a significantly higher value for E'_a. The value of the modulus will be temperature dependent as discussed in the previous chapters and can be mapped by comparison with the temperature dependence of the modulus as measured using dynamic mechanical thermal analysis. The main temperature-dependent component is the T_g effect and the modulus should be fairly constant for a structural adhesive up to within about 40°C of the T_g.

7.4.9 Single lap joints

As indicated in Table 7.6 there are a variety of lap joints which have been proposed for testing. Lap joints are easy to fabricate and replicate the type of joint which is often used in practical bonding situations. Although in principle the loading is simple, when we look at it in detail, it emerges that it is quite complex. The shear stresses are illustrated in Figure 7.29.

The stress profile depends on the rigidity of the substrate. If the substrate is rigid then a fairly uniform deformation of the adhesive will be induced. However, if elasticity is possible in the substrate, as would be the case with a plastic material, then a significantly nonlinear stress distribution can arise in the adhesive. The stresses at the edges are greater than at the centre of the bond, making the edges prone to crack initiation. If the substrate is elastic then the stress distribution will be even more extreme than that if there is a rigid substrate. The effect of flexibility in the substrate is to allow deformation and hence the true profile will involve not only deformation of the adhesive, but also of the substrate in the area of the joint. The extent to which this deformation occurs depends on the rigidity of the substrate. To avoid this problem, thick substrates are often used (see Table 7.6).

7.4.10 Fracture mechanics

Generally, the application of fracture mechanics to the analysis of adhesive joints will involve linear elastic fracture mechanics (LEFM). The analysis is usually based on either energy analysis or

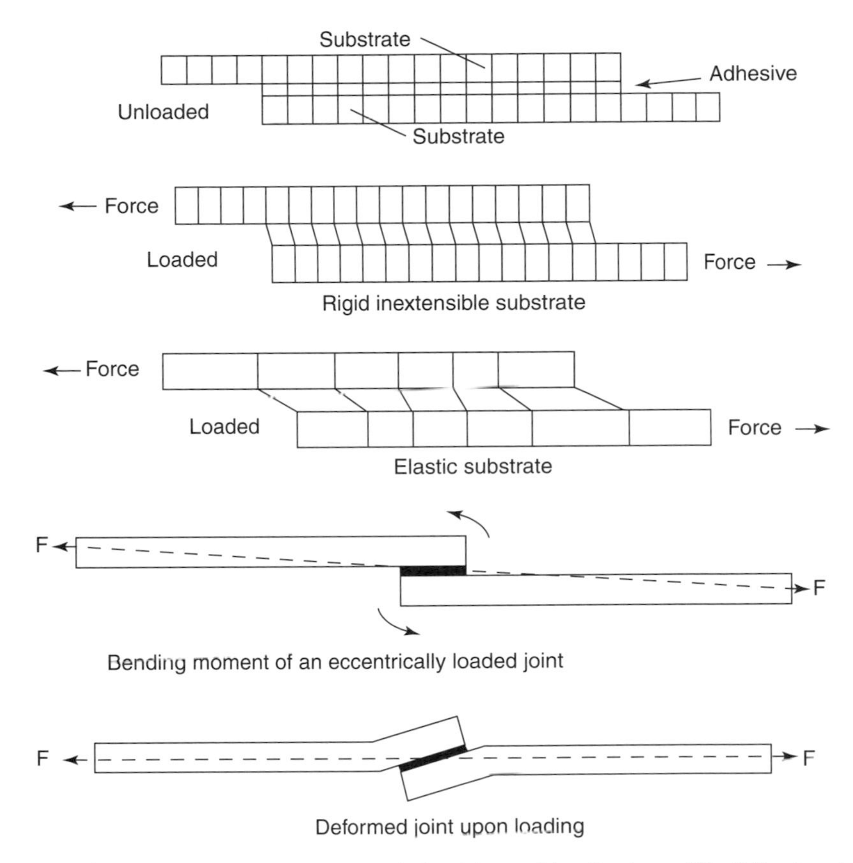

Figure 7.29 Stress profile of loading a simple lap joint and implications of flexibility in substrate.

localised stress field analysis or a combination of both. The aim of the analysis is to create values for the fracture energy which are independent of the geometry used for the analysis.

7.4.10.1 *Energy balance approach*

The energy balance approach, commonly termed the G approach, is based upon the work of Griffith. The quasicrack created in the adhesive is assumed to propagate with the conversion of work done by the external force, F, into elastic energy stored in the material and will ultimately result in the creation of new surfaces which have an energy, γ. The process can be written as:

$$\partial(F-U)/\partial a \geq \gamma \partial A/\partial a \tag{7.11}$$

where ∂A is the increase in surface area for a crack growth increment of ∂a. For a crack growth in a plate of thickness b and substituting $\partial A = 2b\partial a$, Equation (7.11) becomes:

$$(1/b)\partial(F-U)/\partial a \geq 2\gamma \tag{7.12}$$

We assume that the energy dissipation around the crack is independent of the specimen's geometry and loading conditions. The right-hand side of Equation (7.12) can be replaced by G_c as follows:

$$G_c \leq (1/b)\partial(F-U)/\partial a \tag{7.13}$$

where G_c is the energy required to increase the crack by a unit length in a unit width specimen. Using the first law of thermodynamics and neglecting dynamic effects it follows that:

$$G_c = G_0 + \psi \tag{7.14}$$

where ψ is the energy dissipated through viscoelastic and plastic deformations at the crack tip and G_0 is the intrinsic fracture energy. It is found that for cross-linked rubbery adhesives:

$$\psi = G_0 f\left(\dot{a}, T, \varepsilon\right) \tag{7.15}$$

where f is a function which depends upon the crack growth rate $\dot{a}$, temperature T and strain ε. If we suppose the following is appropriate:

$$\Phi\left(\dot{a}, T, \varepsilon\right) = 1 + f\left(\dot{a}, T, \varepsilon\right) \tag{7.16}$$

and combining Equations (7.14)–(7.16) we obtain:

$$G_c = G_0 \Phi\left(\dot{a}, T, \varepsilon\right) \tag{7.17}$$

If the adhesive is rigid and the viscoelastic and plastic energy losses are negligible, such that $\Phi\left(\dot{a}, T, \varepsilon\right) \rightarrow 1$ or $f\left(\dot{a}, T, \varepsilon\right) \rightarrow 0$, then the fracture energy is equivalent to G_0. This argument implies that the bonding energy is directly accessible through the measurement of the fracture energy.

The alternative approach is to use LEFM. This approach assumes that the material behaves as a Hookian solid. For a specimen of thickness b, loaded by applying a force P, the stored elastic energy U_1 is equivalent to the area under the displacement curve and is as follows:

$$U_1 = (1/2) P \Delta \tag{7.18}$$

where Δ is the displacement. If the crack grows by an increment ∂a, the stiffness of the samples decreases, which produces a change in load ∂P and a deflection $\partial \Delta$. The stored elastic energy is therefore:

$$U_2 = 1/2\left(P + \partial P\right)\left(\Delta + \partial \Delta\right) \tag{7.19}$$

The change in stored energy as a consequence of the deformation is therefore:

$$\partial U = U_2 - U_1 \tag{7.20}$$

The external work done is therefore:

$$\partial F = P \partial \Delta + 1/2(\partial P \partial \Delta) \tag{7.21}$$

or

$$\partial F = \left(P + \partial P / 2\right) \partial \Delta \tag{7.22}$$

Substituting Equations (7.19) and (7.20) into Equation (7.21) and combining with Equation (7.22) we have:

$$\partial\left(F - U\right) = 1/2\left(P \partial \Delta - \Delta \partial P\right) \tag{7.23}$$

Substituting Equation (7.23) into Equation (7.17) we have the following condition for crack growth:

$$G_c = \left(1/2b\right)\left((P \partial \Delta / \partial a) - (\Delta \partial P / \partial a\right) \tag{7.24}$$

The compliance of the system can be defined as:

$$C = \Delta / P \tag{7.25}$$

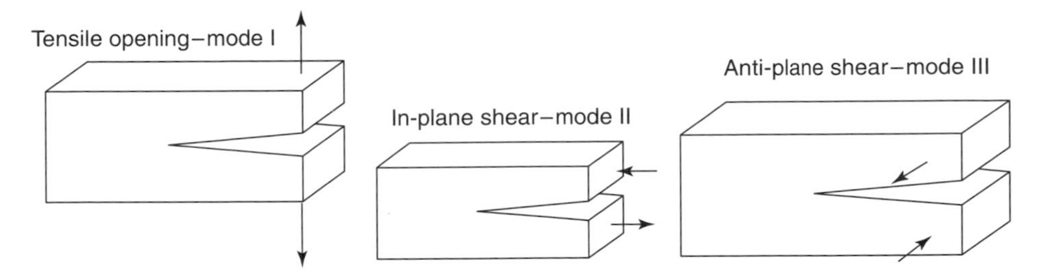

Figure 7.30 Loading modes for measurement of G_c.

Such that:

$$\partial\Delta = P\partial C + C\partial P \quad (7.26)$$

By substituting Equation (7.24) into Equation (7.26) we obtain the following relationship:

$$G_c = P_c^2 \partial C / 2b\partial a \quad (7.27)$$

This equation is the basis of most calculations of G_c which require the load, specimen width and variation of specimen compliance with crack length. The latter is determined by analytical and experimental methods. Typically there are three modes of G_c corresponding to the three basic modes of loading (see Figure 7.30). We can therefore express G_c as fracture energies corresponding to the three modes of loading with G_{Ic}, G_{IIc} and G_{IIIc}. It is usually the mode G_{Ic} which is obtained from experiments.

7.4.10.2 *Shear stress analysis theory*

The usual analysis predicts the nominal in-plane adhesive shear stress distribution by neglecting the out-of-plane effects. Therefore, the only loads required for solution are those in the plane of the adherents.

Consider an element of length Δx which is subjected to loads P_1 and P_2 per unit width as shown in Figure 7.30. At equilibrium the condition for the two adherents is:

$$\partial P_1 / \partial x + \tau = 0 \qquad \partial P_2 / \partial x - \tau = 0 \quad (7.28)$$

where:

$$P_1 = P - P_2 \qquad \text{and} \qquad \sigma_1 = P_1 / t_1 \quad \sigma_2 = P_2 / t_2 \quad (7.29)$$

If we consider displacements in the x-direction, denoted u_1 and u_2, then the shear strain γ in the adhesive is:

$$\gamma = (u_1 - u_2) / t_a \quad (7.30)$$

where t_a is the thickness of the adhesive layer. The shear stress, τ, is then given by:

$$\tau = (G_a / t_a)(u_1 - u_2) \quad (7.31)$$

where G_a is the adhesive shear modulus. The adherent strains on the element in the x-direction are:

$$\varepsilon_1 = \partial u_1 / \partial x = \left((1 - v_1^2) / E_1 t_1\right)(P - P_2) \quad \text{and} \quad \varepsilon_2 = \partial u_2 / \partial x = \left((1 - v_2^2) / E_2 t_2\right) P_2 \quad (7.32)$$

where t_1, t_2, v_1, v_2, E_1 and E_2 are, respectively, the thickness, Poisson's ratio and Young's modulus for adherents 1 and 2 . Combining Equations (7.32) and (7.31) yields:

$$\partial^2 P_2 / \partial x^2 - \alpha^2 P_2 = \beta P \quad (7.33)$$

where:

$$\alpha^2 = (G_a / t_a)\left(\left((1-\nu_1^2)/E_1t_1\right) + \left((1-\nu_2^2)/E_2t_2\right)\right)$$

$$\beta = (G_a / t_a)\left(\left((1-\nu_1^2)/E_1t_1\right)\right) \quad (7.34)$$

The distribution of adhesive shear stress can then be expressed as:

$$\tau = -\frac{B}{A^2}\alpha\left(-\frac{\cosh(l-x)}{\sinh(\alpha l)} + \left(1-\frac{A^2}{B}\right)\frac{\cosh(\alpha x)}{\sinh(\alpha l)}\right)P \quad (7.35)$$

where

$$A = \alpha\sqrt{t_a} \quad \text{and} \quad B = -\beta t_a \quad (7.36)$$

These equations describe the stress distribution in an adhesive joint, which is determined by the geometry and material properties of both the adherent and adhesive material. However, it is interesting to understand the behaviour of the shear stress distribution, τ, within the joint.
For $t_a > 0$, τ is a bounded function. For $t_a \rightarrow 0$, α becomes large and τ would have the following asymptotic form:

$$\tau = -(B / A^2)\alpha\left(-e^{-\alpha x} - 0\left(e^{-\alpha(2l-x)}\right) + \left(1-(A^2 / B)\right)\left(e^{-\alpha(l-x)} + 0\left(e^{-\alpha(l+x)}\right)\right)\right)P \quad (7.37)$$

Thus from Equation (7.37) it is clear that for:

$$\lim_{\alpha\rightarrow\infty} \alpha e^{-c\alpha} = \begin{Bmatrix} 0, c > 0 \\ \infty, c = 0 \end{Bmatrix} \quad \text{and} \quad t_a \rightarrow 0, \alpha \rightarrow \infty$$

then:

$$\tau = \begin{Bmatrix} 0, 0 < x < l \\ \infty, x = 0, x = l \end{Bmatrix} \quad (7.38)$$

In other words as t_a drastically decreases, the interface shear stress will be concentrated at both end points, $x = 0$ and $x = l$, and will be zero elsewhere. These two concentrated shear forces, T_1 and T_2, satisfy the equilibrium condition $T_1 + T_2 = P$. In the case of identical adherents of thickness t, Poisson's ratio ν and Young's modulus E, the constants A and B become:

$$A^2 = \alpha^2 t_a = 2G_a((1-\nu^2)/Et) \quad \text{and} \quad B = -\beta t_a = G_a\left((1-\nu^2)/Et\right) = 1/2A^2 \quad (7.39)$$

and therefore, for $t_a \rightarrow 0$, the forces T_1 and T_2 become:

$$T_1 = T_2 = P/2 \quad (7.40)$$

Figure 7.31 represents a typical shear stress distribution in the adhesive layer for identical adherents and relatively thick adhesive.

7.4.10.3 *Implications of stress distribution for adhesive bonding*

The above analysis highlights the problem of the stress being high at the edges of the joint. The basic design, in which the edges are perfectly square, is not optimal. Alternative designs are presented in Figure 7.32. The introduction of a taper or a fillet changes the stress profile within the joint and decreases the stress gradient at the edges.

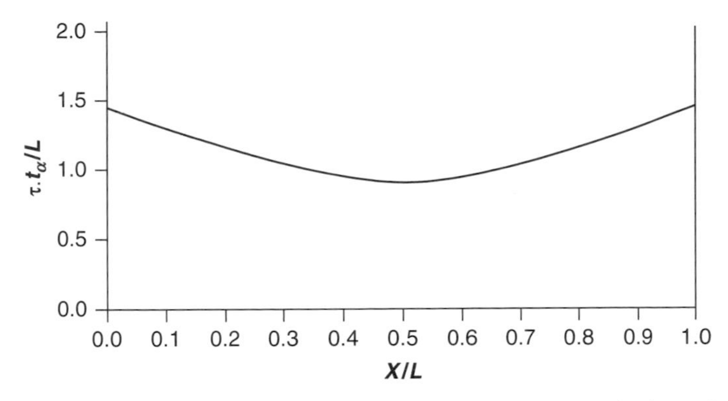

Figure 7.31 Typical shear stress distribution in adhesive bond line with identical adherents.

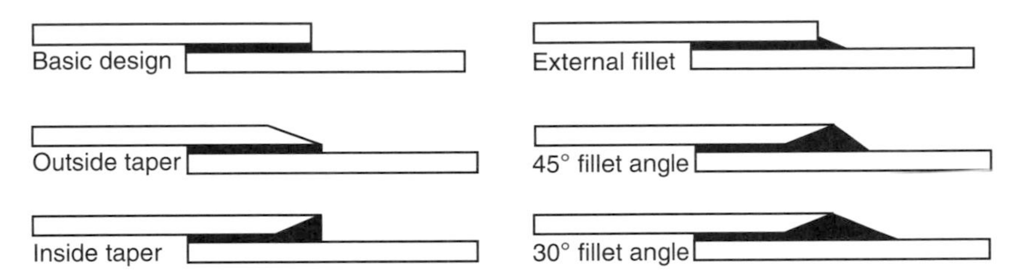

Figure 7.32 Designs of various joints indicting how edge changes influence strength.

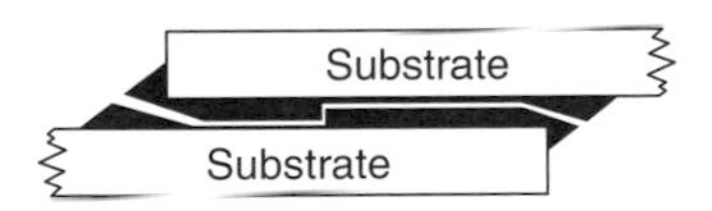

Figure 7.33 Profile of a typical cohesive fracture of a good joint.

The basic design gives the greatest stress gradient at the edge. Introduction of a taper into the edge will also reduce the stress profile as will the introduction of a fillet. As the fillet angle decreases so this will reduce the stress profile in the joint. A cohesive failure will result in a fracture which will vary as shown in Figure 7.33. In short joints the failure can move very close to the surface of the substrate, but there will usually be a thin layer of adhesive left over the surface on both substrates.

7.4.10.4 *Peel test on joints*

Joints can fail at lower stress as a consequence of peel forces than if the structure is subjected to shear or tension. The peel forces become significant when one of the substrates is flexible and the other is very rigid (see Figure 7.34). The tensile or cleavage stress σ_{11} at a distance, x, in the adhesive layer is given by:

$$\sigma_{11} = \sigma_P \left(\cos \beta_P x + \kappa_P \sin \beta_P x\right) \exp\left(\beta_P x\right) \tag{7.41}$$

where:

$$\beta_P = \left(\left(E_a b / \left(4 E_s I_e h_a\right)\right)\right)^{1/4} \quad \text{and} \quad \kappa_P = \left(\left(\beta_P M_P\right) / \left(\beta_P M_P + \sin \alpha\right)\right) \tag{7.42}$$

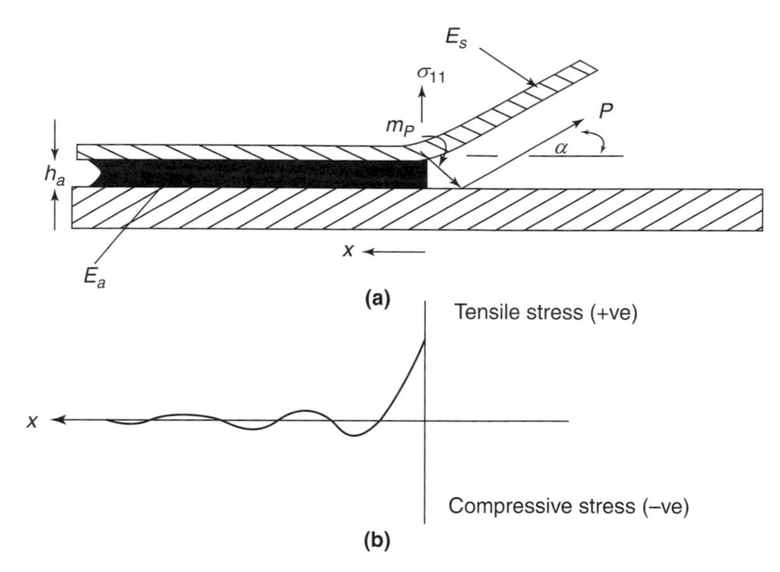

Figure 7.34 Peeling: (a) schematic diagram of peel test; (b) stress distribution in adhesive layer ahead of advancing peel front.

where σ_P is the boundary cleavage stress in the adhesive at $x = 0$, i.e. at the peel front, I_e is the moment of inertia of the flexible substrate cross-section, M_P is the moment arm of the peel force (peel energy × distance, Pm_p), α is the peel angle and β_P is a constant.

The predicted stress profile goes into compression as well as tension. This simple model is not strictly correct but it does illustrate that the stress profile depends on the flexibility of the upper substrate. The adhesives which are used to attach labels to rigid substrates are usually pressure sensitive. Essentially, they are plasticised polymers and exhibit significant viscoelastic effects, which are not included in Equation (7.42). The consequence of the viscoelasticity is that the energy dissipation and the effects of stringing across the gap can make a significant contribution to the fracture energy.

7.4.11 Service life of an adhesive joint

In the case of welded or riveted structures, the strength of the joint is determined by the corrosion and fatigue of the metal structure. In the case of bonded structures, it is the durability of the joint which is the controlling factor. The fracture strength is changed as a consequence of changes in the mechanical properties of the adhesive and of the interface. The service life is very dependent on the following factors:

- the period of the varying load or stress to which the joint is subjected
- the range of stresses experienced by the joint
- the frequency of the load or stress variation
- the ambient temperature
- the environment to which the joint is subjected

Below the glass transition temperature, T_g, the modulus and the energy dissipation are often fairly constant. In the context of an adhesive, the implication is that provided the joint does not approach T_g the strength of the joint will not vary significantly. Similarly since the energy dissipation is tied to T_g, the fracture strength will not vary significantly with temperature or frequency. However, as the glass transition point approaches, the possibility that the fracture will be temperature- or frequency-dependent arises. The major problem for joints which are exposed to a normal atmosphere is the ability of moisture to enter the resin system. The initial absorption of moisture will plasticise the resin; this has the advantage of plasticising the resin and increasing the energy involved in the fracture. However, the moisture can permeate to the interface and here it is disadvantageous and can aid corrosion and alter the interfacial energy. To achieve durable joints, careful preparation of the surface is critical. One of the most common applications is the bonding of aerospace structures. Many aircraft are created using aluminium alloys which are clad with pure aluminium in order to suppress corrosion. The alloys are used to impart ductility to the aluminium and to allow the fabrication of complex shapes. The surfaces to be bonded are usually etched to remove possible contamination and then a rough surface to aid bonding is created by anodising the surface oxide layer. The surface which is created depends on the nature of the anodisation process. Phosphoric and chromic–sulfuric acid processes (see Figure 7.35) create a surface oxide structure. Both the chromic–sulfuric acid and phosphoric acid treatments create an artificial oxide layer which both increases the area of contact between the oxide and resin and creates 'hollows', structures which are filled with resin and add to the lock and key contribution to the interfacial strength.

The phosphoric acid etch is used by the aerospace industry to improve the durability of the joint. In the case of aluminium, considerable effort has been directed at improving the durability of adhesive joints and special pretreatments have been developed to assist the creation of good bonds and improve durability. The pretreatments are of two types: the first type is purely protective and usually involves the use of a low viscosity reactive resin to seal the delicate oxide created using the etching and anodisation process. The second type of process involves the use of various formulations which will either react with or modify the surface of the oxide. Various organic silane compounds can be used which will react with the surface and produce reactive functions

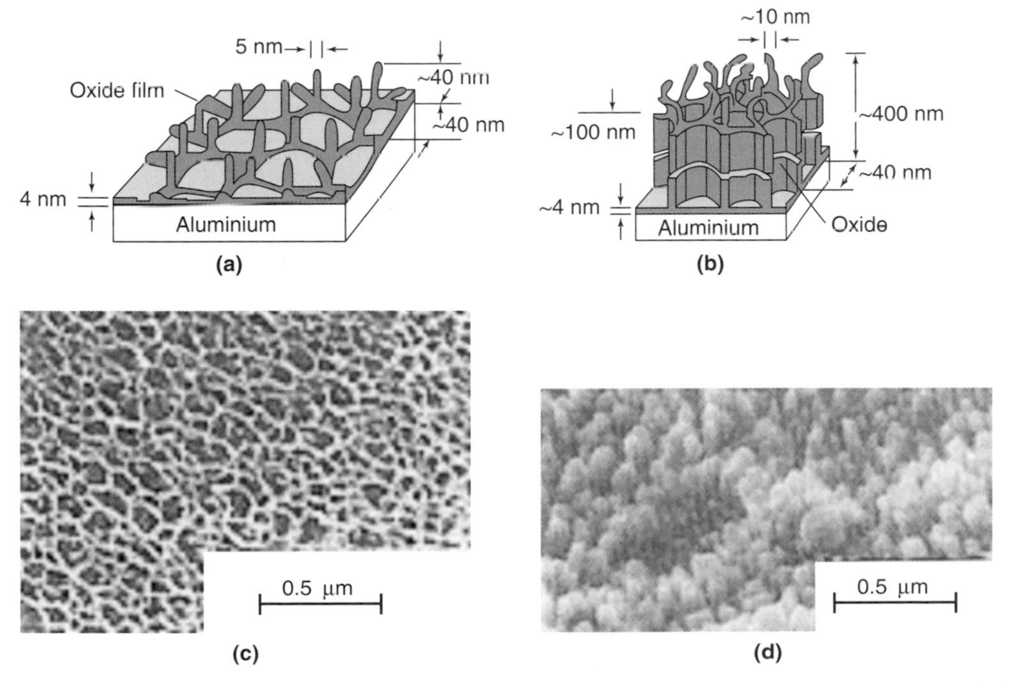

Figure 7.35 Surface structures: (a) schematic of surface structure created using chromic–sulfuric acid; (b) schematic of surface structure created using phosphoric acid; (c) etching of aluminium surface; (d) electron micrograph of surface.

which allow direct bonding between the resin and the oxide. Other formulations involve the use of zirconium and titanium compounds, which act in a similar fashion to the silanes. Whilst the simplistic view is that there is direct coupling between the resin and the oxide, in practice complex sol-gel reactions occur which create a textured surface which aids bonding and protects the oxide from hydrolytic reactions and toughens the resin.

7.4.12 Toughening of adhesives

Structural epoxy resins are intrinsically brittle and are prone to fracture. Many adhesives and resins are toughened by the incorporation of rubber particles into the matrix. Carboxy terminated butadiene acrylonitrile (CTBN) is initially dispersed in the epoxy resin to form a homogeneous mixture, but as polymerisation occurs so the CTBN phase separates to form micrometre-size rubber particles within the matrix. The rubber particles are able to act as stress release centres and increase the fracture toughness of the resin. An electron micrograph of a typical CTBN modified epoxy resin showing the detachment of the rubber from the resin matrix (see Figure 7.36) illustrates how the toughening mechanism operates. In the electron micrograph the holes in the matrix reflect the structure created by rubber particles being pulled out of the matrix. As the crack propagates the matrix squeezes the rubber particles and energy is dissipated by the distortion of the interface. CTBN levels of 2–6% are typically used. Recently, nanosilica particles dispersed in the epoxy matrix have been used for toughening. The nanoparticles are able to suppress crack growth. Toughening can also be achieved by incorporating nanosilica particles. Another very successful approach is to use a thermoplastic dispersed in the thermoset. The dispersion of polyethersulfone as small, hard, spherical particles is found to significantly enhance the fracture toughness of amine cured epoxy resins. This latter approach is used in the toughening of the resin in the carbon fibre composite used to create the tail of the Boeing 777 aircraft.

7.5 Polymers in corrosion protection

Corrosion is not a glamorous scientific topic and often receives little attention. However, the replacement of metals and other materials due to corrosion typically involves 4% of the gross national product of the manufacturing-based economies and is a particularly important topic for nanostructured systems. Of potentially greater interest is the ability to design coatings with special corrosion-related properties. The ability to design coatings to replace chromate-based corrosion protection films that are widely used in aircraft and other critical areas is of particular environmental importance. It is now established that conducting polymers, such as doped polyaniline (PANI), are effective at lowering the corrosion rates for metals exposed to aqueous solutions. But, the ability to add the doped PANI to normal coating formulations is limited because of its low solubility in water and many organic solvents. However, quaternisation of the PANI with lignosulfonates enhances the solubility of inherently conducting polymers in aqueous systems

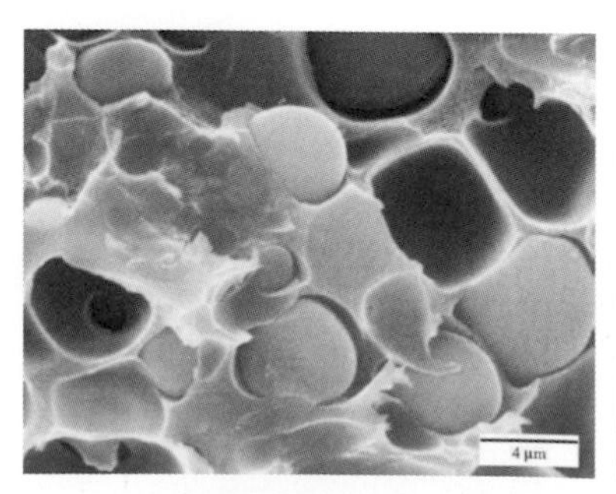

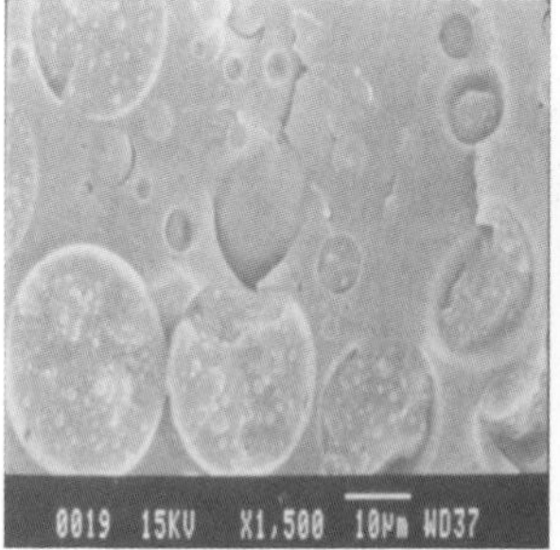

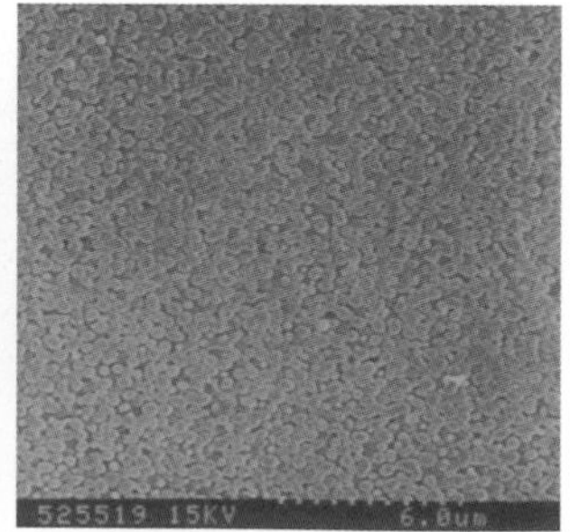

Figure 7.36 Electron micrograph of a fracture surface showing CTBN particles.

and allows the polymers to be included in water-based coating formulations. Another approach to the formation of a nanocomposite protective film involves a two-stage sol-gel process where acid-catalysed hydrolysis and condensation of tetramethoxysilane and glycopropyltrimethoxysilane is used to form organosilica nanoparticles with peripheral epoxy-functional groups. The nanoparticles are cross-linked through a conventional chemical reaction between amino- and epoxy-functionalities. The cross-linking and self-assembly occur as the solvent evaporates when a solution of particles, cross-linking species and a solvent are deposited on the surface to be coated. This self-assembled nanophase particle (SNAP) coating process forms a stable continuous and highly adherent protective film on the metal surface and has been used to engineer nanostructured composite ceramic coatings for aluminum aerospace alloys. The adhesion and structure of these organic/inorganic nanocomposite coatings can be tailored by changing the chemistry of the SNAP process.

More conventionally, glass-filled polymer coatings can be produced by incorporating glass platelets into the polymer matrix (see Figure 7.37). The platelets are able to align in the coating and reduce the percolation of gases and, most importantly, moisture through the coatings. Glass-filled polyester and isophthalate coatings have been used extensively for corrosion protection and can give a durability which approaches 25 years. The coatings are thermoset and the glass is incorporated at a level which achieves the type of barrier shown in Figure 7.37.

The glass platelets form a tortuous path for the moisture diffusion and the paths which the molecules have to take will be significantly longer than if they simply had to diffuse through the polymer coatings. The glass flake can improve the physical properties of the polymer matrix, imparting reinforcement and increased dimensional stability. The glass flake can increase tensile/flexural strength modulus, reduce shrinkage and warpage and increase the impact strength. The overlapping platelets can provide a barrier to the ingress of chemicals, increasing the chemical resistance, reducing liquid and vapour permeation and providing an in-situ barrier to oxygen permeation. Because the glass produces a physical barrier to oxygen ingress it can improve the fire resistance, aid char formation, produce a stable char and reduce the surface spread of flame and smoke emissions. The aligned glass flakes have been found to influence the tribological properties of the polymer composite, resulting in improved wear and abrasion resistance, improved thermal properties and an increase in the value of the T_g. If the glass flake is pretreated with silane, that will also enhance the interaction between the flake and the polymer and will lead to improved mechanical properties. Coatings in which flaked glass is dispersed are widely used as protective coatings where hazardous environments are experienced. Coatings containing glass flake are used as liners for oil pipelines, water distribution systems, tanks in ships, offshore drilling rigs, the Thames Barrier in London, the Forth Rail Bridge in Scotland, and a variety of containers where attack by the fluids of the metal structure would compromise the structures.

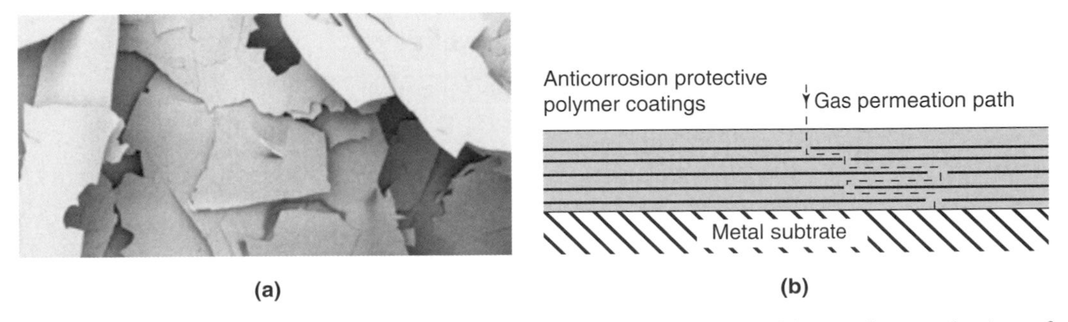

Figure 7.37 Corrosion resistant coatings: (a) electron micrograph of glass flakes used in production of corrosion resistant coatings; (b) schematic of a coating.

7.6 Gas diffusion through polymer matrices

The permeability of a gas through a polymer membrane, P, is the product of two quantities: the diffusion coefficient, D, and the solubility coefficient, S. The solubility of a gas in a membrane depends on the functionality of the polymer. If the polymer is polar it will be able to absorb polar molecules such as water more easily than a nonpolar gas such as nitrogen or oxygen. On the other hand, the diffusion coefficient depends on the ability of the gas molecules to find a path through the matrix. The path is created by motion of the chains, which allows movement of the gas molecules. The diffusion path is influenced by the crystallinity in a polymer such as polyethylene and the T_g in an amorphous polymer. The higher the value of the T_g the lower the diffusion coefficient.

The large variation in the permeability for a range of amorphous and crystalline polymers is illustrated in Figure 7.38. By varying the chemical structure and hence solubility, differences are observed in the permeability. Gas separation is achieved by the use of selective combinations of the polymer materials. Gas membranes are usually constructed using a rigid, open, cellular polymer matrix over which a thin more selective polymer membrane is deposited. The cellular structure only provides the support and the separation is achieved by the thin membrane layer.

The permeability of a gas through a matrix which includes aligned platelets will be described by a tortuous path model:

$$P_{tort} = P_0\left\{\left(1/\left[1+\left(\alpha^2\varphi^2/(1-\varphi)\right)\right]\right)\right\} \tag{7.43}$$

where α is the aspect ratio of the platelets and ϕ is their volume fraction. In the case of clay or a glass-filled system the aspect ratio may have values which are of the order of 100–1000. The net effect is that the permeability of the material to the gas is reduced by a factor which can be of the order of 10–100 (see Figure 7.39). The data points in Figure 7.39 represent the results obtained with different levels of platelets for different polymer matrices.

7.7 Selection of polymeric materials for particular applications

There are a number of computer programmes and websites which can aid the engineer in the selection of the correct material for a particular purpose. One such compilation had been provided by Waterman and Ashby (1997). These aids for materials selection are very useful and will help sort out the obviously incorrect materials from those which could possibly be used. As an example we will

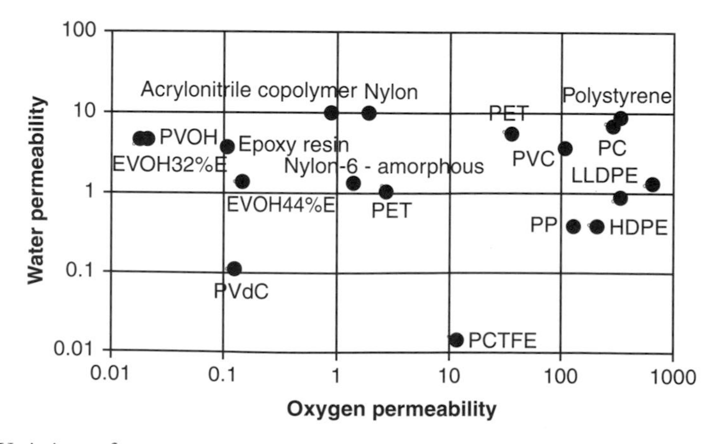

Figure 7.38 Variation of water permeability plotted against oxygen permeability for some common polymers.

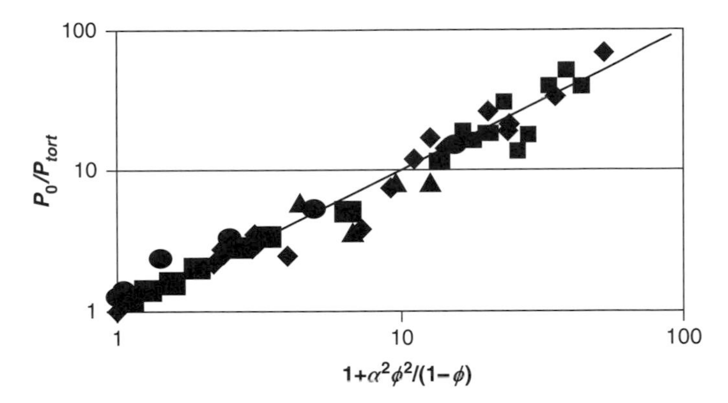

Figure 7.39 Plot of ratio of permeability against volume fraction of platelets in matrix.

consider the selection of an electrical connector housing for a car. The under bonnet environment is one in which extremes of temperature may be experienced: as low as −20°C when the car is parked in the open and in excess of 100°C when the car is in traffic on a very warm day. In fact, in areas close to the engine, the temperatures might approach 130–140°C. The electrical connector has to be produced cheaply, at high volumes and has to be 100% reliable. The implication is that the preferred method of manufacture would be injection moulding and therefore a thermoplastic would be preferred rather than a thermoset material. From the point of view of recycling, a thermoplastic would also be preferred as there are fewer problems when attempting to recover the material from scrap.

Because of the requirement for the plastic to be used over a significant temperature range, and recognising that the fixture will probably be attached to a metal chassis, stresses would be kept to a minimum if the thermal expansion coefficient were to have a value of 5×10^{-5}. The materials which have values of this order include: nylons, polyphenylene oxide, polycarbonates, polyesters, polyetherimides, some polypropylene polymers, polyphenylene sulfides, some PVCs and polysulfones.

The second requirement to be considered is the temperature of the engine compartment. For the unit to have mechanical stability it is desirable to have a glass transition temperature which is well above the operational temperature. A rule of thumb would be to have the glass transition temperature 20–30°C above the maximum operating temperature. In this illustration we will consider that an appropriate temperature would be 180°C. Of the above list of potential materials only nylon 6,6 (glass-filled), polyester PBT (30% glass-filled), polyethersulfone and polyphenylenesulfide match the criteria.

The electrical housing is a snap assembly and hence there are critical requirements in terms of flexibility and an appropriate value for the flexural modulus would be 4 GN m^{-2}. This criterion eliminates all the glass-filled materials and leaves polyethersulfone and polyarylsulfone as potential candidates.

The unit will be operated in an environment in which a high level of aliphatic hydrocarbons will be present for most of the time, hence the final selection can be made on the basis of resistance to ESC when exposed to hydrocarbons. In this context the polyethersulfone exhibits slightly better characteristics.

Having identified a candidate, it is always important to check that there are no adverse properties which might exclude that material from the proposed application. Polyethersulfone can be operated to −40°C, it is creep resistant, it resists oxidative degradation and retains its electrical resistance when operated at high temperatures for long periods. It is the appropriate material for the job.

The usual mistake in materials selection is to forget to include one or more of the critical factors in the selection list. The ability of the engineer to obtain the right material for the job depends critically upon the engineer's ability to identify all the possible factors which could influence the operation of the device. Having selected polyethersulfone as a candidate material it is important to check how many grades are available and whether or not they all have the desired characteristics. It is not unusual to find that the materials which are most easily moulded have slightly inferior physical characteristics since they will contain a higher proportion of lower molar mass material. Postfailure analysis has often shown that the generic material may have been correctly selected, but the wrong grade was used and thus failure ensued. A correctly selected plastic can often be the most appropriate material, providing the right strength, flexibility weight balance for many critical applications.

Brief summary of chapter

- This chapter has attempted to show how knowledge of the basic polymer science presented in the introductory chapters can be applied to help understand issues which will be encountered by practising engineers.

References and additional reading

Asby M.F. and Jones D.R.H. *Engineering Materials: An Introduction to Their Properties and Applications*, Pergamon Press, Oxford, 1980.

Kinloch A.J. *Adhesion and Adhesives*, Chapman and Hall, London, UK, and New York, NY, USA, 1987.

Kinloch A.J. *Developments in Adhesives*, Applied Science Publishers, London, UK, 1981.

Pocius A.V. *Adhesion and Adhesives Technology*, 2nd edn., Hanser, Munich, Germany, 2002.

Wright D.C. *Environmental Stress Cracking of Plastics*, RAPRA, Shawbury, UK, 1996.

Pethrick R.A. and North A.M. *Chemistry World*, 2008, **5**(12), 34.

Waterman N.A and Ashby M.F. (Eds.) *The Materials Selector Vols 1–3*, 2nd edn., Chapman and Hall, London, UK, 1997.

8

Polymer chemistry and synthesis

8.1 Introduction

The synthesis of high molar mass polymer materials is achieved by the formation of a chemical bond between low molar mass entities. The physical properties of the polymers formed depend on their chemical structure and molar mass. How quickly a chain is created will depend critically upon the detail of the chemical processes involved in the synthetic process. If the polymer is created from a vinyl monomer, the product is likely to be different from that obtained from a condensation reaction, which builds the polymer chain via a series of stepwise processes. Polymers can be formed by opening the ring of strained ring structures and, depending on the species attacking the ring, the process may either resemble the chain-like process typical of vinyl polymerisation or the stepwise process of condensation polymerisation.

8.2 Condensation polymerisation

This process is typically used for the creation of polyesters and nylon. The detailed synthesis involves the use of a catalyst but the overall kinetics of the reaction can be described by a general reaction scheme. It is assumed that the rate of the reaction is independent of chain length and is purely defined by the functional groups involved in the reaction. But as polymers are formed the viscosity will increase and there will be a time required for the functional groups to 'find' each other and the processes are considered to be *diffusion* controlled. The rate of chain growth is essentially equal to the rate of esterification, which is in turn proportional to the rate of the functional groups consumption. As an illustration we will consider the formation of polyester when the process is acid catalysed but no added acid is present:

H H H H O O HO OH H H + HO H H H H H H OH H H H H H H

k_1 → H H H H O O H H H H H H HO H H O OH H H H H H H

The rate of the reaction is designated by the rate constant k_1. The rate of esterification is therefore:

$$-\partial[COOH]/\partial t = -\partial[OH]/\partial t = k_1[OH][COOH]^2 \qquad (8.1)$$

In this equation, the term $[COOH]$ is raised to the power two, indicating that the process requires the presence of acid to proceed. However, if a strong acid is added then the equation has the form:

$$-\partial[COOH]/\partial t = -\partial[OH]/\partial t = k_1[OH][COOH][ACID] \qquad (8.2)$$

where [*ACID*] represents the concentration of a strong acid which is added to catalyse the process. However, since this acid is not incorporated into the polymer its concentration remains constant and can be incorporated into a new rate constant k_1' which is related to the original via:

$$k_1\left|H^+\right| = k_1' \tag{8.3}$$

For simplicity we will also assume that the initial concentrations of the [COOH] and [OH] groups are the same and equal to *c*. The rate of esterification for the case where there is no added acid will then have the form:

$$-dc / dt = k_1 c^3 \tag{8.4}$$

and for the acid catalysed case:

$$-dc / dt = k_1' c^2 \tag{8.5}$$

Integration of Equation (8.4) yields:

$$-\int_{C_0}^{C} dc \Big/ C^3 = \int_0^t k_1 dt$$

$$\left(1 / C^2\right) - \left(1 / C_0\right) = 2k_1 t \tag{8.6}$$

where C_0 is the initial concentration and C is the concentration after some time t. In the case of the acid catalysed system, integration of Equation (8.5) yields:

$$-\int_{C_0}^{C} \frac{dc}{C^2} = \int_0^t k_1' dt$$

$$\left(1 / C\right) - \left(1 / C_0\right) = k_1' t \tag{8.7}$$

The growth of the polymer chain is related to the extent of the reaction which is designated by the symbol p, and defined as:

$$C = C_0\left(1 - p\right) \tag{8.8}$$

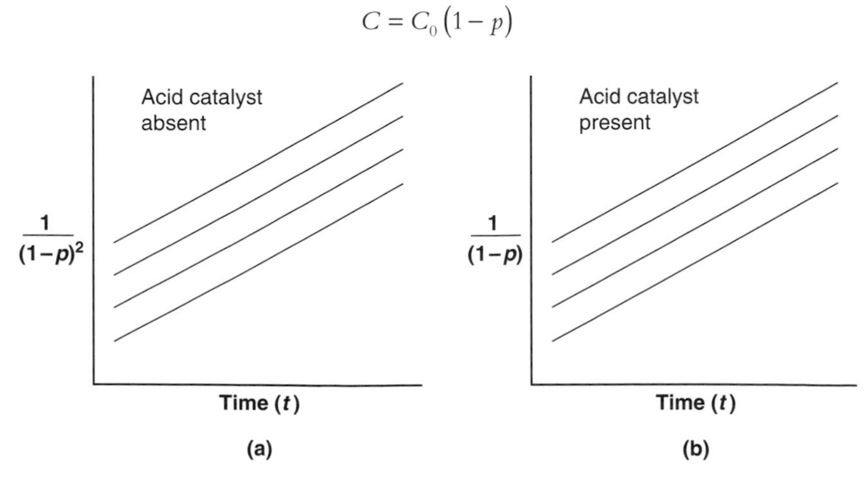

Figure 8.1 Kinetic plots for esterification process: (a) without strong acid catalyst; (b) with strong acid catalyst. Parallel lines represent measurements taken at different temperatures.

The concentration C is therefore the number of functional groups which are remaining at the time t. Combining Equations (8.6) and (8.8) yields:

$$C_0^2 / C^2 - 1 = C_0^2 2k_1 t$$

$$\left(1/(1-p)^2\right) - 1 = C_0^2 2k_1 t \tag{8.9}$$

Thus a plot of $1/(1-p)^2$ against t should produce a straight line from which k_1 can be calculated (see Figure 8.1).

8.2.1 Degree of polymerisation and molar mass

The mean degree of polymerisation $\overline{DP}_n$ is the number of monomer units in a polymer chain. Therefore:

$$\overline{DP}_n \times \text{molar mass of monomer} = \text{number average molar mass}$$

If we define the number of structural units (monomers) present as N_0 and the number left after time t as N then:

$$N = N_0\left(1-p\right) \tag{8.10}$$

If we equate $N_0 \equiv C_0$ and $N \equiv C$, and recognise that N is the number of polymer chains at time t, so that the average number of units in each chain is:

$$\overline{DP}_n = N / N_0 = \left(N_0\right) / N_0\left(1-p\right) \tag{8.11}$$

then:

$$\overline{DP} = 1/\left(1-p\right) \tag{8.12}$$

In practice, if we plot a graph of $\overline{DP}$ against t it is often highly linear up to a value of 100. If the molar mass of the monomer is ~100, this implies that the above approximation holds up to a molar mass of 10,000, which is a typical value for a condensation polymer. Table 8.1 summarises the variation of the molar mass with the extent of reaction.

Table 8.1 indicates that when half of the reactive groups present have been consumed the average molar mass is only twice that of the original starting monomer. In other words, the process builds polymer chains very slowly. Even when 90% of the groups have been consumed, the molar mass is only 1,000, which would be considered an *oligomer* and not really a polymer. It requires 99.9% of the groups to have reacted to produce a reasonable chain length. If there are even a small number of monomers which only have one functionality (impurities) then it may be impossible to create a high molar mass material.

8.2.2 Molar mass distribution

Not all polymer chains will be of the same length at any given point in time and we would expect to find a distribution of chain lengths in any sample. The probability of finding a particular

Table 8.1 Illustration of way in which molar mass grows with degree of reaction

Reaction (%)	p	$\overline{DP}_n$	Average molar mass*
0	0	1	100
50	0.5	2	200
90	0.9	10	1000
99	0.99	100	10,000
99.9	0.999	1000	100,000

* Assumes molar mass of monomer is 100.

chain length will depend on the statistical probability of reaction having occurred. It is assumed that all groups are equally likely to undergo reaction. The extent of the reaction p is defined as the probability that a functional group has reacted at time t. The definition is effectively the equivalent of that presented above and the probability of finding a functional group *unreacted* is $1 - p$.

The probability that a given molecule selected at random contains x monomer units containing a reactive group is by definition p. Therefore the probability of finding $x-1$ such reacted groups in a molecule is:

$$p^{(x-1)} = (\text{Prob1 Prob2 Prob3} \ldots)$$

where the integers indicate the number of groups incorporated in the growing chain. The presence of an unreacted group, i.e the probability that a single group has not been incorporated in a chain, is $1 - p$. Hence the total probability of finding the complete molecule is:

$$p^{x-1}(1-p) \tag{8.13}$$

and is also equal to the fraction of all the polymers which have x-mers in the chain. If there are N polymer molecules present and the number of x-mers is N_x then:

$$N_x/N = p^{x-1}(1-p) \tag{8.14}$$

Now if the total number of monomers present was originally N_0 then:

$$N/N_0 = (1-p) \tag{8.15}$$

or

$$N_x = N_p^{x-1}(1-p) = N_0(1-p)^2 p^{x-1} \tag{8.16}$$

The N_x is the number average distribution function of a linear condensation polymerisation and when mapped as an extent of reaction has the form shown in Figure 8.2. Thus on a number basis monomer units are always more plentiful than any other x-mer even at the highest extent of

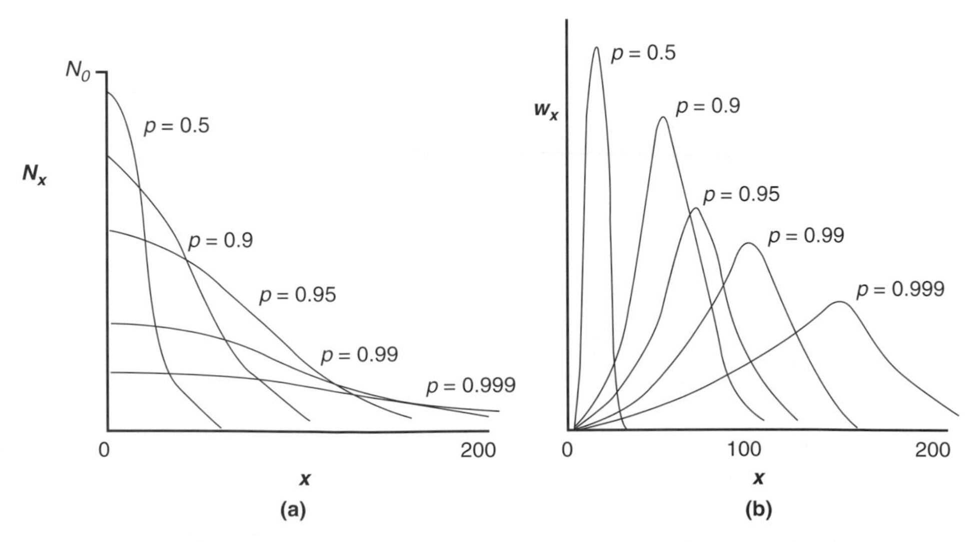

Figure 8.2 Variation of N_x with x for various extents of reaction (a) and the weight distribution function w_x (b).

reaction. However, a better measure of the distribution is to consider the *weight distribution function*, w_x, since this takes into account the weight of each x-mer in the sample:

$$w_x = (xN_x)/N_0 \tag{8.17}$$

where xN_x is the relative weight of x-mers multiplied by the number of x-mers present and N_0 is the weight of all monomer units initially present. Equation (8.17) can be modified by incorporation of Equation (8.16) to give:

$$w_x = \left[(xN_0(1-p)^2 p^{x-1})/N_0\right] = x(1-p)^2 p^{x-1} \tag{8.18}$$

If we plot the weight distribution function, we obtain Figure 8.2(b). The maximum in the weight fraction moves to higher molar mass as p approaches 1. It is possible to calculate the weight average degree of polymerisation $\overline{DP_w}$, which has the form:

$$\overline{DP_w} = (1-p)/(1+p) \tag{8.19}$$

Hence the molar mass distribution $\overline{DP_w} / \overline{DP_n}$ is given by:

$$\overline{DP_w} / \overline{DP_n} = ((1+p)(1-p))/(1-p) = 1+p \tag{8.20}$$

Thus as p tends to 1, i.e. 100% of the monomers have reacted, then the molar mass distribution will tend to 2; the breadth of the distribution increases as the reaction proceeds.

8.2.3 Molar mass control

The molar mass can be limited by simply cooling a reaction mixture at an appropriate time; however, on a large scale this is an inconvenient method of controlling the reaction. A cleaner method is to introduce into the reaction mixture a certain amount of a monofunctional reagent, which will end cap the growing polymer. Suppose we have N_A hydroxyl groups and N_B carboxylic groups and $N_B > N_A$ and the ratio $r = N_A / N_B$. The total number of monomers present will be equal to $(N_A + N_B)/2 = N_A(1+(1/r))/2$. The fraction of A groups reacted is equal to p and the fraction of B groups reacted will be rp. Then the fraction of A groups remaining will be $N_A(1-p)$ and the number of B groups remaining will be $N_B(1-rp)$. Therefore the total number of groups remaining will be:

$$N_A(1-p) + N_B(1-rp) = N_A((1-p)+(1-rp)/r) \tag{8.21}$$

At this point the number of polymer molecules present is therefore:

$$N_A / 2((1-p)+(1-rp)/r) \tag{8.22}$$

The factor of 2 reflects that two groups join to form a single bond. Thus $\overline{DP_n}$ is equal to the total number of monomer units divided by the number of polymer chains:

$$\overline{DP_n} = \frac{N_A(1+(1/r))/2}{N_A(1-p+(1-rp)/r)/2} = \frac{1+r}{1+r-2rp} \approx \frac{1+r}{1-r} \quad \text{as } p \to 1 \tag{8.23}$$

Thus if 1 mol % of a monofunctional acid is added to the mixture then the maximum number average molar mass achievable is:

$$\begin{aligned} &= (1+(100/101))(1-(100/101)) \times \text{molar mass monomer} \\ &= 201 \times \text{molar mass monomer} \end{aligned} \tag{8.24}$$

This calculation illustrates how the absence of impurities is crucial if a high molar mass is to be obtained. In particular, the molar ratio between the reacting groups must be very close to unity otherwise the chain growth is limited. In practice, there is always the possibility of the two ends reacting to give a cyclic product, although usually once a significant chain length has been achieved then the probability of this occurring becomes statistically less. Polymerisation in solution can, however, enhance the probability of finding cyclic polymers.

8.3 Vinyl polymerisation

A chain reaction has three essential steps: initiation, propagation and termination. The initiation process involves the creation of a species that can interact with a monomer to produce a growing chain which contains an active centre. In the case of vinyl monomers, the species adding to the monomer may either create a radical, anion or cation. Depending upon the nature of the active centre, so there will be different possible chemical processes which can terminate the reaction. However, it is possible to describe the general processes which are occurring in terms of some simple mathematical relationships.

8.3.1 Initiation

The initiation process will involve the creation of the *active centre*. The most commonly used initiators for vinyl polymerisation are *free radicals*. Free radicals are usually created by the dissociation of a compound which contains a weak bond. The sources of free radical commonly used are peroxides or azo compounds (see Table 8.2). Changing the structure of the groups attached to the weak bond can significantly influence the temperature at which the molecule decomposes and hence is able to initiate reaction. In practice, it is not unusual for manufacturers to use a cocktail of various molecules so as to achieve a smooth initiation of the reaction process. In many reactions, there is a significant amount of energy created by the chemical reaction and this exothermal energy can be used to trigger subsequent chemistry. The peroxides will decompose to produce reactive peroxide radicals, whereas the azo compounds will decompose to give radicals with the evolution of nitrogen. The gas evolved can be an advantage if foaming is required but can be a disadvantage when a consolidated resin is required as the final product.

The initiators, summarised in Table 8.2, illustrate the way in which change in the structure of the peroxide will change its reactivity. The self-accelerating decomposition temperature (SADT) is an indication of the storage problems which might be encountered with a particular system. Once this temperature has been exceeded the molecule will spontaneously decompose. The SADT is a reflection of the bond strength, the lower the SADT the weaker the bond, and can be an issue if the initiator is to be transported as would be the case in coating applications. A system with a low SADT cannot be safely transported in unrefrigerated containers or by air, which may create difficulties with onsite polymerisation processes.

8.3.2 Kinetics of vinyl or addition polymerisation

The kinetics of the radical chain process can be summarised in terms of three steps:

- *Initiation*: Generation of the free radicals which will produce polymerisation.
- *Propagation*: Growth of the polymer chains by addition of monomers to the free radical *active centre*.
- *Termination*: Processes whereby the free radical centre is converted into a stable entity and the free radical is destroyed.

Table 8.2 Some commonly used peroxide and azo initiator systems

Molecule	Chemical structure	SADT (°C)
Methylethylketonepreoxide	$HOO-C(CH_3)(C_2H_5)-O-O-C(CH_3)(C_2H_5)-OOH$	60
2,5-Dimethyl-2,5-ditertbutylperoxylhexane	$CH_3-C(CH_3)_2-O-O-C(CH_3)_2-CH_2-CH_2-C(CH_3)_2-O-O-C(CH_3)_2-CH_3$	80
Cumylhydroperoxide	$C_6H_5-C(CH_3)_2-O-OH$	55
Ditertbutylperoxyisopropylbenzene	$CH_3-C(CH_3)_2-O-O-C(CH_3)_2-C_6H_4-C(CH_3)_2-O-O-C(CH_3)_2-CH_3$	80
Di-2-ethylhexylperoxydicarbonate	$CH_3-(CH_2)-CH(C_2H_5)CH_2O-C(=O)-O-O-C(=O)-OCH_2CH(C_2H_5)-(CH_2)-CH_3$	5
2,2′-Azobisisobutyronitrile	$CH_3-C(CH_3)(CN)-N{=}N-C(CH_3)(CN)-CH_3$	50
2,2′Azodi-2-methylbutyronitrile	$CH_3CH_2-C(CH_3)(CN)-N{=}N-C(CH_3)(CN)-CH_2CH_3$	45

8.3.2.1 *Initiation*

In the simplest illustration, initiation may be achieved by the decomposition of a molecule such as benzoyl peroxide or azobisisobutyronitrile (AIBN).

Alternatively, initiation can be achieved at room temperature or low temperature by irradiation with UV light, usually using a mercury lamp operating at 360 nm.

$$RO{-}OR \xrightarrow{h\upsilon} 2RO^{\bullet} \quad \text{or} \quad RN{=}NR \xrightarrow{h\upsilon} 2R^{\bullet} + N_2$$

Yet another convenient source of radicals is oxidation/reduction reactions.

$$Fe^{2+} + ROOR \longrightarrow Fe^{3+} + RO^{\ominus} + RO^{\bullet}$$

8.3.2.2 *Propagation*

Initiating or primary radicals react stepwise with monomers in propagation reactions

$$R^{\bullet} + CH_2{=}CH(C_6H_5) \longrightarrow RCH_2{-}CH^{\bullet}(C_6H_5)$$

$$RCH_2{-}CH^{\bullet}(C_6H_5) + CH_2{=}CH(C_6H_5) \longrightarrow RCH_2{-}CH(C_6H_5){-}CH_2{-}CH^{\bullet}(C_6H_5)$$

The radical chain grows quickly and therefore high molar mass polymer is produced soon after the reaction starts. Primary radicals are being generated all the time so that new chains are created continuously. Thus, unlike in condensation polymerisation, we have a great variety of chain lengths soon after polymerisation starts when the total monomer loss or conversion to polymer is ~5%.

8.3.2.3 *Transfer*

If there are molecules in the reaction mixture which can undergo a chemical reaction with the radicals then *transfer* reactions can occur. The reaction will usually involve the transfer of a proton to the growing polymer to produce a stable molecule and generate another radical which can initiate further polymerisation. If the transfer involves a monomer molecule, it is known as a *monomer* transfer.

$$RCH_2{-}CH^{\bullet}(C_6H_5) + CH_2{=}CH(C_6H_5) \longrightarrow RCH{=}CH(C_6H_5) + CH_3{-}CH^{\bullet}(C_6H_5)$$

As a general rule transfer reactions do not influence the rate of consumption of monomer, since radicals are not lost in the process. If the molecule is either an impurity or purposely added then a similar reaction may occur.

$$RCH_2{-}CH^{\bullet}(C_6H_5) + HCCl_3 \longrightarrow RCH{-}CH_2(C_6H_5) + CCl_3^{\bullet}$$

The CCl_3 radical can then initiate polymerisation.

8.3.2.4 *Termination*

The growing radicals are terminated (destroyed) by several processes. The main processes are disproportionation and recombination. The termination can involve an added species which 'mops up' radicals, either because it is a radical itself or because it can form stable radicals.

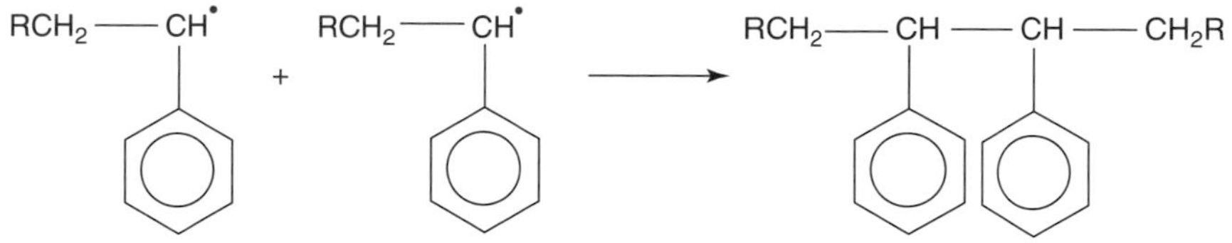

$$RCH_2{-}\dot{C}H(C_6H_5) + RCH_2{-}\dot{C}H(C_6H_5) \longrightarrow RCH_2{-}CH(C_6H_5){-}CH(C_6H_5){-}CH_2R$$

Recombination

$$RCH_2{-}\dot{C}H(C_6H_5) + RCH_2{-}\dot{C}H(C_6H_5) \longrightarrow RCH_2{-}CH_2(C_6H_5) + CH(C_6H_5){=}CHR$$

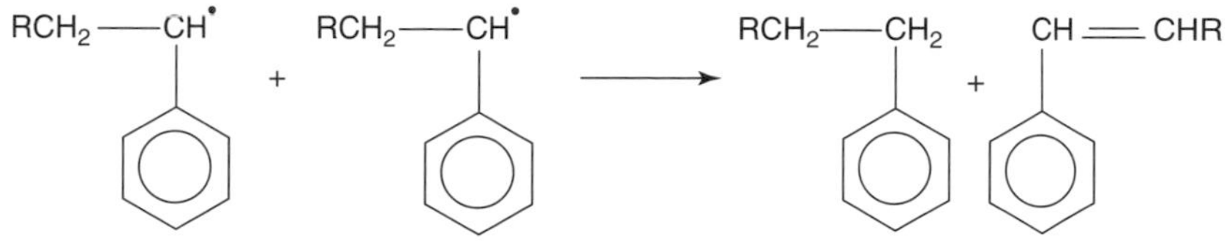

Disproportionation

The most important of all is the reaction of polymeric radicals with dissolved oxygen. The latter has two unpaired electrons and readily forms peroxyl radicals

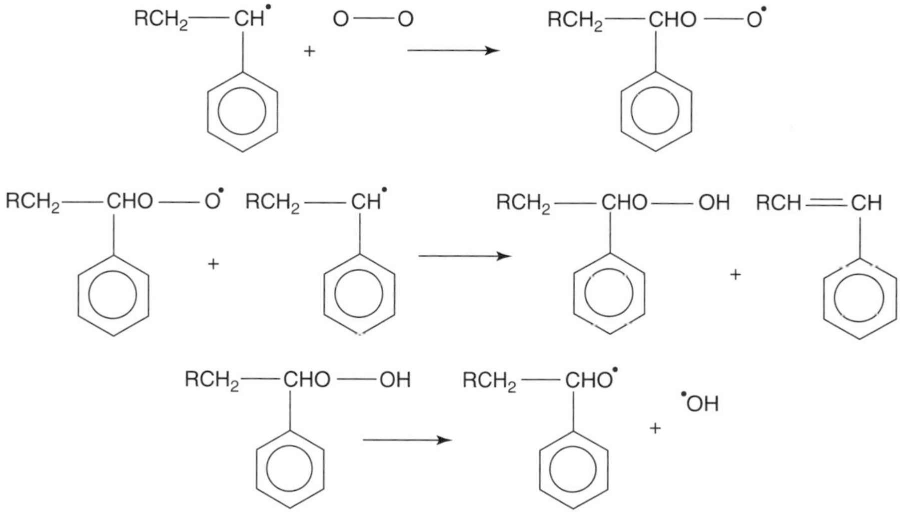

$$RCH_2{-}\dot{C}H(C_6H_5) + O{-}O \longrightarrow RCH_2{-}CH(C_6H_5)O{-}\dot{O}$$

$$RCH_2{-}CH(C_6H_5)O{-}\dot{O} + RCH_2{-}\dot{C}H(C_6H_5) \longrightarrow RCH_2{-}CH(C_6H_5)O{-}OH + RCH{=}CH(C_6H_5)$$

$$RCH_2{-}CH(C_6H_5)O{-}OH \longrightarrow RCH_2{-}CH(C_6H_5)\dot{O} + \dot{O}H$$

For fast free radical polymerisation, oxygen must be excluded by using either a nitrogen atmosphere to blanket the reaction mixture or a vacuum over the reagents.

8.3.3 Kinetics of free radical polymerisation

The polymerisation process:

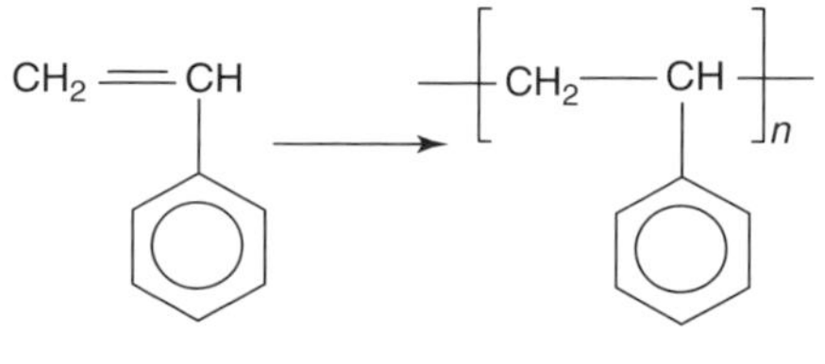

$$CH_2{=}CH(C_6H_5) \longrightarrow \left[CH_2{-}CH(C_6H_5) \right]_n$$

has an unfavourable standard entropy change; however, it is also a highly exothermic process: ΔH^0 and the resultant standard free energy change ΔG^0 are sufficiently negative to regard the reaction as going to completion in favourable circumstances. The elementary reactions can be treated similarly

$$I_2\left(initiator\right) \rightarrow 2I^*\left(radicals\right) \quad k_i \text{ initiation}$$

The initiator I_2 decomposes to produce two radical species with a rate constant k_i. The initiator radical reacts with the monomer M to produce a radical species which can propagate polymerisation.

$$\left.\begin{aligned} I^* + M &\rightarrow IM^* \ k_1 \\ IM^* + M &\rightarrow IM_2^* \ k_2 \\ IM_2^* + M &\rightarrow IM_3^* \ k_3 \end{aligned}\right\} \quad \text{Propagation}$$

$$RM^* + RM^* \rightarrow -RMMR \ k_{tc} \quad \text{Recombination}$$

$$RM^* + RM^* \rightarrow RM(+H) + RM(-H) \ k_{td} \quad \text{Disproportionation}$$

$$RM^* + M \rightarrow RM(+H) + M(-H) \ k_{tm} \quad \text{Monomer transfer}$$

where $RM(-H)$ and $RM(+H)$ are, respectively, a radical minus a hydrogen atom and a radical plus a hydrogen atom. Both species will be stable whereas $M(-H)$ is a radical and can initiate further polymerisation. We can make a number of assumptions:

- The reactivity of the radicals is independent of chain length k_p.
- The most dominant processes are propagation, initiation, transfer and termination.
- The radical concentration remains constant during the polymerisation process. Note that the rate of loss of I is: $I = -d[I]/dt = k_i\left[I\right]$ but the rate of formation of radical I is given by $d[R^*]/dt = 2k_i\left[I\right]$ since $2I$ are formed from one I_2. The rate at which radicals are formed is equal to the rate of destruction, thus the rate of formation of primary radicals is:

$$2k_i\left[I\right] = R_i \quad \left(\text{rate of initiation}\right)$$

- The rate of destruction of polymeric radicals $= k_t\left[R_n^*\right]^2$; the reaction will remove two radicals per reaction. There is no need to distinguish between recombination and disproportionation since both processes remove two radicals per reaction.

According to the steady-state assumption:

$$2k_i\left[I\right] = 2k_t\left[R_n^*\right]^2 \tag{8.25}$$

The rate of initiation must be balanced by the rate of termination. Rearranging Equation (8.25) gives:

$$\left[R_n^*\right] = \left(\left(k_i[I]\right)/k_t\right)^{1/2} = \left(k_i/2k_t\right)^{1/2} \tag{8.26}$$

The rate of reaction or polymerisation R_p is the rate at which monomer is consumed and polymer formed, i.e. $-d\left[M\right]/dt$. Thus from the above assumptions we can write:

$$-d\left[M\right]/dt = k_p\left[M\right]\left[R_n^*\right] \tag{8.27}$$

or

$$R_p = k_p \left(R_i / 2k_t\right)^{1/2} \left[M\right] \tag{8.28}$$

Substituting for $\left[R_n^*\right]$

$$-d\left[M\right]/ dt = k_p \left(k_i / k_t\right)^{1/2} \left[I\right]^{1/2} \left[M\right] = R_p \tag{8.29}$$

8.3.4 Experimental measurement of polymerisation kinetics

The particular method has to be tailored to the particular monomer system but there are some methods which are generally applicable for most systems.

- *Spectroscopic methods*: The polymerisation process usually will involve the disappearance of some chemical grouping and the appearance of another species. It is often possible to follow these changes using infrared or nuclear magnetic resonance (NMR) spectroscopy. Infrared spectroscopy is ideal for the study of isocynanate being converted to polyurethane using the disappearance of the –C≡N stretching mode, or in the case of the epoxy resin the disappearance of the ring stretching mode at 915 cm^{-1}. Unfortunately the –C=C– is often not infrared active and therefore is not easily studied with this technique, but it can be investigated using Raman scattering spectroscopy. In favourable cases, the disappearance of double bonds can be studied using NMR spectroscopy and some changes in the structure can produce spectral shifts which allow measurements of the kinetics.
- *Gravimetric methods*: The most direct method of following the rate of polymerisation is to withdraw samples as a function of time from the reaction mixture and determine the amount of polymer which has been created by precipitation of the polymer solid. However, this method is rather tedious.
- *Dilatometry*: Since the density of the monomer and the polymer are usually different there is a change in volume as the polymerisation occurs. In most cases the volume decreases as the density increases. From the known densities of the monomer and the polymer, the volume contraction can be converted into an extent of reaction versus time plot (see Figure 8.3).

Analysis of the initial slope of the line in Figure 8.3 provides the quantity $-d[M]/dt$ in Equation (8.29).

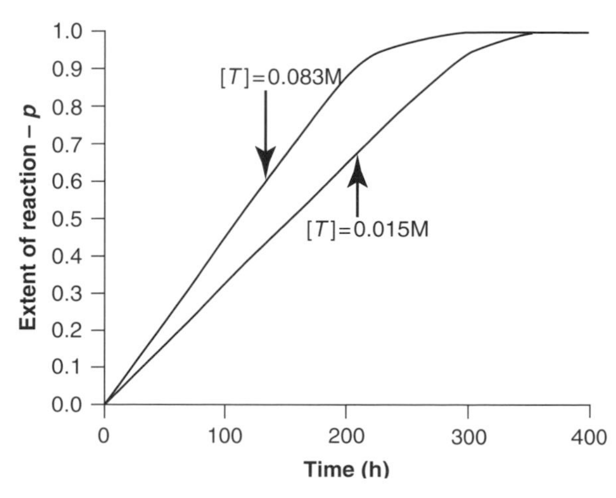

Figure 8.3 Plot of extent of reaction against time for a vinyl polymerisation process carried out with different levels of catalysts.

In the simplest case presented above the slope will vary as $[I]^{1/2}$ and it is possible from a series of experiments to determine the combined rate constant $k_p(k_i/k_t)^{1/2}$. If the rates of reaction are determined at various temperatures, it is possible to measure the activation energy from the polymerisation process. The activation energy for the propagation process is necessarily very low and the temperature dependence is usually determined by the energy required for the initiation of the reaction: the bond energy of the initiator. The activation energy for the termination process is usually very low and does not play a significant role in influencing the overall activation energy. There are, however, cases where k_t can have an influence, but these will not be discussed here.

8.3.5 Molecular weight and $\overline{DP_n}$

The $\overline{DP_n}$ in the absence of transfer reactions can be defined as:

$$\overline{DP_n} = \frac{\textit{rate at which monomer is consumed}}{\textit{rate at which stable polymer is formed}} \tag{8.30}$$

which is also equal to:

$$\overline{DP_n} = \frac{\textit{rate of propagation}}{\textit{rate of polymer formation by termination}} \tag{8.31}$$

For termination by recombination:

$$R^* + R^* \rightarrow -R-R- k_{tc} \tag{8.32}$$

The rate of polymer production is defined by:

$$d[polymer]/dt = k_{tc}\left[R_n^*\right]^2 \tag{8.33}$$

where R_n^* is the growing polymer chain, which is a radical and contains n monomer units. Therefore:

$$\overline{DP_n} = \left[k_p[M]\left[R_n^*\right]\right] / \left(k_{tc}\left[R_n^*\right]^2\right) = \left(k_p[M]\right) / \left(k_{tc}\left[R_n^*\right]\right) \tag{8.34}$$

but

$$\left[R_n^*\right] = \left(\left(k_i[I]\right)/k_{tc}\right)^{1/2} \tag{8.35}$$

Thus:

$$\overline{DP_n} = k_p[M] / \left(\left(k_i k_{tc}\right)^{1/2}[I]^{1/2}\right) \tag{8.36}$$

For termination by disproportionation:

$$-R^* + -R^* \rightarrow -R(+H) + -R(-H)\ k_{td} \tag{8.37}$$

$$d[polymer]/dt = 2k_{td}\left[R_n^*\right]^2 \tag{8.38}$$

The factor of 2 in Equation (8.38) reflects the creation of two stable polymer chains as a result of this process.

$$\overline{DP_n} = \left[k_p[M]\left[R_n^*\right]\right] / \left(2k_{tc}\left[R_n^*\right]^2\right) = \left(k_p[M]\right) / \left(2(k_i k_{tc})^{1/2}[I]^{1/2}\right) \tag{8.39}$$

In the presence of a transfer agent, T, the reaction will have the form:

$$-R_n^* + T \rightarrow -R(+H) + T^*(-H) \quad k_{tT}$$

and in the case of transfer to monomer M, the reaction has the form:

$$-R_n^* + M \rightarrow -R(+H) + M^*(-H) \quad k_{tM}$$

These reactions will form stable polymers and hence reduce the degree of polymerisation of the material. The rate for formation of polymer $d[polymer]/dt$ then becomes:

$$d[polymer]/dt = k_{tc}\left[R_n^*\right]^2 + k_{tM}\left[R_n^*\right][M] + k_{tT}\left[R_n^*\right][T] \tag{8.40}$$

where k_{tM} and k_{tT} are, respectively, the rates of termination by monomer transfer and by transfer to an added reagent. Equation (8.40) assumes the dominant process is termination by recombination. The degree of polymerisation will then be:

$$\overline{DP_n} = \left(k_p[M]\left[R_n^*\right]\right) \Big/ \left(k_{tc}\left[R_n^*\right]^2 + k_{tM}\left[R_n^*\right][M] + k_{tT}\left[R_n^*\right][T]\right)$$

$$= k_p[M] \Big/ \left((k_{tc}k_i)^{1/2}[I]^{1/2}\right) + k_p / k_{tM} + k_p[M] / k_{tT}[T] \tag{8.41}$$

In the absence of transfer, Equation (8.41) has the same form as Equation (8.36). In the above theory, it is assumed that all radicals that have been formed have the same reactivity as the original propagating radical. If that is not the case then retardation or inhibition of the polymerisation reaction may result.

In the case where monomer transfer occurs, we cannot control the $\overline{DP_n}$ and the maximum achievable is controlled by k_{tM} so that Equation (8.41) becomes:

$$\overline{DP_n} = \left(k_p[M] \Big/ \left((k_{tc}k_i)^{1/2}[I]^{1/2}\right)\right) + k_p / k_{tM} \tag{8.42}$$

and by judicious choice of $[I]$ and $[M]$ the first term can be made small compared with the second so that:

$$\overline{DP_n} \approx k_p / k_{tM} \tag{8.43}$$

8.3.6 Inhibition and retardation

Inhibitors are often added to reaction mixtures to stabilise the system for shipping etc. This practice is common for coatings, where the formulation may be shipped in an activated form and when applied to a hot substrate will then polymerise to form a hard coating. To avoid the coating 'going off' it is desirable to reduce the rate of polymerisation to zero. This is achieved by adding an inhibitor, which will typically be a radical trap. Molecules such as quinone can effectively scavenge any radicals that are formed and will inhibit the propagation process. The amount of inhibitor added has to be much less than the amount of initiator. If there is prolonged storage then it is possible that all the inhibitor will be consumed and polymerisation will then occur.

8.3.7 Determination of absolute rate constants

From the rate of polymerisation R_p we can deduce $k_p(k_i/k_t)^{1/2}$ since:

$$R_p = k_p\left(k_i / k_t\right)^{1/2}[I]^{1/2}[M] \tag{8.44}$$

From measurements of the molar mass we obtain $\overline{DP_n}$ and taking the simplest case where disproportionation is dominant we can therefore deduce $k_p/(2(k_ik_{td})^{1/2})$ from

$\overline{DP_n} = k_p[M]/(2(k_i k_{td})^{1/2}[I]^{1/2})$, i.e. determine the $\overline{DP_n}$ of polymer obtained at low concentration and conversion. In the case where recombination is dominant, the parameter measured will be $k_p/(k_i k_{td})^{1/2}$. Both these termination rate constants involve the ratio $k_p / k_t^{1/2}$ so that even if k_i is known, e.g. from inhibition experiments, the separate values of k_p and k_t cannot be obtained from these data. It should be noted that k_i can be extracted from these data and compared with values from other methods. Using the rate of polymerisation:

$$R_p = \left(k_p k_i^{1/2} / k_t^{1/2}\right)[M][I]^{1/2} \tag{8.45}$$

and the degree of polymerisation:

$$\overline{DP_n} = \left(k_p[M]\right) / \left(k_{tc}^{1/2} k_i^{1/2} [I]^{1/2}\right) \tag{8.46}$$

Dividing Equation (8.45) by Equation (8.46) gives:

$$\left(R_p / \overline{DP_n}\right) = k_i[I] \tag{8.47}$$

To determine k_p and k_t separately, we must carry out experiments involving *nonstationary state* concentrations of radicals. In developing the above theory, it was always assumed that a stationary state existed, i.e. that the creation and destruction of radicals was always in equilibrium. In a nonstationary state experiment, the concentration of radicals present is varied in a controlled manner.

8.3.7.1 *Rotating sector method*

The most common method of varying the radical concentration in a nonequilibrium manner is to use a photo initiator and arrange for the illumination to be turned on and off in a regular manner, as shown in Figure 8.4.

The apparatus consists of a suitable light source which is collimated and chopped by a rotating sector. The sector is designed to provide an appropriate period of illumination and usually will have equal areas cut away. As the sector rotates, light is flashed on and off and the sample is contained in a thermostatted bath. For analysis, it is often simpler to make the period of

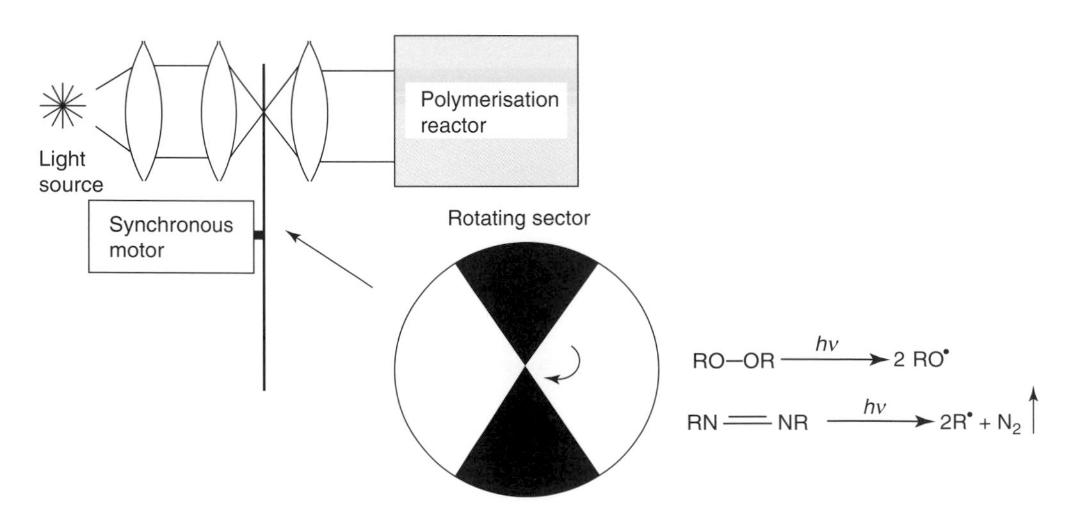

Figure 8.4 Rotating sector polymerisation apparatus and initiator creation chemistry.

illumination and dark of equal length. In general, the rate of initiation of a photochemical radical generation reaction is proportional to the intensity of the light – F. Therefore:

$$R_i = 2k_i[I] \propto F \tag{8.48}$$

In the expression for the rate of polymerisation, P_{ry}, R_i appears as $R_i^{1/2}$, hence

$$R_p \propto F^{1/2} \tag{8.49}$$

If the light is flashed on and off slowly, the stationary state concentration of radicals is quickly reached during each light period and the rate of polymerisation R_p is simply proportional to the fraction of time the light is on multiplied by the intensity, i.e.:

$$R_p \propto 1/2F^{1/2} \quad \text{for slow rotation} \tag{8.50}$$

The factor of 1/2 appears in Equation (8.50) to reflect the fact that the light is on for equal periods with the dark sectors (see Figure 8.5). On the other hand, if the light is flashed on and off very quickly the concentration of radicals changes little during a flash, however, the effect is the same as reducing the overall intensity by a constant factor but leaving it on all the time, in our case equivalent to $F/2$.

Thus $R_p \propto (F/2)^{1/2}$ for fast rotation and for constant illumination $R_p \propto F^{1/2}$; for fast chopping $R_p \propto 1/\sqrt{2}F^{1/2}$ and for slow chopping $R_p \propto 1/2F^{1/2}$. If we investigate the rate of polymerisation as a function of the rate of rotation ω then we observe a variation of the type shown in Figure 8.6. Obviously there must be a speed of rotation of the sector where we undergo a change from the slow to fast limiting behaviour and this can be related to the mean lifetime τ_R of the free radical and is defined as:

$$\tau_R = \frac{\text{number of radicals}}{\text{number of radicals disappearing per unit time}} \tag{8.51}$$

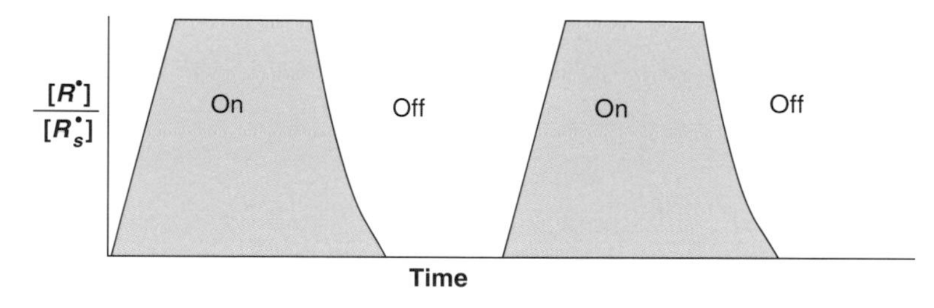

Figure 8.5 Variation of radical concentration with time in a rotating sector experiment.

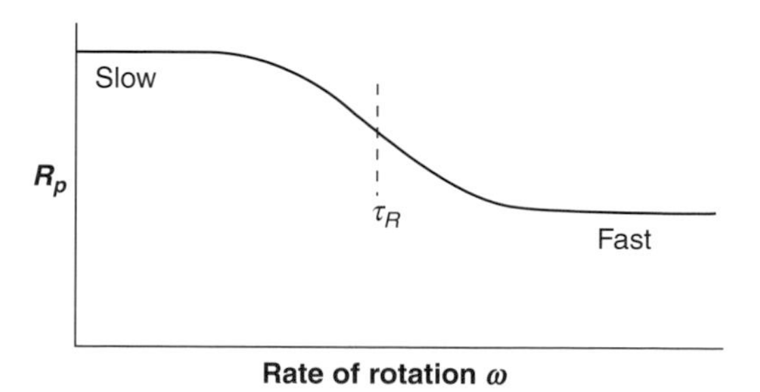

Figure 8.6 Schematic of variation of rate of polymerisation against rotation sector speed.

which is equal to:

$$\tau_R = \left[R_n^*\right] / \left(2k_t \left[R_n^*\right]^2\right) = 1 / \left(2k_t \left[R_n^*\right]\right) \tag{8.52}$$

Substituting for the radical concentration: $[R_n^*] = R_p / k_p[M]$ and $R_p = -d[M]/dt = k_p[M][R_n^*]$ yields:

$$\tau_R = k_p \left[M\right] / \left(2k_t R_p\right) \tag{8.53}$$

Equation (8.53) is true for all polymer systems. Thus, we can measure τ_R and obtain R_p from conversion versus time plots of the form shown in Figure 8.3, and we can deduce k_p/k_t. The value of τ_R is obtained from Figure 8.6.

8.3.7.2 *Transfer constants*

The transfer constants k_{tM} and k_{tT} can now be obtained. Equation (8.41) can be rewritten as:

$$1/\overline{DP_n} = \left(\left(k_{tc}^{1/2} k_i\right)^{1/2} \left[I\right]^{1/2}\right) / k_p \left[M\right] + k_{tM} / k_p + k_{tT} \left[T\right] / k_p \left[M\right] \tag{8.54}$$

If we plot $1/\overline{DP_n}$ for a polymerisation carried out at low conversion against $[I]^{1/2}/[M]$, keeping the ratio of $[T]/[M]$ fixed, we obtain plots of the type shown in Figure 8.7.

If no transfer agent is present then the intercept will be equal to k_{tm}/k_p, and knowing k_p, k_{tM} can be calculated. If a transfer agent T is then added, k_{tT} can be calculated. Often k_{tT}/k_p is left as a combined constant C_s, and is called the transfer constant. Table 8.3 indicates typical values of some chlorinated molecules. The constant increases with the lability of the chlorine atom.

In the case of CB_4, the transfer process is comparable with the rate of propagation and the net effect is that only oligomers, i.e. dimers, trimers, etc., are formed. Experiments on a range of molecules give

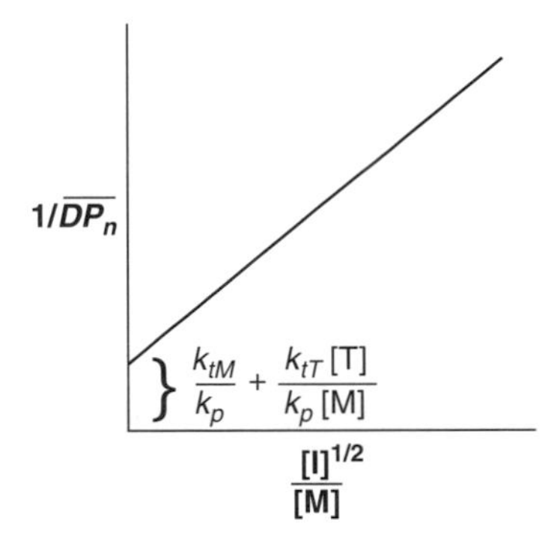

Figure 8.7 Plot of inverse of degree of polymerisation against ratio of square of initial initiator concentration divided by monomer concentration.

Table 8.3 Typical values for combined transfer–propagation constant, C_s

Transfer constant	$C_s \times 10^4$ measured at 60°C		
	Monomer system		
Transfer agent	**Styrene**	**Methylmethacrylate**	**Vinylacetate**
$C_6H_5CH_3$	0.125	0.20	2.2
CH_2Cl_2	0.15	-	-
$CHCl_3$	0.5	1.77	150
CCl_4	90	2.40	9600
CBr_4	22,000	2,700	29,000

Table 8.4 Values of propagation and termination rate constant for common monomers measured at 60°C

Monomer	**Propagation rate k_p ($M^{-1}s^{-1}$)**	**Termination rate constant $10^7 \times k_t$ ($M^{-1}s^{-1}$)**
Styrene	210	115
Methylmethacrylate	410	25
Methylacrylate	1000	3.5
Vinylacetate	2600	110
Vinychloride	11000	2100

values for the propagation and termination rate constants (see Table 8.4). The values of $k_t >> k_p$ and it is not surprising that two radicals react quickly with each other. Typical values for k_i are 10^{-6} to 10^{-7} s^{-1}.

8.3.7.3 *Effect of temperature on free radical vinyl polymerisation*

Each of the elementary reactions can be described by an Arrhenius-type of temperature dependence with a characteristic activation energy E_i and pre-exponential factor, A_i:

$$k_i = A_i \exp(-E / RT) \tag{8.55}$$

It is useful to consider two specific cases. First, we consider photochemically initiated polymerisation. In this case the production of the initiating radicals is the result of the absorption of light. This process is not influenced by the temperature at which the reaction is carried out, however, the other processes will be influenced by temperature. In this instance, the rate constants which are influenced by temperature will be k_p and k_t. From Equation (8.45), we see that the temperature sensitivity will be reflected in the ratio of the rate of propagation to termination:

$$k_p/k_t^{1/2} = A_p/A^{1/2} \exp\left(-\left(E_p - (E_t/2)\right)/RT\right) \tag{8.56}$$

A plot of $\log\left(k_p / k_t^{1/2}\right)$ against $1/T$ enables us to calculate $A_p/A_t^{1/2}$ and $\left(E_p - (E_t / 2)\right)$. For a typical polymerisation reaction, the activation energy for propagation, E_p, has a value of ~28 kJ mol^{-1} and the activation energy for termination, E_t, is ~12 kJ mol^{-1}. The overall activation energy for the reaction will therefore have a value in the range 20–24 kJ mol^{-1}, and this is reflected in a 30% increase in $k_p / k_t^{1/2}$ for a 10°C rise in temperature.

Scondly, we consider thermal initiation. In this case catalysts may or may not be present. The principal difference between this case and that illustrated in our first example is that the first step, the initiation, is temperature dependent. Inspection of Equation (8.45) indicates that the overall temperature dependence is dependent on the ratio $\left(k_p k_i^{1/2}\right) / k_t^{1/2}$ and therefore:

$$\left(k_p k_i^{1/2}\right)/k_t^{1/2} = \left(A_p A_i^{1/2}\right)/A_t^{1/2} \exp\left(-\left(E_p + E_i / 2 + E_t / 2\right)/RT\right) \tag{8.57}$$

The breaking of a bond usually involves a high energy and is typically of the order of 120 kJ mol^{-1}. Because of the high value of the activation energy for initiation, increasing the temperature can have a marked effect on the speed of polymerisation. The overall activation energy for the process becomes $= E_p + E_i / 2 - E_t / 2 \approx 28 + 120 / 2 + 12 / 2 \approx 82$ kJ mol^{-1}. Collision theory of chemical reactions gives $A = kT/h \exp(\Delta S/R)$, which leads to estimates for A in the range 10^{11}–10^{12} for a bimolecular reaction and 10^{13}–10^{14} for a unimolecular reaction. Substituting appropriate values into $\left(A_p A_i^{1/2}\right)/A_t^{1/2}$ gives an estimate for the overall pre-exponential factor in Equation (8.57) in the range 10^{11}–10^{13} $M^{-1}s^{-1}$. If we examine the expression for the degree of polymerisation:

$$\overline{DP_n} = \left(k_p/\left(2\left(k_i k_{td}\right)^{1/2}\right)\right)\left([M]/[I]^{1/2}\right) \tag{8.58}$$

we see that the temperature dependence is controlled by the ratio of $k_p/(k_i k_t)^{1/2}$ and hence the effective activation energy influencing the degree of polymerisation is:

$$\overline{DP_n} = E_p - E_i / 2 - E_t / 2 = 28 - 60 - 6 = -38 \text{ kJ mol}^{-1} \tag{8.59}$$

The overall effect is therefore negative and reflects the way in which the various factors influencing the reaction process control the overall process. The implication is that as the temperature of reaction increases the average molar mass of the polymer decreases. This has important implications if one is attempting to produce a high molar mass polymer for a particular application. In the case of photochemically initiated polymerisation, the DP_n becomes analogous to the dependence of $k_p / k_t^{1/2}$ and the effective activation energy will be $= E_p - E_t / 2 = 28 - 6 \text{ kJ mol}^{-1}$. The activation energy is now positive and the molar mass increases with increase in temperature along with the rate. When transfer reactions dominate DP_n then we have:

$$\overline{DP_n} \propto k_p/k_{tM} \quad \text{or} \quad k_p / k_{tM} \left(k_p/k_{tM}\right) \tag{8.60}$$

Using the approach outlined above, the effective activation energy is:

$$= E_p - E_{tM} \quad \text{or} \quad = E_p - E_{tT} \tag{8.61}$$

Since the effective energy can have values of ~20–60 kJ mol^{-1}, the overall activation energy can become negative, hence an increase in temperature tends to decrease the molar mass.

8.4 Free radical copolymerisation

In order to be able to design a polymer material to fit a particular application, it may be advantageous to incorporate a small amount of another monomer into the chain. These polymers, which contain more than one monomer in the chain, are known as *copolymers*. The kinetics presented above assumes that there is only one monomer type present and is usually referred to as *homopolymerisation*. Many commercial polymers contain mixtures of monomers, one species usually being present in small proportions but having a very advantageous effect on the properties of the material.

Modified polyethylene

For certain applications it is desirable to suppress crystallinity and increase transparency but still retain a high modulus material. Polyethylene can be modified to incorporate acrylic comonomers. The polymers used in certain packaging applications incorporate 1–10% of methacrylic acid in the polymerisation mixture. The resulting material is physically tougher than ordinary polyethylene as a result of hydrogen bonding and ionomer interactions. If treated with alkaline solution, the polymer will become charged and the electrostatic interactions can enhance the bonding between neighbouring polymer chains and increase the rigidity of the films.

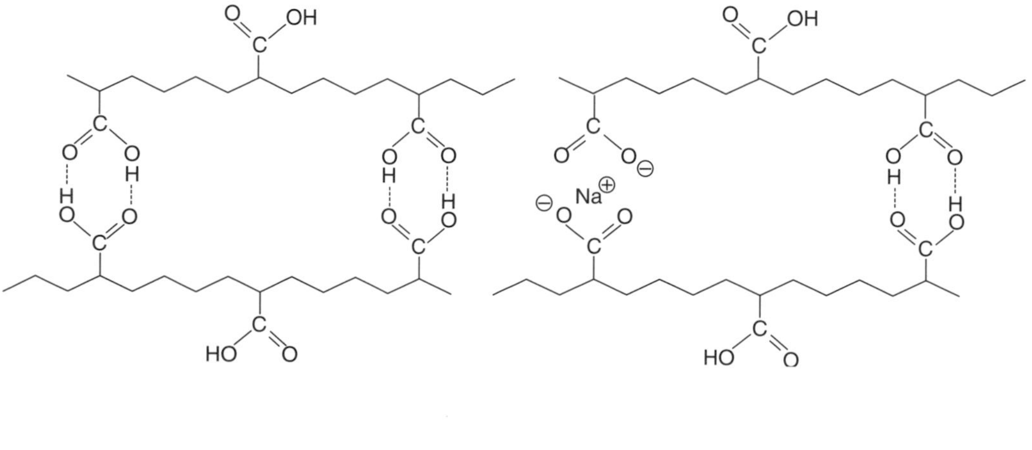

Polyethylene–propylene copolymers

Incorporation of about 35% of propylene into a polyethylene will dramatically reduce the crystallinity and results in a rubbery material.

Styrene acrylonitrile copolymers

Polystyrene is an intrinsically brittle material at room temperature; however, the incorporation of 20–30% of acrylonitrile produces a material which has a higher softening point and improved impact strength relative to the unmodified material.

8.4.1 Kinetics of copolymerisation

Consider the case of two vinyl monomers, A and B, being polymerised using a free radical initiator. For simplicity we make certain assumptions:

- It is assumed that A^* can react with A and with B, and B^* can react with A and with B.
- It is assumed that the radical reactivity is independent of the chain length.
- The radical reactivity is determined only by the terminal unit of the chain and the nature of the sequence to which it is connected has no effect on the reactivity.

There are four possible reactions which can occur:

$$\sim A^{\bullet} + A \xrightarrow{k_{pAA}} \sim AA^{\bullet}$$

$$\sim A^{\bullet} + B \xrightarrow{k_{pAB}} \sim AB^{\bullet}$$

$$\sim B^{\bullet} + A \xrightarrow{k_{pBA}} \sim BA^{\bullet}$$

$$\sim B^{\bullet} + B \xrightarrow{k_{pBB}} \sim BB^{\bullet}$$

The rate of disappearance of the two monomers expressed in the normal way will be:

$$-d[A]/dt = k_{pAA}\left[A^*\right][A] + k_{pBA}\left[B^*\right][A] \tag{8.62}$$

and

$$-d[B]/dt = k_{pAB}\left[A^*\right][B] + k_{pBB}\left[B^*\right][B] \tag{8.63}$$

We need to determine the composition of the copolymer

$$\frac{A_{pol}}{B_{pol}} = n = \frac{\text{number of } A \text{ units in polymer}}{\text{number of } B \text{ units in polymer}} = \left(-d[A]/dt\right)/\left(-d[B]/dt\right) \tag{8.64}$$

Combining Equations (8.62) and (8.63) with Equation (8.64) yields:

$$n = \frac{[A]\left(k_{pAA}\left[A^*\right] + k_{pBA}\left[B^*\right]\right)}{[B]\left(k_{pAB}\left[A^*\right] + k_{pBB}\left[B^*\right]\right)} \tag{8.65}$$

where $[A]$ and $[B]$ are the monomer concentrations at time t. If we assume stationary state concentrations of A^* and B^* as in the case of homopolymerisation, then:

$$-d\left[A^*\right]/dt = k_{pBA}\left[B^*\right][A] - k_{pAB}\left[A^*\right][B] = 0 \tag{8.66}$$

The rate of loss of A^* is balanced by the gain in A^* through reaction of monomer A with B^* and the loss of A^* through reaction with monomer B. Note that there will be no change in A^* in the first equation, when A^* reacts with A. Similarly:

$$-d\left[B^*\right]/dt = k_{pAB}\left[A^*\right][B] - k_{pBA}\left[B^*\right][A] = 0 \qquad (8.67)$$

From combining Equations (8.66) and (8.67) we obtain:

$$\left[A^*\right] = \left(k_{pBA}/k_{pAB}\right)\left([A]/[B]\right)\left[B^*\right] \qquad (8.68)$$

Substituting in Equation (8.65) we obtain:

$$n = \left([A]/[B]\right)\left(k_{pAA}\left(k_{pAA}[A]\left[B^*\right]/k_{pAB}\right) + k_{pBA}\left[B^*\right]\right)/\left(\left(k_{pAB}\left(k_{pBA}[A][B]\right)/k_{pAB}\right) + k_{pBB}\left[B^*\right]\right) \qquad (8.69)$$

If we divide through by k_{pBA}:

$$n = \left([A]/[B]\right)\left(\left(k_{pAA}/k_{pAB}\right)[A] + [B]\right)/\left(\left(k_{pBB}/k_{pBA}\right)[B] + [A]\right) \qquad (8.70)$$

If we multiply through by $[B]$:

$$n = \left([A]/[B]\right)\left(\left(\left(k_{pAA}/k_{pAB}\right)[A] + [B]\right)/\left(\left(k_{pBB}/k_{pBA}\right)[B]\right) + [A]\right) \qquad (8.71)$$

Equation (8.71) describes the *instantaneous composition* of the copolymer. We now define the *reactivity ratio* of A as r_A:

$$r_A = \frac{k_{pAA}}{k_{pAB}} = \frac{\text{rate constant of } A^* \text{with } A}{\text{rate constant of } A^* \text{ with } B} \qquad (8.72)$$

and similarly r_B

$$r_B = \frac{k_{pBB}}{k_{pBA}} = \frac{\text{rate constant of } B^* \text{ with } B}{\text{rate constant of } B^* \text{ with } A} \qquad (8.73)$$

A number of these constants have been measured and are tabulated in Brandrup *et al.* (1975). Combining Equations (8.71)–(8.73) we obtain:

$$n = \frac{[A]}{[B]}\frac{\left(r_A[A] + [B]\right)}{\left(r_B[B] + [A]\right)} \qquad (8.74)$$

Note that in Equation (8.74) the concentration of the monomers refers to the values at any instant in time. As the reaction proceeds, so the value of this ratio will change, reflecting the consumption of one monomer relative to another.

To obtain the variation of n with time during a polymerisation, we need to integrate the equations. However, in practice it is desirable to obtain a polymer with a constant composition and this is achieved by maintaining the composition constant by feeding in the monomer which is most rapidly depleted. If we plot the fraction of A in the feed, F_A, against the fraction of A incorporated in the polymer, a plot of the type shown in Figure 8.8 can be obtained. In Figure 8.8, the curves correspond to the conditions:

curve i $r_A > 1 \quad r_B < 1$ curve ii $r_A > 1 \quad r_B > 1$ curve iii $r_A = r_B$

curve iv $r_A < 1 \quad r_B < 1$ curve v $r_A < 1 \quad r_B > 1$

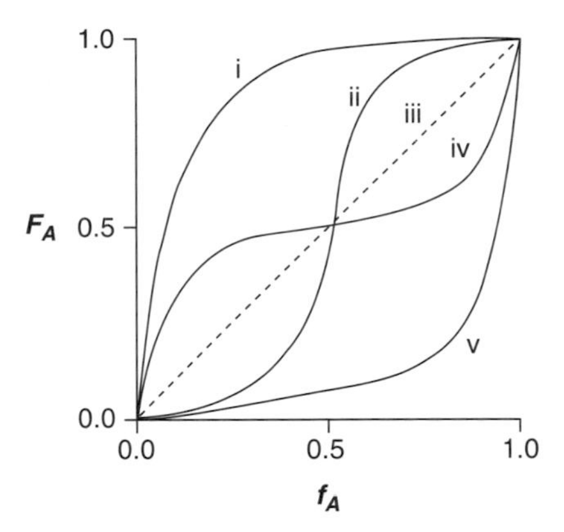

Figure 8.8 Variation of fraction of A in polymer, F_A, against fraction of monomer present in feed stock f_A.

To use the curves to obtain a polymer of the desired composition one decides on a particular value of F_A, looks up the values of r_A and r_B and then reads off the feed value needed to produce the polymer. Note that if $[A]_{pol}/[B]_{pol} = [A]_{feed}/[B]_{feed}$ then:

$$\left(r_A[A]+[B]\right)/\left(r_B[B]+[A]\right)=1 \quad \text{or} \quad \left(1-r_B\right)/\left(1-r_A\right)/=[A]/[B] \tag{8.75}$$

This equation indicates that only specific values of $[A]/[B]$ can give the same composition in the polymer; this is known as the *azeotropic* composition. From the graph, when $r_A > 1$ and $r_B < 1$ and $r_A < 1$ and $r_B > 1$ there is no value of $[A]/[B]$ that gives an azeotopic condition. For $r_A > 1$ and $r_B > 1$ and $r_A < 1$ and $r_B < 1$ there is one condition which gives the azeotropic condition.

8.4.2 Mean sequence length

The physical properties of the polymer formed can depend very much on the sequence structure which is produced. If one monomer has a tendency to react with itself then the polymer which is produced will have *blocks* of that monomer and this will give it very different characteristics from one in which the monomers are incorporated in a random fashion. The tendency to form blocks is controlled by the relative values of r_A and r_B. The probability that A^* will propagate by reacting with A is P_{AA} and is:

$$P_{AA}=\frac{\text{rate of reaction of } A^* \text{ with } A}{\text{rate of reaction of } A^* \text{ with } A \text{ and B}}$$

$$=\frac{k_{pAA}\left[A^*\right][A]}{k_{pAB}\left[A^*\right][B]+k_{pAA}\left[A^*\right][A]}=\frac{r_A[A]}{[B]+r_A[A]} \tag{8.76}$$

also

$$P_{AB}=1-P_{AA}=[B]/\left(r_A[A]+[B]\right) \quad \text{and} \quad P_{BA}=[A]/\left(r_B[B]+[A]\right) \tag{8.77}$$

The average sequence length of A units in the polymer is $\overline{S_A}=1/P_{AB}$ since as soon as A^* reacts with B, the sequence of A units is broken:

$$\overline{S_A}=r_A(1/2)\left([A]/[B]\right)+1 \tag{8.78}$$

and similarly

$$\overline{S_B}=1/P_{BA}=r_B\left([B]/[A]\right)+1 \tag{8.79}$$

Values of r_A and r_B

The values of the reactivity ratios are determined by analysis of polymers produced in copolymerisation stopped at low concentration so that $[A]/[B]$ is equal to $[A]_0/[B]_0$ and is approximately constant over the time scale of the polymerisation process. In practice we find that r_A and r_B are both <1, or that $r_A > 1$ and r_B is small, or $r_B > 1$ and r_A is small. It is very unusual for r_A and r_B to be both >1. The latter would mean that each radical type reacts faster with its own monomer. If one ratio is larger than 1, this indicates a very reactive monomer. Hence $r_A > 1$ and r_B is generally <1 and *vice versa*. When r_A and r_B are about the same (~1) we obtain fairly random incorporation of monomer and the radical sees each monomer to be equally attractive in reaction. When $r_A > 1$ and r_B is small or *vice versa* then the trend is to obtain large blocks of one monomer, particularly of A units broken by smaller runs of B. When $r_A < 1$ and $r_B < 1$, each radical prefers to react with the opposite monomer, hence there is a tendency to produce a 1:1 alternating copolymer. The product $r_A r_B$ is often taken as a measure of the tendency to form an alternating copolymer and the tendency increases as the product tends towards zero. The important point to take from this discussion is that the detailed polymer structure will depend on the reactivity of the various radicals.

As an illustration we can consider the values for some common monomers. Acrylonitrile has a value of 0.15 when reacting with methylmethacrylate (MMA), which has a value of 1.65, and the copolymer formed will have blocks of MMA. Styrene with a reactive ratio of 0.52 reacts with MMA to produces a fairly random copolymer. Acrylonitrile reacting with styrene has a value of 0.03 and styrene a value of 0.52, ($r_{AN} r_{St}$ ~0.015), and the resulting polymer has a high tendency to form an alternating structure.

8.5 Methods of polymerisation

The polymerisation process is often highly exothermic, of the order of 80 kJ mol^{-1}, and it is essential that the heat produced by the reaction is effectively removed from the reaction vessel. It can be dangerous to attempt to carry out a polymerisation on a neat monomer system as the temperature rise can produce an explosion! Several approaches are usually adopted:

- dispersion of the monomer in an inert solvent
- carrying out the polymerisation as either a suspension of monomer in an inert solvent (which will often be water)
- polymerisation of the monomer dispersed in the liquid as an emulsion

Suspension polymerisation will usually produce larger beads than those produced by emulsion polymerisation. Depending on which process is used so the product can have slightly different processing characteristics.

- *Solution polymerisation*: The product of solution polymerisation will usually be a free-flowing powder. The removal of solvent from the polymerisation solution can be a major issue in the case of industrial synthesis and this approach is rarely used for bulk polymer production unless an alternative method is not available.
- *Bulk polymerisation*: In certain situations, such as the production of slabs of polymer, it is possible to polymerise the neat resins. The main problem which is encountered with this process is the effective removal of the heat generated during the polymerisation and the ability to incorporate all the monomers in the final slab stock. It is undesirable to have trapped monomer left in the slab stock as this will slowly leach out and can have very disadvantageous effects. Rate control agents can be used.
- *Suspension polymerisation*: The monomer is dispersed with surfactant in an inert medium, usually water, and each micelle contains a small concentration of initiator. On heating, the polymerisation is activated and the monomer is converted to polymer. The result of this

process will often be polymer beads of size 5–20 μm and these are very suitable for subsequent blending and combing into polymer product. A minor issue is that the surface of the beads will be contaminated due to the surfactant used to create the initial dispersion.

- *Emulsion polymerisation*: This process is similar to that for suspension polymerisation but with some very important differences. The initiator is dispersed in the surfactant and polymerisation occurs in the micelles. However, the monomer diffuses to the micelle from the surrounding solution and the size of the micelle grows as the polymerisation takes place. The net result is that the particles increase in size from almost a nanoscopic scale to micron size as the polymerisation proceeds. This process is often used to disperse polymers in a solvent. The simplest example of this type of material is a household paint formulation. The drying process is the evaporation of an aqueous solvent mixture and the paint film is created by the fusion of these small polymer emulsion particles.

8.6 Specialist chemical reactions

With the desire to tailor make the polymer to fit a particular application there has been considerable interest in discovering new methods of polymerisation. The majority of new methods start with a vinyl monomer and attempt to create a polymer chain in a more controlled manner. The above summary of the kinetics of polymerisation did not consider some of the side reactions which can lead to problems with defining the structure of the product that is formed. Although the steps involved in these processes are different from those outlined above, the overall kinetics and the factors controlling the polymerisation process are similar.

Why do we wish to have better control of the chemistry?

A number of side reactions can occur in a typical free radical reaction which will significantly influence the nature of the polymer created. To illustrate this we will consider the polymerisation of polyethylene (see Figure 8.9). There is a finite possibility that the radical at the end of the chain will move into a conformation which allows it to abstract a proton from a point further back along the chain. The removal for the proton creates a radical which is very active and readily adds ethylene to start the growth of side chains. The resulting polymer, instead of being totally

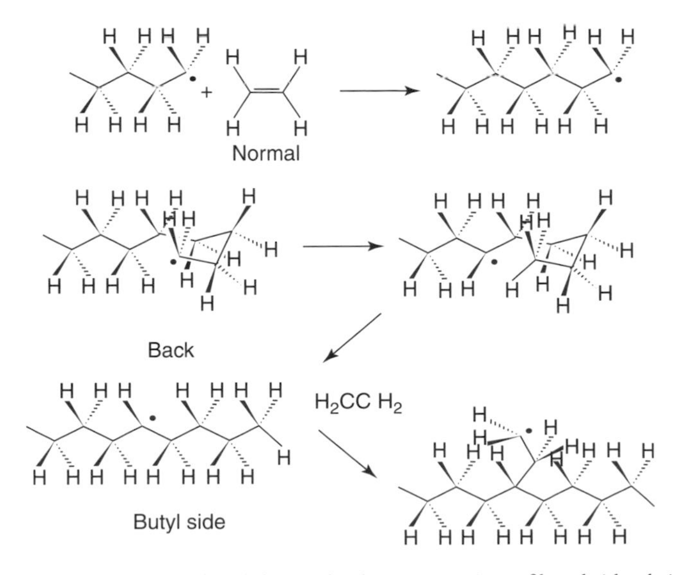

Figure 8.9 Effects of back biting, leading to creation of butyl side chains.

Table 8.5 Examples of polymer systems in which stereochemical control is used to produce a material with desired characteristics for a particular application

Polymer	Stereochemistry	Typical uses
HDPE	–	Bottles, drums, pipes, conduits, sheets, films, wire and cable insulation
Polyethylene, ultrahigh molar mass	–	Surgical prostheses, machine parts, heavy duty applications
Polypropylene	Isotactic	Automobile and appliance applications, rope, cordage, webbing, carpeting, films
Poly-1-butene	Isotactic	Films, pipes
Poly-4-methyl-1-pentene	Isotactic	Packaging, medical supplies, lighting
Polystyrene	Syndiotactic	Speciality applications
1,4-Polybutadiene	*Trans*	Metal can coatings, potting compounds for transformers
1,4-Polyisoprene	*Trans*	Golf ball covers, orthopaedic devices
1,4-Polybutadiene	*Cis*	Tyres, conveyor belts, wire and cable insulation, footware
1,4-Polyisoprene	*Cis*	Tyres, footware, adhesives, coated fabrics
Ethylene–propylene block copolymer	Isotactic	Food packaging, automobile trims, toys, bottles, films, heat-sterilisable containers

linear, will have a series of branches which contain C_2, C_3 or C_4 fragments. These branched chains have the effect of disrupting the crystal growth process and are usually incorporated into the amorphous phase. The lowering of the crystallinity will result in a softer polymer material. A polymer with a high linear content has a high degree of crystallinity and hence density and is known as a high density polyethylene (HDPE). As the branched chains content is increased, so the density is decreased and the product is termed low density polyethylene (LDPE). Because of the considerable interest in polyethylene, detailed studies have been undertaken to understand how changes in the polymerisation conditions will influence the degree of branching and the density of the resulting solid.

It is desirable to be able to control the structure of a polymer and hence instead of allowing the polymerisation to produce side chains in a random manner, it may be appropriate to incorporate these by adding monomers such as 1-butene, 1-hexene or higher 1-alkenes to the reaction mixture and hence produce a more controlled polymer.

In the 1950s, Karl Zeigler and Giulio Natta explored the possibility of using coordination catalysts to achieve polymerisation. The so-called *Zeigler–Natta* catalysts allowed a degree of control of the polymerisation to be achieved that was not possible with the radical process (see Table 8.5). A range of such catalysts have since been discovered, allowing the design of a polymer structure which is fit for purpose. Examples of commercial polymers which are produced using special catalysts are summarised in Table 8.5. Inspection of the polymers listed in Table 8.5 indicates that the polymerisation process has been achieved by stereochemical control at the reaction site and this requires the use of a coordination rather than a free radical process.

What other factors can we control?

Other than sterochemical control there are two factors which are desirable to control. First, the length of the chain, the molar mass is a very important parameter which will have a significant effect on the rheology of the polymer melt and the ultimate mechanical properties. Secondly, whilst the reactivity ratio will produce blocky copolymers it is sometimes desirable to produce polymers with a definite block copolymer structure. Coordination catalysts have the ability to control both these factors and hence are used extensively industrially to produce polymers.

8.7 Heterogeneous catalysis

The best known of the industrial heterogeneous catalysts are *Zeigler–Natta* catalysts. The catalyst is prepared by mixing in a dry, inert solvent in the absence of oxygen titanium trihalides and tetrahalides with trialuminium compounds.

The chemical processes which are believed to take place are summarised as follows:

$$AlR_3 + TiCl_4 \rightarrow AlR_2Cl + TiRCl_3$$

$$AlR_2Cl + TiCl_4 \rightarrow AlRCl_2 + TiRCl_3$$

$$AlR_3 + TiRCl_3 \rightarrow AlR_2Cl + TiR_2Cl_2$$

$$TiRCl_3 \rightarrow TiCl_3 + R^*$$

$$TiR_2Cl_2 \rightarrow TiRCl_2 + R^*$$

The radicals formed in these reactions may be removed by recombination, disproportionation or reaction with solvent. While such reactions occur in the catalyst during their formation process and possibly during ageing, they are not the initiation mechanism. The catalysts formed by the above reactions can undergo reduction as follows:

$$TiRCl_2 \rightarrow TiCl_2 + R^*$$

$$TiRCl_3 \rightarrow TiCl_2 + RCl$$

and the equilibrium always contains:

$$TiCl_4 + TiCl_2 \leftrightarrow 2TiCl_3$$

The catalyst is precipitated from solution and can have a variety of different crystalline forms: α, β, γ or δ forms of $TiCl_3$. All of these forms give stereoregular polymers. The β form gives much more of the atactic polymer structure. Being a heterogeneous material, it is possible that there will be a range of structures present, each of which can initiate and control the polymerisation in a different way. There will be catalytic sites on the solid, similar to those shown in Figure 8.10. The titanium will act as an acceptor site for the π electrons of the approaching vinyl monomer and the initial step is π coordination of the double bond to the titanium. The π coordinated bond is converted into a formal bond by the attack of the R group which is located in one of the arms of the catalyst and there is a 90° shift in the bond pattern.

The overall result is to create a new vacant site at which π coordination can occur. The process can occur until a group is added to the reaction mixture which allows the reaction to terminate.

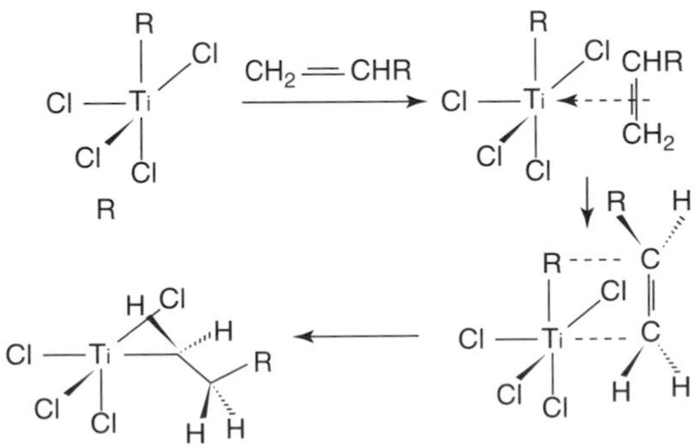

Figure 8.10 Simulation of active site in Zeigler–Natta catalyst.

Because the addition is to a specific site, the orientation of the approaching groups is controlled by the stereochemistry of the site and stereoregular addition occurs with the production of a stereopolymer.

The active site in the case of the Zeigler–Natta catalyst can be considered to be of the form shown in Figure 8.10. The vinyl bond attaches itself to the vacant site, first via a π bond complex, which is then converted into a σ bonded complex by the movement of the R group which is already attached to the site. As a consequence the addition process is influenced by the steric constraints placed in the site by the chlorine and alkyl groups. Since the monomer is always in contact with the site of the catalyst it is possible to achieve a high degree of steric control. The most important monomer to have been polymerised by this catalyst is polypropylene. The product of the synthesis is a highly isotactic polymer which has a high melting point. The syndiotactic and atactic polymers are soft materials which are not as useful for fabrication purposes as the isotactic material.

The above Zeigler–Natta catalyst is a heterogeneous catalyst, i.e. it acts as a suspension in the reaction media, and, in effect, the polymerisation process occurs at a solid surface. The location of the active site on the solid surface will influence the activity and introduces control of the manner in which the approaching monomer can attach itself to the site. In practice, the solid surface helps to dictate the stereochemistry of the product. As noted above, changes in the lattice structure of the $TiCl_4$ will influence the stereochemistry of the product produced. The reactivity of various monomers is also influenced by their ability to approach the active site. Increasing the bulk of the groups next to the double bond will influence the speed at which the reaction will proceed; therefore the observed order of reactivity for several 1-alkenes is:

$$CH_2{=}CH_2 > CH_2{=}CHCH_3 > CH_2{=}CHCH_2CH_3 > CH_2{=}CHCH_2CH(CH_3)_2 >$$

$$CH_2{=}CHCH(CH_3)_2 > CH_2{=}CHCH(CH_2CH_3)_2 > CH_2{=}CHC(CH_3)_3$$

The termination of the reaction can occur in several ways, as shown in Figure 8.11. The various processes which can produce termination all involve hydrogen transfer to the monomer either from a group attached to the surface (internal hydride transfer), transfer to a cocatalyst or to an added alkyl metal compound and transfer to an added hydrogen. Saturated chain ends are normally more prevalent in Zeigler–Natta polymers.

Hydrogen is the preferred transfer agent for controlling molecular weight because it reacts cleanly, leaves no residue and is low in cost. Without the application of transfer reagents, the

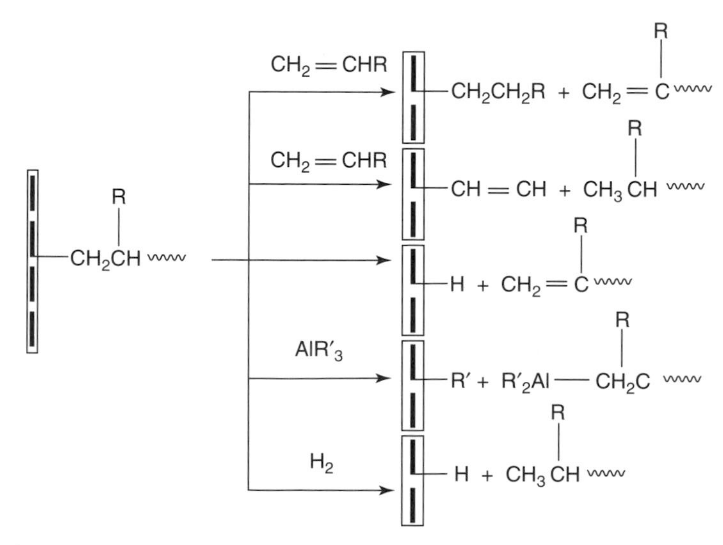

Figure 8.11 Termination mechanism for Zeigler–Natta catalysed polymers.

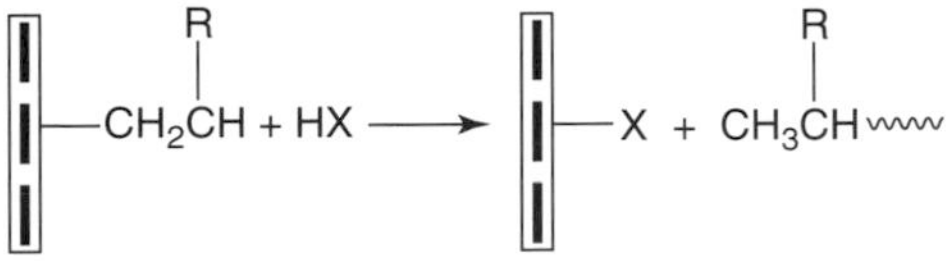

$$[\text{M}]\text{—}CH_2CH(R) + HX \longrightarrow [\text{M}]\text{—}X + CH_3CH(R)\sim\sim$$

Figure 8.12 Common termination reaction.

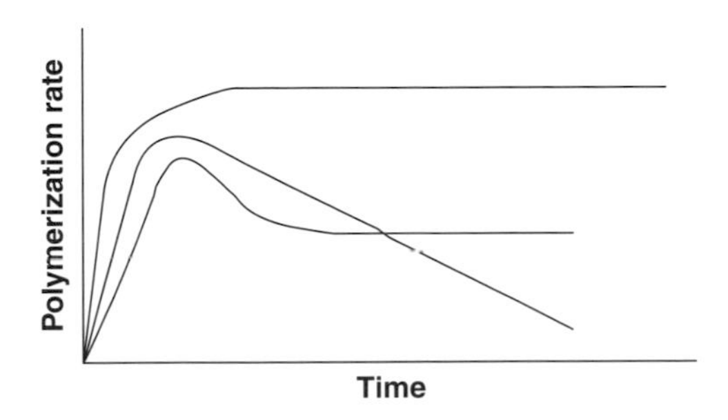

Figure 8.13 Plots of polymerisation rate against time.

molecular mass would in most instances be too high for commercial use. A true termination reaction is brought about by compounds containing active hydrogen (see Figure 8.12).

It is difficult to write down all the kinetic equations associated with all the possible reactions. If the rate of polymerisation is monitored as a function of time, the types of plots observed are summarised in Figure 8.13. There are three types of curves: rising to a constant rate, rising and then falling to a constant value, and rising and then falling continuously until it stops. Of the three, the decaying rate type is the most common. Rate decay has been attributed to such factors as structural changes that reduce the number or activity of the active centres and encapsulation of active centres by polymer, which prevents approach by monomers. Molar mass distributions are generally broad when insoluble catalysts are used and much narrower when soluble catalysts are used. In the former case, the broad distribution may arise from the decay of catalyst activity or activity from the presence of sites of variable activity.

8.8 Homogeneous catalysis

8.8.1 Homogeneous metallocene catalysts

In the early 1990s, an alternative system of catalysts became available and has had a major impact on the range of polymer materials which are available with tight stereochemical control. The metallocene catalysts are based on the complexes formed between a metal cation and cyclopentadiene (Cp). The earliest metallocene catalysts were biscyclopentadienyl titanium dichloride used in combination with dialkylaluminium chloride compounds. These catalysts exhibited low catalytic activity towards ethylene and were generally unreactive towards propylene. The active centre of the catalysts is represented as:

$$Cp_2TiCl_2 \quad R_2AlCl$$

It was later found that addition of water substantially increased the activity. The increase was the result of a reaction between the water and the alkylaluminium cocatalysts to form complex alkylalumoxanes. Subsequently, it was shown that especially high activities were realised if methylalumoxanes (MAOs) were used in conjunction with metallocene catalysts. MAO, formed by controlled hydrolysis of trimethylaluminium, has a complex oligomeric structure with molar masses of 1,000 and 1,500, most likely consisting of methyl-bridged aluminium atoms alternating with oxygen (see Figure 8.14).

MAO is now used with a wide variety of metallocenes having the general structure shown in Figure 8.15, in which the transition metal (M) is usually Zr, Ti, Hf or Sm; X is Cl or alkyl; Z is an

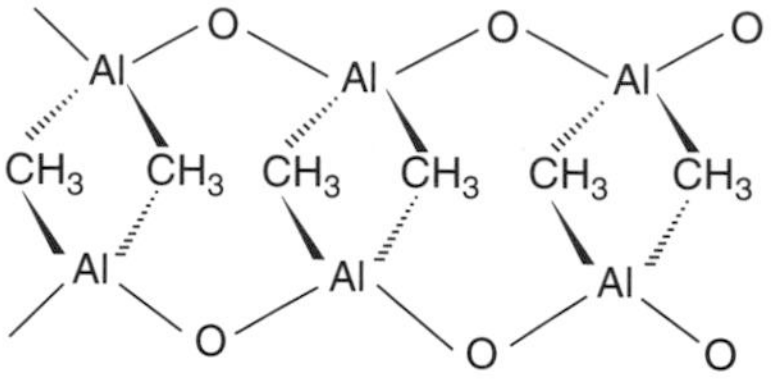

Figure 8.14 Proposed structure of alkylalumoxanes.

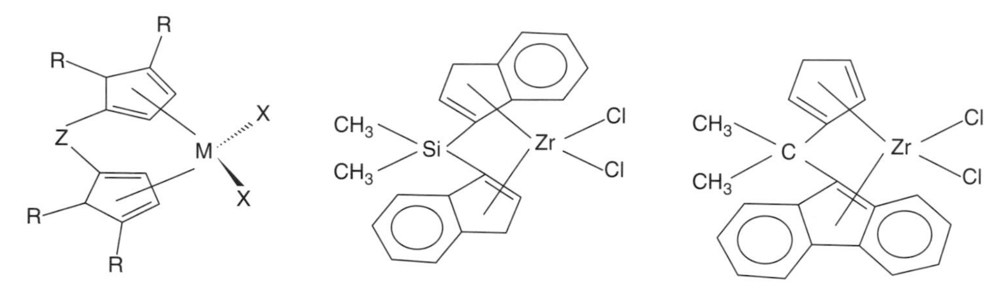

Figure 8.15 Proposed structures for some metallocene catalysts.

optional bridging group, usually $C(CH_3)_2$, $Si(CH_3)_2$ or CH_2CH_2; and R is H or alkyl. Fused ring systems are also found in place of the cyclopentadiene ligands.

The above catalysts are able to produce isotactic and syndiotactic polypropylene. The metallocene catalyst is often referred to as a *single site* catalyst to differentiate it from the classic Ziegler–Natta catalyst. A proposed mechanism for formation of the active site in a zirconium catalyst, L_2ZrCl_2 (where L represents the π ligand), involves initial complexation between MAO and the catalyst, followed by Cl–CH_3 exchange to form $L_2Zr(CH_3)_2$. The methylated zirconium reacts further with MAO to form the active species (see Figure 8.16).

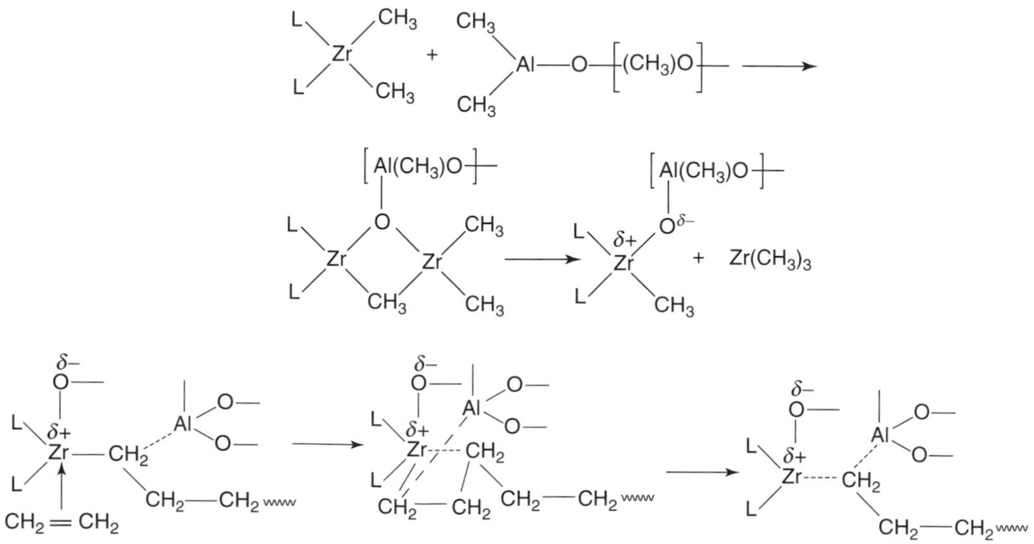

Figure 8.16 Proposed polymerisation mechanism for ethylene.

Polymers prepared with metallocene catalysts have narrow molar mass distributions compared with those produced by heterogeneous catalysts and consequently have better defined physical properties. In the case of polyethylene, the narrow molar mass polymers will be more easily processed, but may have inferior properties if the desired physical properties rely on the presence of a small fraction of higher molar mass material to produce entanglement or tie molecules in crystalline phases.

8.8.2 Atom transfer radical polymerisation

The atom transfer radical polymerisation (ATRP) process produces a living free radical polymerisation, as exemplified by styrene, using 1-chloro-1-phenylethane or the bromo- analogue as initiator in the presence of a copper(I) bipyridyl (byp) complex. Initiation occurs when a halogen atom is transferred from the 1-halo-1-phenylethane to the complex. The resultant 1-phenylethyl radical in turn adds to a styrene molecule. The halogen atom is then reversibly transferred to the styryl radical, thus preventing radical termination reactions from occurring while allowing the propagation reaction to occur (see Figure 8.17) An important consequence of living polymerisation is that the average degree of polymerisation is simply equal to the ratio of the initial monomer concentration to the initiator concentration:

$$\overline{DP} = [M]_0/[I]_0 \qquad (8.80)$$

All the chains are initiated at about the same time and there are no chain terminations or transfer reactions; the chains all grow to approximately the same length. The low polydispersity is a characteristic of living polymerisation and values of 1.05 can be achieved in favourable cases.

8.8.3 Group transfer polymerisation

In the 1980s, a new method of polymerisation of acrylic monomers was developed. Unlike conventional anionic polymerisation, *group transfer polymerisation* (GTP), affords low polydispersity living polymer at room temperature or above. Typically, an organosilicon compound is used to initiate polymerisation in solution in the presence of an anionic or Lewis acid catalyst. To illustrate the process we will consider methyltrimethylsilyl acetal of dimethylketone as the initiator and bifluoride ion as the catalyst. In each propagation step, the SiR_3 group is transferred to the

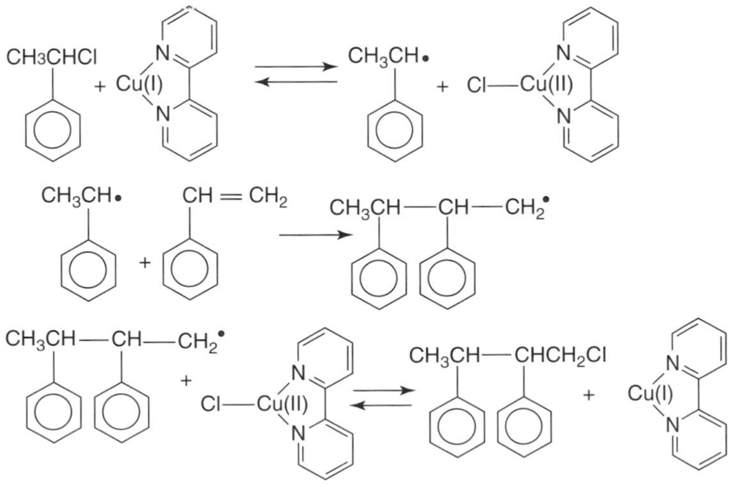

Figure 8.17 Schematic of ATRP polymerisation process.

carbonyl oxygen of the incoming monomer, hence the name GTP. If a difunctional initiator is used, the chain propagates from each end:

$$R_2CHCO_2R \xrightarrow{R'_2N^{\ominus}Li^{\oplus}} \left[R_2\bar{C}CO_2R \longleftrightarrow R_2C{=}C(O^{\ominus})(OR)\right] \xrightarrow{R_3SiCl} R_2C{=}C(OR)(OSiR_3)$$

$$(R = CH_3)$$

$$R_2C{=}C(OR)(OSiR_3) + CH_2{=}C(CH_3)CO_2CH_3 \xrightarrow{HF_2^{\ominus}} RO_2C{-}CR_2{-}CH_2C(CH_3){=}C(OSiR_3)(OCH_3)$$

$$\xrightarrow{n\,CH_2{=}C(CH_3)CO_2CH_3} RO_2CCR_2{-}\left[CH_2C(CH_3)_2\right]_n{-}CH_2C(CH_3){=}C(OSiR_3)(OCH_3)$$

and

$$\begin{array}{l} CHSSiCH_3 \\ | \\ CHSSiCH_3 \end{array} + CH_2{=}CHCO_2R \xrightarrow{ZnI_2} \begin{array}{l} CHS{-}\left[CH_2CH(CO_2R)\right]_n{-}CH_2CH{=}C(OSiCH_3)OR \\ | \\ CHS{-}\left[CH_2CH(CO_2R)\right]_n{-}CH_2CH{=}C(OSiCH_3)OR \end{array}$$

Once the monomer is consumed, a different monomer may be added to form a block copolymer or the chain can be terminated by removal of catalyst, protonation or alkylation.

$$\sim\!CH_2C(CH_3){=}C(OSiR_3)(OCH_3) \xrightarrow{CH_3OH} \sim\!CH_2C(CH_3){=}C(OSiR)(OCH)$$

$$\sim\!CH_2C(CH_3){=}C(OSiR_3)(OCH_3) \xrightarrow{C_6H_5CH_2Br} \sim\!CH_2C(CH_3)(CO_2CH_3)CH_2C_6H_5$$

This approach has been used to produce acrylic block copolymers which are designed for use as lacquer in the automobile industry. By changing the nature of the elements of the block copolymer it is possible to improve the gloss, adhesion and ability to achieve flow when the paint is being baked.

8.8.4 Cobalt-catalysed polymerisation

Cobalt-catalysed polymerisation is used to convert styrene polyester and isophthalate mixtures, which form the basis of much of composite manufacture, to the corresponding solid matrix.

The cobalt catalytic process allows polymerisation to be initiated at ambient temperatures and below using higher temperature peroxide catalysts. The initiation process can be summarised as follows:

$$Co^{2+} + ROOH \rightarrow Co^{3+} + RO^{\bullet} + HO^{\ominus}$$

$$Co^{3+} + ROOH \rightarrow Co^{2+} + ROO^{\bullet} + H^{\oplus}$$

$$2RO_2^{\bullet} \rightarrow 2RO^{\bullet} + O_2$$

$$RO^{\bullet} + ROOH \rightarrow 2RO_2^{\bullet} + ROH$$

The cobalt is assumed to oscillate between the +2 and +3 oxidation states and in so doing produces an alkoxy and an alkylperoxy radical from two molecules of hydroperoxide. An alternative scheme has been proposed in which dicobalt complexes are involved. The main difference between the two schemes is that the overall transformation does not produce an alkoxyradical:

$$3ROOH \rightarrow ROH + 2ROO^{\bullet} + H_2O$$

The cobalt is often used as cobalt(II) 2-ethylhexonate and these complexes are believed to act in a similar manner as in the dicobalt system illustrated below. However, the process is more complex than the above scheme would suggest as the reaction is accelerated by the addition of tertiary amines. Molecules such as diethylaniline can enhance the speed of the reaction. Tertiary amines can themselves help enhance the dissociation of peroxides and it is apparent that the amine can help with enhancing dissociation in addition to changing the coordination of the cobalt atom. These systems are widely used to cure a range of resin systems and have the advantage of operating over a broad temperature range. In practice, atmospheric oxygen can play a role in the process and the phenomenon of styrene inhibition, in which polymerisation is inhibited, is associated with benzaldehyde formation, which inhibits the action of the cobalt catalyst.

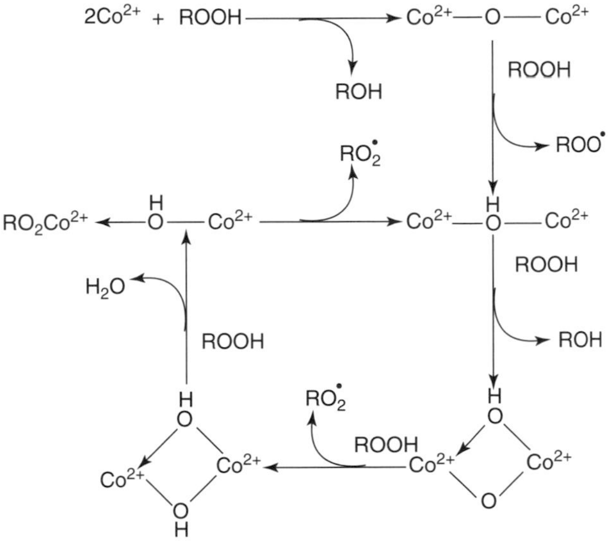

$2Co^{2+}$ + ROOH → Co^{2+}—O—Co^{2+}
ROH
ROOH
$ROO^{\bullet}$
$RO_2^{\bullet}$
RO_2Co^{2+} ← H–O—Co^{2+} → Co^{2+}—O(H)—Co^{2+}
ROOH
H_2O
ROOH
ROH
$RO_2^{\bullet}$
ROOH

8.9 Polymer degradation

At high temperatures polymers tend to undergo thermal or oxidative degradation. Thermal degradation occurs when a polymer is heated in an inert atmosphere. In the melt phase, the polymer may undergo what is essentially the reverse of the polymerisation process. In the presence of oxygen, chemical reactions can occur which will change the degradation process. There are a number of causes of degradation in addition to thermal effects, they include photodegradation, hydrolytic degradation and oxidative degradation. The tendency to degradation depends on the chemical structure of the polymer backbone. Polymer degradation can occur at temperatures lower than the ceiling temperature and result in changes in a variety of physical properties: tensile strength, colour, shape, etc. These changes are usually associated with a lowering of the molar mass, the oxidative attack of environmental stress cracking.

8.9.1 Analysis of polymer degradation: thermogravimetric analysis

The simplest and most common method of assessing polymer degradation is *thermogravimetric analysis* (TGA) (see Figure 8.18). The apparatus consists of a very sensitive balance from which is suspended via a quartz thread a small pan containing the material to be analysed (see Figure 8.18(a)). The analysis involves taking a small piece (~100 mg) of polymer and subjecting it to either isothermal or ramped heating. Typically, ramped heating will be used initially to determine the onset of degradation. Differentiation of the plot of the weight loss against time yields a peak which determines the temperature at which maximum weight loss occurs. Whilst for a simple system the point at which maximum weight loss is occurs is easy to identify, in more complex systems taking the differential of the mass lost can produce peaks which help to identify specific processes. Many polymers will produce a char residue and this can be identified as a high temperature plateau in the weight loss. Observation of such a plateau may be an indication that the sample contains inorganic filler.

8.9.2 Kinetics of polymer degradation: the random scission model

For simple thermal degradation, it is normally assumed that the chain breaks in a random fashion and the process which occurs is the reverse of polymerisation. This process occurs once the *ceiling temperature* has been exceeded. Being a kinetic process, it is relatively slow at the *ceiling temperature*, but its rate increases as the temperature is increased above this value. To illustrate this process we will consider polyethylene (see Figure 8.19).

Consider a polymer chain that has an average initial chain length of P_o. After a time t scission has reduce the chain length to a value of P_t as a result of S scissions per molecule, then:

$$P_t = P_o/(S + 1) \tag{8.81}$$

Therefore:

$$S = P_o/P_t - 1 \tag{8.82}$$

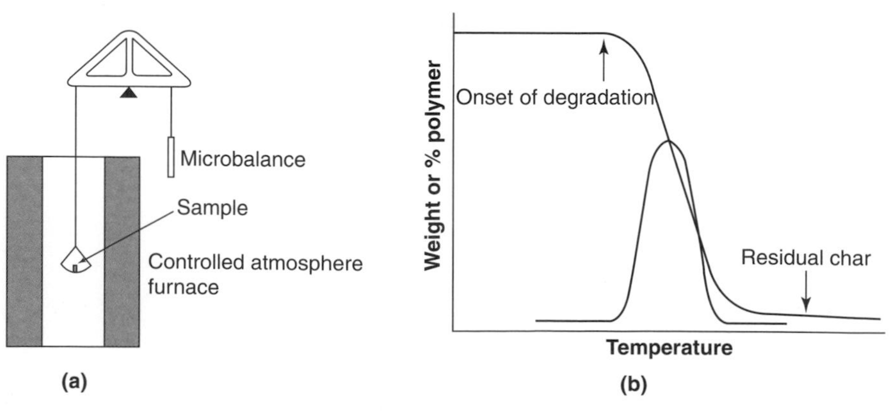

Figure 8.18 (a) Schematic of TGA apparatus; (b) idealised trace showing weight loss as a function of temperature and differential of peak.

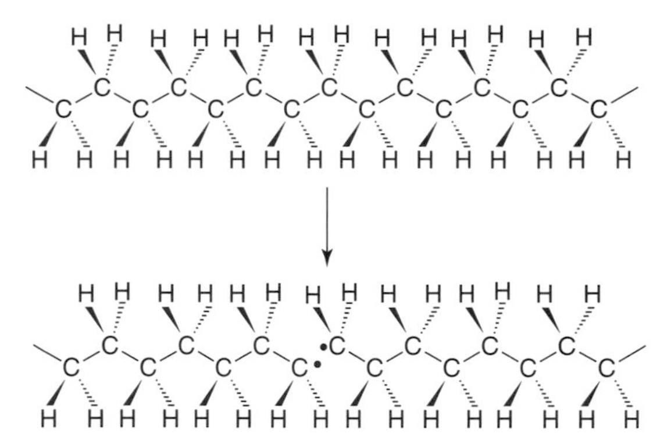

Figure 8.19 Thermal degradation of polyethylene.

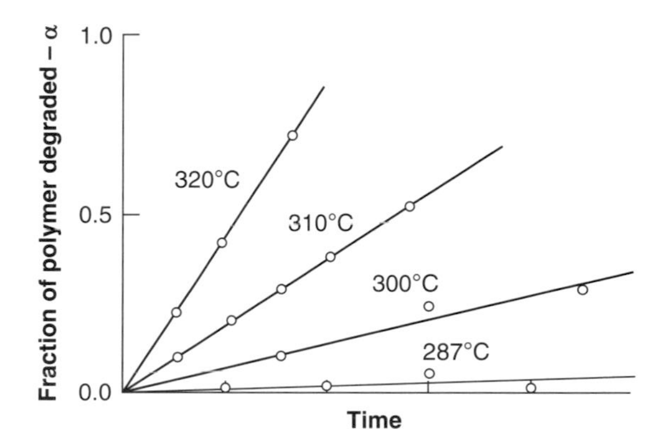

Figure 8.20 Thermal degradation of anionically polymerised polystyrene at different temperatures.

If α is the fraction of bonds broken at time t then:

$$\alpha = S/(P_o - 1) \approx S/P_o = 1/P_t - 1/P_o \tag{8.83}$$

Now if scission is really a random process then:

$$\alpha = kt \tag{8.84}$$

Therefore a plot of α versus t should be linear and pass through the origin. Data for the thermal degradation of a M_n 229,000 anionic polystyrene exposed to a range of different temperatures are presented in Figure 8.20 and illustrate the linearity of the scission process as a function of time. These data imply that the simple assumption about random scission is a reasonable approximation for these data.

8.9.3 Degradation of polyethylene

Before considering the complexities of the degradation process in a range of polymers it is appropriate to look at the simplest system: polyethylene (see Figure 8.21). The thermal degradation will produce two chains, each with a radical termination. The simplest process is then for the electron rearrangement to generate the ethylene monomer and essentially unzip the polymer chain. However, there is always the possibility of hydrogen migration via an intramolecular transfer, which will move the radical to a location away from the chain end. This process will involve the chain having adopted a *gauche* conformation and leads to a lower energy process compared with that for

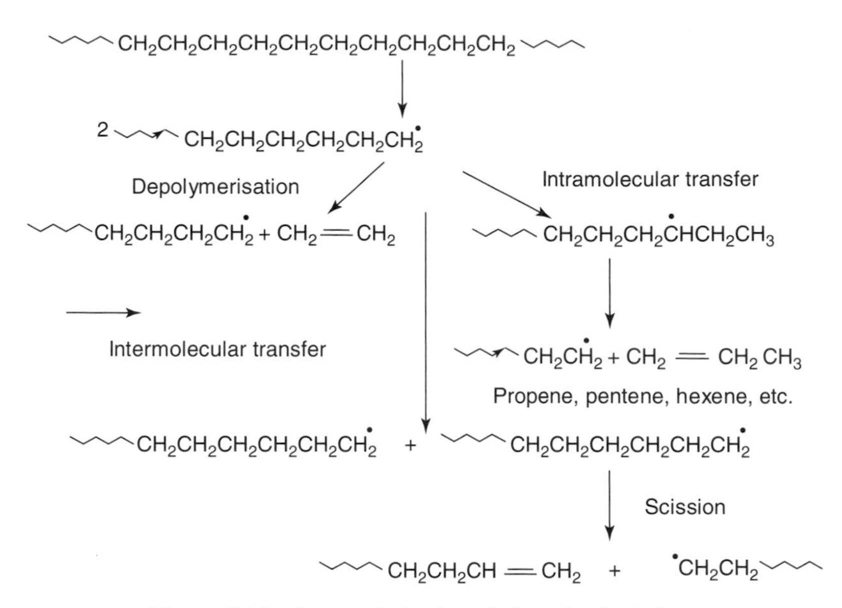

Figure 8.21 Schematic for degradation of polyethylene.

the all-*trans* conformation. The slight lowering of the total energy is very small compared with that for the dissociation process, but helps to explain why analysis of the degradation products reveals that there are significant amounts of propene, pentene, hexane, etc. found in the gas phase.

The radical formed by the intramolecular process loses the low molar mass unsaturated molecule and once more forms the terminally substituted radical chain. Hydrogen subtraction can occur from the carbon atom β to the radical and a terminal unsaturated bond is formed. This scheme illustrates the fact that even in a simple system such as polyethylene there can be a complex array of products formed. The above process is carried out in an inert atmosphere. If oxygen is present then the products formed can contain oxygen and the degradation scheme becomes very complex. For the simple case of thermal degradation a general mechanism can be devised.

8.9.4 General mechanism of radical depolymerisation

The first step is random scission:

Random initiation $$M_n \rightarrow M^*_j + M^*_{n-j} \quad \_k_{r1} \tag{8.85}$$

Terminal initiation $$M_n \rightarrow M^*_{n-1} + M \quad \_k_{t1} \tag{8.86}$$

Depropagation $$M_i \rightarrow M^*_{i-1} + M \quad \_k_{d1} \tag{8.87}$$

Intramolecular transfer and scission $$M_i \rightarrow M^*_{i-z} + M_z \quad \_k_{is1} \tag{8.88}$$

Intermolecular transfer $$M^*_i + M_n \rightarrow M_i + M^*_n \quad \_k_{i1} \tag{8.89}$$

Scission $$M^*_n \rightarrow M_j + M^*_{n-j} \quad \_k_{s1} \tag{8.90}$$

Termination $$M^*_i + M^*_j \rightarrow M_i + M_j \quad \_k_{t2} \tag{8.91}$$

$$M^*_i + M^*_j \rightarrow M_{i+j} \quad \_k_{t3} \tag{8.92}$$

The kinetics of such a complex process are not easily reduced to a simple mathematical relationship; however, we can get a feel for the rate of the process by considering the rate of disappearance of polymer, M_n

Table 8.6 Summary of percentage of monomer resulting from thermal degradation of some common polymers

Polymer	Monomer (%)	Polymer	Monomer (%)
Methylmethacrylate	100	Methylacrylate	42
Styrene	42	α-Methylstyrene	100
m-Methylstyrene	52	Ethylene	<1
Methylacrylonitrile	100	Vinylidenechloride	100
Isobutene	32	Propylene	2
Butadiene	1.5	Isoprene	12

$$\frac{\partial[M_n]}{\partial t} = k_{r1}M_n + k_{t1}M_n + k_{d1}M_n + k_{is1}M_n + k_{i1}M_iM_n + k_{s1}M_n + k_{t2}M_iM_j + k_{t3}M_iM_j \quad (8.93)$$

In the above equation the rate constant for the scission process is expected to be the rate determining step so that $k_{s1} \sim k_{r1}$ and these rate constant will be similar in value. Equation (8.93) can be reduced to:

$$\partial[M_n]/\partial t = k^*M_n \quad (8.94)$$

where k^* is a pseudorate constant which incorporates the rates for the primary scission processes. Study of a range of different polymeric materials indicates the extent to which the process is a simple chain scission. The loss of a monomer depends on the chemical structure of the polymer chain (see Table 8.6).

It is noticeable that many polymers do not simply depolymerise and the monomer yield appears to change dramatically for relatively small changes in the backbone structure. For instance, changing from MMA to polymethylacylate (PMA) leads to a reduction in the monomer yield from 100 in the case of polymethylmethacrylate (PMMA) to a value of 42% in PMA. Similarly the monomer yield in the case of styrene is only 42%, whereas in α-methylstyrene the yield is 100%. In other aliphatic materials low yields are observed.

8.9.5 Depolymerisation versus transfer

In order to understand the differences highlighted in Table 8.6, it is appropriate to consider how the changes in structure impact on the stability of the radicals which are formed by scission.

$$\left[-CH_2-\overset{X}{\underset{Y}{C}}- \right]_n$$

The following generalisation can be made:

- For $X + Y = H$, no monomer is produced and extensive H transfer via H abstraction occurs, e.g. polyethylene.
- For $X = H$, the amount of monomer produced depends on the nature of Y and the resultant stability of the radical which is formed. In the case of polypropylene, the radical is very unstable and little monomer is formed, whereas in the case of styrene the radical is stabilised and more monomer is generated.
- For $X \neq H$, $Y \neq H$, a large amount of monomer (up to 100%) can be formed as illustrated in the case of PMMA.

In summary, the stability of the terminal radical is a determining factor for the amount of the monomer which will be recovered in the case of thermal degradation. Before considering specific polymer systems it is appropriate to consider degradation when oxygen is present.

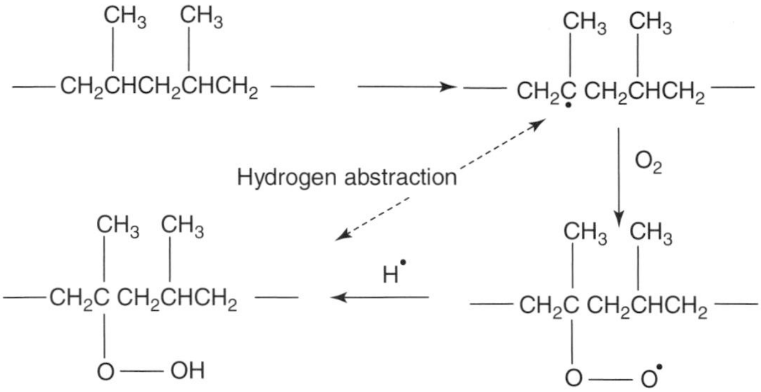

$$-CH_2CH(CH_3)CH_2CH(CH_3)CH_2- \longrightarrow -CH_2\dot{C}(CH_3)\,CH_2CH(CH_3)CH_2-$$

$$\xrightarrow{O_2} -CH_2C(CH_3)(O-O^{\bullet})\,CH_2CH(CH_3)CH_2- \xrightarrow{H^{\bullet}} -CH_2C(CH_3)(O-OH)\,CH_2CH(CH_3)CH_2-$$

Hydrogen abstraction

The oxygen can either be in an excited state (singlet oxygen), which can directly abstract the hydrogen, or it can react with a radical site created by chain scission. The peroxide which is created will readily undergo decomposition to produce a hydroxyl radical which can undergo further hydrogen abstraction producing water and more free radical sites on the polymer backbone. The rearrangement of the oxygen bearing radicals can create ketones and eventually aldehydes or acids. It is appropriate to consider the degradation of some specific polymer systems.

8.9.6 Degradation of polyvinylchloride

PVC widely used in domestic and industrial applications. Most of the materials which are encountered are stabilised to avoid photodegradation and are often stabilised to inhibit thermal and oxidative degradation. The characteristics of degradation of PVC are:

- Production of HCl at low temperatures.
- Polyene sequences develop in residues (develops increasingly darker colours as unsaturation increases).
- Residue begins to cross link.
- Benzene detected as a volatile product.
- Large quantity of carbonaceous char produced as residue.
- Autocatalysis of degradation.

The degradation of PVC does not necessarily lead to chain scission but initially involves the loss of HCl and the creation of linked unsaturated sequences and observation of characteristic changes in colour: yellow to red to black. The process is shown in Figure 8.22.

In order to understand the dehydrochlorination process it is appropriate to examine the decomposition temperatures of a range of related small molecules (see Figure 8.23). The decomposition temperature reflects the stability of the radical which is formed when the C–Cl bond is broken. In the case of (vii) with a decomposition temperature of 180°C, the high steric strain of the three ethyl groups attached to the carbon atom which contains the chloride produces a bond which is easily broken, a radical which combined with *trans* elimination

$$\sim\!CH_2CH(Cl)CH_2CH(Cl)CH_2CH(Cl)CH_2CH(Cl) \longrightarrow \sim\!CH{=}CHCH{=}CHCH{=}CH$$

Figure 8.22 Dehydrochlorination reaction for PVC.

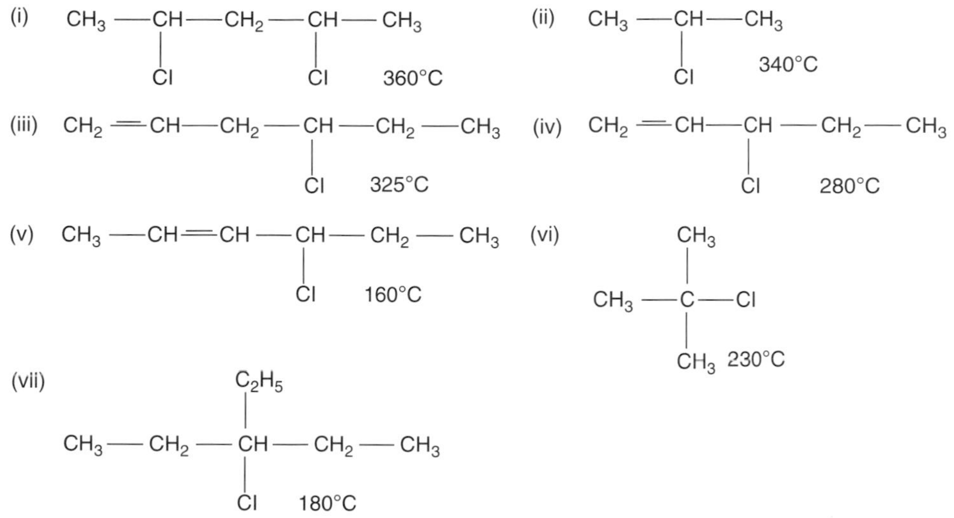

(i) $CH_3-CH(Cl)-CH_2-CH(Cl)-CH_3$ 360°C

(ii) $CH_3-CH(Cl)-CH_3$ 340°C

(iii) $CH_2=CH-CH_2-CH(Cl)-CH_2-CH_3$ 325°C

(iv) $CH_2=CH-CH(Cl)-CH_2-CH_3$ 280°C

(v) $CH_3-CH=CH-CH(Cl)-CH_2-CH_3$ 160°C

(vi) $(CH_3)_3C-Cl$ 230°C

(vii) $CH_3-CH_2-C(C_2H_5)(Cl)H-CH_2-CH_3$ 180°C

Figure 8.23 Decomposition of a range of related small molecule systems.

of a proton from the β carbon atom produces a very facile reaction. If a double bond already exists next to the C–Cl bond as in (v), the decomposition temperature is lowered to 160°C, the loss of chlorine being assisted by the electron donation from the methyl group attached to the double bond. If the methyl group is absent, as in (iv), then the decomposition temperature is increased to 280°C. The effects of steric hindrance are also shown in the case of (vi) where the decomposition is again lowered to a value of 230°C, but the donation ability of the methyl groups is clearly evident relative to the lower decomposition temperature of (vii). The less steric hindrance in (iii) and the stabilising influence of moving the double bond one atom further from the chlorine increases the decomposition temperature to 325°C. Similarly, removal of one methyl group reduces the steric strain and the decomposition temperature of (ii) at 340°C is significantly higher than that of form (vi). The balance of reduction of steric hindrance and stabilisation of the methyl group is further exemplified in (i), which has a decomposition temperature of 360°C. It would appear from the study of these model compounds that PVC should be relatively stable, but it has a tendency to become unsaturated at room temperature if action is not taken to stabilise it. The dehydrochlorination process is summarised in Figure 8.24. The first step is usually considered to be scission of the C–Cl bond. Whilst it is usually considered that the polymerisation produces head–tail sequences, there is always the possibility of head–head sequences being formed at a very low level along the polymer backbone.

In these head-to-head sequences the neighbouring chlorine atoms will be very sterically crowded and therefore it is not surprising that one of the chlorine atoms is easily lost to give the required radical to initiate degradation. The hydrogen *trans* to the chlorine tends to be eliminated more easily leading to molecular elimination of HCl.

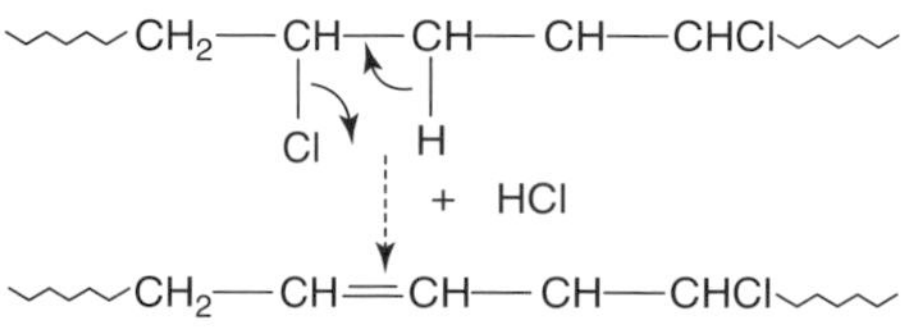

$$\sim CH_2-CH(Cl)-CH(H)-CH-CHCl\sim$$

$$\downarrow + HCl$$

$$\sim CH_2-CH=CH-CH-CHCl\sim$$

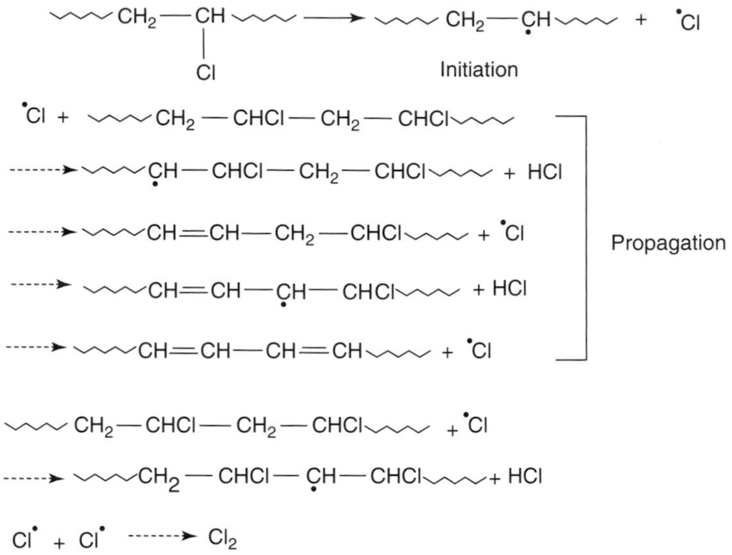

$$\sim CH_2{-}CH(Cl)\sim \longrightarrow \sim CH_2{-}\dot{C}H\sim + \dot{C}l$$

Initiation

Propagation:

$$\dot{C}l + \sim CH_2{-}CHCl{-}CH_2{-}CHCl\sim$$

$$\dashrightarrow \sim \dot{C}H{-}CHCl{-}CH_2{-}CHCl\sim + HCl$$

$$\dashrightarrow \sim CH{=}CH{-}CH_2{-}CHCl\sim + \dot{C}l$$

$$\dashrightarrow \sim CH{=}CH{-}\dot{C}H{-}CHCl\sim + HCl$$

$$\dashrightarrow \sim CH{=}CH{-}CH{=}CH\sim + \dot{C}l$$

$$\sim CH_2{-}CHCl{-}CH_2{-}CHCl\sim + \dot{C}l$$

$$\dashrightarrow \sim CH_2{-}CHCl{-}\dot{C}H{-}CHCl\sim + HCl$$

$$Cl^{\bullet} + Cl^{\bullet} \dashrightarrow Cl_2$$

Figure 8.24 Dehydrochlorination reactions for PVC.

The HCl which is created by this initial step can act as a catalyst for further degradation.

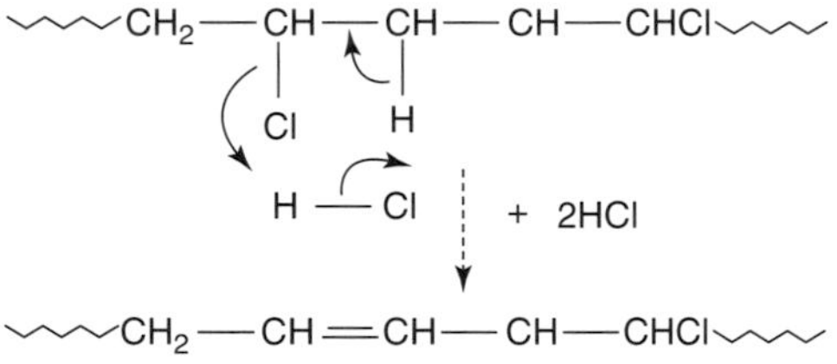

$$\sim CH_2{-}CH(Cl){-}CH(H){-}CH{-}CHCl\sim \quad H{-}Cl$$

$$\dashrightarrow \; + 2HCl$$

$$\sim CH_2{-}CH{=}CH{-}CH{-}CHCl\sim$$

It is believed that at low temperatures the HCl catalysed degradation is fairly efficient and the main cause of the discolouration of PVC. Other reactions which are possible are cross-linking and benzene elimination.

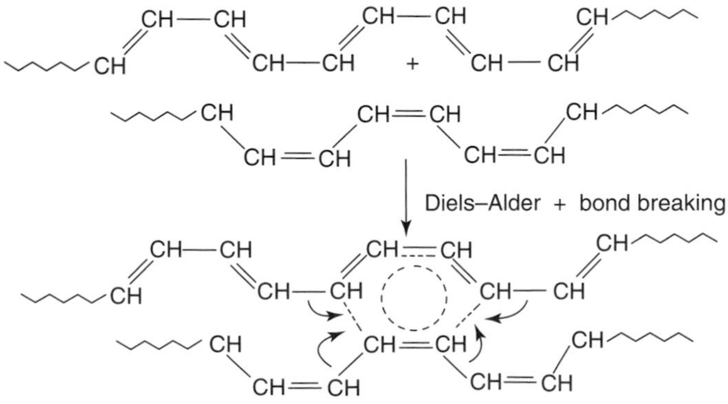

$$\sim CH{=}CH{-}CH{=}CH{-}CH{=}CH{-}CH{=}CH\sim \; + \; \sim CH{-}CH{=}CH{-}CH{=}CH{-}CH{=}CH{-}CH\sim$$

Diels–Alder + bond breaking

$$\sim CH{=}CH{-}CH{=}CH{-}CH{=}CH{-}CH{=}CH\sim \; \text{(cyclic adduct)} \; \sim CH{-}CH{=}CH{-}CH{=}CH{-}CH{=}CH{-}CH\sim$$

The Diels–Alder addition forms a cross-link, which is followed by rearrangement and elimination of benzene. If these reactions are suppressed then the possibility of chain scission becomes a possible route to degradation.

8.9.7 Polyvinyl acetate

One of the products of degradation is acetic acid and this is generated in a reaction analogous to the elimination of HCl from PVC. The elimination reaction creates a double bond in the polymer backbone and as a consequence a yellowing of the polymer.

8.9.8 Polymethylmethacrylate

The normal degradation is a two-step process (see Figure 8.25). However, the first step is completely absent in anionically polymerised polymer. Monomer is virtually the only product. The monomer is produced by the reverse of the propagation process, *unzipping*. The unzip length is approximately 200 units and the process is characterised by the pungent smell of the MMA monomer. From the point of view of potential recycling, in principle it is possible to heat PMMA and recover monomer, which can then be repolymerised to once more produce the polymer form.

In practice, PMMA can be recycled to a limited extent if dissolved in monomer and then polymerised. At high temperatures, the process is simply the unzipping consequent upon random scission of the backbone. In radically prepared PMMA, ~50% of chain ends are unsaturated, the termination process being dominantly disproportionation.

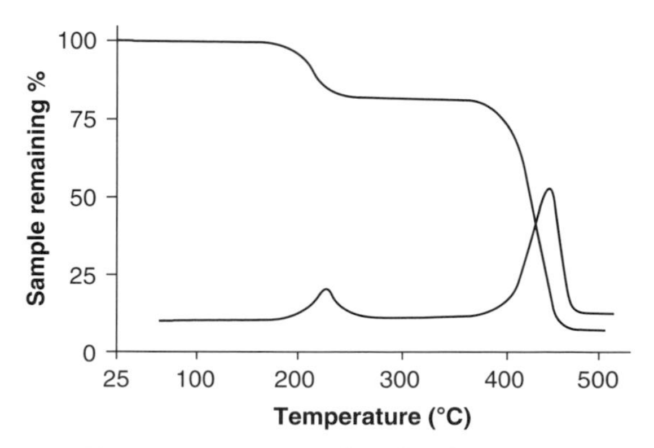

Figure 8.25 TGA analysis data for PMMA.

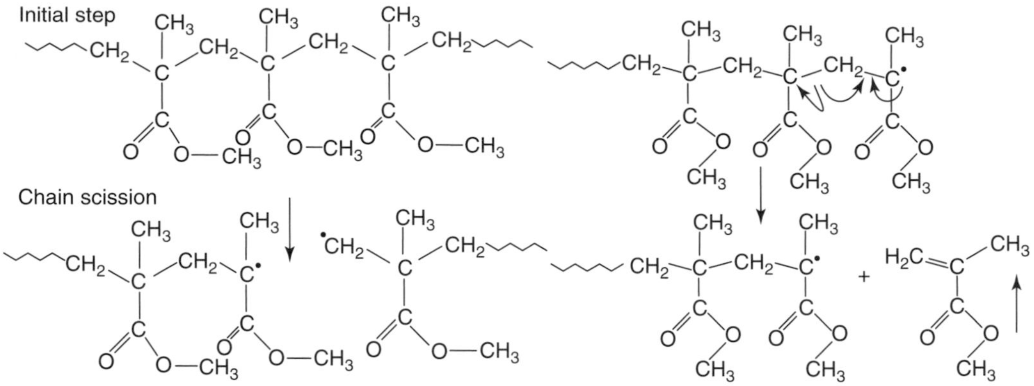

Initial step

Chain scission

The double bond destabilises the bond at the β position; scission is promoted at lower temperatures.

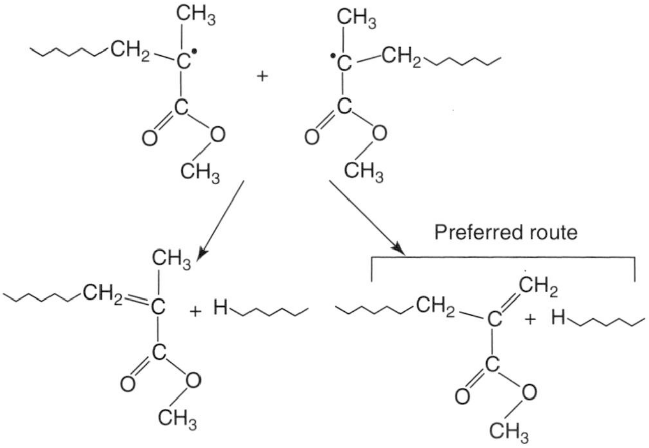

Preferred route

The unzip length is about 200 units. Fifty per cent (i.e. those with terminal unsaturation) of the chains will start to depropagate at low temperatures. If the molecular weight of the unzipping chain is less then the zip length, the chain will unzip completely to monomer, thus for $M_n < 20{,}000$, approximately 50% of the polymer will degrade via chain end scission. A side reaction which can alter the depolymisation process is associated with the formation of an anhydride ring and occurs especially when the polymer is a methylmetharylate–methacryclic acid copolymer. The anhydride can unzip, but the subsequent depolymerisation is blocked.

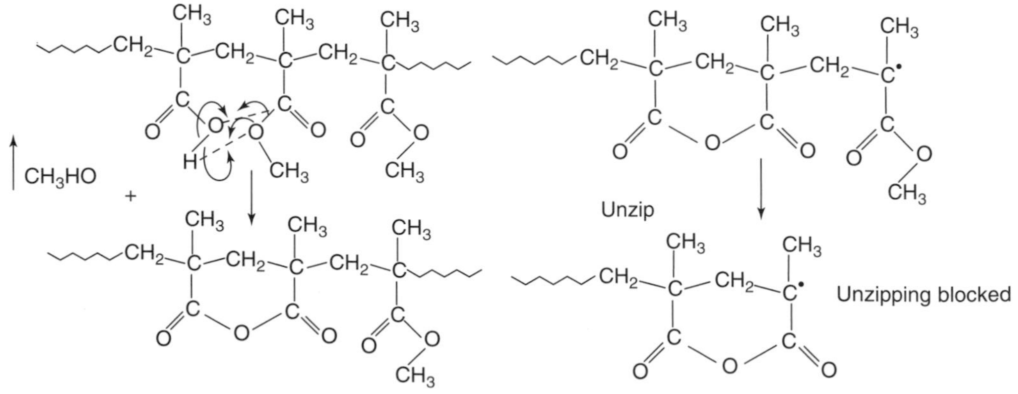

Unzip

Unzipping blocked

8.9.9 Degradation routes for alkylmethacrylate polymers

A study of the degradation of a series of alkylmethacrylate polymers indicates the role which the side chain can play in the degradation process. The product of the degradation depends on two factors: the number of β hydrogens and the size of the ester grouping, as shown in Table 8.7.

If the ester group is sterically hindered then there is a high probability that scission will occur in the side chain leaving an acid function attached to the backbone (see Figure 8.26). Alternatively, if there is a β hydrogen present then elimination can occur, forming an unsaturated bond in the backbone. Both these processes are in addition to the normal type of degradation found in PMMA (see Figure 8.26).

8.9.10 Degradation of polyethyleneterephthalate

The melting point of polyethyleneterephthalate (PET) is 250°C and hence high temperatures are required for processing. Degradation starts at ~250°C with a drop in the molar mass being observed: volatile production only becomes rapid above 350°C. The principal degradation reaction is ester scission as shown in Figure 8.27. Major products that are ultimately formed from PET include: terephthalic acid, acetaldehyde and carbon monoxide. The minor products that are ultimately produced include: anhydrides, benzoic acid, vinyl benzoate, water, ethane and ethyne.

Table 8.7 Number of β-hydrogen atoms in alkylmethacrylate polymers

Ester group	No. of β-hydrogens	Comment	Ester group	No. of β-hydrogens	Comment
Methyl	0		n-Pentyl	0	
Isobutyl	1	Unzipping	Ethoxyethyl	2	
Ethyl	2		n-Propyl	2	
n-Butyl	2	Mostly	n-Hexyl	2	Unzipping
n-Heptyl	2		n-Octyl	2	
Isopropyl	6	Ester degrade	*s*-Propyl	5	Ester degrade
t-Butyl	9				

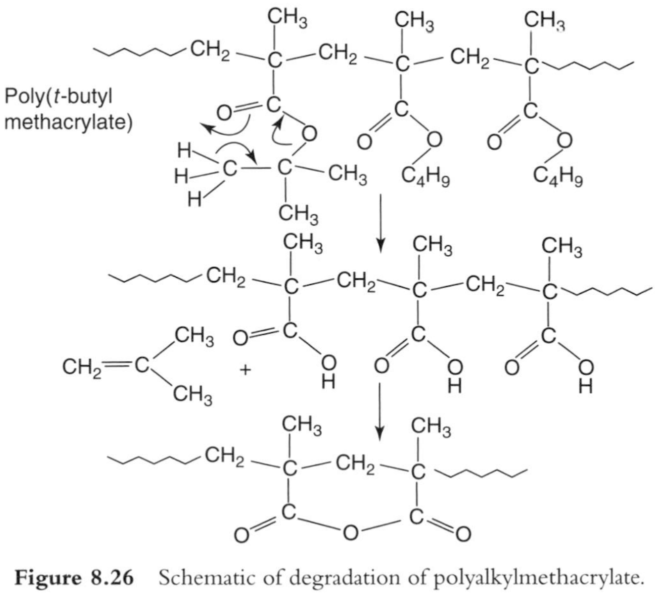

Figure 8.26 Schematic of degradation of polyalkylmethacrylate.

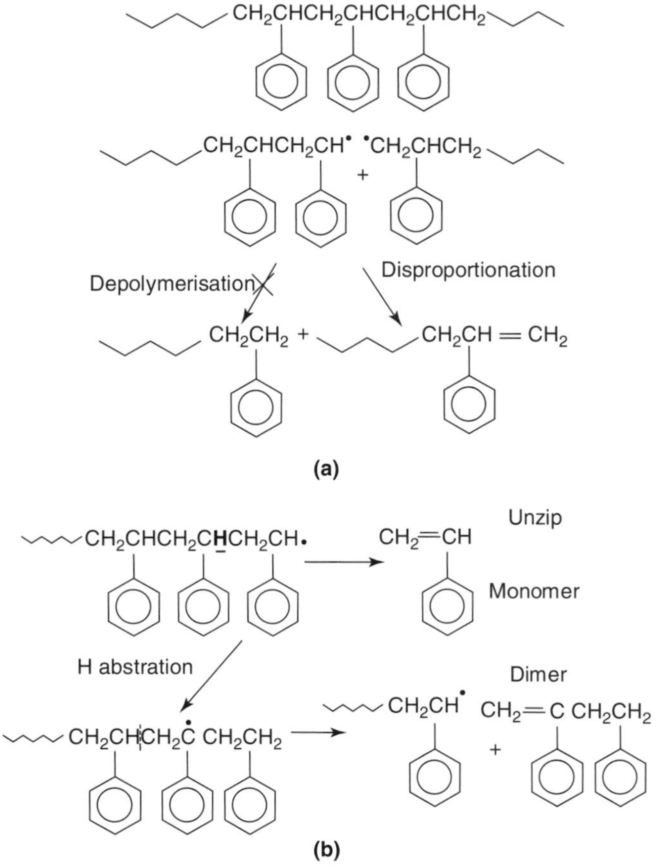

Figure 8.27 Schematic of degradation processes: (a) below 300°C; (b) above 300°C.

The equilibrium water content is typically 0.3% at 25°C and at 50% relative humidity. Hydrolysis is quantitative (one molecule will result in one chain scission) and a rapid drop in molar mass is observed. Injection moulding of preforms for bottles requires the water content to be reduced to 0.003%. Drying requires a balance between the rate of diffusion of water from the polymer and the rate of hydrolysis. The compromise is to carry out the drying process at 170°C for 4 hours.

Many polymers, especially step-growth polymers, are degraded by specific chemicals such as strong acids and strong alkali. Condensation polymers can degrade by the reverse of the method by which they have been made. Other degradation routes involve interaction with strong oxidising agents and interaction with UV radiation.

8.9.11 Polystyrene

For degradation of polystyrene below 300°C, the process is chain scission which occurs randomly along the chain, and there are no volatile products. Above 300°C, the products are monomer, and approximately 4–50% dimer and trimer. The mechanisms of degradation are summarised in Figure 8.28. Above 330°C, the abstraction reaction now becomes more significant and dimer, trimer etc. become more prevalent. However, with poly(α-methylstyrene) there is no tertiary hydrogen and no abstraction can occur, thus a 100% monomer is obtained.

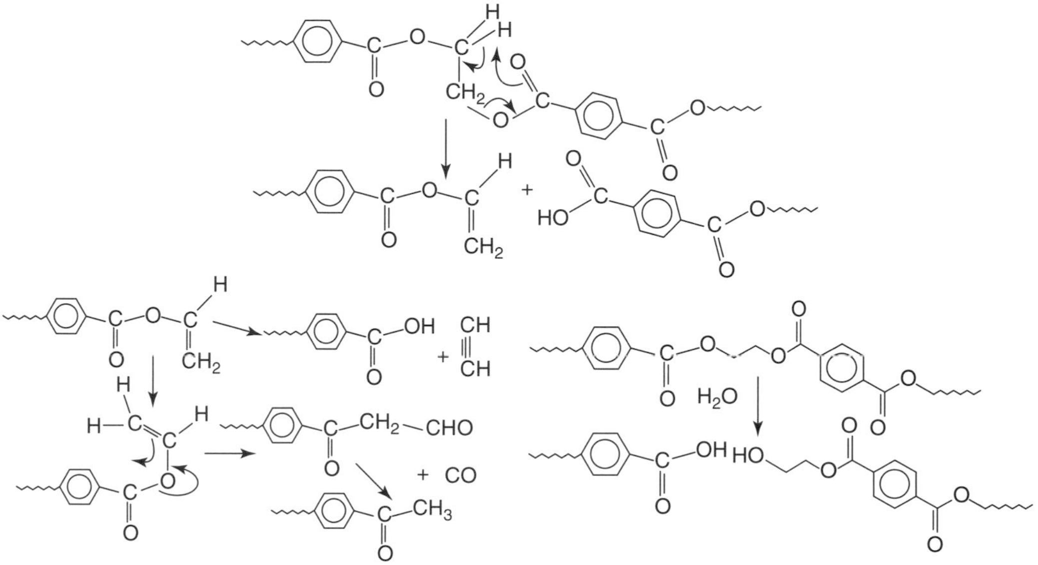

Figure 8.28 Schematic of degradation of polyethyleneterephthalate.

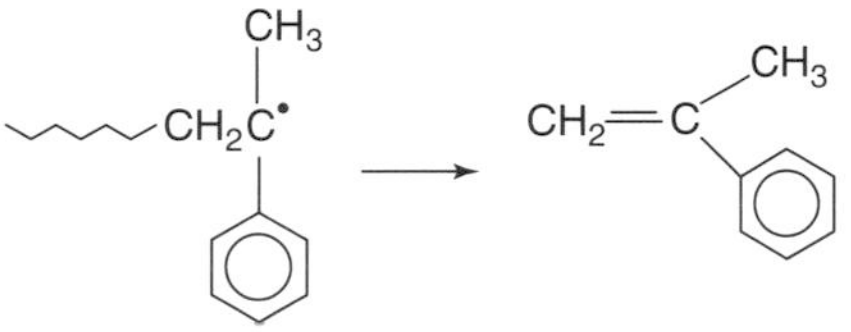

In the degradation of α-methylstyrene, the monomer is 100% recovered because there is no possibility of any reaction other than unzipping occurring.

8.9.12 Hydrolysis

Nylon is sensitive to degradation by acids and a nylon moulding will crack when attacked by strong acids. A fuel pipe fractured when a small drip of 40% sulfuric acid from a nearby battery fell onto a nylon 6,6 moulded connector in a diesel line. The crack grew with time until it penetrated the interior, so initiating a slow leak of diesel. The crack continued to grow until final separation occurred, and diesel fuel poured into the road.

8.9.13 Importance of β hydrogen in degradation

If there is a β hydrogen next to the site where the radical is formed then there is a high probability of hydrogen transfer reactions and hydrogen migration. The stability of the radical formed will depend on the nature of the functional groups to which the radical is attached. Hence, in the degradation of polystyrene, the radical which is formed is stabilised by interaction with the phenyl group.

8.10 Polymers and fire

Today there are primarily six commodity polymers in use: polyethylene, polypropylene, PVC, PET, polystyrene and polycarbonate. These make up nearly 98% of all polymers and plastics encountered in daily life. Each of these polymers has its own characteristic modes of degradation and resistances to heat, light and chemicals. Polyethylene and polypropylene are sensitive to oxidation and UV radiation, while PVC may discolour at high temperatures due to loss of hydrogen chloride gas, and become very brittle. PET is sensitive to hydrolysis and attack by strong acids, while polycarbonate depolymerises rapidly when exposed to strong alkalis.

Polymers do not burn: it is the degradation products which are combustible, i.e. under the influence of heat, the large polymer molecules break into small, volatile molecules which escape from the polymer into the air around the polymer, where they ignite, generating more heat. As materials, polymers offer particular problems in a fire:

- *Low thermal conductivity*: Under radiant heat loads (common in a fire), surface temperatures will quickly rise above the decomposition temperature of the polymer, as heat cannot be easily dispersed throughout the bulk of the polymer.
- *Toxic and corrosive products*: Most damage to buildings and almost all fatalities result from the generation of toxic, irritant or corrosive gases.
- *Dense smoke*: Even when the smoke is not toxic, high smoke density will reduce the likelihood of escape. Even relatively low smoke densities have been shown to induce panic.
- *High surface area*: As a general rule, the higher the surface area, the greater the rate of combustion, all other factors being equal. Foams, often used as insulation, have particularly high surface areas.
- *'Drip' factor*: Poor mechanical properties at elevated temperatures mean that structural integrity is lost. In the worst scenario, burning polymer can drip from suspended structures, increasing the rate of fire spread and the likelihood of injury.

8.10.1 Cone calorimeter

The *cone calorimeter* has been devised to provide a quantitative method of assessing the susceptibility to heat and fire of materials. The device, shown in Figure 8.29, consists of a device for measurement of the weight loss as a function of time (a load cell and a method of extraction of the gases created during the combustion process).

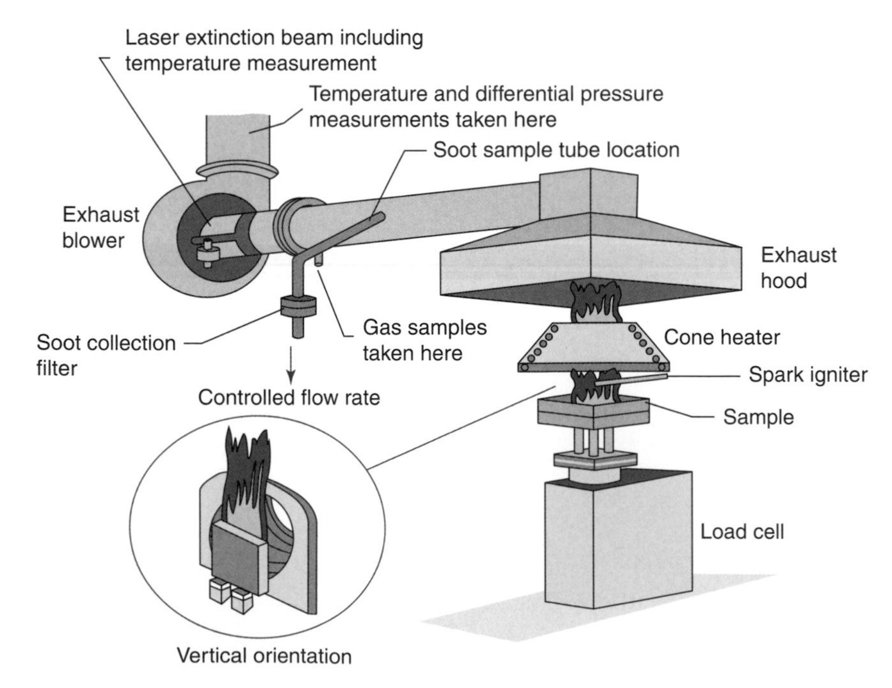

Figure 8.29 Cone calorimeter.

The gases are subjected to a range of analyses, depending on the nature of the instrument. A laser interferometer allows examination of the temperature of the flame and the density of the smoke particles which are generated. The temperature and pressure of the gas are also measured, allowing the thermodynamics of the combustion process to be studied. A soot sample tube allows quantitative measurement of the particulate content of the gases. Burning can be carried out in a horizontal or vertical mode. The choice of the mode of burning can simulate different potential fire situations but also influences the supply of oxygen to the sample. The cone calorimeter is a very popular method for the analysis of the factors which influence combustion. Heat release is the most important single factor in a fire and can be described as the driving force of a fire and is measured by the cone calorimeter.

8.10.2 Experimentally measurable parameters

One of the most important parameters to determine in relation to the susceptibility of a polymeric material to undergo combustion is the temperature at which spontaneous combustion occurs. In this context the *limiting oxygen index* (LOI), is the oxygen content (in %) of an O_2/N_2 mixture required to just sustain the combustion of the polymer. Air contains 21% oxygen and in normal conditions a polymer with an LOI higher than 21% should be combustible in air. At the higher temperatures typical of a fire, the LOI value falls and thus a value of >27% is required before a polymer can be considered to be fire retardant. Typical values are listed in Table 8.8.

Another important parameter in determining whether a polymer will burn is the *heat of combustion*. The heat of combustion is a measure of the heat load contribution to the fire from the combusting polymer. Note that the heat of combustion does not correlate with the LOI, although it is an important parameter in its own right, as it is a measure of the fuel load that the polymer will contribute to the fire. A high heat of combustion will melt a considerable amount of polymer and hence accelerate the fire.

The *char yield* is the amount of solid carbonaceous char remaining upon combustion (usually measured by heating to some arbitrary temperature, e.g. 900°C). Char can seal the surface of the polymer and prevent the escape of gaseous products or the ingress of oxygen. This is an important parameter, which can determine the difference between a material burning or being self-extinguishing. If the char insulates the polymer surface then the material will cease to burn.

The *smoke density* rating is also an important parameter. In a fire the amount of smoke which is generated and its density can determine the extent to which a person trapped in the fire becomes confused and is unable to escape. Polyolefins decompose to aliphatic molecules that burn cleanly and so they have a low smoke yield. However, they produce char. Polystyrene produces no char as it decomposes almost quantitatively to monomer, dimer and trimer. Furthermore, as carbon-rich aromatic compounds, these decomposition products burn with a smoky flame. Thus polystyrene is particularly troublesome in a fire, producing a heavy black smoke.

Other factors which may be important in a fire are the extent to which the smoke contains toxic components. For instance, PVC burns with the emission of toxic acidic vapours, polyurethanes can release isocyanides, polytetrafluororethylene (PTFE) can release hydrofluoric acid. Unfortunately there is no easy answer to the choice of a polymer which will exhibit high

Table 8.8 Typical values for LOI for a number of common polymers

Polyurethane foam	16.5	Polymethylmethacrylate	17.3
Polyethylene	17.4	Polypropylene	17.4
Polystyrene	17.8	Polyester (PET)	20.0
Polycarbonate	22–28	Nylon 6,6	24
Neoprene rubber	26	Phenolic thermoset	35
UPVC	45	PTFE	95

fire retardant characteristics. Polyphenyleneoxide has a high char yield but is associated with a high smoke density. PVC is often used with high levels of plasticisers that lower the LOI.

Fire retardants only go part of the way toward solving the problem and they often present their own problems. Halogen compounds are increasingly viewed with disfavour because of the corrosive gases they produce. Tricresylphosphate, often used in PVC formulations, contributes to a very high smoke density. Other additives are themselves highly toxic, either in the fire or maybe even in use. Certain additives have been thought to cause the health problems with cot mattresses. The combustion cycle is:

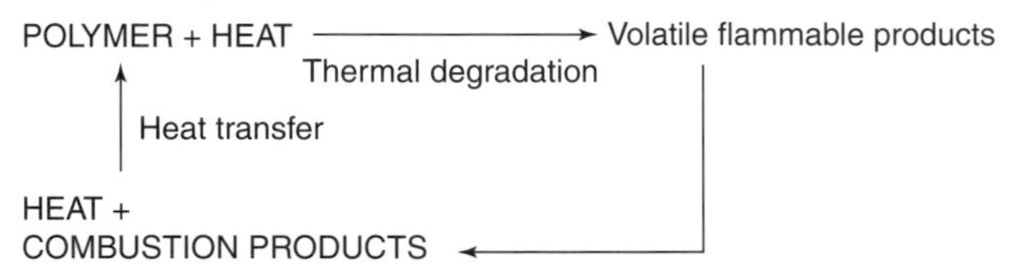

The above cycle can be broken by modifiying the degradation behaviour, quenching the flame, and reducing the heat supply.

There are a number of factors which can improve the fire resistance of a material:

- By degradation to low molecular weight volatile and flammable products which are not favoured (e.g. polypropylene).
- Remove weak C–C and C–H bonds. Stronger C=C and aromatic structures are preferred (e.g. polyetheretherketone and polyethersulfone (PES)).
- Increase in molecular weight upon degradation (cross-linking) reduces volatile products and increases char yield (e.g. PVC).
- Include halogen atoms (Cl, Br), as production of HBr/HCl interferes with flame chemistry and can also produce structures which lead to cross-linking (e.g. PVC).

8.10.3 Improved fire retardancy of polymers

The most obvious approach is to reduce the tendency of the material to emit vapours and hence anything which reduces volatility will reduce the susceptibility to produce a fire. Polymer molecules with stronger bonds, such as phenyl-linked structures, produce cross-linked structures in a fire situation and are usually good char formers. Fire retarding elements such as boron, aluminium and phosphorus can promote char formation by changing the degradation chemistry. Incorporation of chlorine, bromine, and antimony is known to have positive effects by interfering with the chemistry of the flame and breaking the combustion cycle.

Additives based on molecules containing aluminium, boron, phosphorus, antimony, bromine, or chlorine are used to reduce flammability and are the basis of flame retardants which are often incorporated in polymer materials at the compounding stage. Fire retardants act by interfering with the fire cycle at different points.

- Some additives purely volatilise, taking energy out of the sample and effectively cooling the material. An example of such materials is aluminium trihydate (ATH). When the sample is heated, the ATH loses water. This process absorbs energy and can suppress the degradation. Melamine, which is used as a fire retardant, acts by absorbing energy as it is sublimed into the vapour phase. The degradation mechanism for molecules containing phosphorus produces species that promote char formation and hence decreases the production of volatile products and consequently reduces the fuel load. Additives which promote charring at the surface (known as intumescent systems) are effective, as a carbonaceous char acts as a good barrier to heat, prevents volatile material escaping from the polymer surface and is surprisingly nonflammable. Those who have tried to light a barbecue briquette will know that this is initially quite difficult.

- Additives based on chlorine and bromine are losing favour for environmental reasons and antimony has been linked with cot deaths. However, the release of a halogen into the flame has the effect of terminating a number of radical processes and as a consequence is very good at suppressing flame chemistry.
- Polymer molecules that are difficult to break into small fragments show good fire resistance, but are generally difficult to process or are very expensive. Thermosetting polymers generally show good fire resistance as a result of the difficulty in breaking down the highly cross-linked network. They also show much better dimensional stability at elevated temperatures. Indeed glass-filled thermosets have been shown to be superior to many metals in real fires.
- Simple mineral fillers can often reduce gross flammability by interfering with heat transfer to the polymer or quite simply by reducing the mass of polymer per unit volume of material.

8.10.4 Flame chemistry

The chemical processes which occur in a flame are complex and often auto-accelerating. In other words, many of the processes will produce new radicals and increase the density of radicals as the flame grows. The preponderance of the radicals in the flame will reflect the amount of energy which is emitted and hence the amount of energy which can be used to melt the polymer and further feed combustion. Flame chemistry is complex but there are two key reactions. The first is the chain branching step:

$$H^{\bullet} + O_2 \rightarrow HO^{\bullet} + O^{\bullet}$$

in which a hydrogen atom reacts with an oxygen molecule and produces an oxygen atom and a hydroxyl radical. The number of radicals is doubled in the flame. The other very important step is the highly exothermic reaction of the hydroxyl radical with carbon monoxide:

$$HO^{\bullet} + CO \rightarrow H^{\bullet} + CO_2$$

However, in the presence of halogen atoms or with HCl present we have:

$$H^{\bullet} + HCl \rightarrow H_2 + Cl^{\bullet} \quad \text{and} \quad HO^{\bullet} + HCl \rightarrow H_2O + Cl^{\bullet}$$

In this reaction the chlorine atom is mopping up the hydrogen and hydroxyl radicals and slows down the flame chemistry and reduces the amount of energy which is released. The halogens mop up highly reactive $H^{\bullet}$ and $HO^{\bullet}$ generate $Cl^{\bullet}$ which cannot propagate flame chemistry and from $Cl^{\bullet}$ regenerate HCl.

Bromine containing fire retardants behave in a similar manner to chlorine. The lability of the weaker H–Br bond leads to greater activity than H–Cl. Lability of the C–Br bond can bring problems of thermolysis during processing, and photolysis in use reduces efficiency, causing discolouration. Bromine compounds are generally more expensive than chlorine containing compounds and only used in specific applications.

Antimony trioxide (Sb_2O_3) is often added to enhance the efficiency of certain fire retardants. It is known as a synergistic agent and is usually combined with chlorinated and brominated agents, e.g. 3% of SB_2O_3 + 5% Br = 14% Br w/w; a 3:1 atomic ratio of Sb:X is generally best. The action of antimony trioxide is associated with the formation of antimony oxyhalide with subsequent decomposition products of the oxyhalide as the active components. The chemistry can be summarised as:

$$\begin{aligned}
Sb_2O_3 + 2HCl &\rightarrow 2SbOCl + H_2O;\\
5SbOCl &\rightarrow Sb_4O_5Cl_2 + SbCl_3;\\
4Sb_4O_5Cl &\rightarrow 5Sb_3O_4Cl + SbCl_3;\\
3Sb_3O_4Cl &\rightarrow 4Sb_2O_3 + SbCl_3;\\
SbCl_3 + H^{\bullet} &\rightarrow SbCl_2 + HCl;\\
SbCl_2 + H^{\bullet} &\rightarrow SbCl + HCl;\\
SbCl + H^{\bullet} &\rightarrow Sb + HCl
\end{aligned}$$

8.10.5 Effect of various fillers on limiting oxygen index

Fillers are often added to plastics either to produce opacity or to reduce the cost by effectively replacing an expensive organic material by a cheaper inorganic material. The negative effect is that some of the fillers will not reinforce the polymer and the material created has inferior mechanical properties compared with those of the pure polymer. Examples of the effect of certain fillers on the LOI are shown in Table 8.9.

All the fillers raise the LOI which, in effect, makes it more difficult to burn the material. The higher the LOI, the more oxygen has to be supplied to the material to sustain combustion. Hence ethylenevinylacetate (EVA), which is normally relatively easily burnt, becomes much more difficult to ignite and becomes self-extinguishing when magnesium or aluminium hydroxide is incorporated into the polymer.

8.10.6 Stabilisers

Many commercial polymer systems will contain stabilisers. Hindered-amine light stabilisers (HALS) stabilise against weathering by scavenging free radicals that are produced by photo-oxidation of the polymer matrix. UV absorbers stabilise against weathering by absorbing UV light and converting it into heat. Antioxidants stabilise the polymer by terminating the chain reaction due to the adsorption of UV light from sunlight.

8.10.7 Final comments on fire issues

Once a fire starts in a room containing flammable material, the heat generated can ignite other materials and hence sustain the fire. As a consequence, the rate at which the fire progresses depends on the susceptibility of the material in the proximity of the fire to combustion. The radiant heat and temperature can rise to an extent that all materials within the room are ignited. This point in time is called the 'flash-over' and leads to a fully developed fire. Escape from the room will then be virtually impossible and the spread of the fire to other rooms is highly probable. When a fire reaches flash-over, every polymer will release roughly 20% of its weight as carbon monoxide, resulting in excess toxic smoke. Consequently, most people die as a result of inhalation of the fumes and not as a consequence of burning. Every year about 5,000 people are killed by fires in Europe and more than 4,000 people in the USA. The total cost of fires is around 1% of the gross domestic product.

There are disadvantages with some of the flame retardant systems. AHT and magnesium hydroxide require a very high portion of the filler to be deployed within the polymer matrix: filling levels of more than 60 weight% are necessary to achieve suitable flame retardancy. High filler loadings often will lead to a loss of those physical properties which make the polymers attractive. In Europe, there are reservations about the use of halogenated compounds as flame retardants.

8.10.8 Use of nanofillers to form nanocomposites

Nanofillers such as clay, graphite and carbon nanotubes have the potential to reinforce the polymer matrix and can also have advantageous effects in terms of fire retardancy. In certain applications *nanocomposites* have been found to be very effective fire retardants. The most commonly used additive is exfoliated

Table 8.9 LOI and ignition times for ethylenevinylacetate (EVA) filler systems (125 phr filler)

Filler	LOI	Ignition time (s)
None	18.5	<2
Calcium hydroxide	24.5	<4
Aluminium hydroxide	30.5	20
Magnesium hydroxide	38.5	20

clay. To aid exfoliation the clay is usually treated with a quaternary ammonium surfactant. The clay platelets can inhibit the release of low molar mass fragments and hence reduce the flow of material to the flame. The platelets impart rigidity to the polymer melt and reduce the tendency of the material to drip.

It has been found that it is relatively easy to disperse clay platelets in polymers like EVA and polyamide to form nanocomposites, while it is considerably more difficult for nonpolar polymers such as polyethylene and polypropylene. For the nonpolar polyethylene it is reported that a blend of polyethylene and MSA-g-polyethylene (a copolymer created by grafting maleic acid onto polymer ethylene and partly forming the anhydride) can be used to create a nanocomposite. Only 3–5% of a nanodispersed organoclay within a polymer matrix is required to improve mechanical properties significantly. Nanocomposites often show reductions of heat release rates up to ~70%. Heat release is the most important single factor in a fire and can be described as the driving force of a fire. The two factors which influence the degradation are:

- During the degradation of the nanocomposite the formation of a char can be observed, which shields the nanocomposite below against the external heat.
- There is a reduction in the permeability of burnable gases from the polymer decomposition to the external flame. Additionally, there is a change of reaction mechanism during the degradation of nanocomposites.

Polymers like PMMA and polystyrene which undergo dominantly depolymerisation form chars in the presence of nanoparticles. Of particular importance to the industry is the fact that often very similar reductions in heat release rates are observed for both intercalated and exfoliated structures. In practice, the flame retardancy of polymers is currently achieved using different traditional additives. Halogen-free flame retardancy, which is most important for Europe, is mainly achieved by using ATH and to a smaller degree by magnesium hydroxide. Nitrogen- and phosphorus-based flame retardants are used. The proportion of ATH needed can be very high, up to 150–175 phr for adequate flame retardancy. The low flexibility of such compounds, only moderate mechanical properties and very often difficulties during compounding and extrusion or injection moulding are frequently reported. Alternatively, the amount of halogen-based flame retardants for polymers is lower. Many of these halogen-based materials are now being restricted because of either their impact on the environment or potential toxicity.

Many of the nanocomposite materials alone do not exhibit sufficient flame retardancy in a polymer matrix, for example for cables or polyurethane foams. However, a combination of ATH or magnesium hydroxide with organoclay and other additives can produce sufficient flame retardancy and simultaneously allows the total filler content to be reduced. Synergistic effects of organoclays and classical flame retardants, such as ammonium polyphosphates and phosphorus esters, as well as brominated flame retardants have been observed.

8.11 Polymer identification

Engineers may often be faced with the problem of attempting to identify an unknown plastic. The following section is designed to provide a simple and practical approach to this problem and does not require sophisticated instrumentation.

8.11.1 *Tests to identify an unknown polymer*

Sophisticated spectroscopic methods exist to determine the nature of a particular plastic. In the field an engineer may need to be able to identify a material and the following schedule will give some guidance as to the nature of the material. Tables 8.10–8.12 will help to identify an unknown polymer.

The first step is to cut a sliver from the edge of the material to be identified. The sample will immediately indicate whether or not it is filled and if the filler is a fibrous material. It is usually fairly obvious if the material is glass-filled or if it contains a more particulate filler.

Table 8.10 Analysis of thermoset materials (instructions are given in italic)

Powdery chips often indicate thermosetting resins *Hold sample in lighted match and smell resultant vapour*		
'Carbolic' smell. Sample usually black or brown in colour. Burns but extinguishes itself when flame removed = phenol formaldehyde	Fishy smell. Sample usually brightly coloured (or white) = urea formaldehyde or melamine formaldehyde	
Self-extinguishing black smoke, sharp acid odour and sample lighter in colour than phenol formaldehyde resin = epoxide	*Try scuffing edge of sample with finger nail*	
	Scuffing = urea formaldehyde	No scuffing = melamine formaldehyde

Table 8.11 Analysis of thermoplastic materials (instructions are given in italic)

Smooth sliver obtained indicates thermoplastic. Application of a hot metal wire or rod will produce melting *Drop sample onto hard surface*				
'Metallic' ring indicates presence of styrene polymer. Burn a small sample and blow out flame. Smell resulting smoke		Absence of 'metallic' ring precludes polystyrene (unless foamed) but may be high impact polystyrene (containing butadiene rubber) *Place sample in water and add a few drops of detergent*		
Characteristic smell of styrene = polystyrene	Smell of styrene and bitter smell = styreneacrylonitrile	If sample floats, indicates a polyolefin (unless foamed polystyrene). *Scratch and burn with match.*	If the sample sinks, not a polyolefin. *Burn small sample. Note ease of ignition and type of flame*	
	Smell of styrene and rubber + ABS copolymer	Glossy surface does not scratch. Burns with smell of paraffin wax = polypropylene	Burns with a clear flame. *Blow out flame and smell vapour*	Difficult to ignite and is self-extinguishing *Burn small sample. Note ease of ignition.*
		Glossy surface, slight scratching. Burns and drips like sealing wax = HDPE	Fruity odour = acrylic (probably PMMA)	Flame has greenish tinge and acid smell = PVC or vinylidene chloride copolymer
		Not particularly glossy. Burns with smell of paraffin wax = LDPE	Smell of burning paper = cellulose acetate or proprionate	Flame is yellow and smells of formaldehyde = polyacetal
			Smell of rancid butter = cellulose acetate butyrate	Indefinite smell with slippery surface and threads easily = nylon
				Light flame, decomposition of material but no charring = polycarbonate

Thermosets

If the slice appears to be a powdery material but is not filled, it is probably a thermoset. Table 8.10 gives a step-by-step procedure for the analysis of a particular polymer material. Caution must be exercised in carrying out all of these tests, especially when smelling the product of the burning, as the vapours that are emitted are potentially toxic. It is important to not inhale these vapours, but rather to let them waft across the nostrils and keep the extent to which the materials are burnt to a minimum. The scuffing with a finger nail may be replaced by using a blunt object. The idea

Table 8.12 Characteristics of polymers when burning

Material	Low flame	Ignited	Colour of flame	Smoke	Smell	Other
Acrylic	Ignites when soft	Continues to burn	Yellow, with clear edges		Fruity smell	Drips
Nylon	Difficult to ignite	Self-extinguishing	Blue–yellow tip	Little black smoke	Burning hair	
Polyester film	Burns readily	Continues to burn	Yellow, smoky		Burnt raspberry jam	Drips
PTFE	Will not ignite in ordinary flame					
Urea formaldehyde	Burns with difficulty	Self-extinguishing	Pale yellow, with light blue–green edges		Pungent odour of formaldehyde, fishy smell	
Melamine formaldehyde	Burns with difficulty	Self-extinguishing			Pungent odour of formaldehyde, fishy smell	
Phenol formaldehyde	Burn with difficulty	Self-extinguishing			Carbolic acid	
Polyester resin (GRP)	Burns readily	Formulated to self-extinguish	Smoky		Fruity	
Polyethylene	Difficult to ignite	Continues to burn	Blue flame with yellow tip	None or little smoke	Burning candle when extinguished	Drips
Polypropylene	Difficult to ignite	Continues to burn	Yellow (blue base)	None	Burning candle when extinguished	Drips
PVC	Burn with difficulty	Self-extinguishing	Yellow (blue–green tinge at base)	Grey	Acid	Drips while burning
Polystyrene	Ignites easily	Continues to burn	Orange–yellow	Dense black smoke	Marigolds, sweetish	Drips and continues to burn
Polycarbonate	Burns with difficulty	Self-extinguishing	Smoky flame		Phenol	
ABS	Burns readily	Usually continues to burn	Orange–yellow	Black smuts	Similar to polystyrene also bitter (acrylonitrile) and rubbery (butadiene)	
Cellulose acetate	Ignites easily	Continues to burn	Dark yellow	Grey	Acid vinegar	
Acetal	Burns readily		Almost invisible pale blue		Pungent odour of formaldehyde	Drips

is that the force being applied to the substance is minimal and allows the tester to differentiate between those materials that are subject to damage and those which will exhibit some resistance.

Thermoplastics
If the sliver is not powdery in appearance, it is probably a thermoplastic and the analysis summarised in Table 8.11 is relevant. The first test involves determining whether or not it is able to absorb energy easily. If the material has a metallic ring then it has a high modulus and little ability to absorb energy. If, however, that material contains groups which are able to absorb energy then a dull ring will be heard. It is useful to have experienced the 'ring' of various materials when attempting to calibrate this test.

8.11.2 *Burning tests*

As indicated above, observation of the way in which materials burn can be a useful method of classifying and identifying them. Table 8.12 summarises observations of burning for a number of common polymer materials. The characteristics of the burning process which are important to identify are:

- How easily does the material ignite?
- When ignited how does it burn and does it drip?
- What is the colour of the flame?
- Does it produce smoke and what is the smell?

Brief summary of chapter

- Using careful control of chemistry it is possible to create polymers which have well-defined structures and hence physical properties.
- Depending on the nature of the polymerisation process, it may be possible to obtain materials which have relatively narrow polydispersity.
- The degradation of a plastic is a complex process but its control can significantly improve the fire resistance.
- Using simple tests to better understand the possible nature of an unknown polymer material.

References and additional reading

Brandrup J., Immergut E.H. and Grulke E.A. *Polymer Handbook*, 4th edn., Wiley, Hoboken, NJ, USA, 1999.
Elias H.G. *Macromolecules: Structure and Properties, Vols. I and II*, Wiley, New York, NY, USA, 1977.
Rodriquez F., Cohen C., Ober K. and Archer L.A. *Principles of Polymer Systems*, Taylor and Francis, New York, NY, USA, 2003.
Stevens M.P. *Polymer Chemistry*, Oxford University Press, Oxford, UK, 1990.

9

Polymer physics: models of polymer behaviour

9.1 Introduction

Our understanding of the nature of polymeric materials has been considerably enhanced by the efforts of polymer physicists modelling various physical phenomena. Sometimes these models appear to be a little crude, since they often lack the detail of the chemical structure; however, they can provide insight into the behaviour of flexible chain structures. They are useful for describing the properties of high molar mass polymer chains in solutions, as melts or in the solid state.

These models of polymer behaviour are essentially of two types: those that attempt to carry out an *ab initio* prediction of the shape of the polymer chain based on a detailed knowledge of the chemical structure; and models that assume that the chemical structure does not play a determining role in dictating the physical properties. The latter models will be referred to as *string* models, describing the structure and properties of the polymer chain on a statistical basis by considering it to be composed of flexible, connected links. The quantum mechanical, *ab initio* calculations can only be carried out for molecular systems that contain several hundred atoms and therefore it is difficult to describe a real polymer system. However, these quantum mechanical calculations can be used with Monte Carlo or other statistical methods (molecular dynamics calculations) to gain a picture of the way in which a polymer moves in space. The semi-empirical *string* models have an important role to play in the development of our understanding of the behaviour of high molar mass polymer systems as they can better handle the long-range effects than the *ab initio* calculations. The former are good at local interactions but cannot be effectively extended to long-range effects.

9.2 Simple statistical models of isolated polymer molecules in solution

Historically, the first attempts to predict the structure of an isolated polymer molecule in solution were carried out in the 1950s and predate the quantum mechanical calculations of many organic molecules. These semi-empirical calculations attempted to describe the size and shape of an isolated polymer chain in solution. Light scattering measurements, carried out on very dilute solutions, established the distinct molecular nature of a polymer and showed that the shape and size changed with temperature and concentration.

The questions which the theoreticians attempted to addresses were:

- How can the shape of a polymer molecule in solution be predicted theoretically?
- Which intrinsic characteristics of the polymer chain need to be considered in order to describe its shape?
- How can we describe elasticity in terms of the changes in the shape of a polymer molecule?

9.3 Freely jointed random coil model

The simplest theory describing the isolated polymer chain is the so-called *freely jointed random coil*, which behaves rather like a necklace (see Figure 9.1). As the name implies, the model assumes that

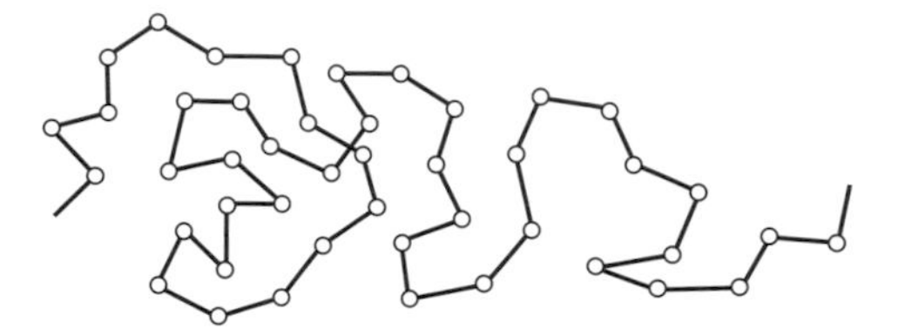

Figure 9.1 Schematic of a freely jointed chain.

the polymer chain is constructed by connecting individual monomers of length l by a completely flexible joint. The flexible joint allows one monomer to be connected to the next monomer in such a manner that it can point in any direction relative to the initial bond direction.

The polymer molecule adopts a statistically averaged structure, which reflects the random possibility of any angle being adopted by each of the bonded units. The completely random selection of the bond directions gives this theory the name of the *random coil* model. If the polymer molecule has a random coil structure then there are certain critical parameters which will describe its size and shape. If we start from one end and map the contour which the chain follows, it is possible to calculate the average distance between the two ends of the polymer chain, the so-called *end-to-end distance*, $\langle r^2 \rangle^{1/2}$. The other parameter which is useful to characterise the shape of the polymer chain is the *radius of gyration*, $\langle s^2 \rangle^{1/2}$, the size of the sphere which would be swept out if the chain were rotating freely in space. These measurements of the shape and size of an isolated polymer chain are accessible from light scattering. If we assume that the bonds are completely flexible then it is easy to see that even for a polymer that only contains about 100 monomer units the possible number of configurations which the bonds can adopt will be very large. If the chain is freely jointed and can adopt any angle between 0 and π then the number of possible configurations can easily exceed 10^4. If the polymer were to contain n monomer units then the length of the string representing the polymer chain would be equal to nl if it were fully extended. However, because the bonds can adopt any bond angle then the *end-to-end* distance and *radius of gyration* will be less than that of the fully extended chain and are determined by random statistics which reflect the effects of the conformation statistics on the shape of the polymer chain.

Since there are no constraints on the directions the bonds can adopt, it is assumed that the path followed by the chain will be completely random, rather like a drunken man attempting to return home from the pub. The end-to-end vector $\boldsymbol{r}$ is the sum of the individual bond vectors r_i:

$$\boldsymbol{r} = \sum_{i=1}^{n} r_i \tag{9.1}$$

The mean square end-to-end distance becomes:

$$\boldsymbol{r}^2 = \sum_{i=1}^{n} r_i \sum_{i=1}^{n} r_j = \sum_{i=1}^{n} r_i^2 + 2\sum_{i=1}^{n-1} \sum_{j=i+1}^{n} r_i r_j \tag{9.2}$$

Equation (9.2) is valid for any polymer chain, regardless of structure, provided that the bonds are linked flexibly (see Figure 9.2). Therefore, if we consider an ensemble of N chains each comprising n segments, the average of the squared end-to-end distance, $\langle r^2 \rangle$, is equal to:

$$\langle r^2 \rangle = \frac{1}{N} \sum_{k=1}^{N} r_k^2 = \sum_{i=1}^{n} \langle r_i^2 \rangle + 2\sum_{i=1}^{n} \sum_{j=i+1}^{n} \langle r_i r_j \rangle \tag{9.3}$$

The scalar product of the arbitrary segment vectors r_i and r_j is:

$$\langle r_i r_j \rangle = l^2 \langle \cos \varphi_{ij} \rangle \tag{9.4}$$

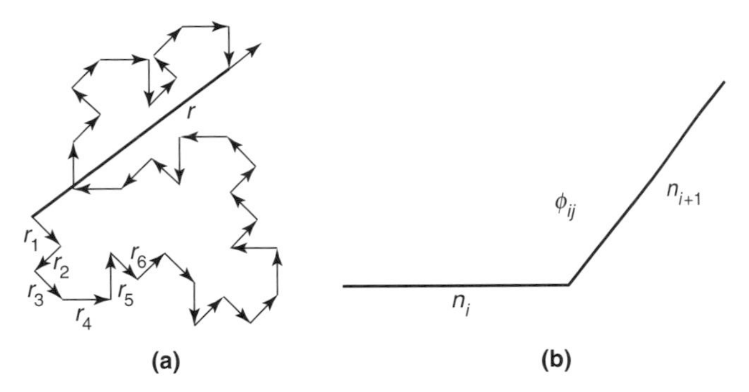

Figure 9.2 Bond vectors: (a) schematic of distribution of bond vectors for a random coil chain; (b) scalar product for two bond vectors.

where ϕ_{ij} is the angle between the two bond vectors. Combining Equations (9.2) and (9.4) we obtain:

$$\langle r^2 \rangle = nl^2 + 2l^2 \sum_{i=1}^{n-1} \sum_{j=i+1}^{n} \langle \cos\phi_{ij} \rangle \tag{9.5}$$

This general formulation is valid for any continuous polymer chain in which no direction is preferred and thus $\langle \cos\phi_{ij} \rangle = 0$ for $i \neq j$ in Equation (9.5) leads to:

$$\langle r^2 \rangle = nl^2 \tag{9.6}$$

In the random walk model, each of the steps has the same size and is equal to the bond length, l. The number of conformations may be expressed by the Gauss formula, where n is the number of units in the chain, l is the length of each unit and L is the total length of the chain. The probability of finding an element of the polymer chain some distance r from the chain end is calculated using a simple statistical distribution function $W(r)$. The probability of finding a chain element at a distance r is obtained by integration of the probability over the sphere of thickness dr and has the form:

$$W(r)\partial r = \left(\beta / \pi^{1/2}\right)^3 \exp\left(-\beta^2 r^2\right) 4\pi r^2 \partial r \tag{9.7}$$

where

$$\beta = \sqrt{3/2\left(n^{1/2} l\right)} \tag{9.8}$$

hence

$$W(r)\partial r = \left(\left(2\pi n l^{1/2}\right)/3\right)^{-3/2} \exp\left(-3r^2/2nl^2\right) 4\pi r^2 \partial r \tag{9.9}$$

where $W(r)$ describes the probability of finding the chain end within the shell ∂r which is a distance r from the origin which contains one of the chain ends (see Figure 9.3). For a polymer chain with 10^4 bonds and each bond equal to 0.25 nm, the *average end-to-end* distance will be about 20 nm, but the distribution indicates that chains as small as 0.5 nm and as large as 40 nm could exist. The value of the *end-to-end distance* is given by:

$$\langle r^2 \rangle_0 = \int_0^\infty r^2 W(r) \partial r \quad \text{where} \int_0^\infty r^2 W(r) \partial r = 1 \tag{9.10}$$

Hence:

$\langle r^2 \rangle = \int r^2 W(r) \partial r = (3/2)\beta^2$, substituting for β gives:

$$\langle r^2 \rangle = nl^2 \tag{9.11}$$

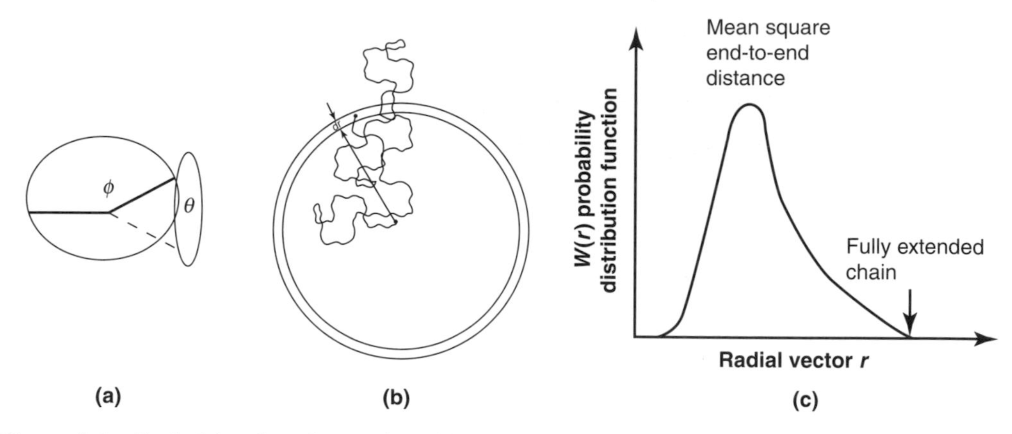

Figure 9.3 Freely jointed random coil model: (a) freely jointed bond indicating polar coordinates defining bond directions θ and ϕ; (b) model showing shell dr over which integration is carried out; (c) form of distribution function.

For a freely jointed chain, the *mean square end-to-end distance* is directly proportional to the number of bonds in the main chain and is directly proportional to the molecular mass.

This is a rather artificial model. However, it is a good basis from which to start to consider the size of a polymer molecule. The *mean square radius of gyration* $<s^2>$ is important in determining the properties of a dilute solution of polymers. For a molecule with a Gaussian distribution of segments, essentially segments distributed in a random fashion, about a centre of gravity then:

$$\langle s^2 \rangle_0 = \langle r^2 \rangle_0 / 6 = 1/4\beta^2 = nl^2/6 \tag{9.12}$$

The fully extended chain length would correspond to r^2 being equal to nl^2, which is a highly improbably situation. Likewise, if all the chains were to be folded very tightly the value of r would be very small. The probability that the coil would be fully collapsed or fully extended is zero. The mean square value corresponds to the most probable size which the chain would occupy. The chain resembles an open wire mesh ball and it is assumed that there will be solvent molecules between the wires. The fact the polymer molecule is dissolved in the solvent implies that the strength of the interaction between solvent and polymer chain is greater than the value between the elements of the polymer chain. If the interactions between the elements of the chain were to become greater than between the chain and the solvent, the polymer would become insoluble and would precipitate from the solution. In fact, this condition is observed when the temperature of the solution is lowered and the polymer is precipitated. The random coil theory tells us nothing about temperature dependence of the dimensions and its predictions are not in good agreement with experimental results. The θ temperature corresponds to the condition when the interaction between solvent and polymer balances the polymer–polymer interactions.

9.4 Valence constrained random coil model

In a real polymer, the bonds will *not* be free to adopt any value of angle relative to one another. The angles are restricted to those allowed by the normal rules of chemical bonding. Thus for two carbon–carbon bonds, the angle between the bonds will be expected to be approximately 110°. The valence constrained random coil model therefore allows free rotation around chemical bonds but restricts the values of ϕ to those allowed by normal chemical bonding rules. The *freely rotating chain* assumes that the bond angle ϕ is constant, but no particular conformation is preferred and the average projection of bond n_j along the direction perpendicular to bond i (see Figure 9.2) is given by:

$$\langle r_i r_{i+1} \rangle = l^2 \cos(180 - \varphi), \quad \langle r_i r_{i+2} \rangle = l^2 \cos(180 - \varphi), \quad \langle r_i r_j \rangle = l^2 \cos(180 - \varphi)^{j-1}$$

Combining Equations (9.3) and (9.5) and substituting the above gives:

$$\langle r^2 \rangle = nl^2 + 2l^2 \sum_{i=1}^{n-1} \sum_{j=i+1}^{n} \left[\cos(180 - \phi)^{j-1} \right] \tag{9.13}$$

The summation can be performed over a single variable (k) by substituting $j - i$ by k:

$$\langle r^2 \rangle = nl^2 \left[1 + \frac{2}{n} \sum_{k=`}^{n-1} (n-k)\alpha^k \right] \tag{9.14}$$

where $\alpha = \cos(180 - \phi)$. Equation (9.14) can be simplified as follows:

$$\langle r^2 \rangle = nl^2 \left[1 + \frac{2}{n} \sum_{k=1}^{n-1} (n-k)\alpha^k \right] = nl^2 \left[1 + 2\sum_{k=1}^{n-1} \alpha^k - \frac{2}{n} \sum_{k=1}^{n-1} k\alpha^k \right]$$

$$= nl^2 \left[1 + \frac{2(\alpha - \alpha^n)}{1-\alpha} - \frac{2}{n}\left(\frac{\alpha(1-\alpha)^n}{(1-\alpha)^2} \right) - \frac{n\alpha^n}{1-\alpha} \right] = nl^2 \left[1 + \frac{2\alpha}{1-\alpha} - \frac{2\alpha(1-\alpha)^n}{n(1-\alpha)^2} \right] \tag{9.15}$$

For an infinitely long chain n = ∞, and:

$$\langle r^2 \rangle = nl^2 \left[1 + \frac{2\alpha}{1-\alpha} \right] = nl^2 \left[\frac{1+\alpha}{1-\alpha} \right] = nl^2 \left[\frac{1+\cos(180-\phi)}{1-\cos(180-\phi)} \right] \tag{9.16}$$

Putting $\phi = 109.5°$, the value of the bond angle for a carbon–carbon bond leads to the mean square end-to-end distance being:

$$\langle r^2 \rangle \approx 2nl^2 \tag{9.17}$$

A further modification of the theory involves the assumption that the units moving are not single chemical bonds but rather small elements (*segments*) of the polymer chain. The chemical bonds which make up a segment of the chain are effectively acting like a flexible bond and are known as a *Kuhn* segment. The Kuhn segment will often contain approximately five to seven real chemical bonds. Using the concept of a Kuhn segment, a better fit between theory and experiment is obtained. The Kuhn segments are often considered to be the 'statistical elements' of the polymer chain. The Kuhn length is designated l' and the number of elements $n' = n/m$, where m is the number of monomers in a Kuhn segment.

The value of m depends on the chemical structure of the polymer chain. Polydimethylsiloxane will have a different value from polyethylene, which will have a different value from polystyrene, etc. It is therefore appropriate to use a more general form of the equation:

$$\langle r^2 \rangle_0 = C_\infty nl^2 \quad \text{where } C_\infty = \langle r^2 \rangle_0 / nl^2 \tag{9.18}$$

where C_∞ is called the *characteristic ratio* and is always greater than 1 for real polymers. Thus C_∞ is a measure of the departure of the real polymer molecule from the predictions of the free jointed random coil model. Whilst the Kuhn *segment* may be seen to be a somewhat arbitrary fudge factor to obtain agreement with experimental results, it does have some real significance. The size of the chain element which moves is in fact about six to eight bond elements. However, the theory does not predict that the shape of the polymer chain might change its size when the temperature is varied.

9.5 Rotational isomeric states model

In reality, the polymer chain will distribute itself between *gauche* and *trans* structures (see Section 1.6.1). The *gauche* are the more hindered, higher energy, higher temperature structures. The rotational isomeric states model (RISM) is based on a statistical mechanical calculation of the distribution of the monomer units between the accessible states for the polymer chain. In practice, this corresponds to working down the chain, selecting an appropriate number of *trans* and *gauche* structures to reflect the energy and hence temperature. The theory once more allows the calculation of the size and shape of the polymer molecule. In Chapter 1, it was pointed out that the conformations are constrained by the short-range interactions which define the *potential energy surface*. The short-range interactions lead to the prediction that the all-*trans* conformation is the lowest energy structure. Rotation about the carbon–carbon bond will change the conformation and its energy will be raised. Because of the very short-range nature of the atom–atom nonbonding interactions, the potential energy surfaces predicted for a normal alkane and for a polymer are almost identical at temperatures when the energy difference between the *trans* and *gauche* becomes comparable to room temperature. A chain with restricted rotation has a preference for certain rotational isomers. In Chapter 1, it was shown that the *trans* state (θ) is the lowest energy state. Using the same arguments as outlined above and incorporating *hindered rotation* with independent torsion angle potentials defined by the angle θ, in Figure 9.3 gives:

$$\left\langle r^2 \right\rangle = nl^2 \left[\frac{1+\cos(180-\phi)}{1-\cos(180-\phi)} \right] \left[\frac{1+\langle \cos\theta \rangle}{1-\langle \cos\theta \rangle} \right] \tag{9.19}$$

The temperature dependence of $\langle r^2 \rangle$ originates from the temperature dependence of $\langle \cos\theta \rangle$ as may be illustrated by the following example. We assume that the three possible rotational isomers *trans*, (+) *gauche* and (–) *gauche* (see Figure 9.4) for the polyethylene chain depend only on the energy levels of the three rotational states of a certain bond and are not influenced by the torsion angles of the neighbouring bonds. Rotation about this virtual bond defines the rotational isomeric potential of polyethylene. The statistical mechanical rotational partition function, z, is a measure of the number of rotational states which the system can adopt at a given temperature. If we consider that the rotational isomeric potential is similar to n-butane then:

$$z = 1 + \sigma + \sigma = 1 + 2\sigma \tag{9.20}$$

where:

$$\sigma = \exp\left(-E_g/RT\right) \tag{9.21}$$

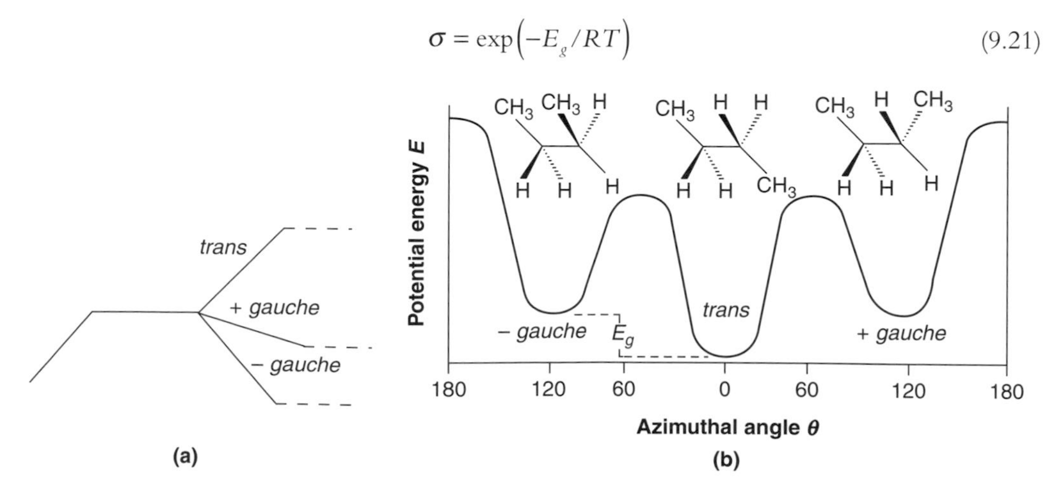

Figure 9.4 Butane: (a) map of bond continuity; (b) potential energy surface and conformations.

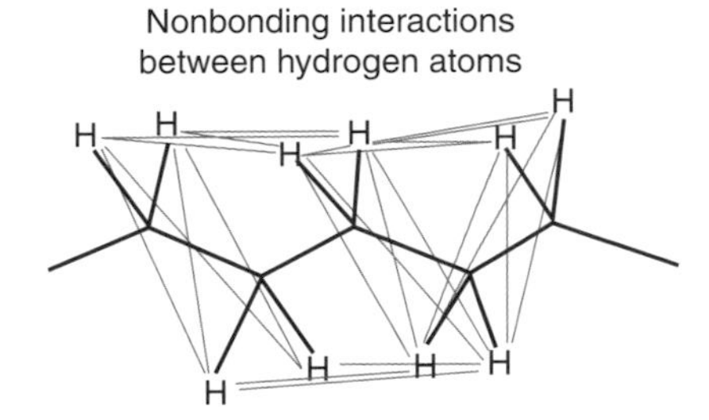

Figure 9.5 Indication of additional interactions involved in defining potential energy surface for a polyethylene chain.

where E_g is the energy difference between the *gauche* and *trans* states. Provided that the rotational potential of the bond is independent of the actual torsion angles of the nearby bond, the partition function $\langle f \rangle$ has the form:

$$\langle f \rangle = \sum_{v} \left(u_v f(\theta_v) \right) / z \tag{9.22}$$

where $\langle f(\theta) \rangle$ is the function describing the distribution between the various possible orientations of the bonds. Thus the function of the angle θ becomes:

$$\langle \cos\theta \rangle = \sum_{v} \left(u_v \cos\theta_v \right) \Big/ z = \frac{1 + \sigma\cos(120°) + \sigma\cos(-120°)}{1 + \sigma + \sigma} = \frac{1-\sigma}{1+2\sigma} \tag{9.23}$$

The factor of two in the denominator of Equation (9.23) arises from the double degeneracy of the higher energy *gauche* conformation. Insertion of Equation (9.23) in Equation (9.19) gives:

$$\langle r^2 \rangle = nl^2 \left[(1 + \cos(180 - \phi) / (1 - \cos(180 - \phi)) \right] \left[(2 + \sigma)/3\sigma \right] \tag{9.24}$$

At 140°C, using $E_g = 2.1\,\text{kJ mol}^{-1}$, $\sigma = 0.54$; the second moment of the end-to-end distance becomes:

$$\langle r^2 \rangle = nl^2 \times 2x \left((2 + 0.54) / (3 \times 0.54) \right) = 3.4nl^2 \tag{9.25}$$

which is lower than the experimentally obtained $(6.7 \pm 0.1)nl^2$. The interactions which alter the potential energy profile are illustrated in Figure 9.5.

Agreement between experiment and theory is obtained by the incorporation of higher-order terms in the partition function, which effectively increases the value of E_g. The calculations are usually based on the potential energy surface for the monomer and then extending the size of the unit until the potential energy surface ceases to change. In practice, the calculations will involve consideration of six to eight bonds before asymptotic values are obtained.

9.6 Long-range interactions: excluded volume

The long-range interaction of elements of the chain can cause the size of the chain to be increased. These long-range interactions are a consequence of the exclusion of a certain volume from being occupied by the polymer chain if there is already an element of polymer chain in that volume (see Figure 9.6). Addition of solvent to the polymer will alter the conformational distribution and hence the size of the polymer chain. If the interactions are weak then the size of the polymer chain will be close to that of the ideal unperturbed value. If, however, the interactions are strong then the conformational distribution will favour the lower energy form and hence

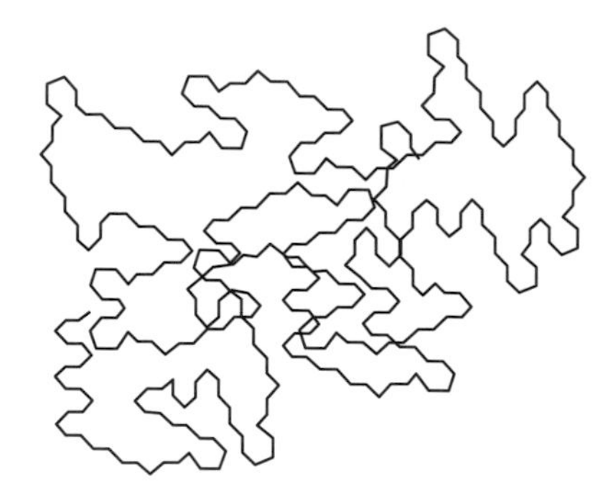

Figure 9.6 As length of chain increases, so probability of it attempting to occupy the same point in space will increase.

the effect will be an expansion in the size of the polymer chain in solution. The real size can be related to the unperturbed value by the relationship:

$$\langle r^2 \rangle = \alpha^2 \langle r^2 \rangle_0 \quad \langle s^2 \rangle = \alpha^2 \langle s^2 \rangle_0 \tag{9.26}$$

where α is the expansion factor and has a value greater than one in good solvents. In bulk amorphous polymer, the interactions between segments on different chains are similar and hence the value of the expansion factor is essentially unity. For high molar mass materials, the exclusion volume effect will always lead to a value of C_∞ that is greater than unity.

9.7 Comparison of the theoretical models

There are several important differences between the prediction of the Kuhn and random coil models. To summarise these differences we can define a factor C_∞, the so-called characteristics ratio which compares the values predicted by the various theories (see Table 9.1).

Table 9.1 indicates that it is only the RISM model that incorporates a temperature dependence of the chain dimensions and gives the correct characteristic ratio. The other, rather counter-intuitive, result which comes from the RISM theory is that increasing the temperature for a polymer such as polyethylene causes the size of the polymer chain to shrink rather than expand as one might have expected. The shrinkage of the polymer coil is a consequence of the higher occupancy of the higher energy conformations and is reflected in the negative temperature coefficient. In the case of polyethylene, the higher energy state is the *gauche* form and is associated with the polymer coiling up. Hence the greater the number of elements of the chain which adopt a higher energy conformation, the more tightly coiled the polymer chain will be. This shrinkage in the size of the polymer chain with increasing temperature is consistent with experimental observation and was one of the major successes of the RISM theory. It is important to note that the size of the polymer chain predicted in dilute solution is very close to the values which are observed in the polymer melt. The solvent interactions in the melt are replaced by interactions from other chains and the net effect is that the size of the chain is similar to that in

Table 9.1 Values of C_∞ and its temperature dependence

Model used in experiment	C_∞	$dC_\infty/dT/10^{-3}$
Experimental	6.7	−1.1
Freely jointed random coil	1	0
Valence restricted random coil	2.1	0
Rotational isomeric states model	6.7	−1.0

Table 9.2 Values of C_∞: size of Kuhn length (b), density (ρ), effective molar mass (M_0) and ratio of M_0/M_n, where M_n is molar mass of a monomer unit

Polymer	Structure	C_∞	B (Å)	ρ (g cm^{-3})	M_0	κ
1,4-Polyisoprene	$-(CH_2CH = CHCH(CH_3))-$	4.6	8.2	0.830	113	2
1,4-Polybutadiene	$-(CH_2CH = CHCH_2)-$	5.3	9.6	0.826	105	2
Polypropylene	$-(CH_2CH(CH_3))-$	5.9	11	0.791	180	4
Polyethyleneoxide	$-(CH_2CH_2O)-$	6.7	11	1.064	137	3
Polydimethylsiloxane	$-(OSi\ (CH_3)_2)-$	6.8	13	0.895	381	5
Polyethylene	$-(CH_2CH_2)-$	7.4	14	0.784	150	5
Polymethylmethacrylate	$-(CH_2C(CH_3)(COOCH_3))-$	9.0	17	1.130	655	6
Polystyrene	$-(CH_2CH(C_6H_6))-$	9.5	18	0.969	720	7

dilute solution. Lowering the temperature will increases the probability of finding an extended sequence of *trans* structures, which is the primary requirement for nucleation of the crystallisation process.

It can be seen from Table 9.2 that the characteristic ratio depends on the chemical structure, as does the Kuhn length and effective molar mass for many common polymers. The bulkier the side groups, the higher the value of C_∞.

In the case of polyisoprene and polybutadiene, the value of M_0 is approximately twice the molar mass of the repeat unit. This implies that the size of the element required to achieve a high degree of flexibility is relatively small, whereas in the case of the more hindered polystyrene, the M_0 value is approximately seven times the monomer mass. In polystyrene, this suggests that motion of the backbone chain involves about seven monomer units. The value of κ indicates the number of bonds which are moving in order to rotate the backbone. The prediction of seven bonds moving in polystyrene is consistent with the experimental results.

9.7.1 Dynamic response of polymer solutions

The rheology of polymer solutions reveals that virtually all polymers behave in a similar manner and that the detailed chemistry of the backbone is not a major factor in defining the behaviour in oscillatory shear. For a dilute polymer solution subjected to a shear field, it is possible to imagine the overall motion of the polymer as being the superposition of a number of modes of motion (see Figure 9.6).

At very low rates of oscillatory shear, the whole polymer molecule will move in the shear field and it can interact with the solvent producing a large loss of energy, $G''(w)$. The viscosity is defined as the rate of energy loss and is defined as $G''(w)/w$. At low frequency, the motion is predominantly diffusion and is designated the zeroth mode. As the frequencies of the oscillations are increased, then the wavelength of the perturbation matches the dimensions of the polymer coil and a condition will be matched where the two extremes of the coil are forced to move in opposite directions, leaving the centre of the polymer coil not moving. The dynamic distortion of the coil has created a node at its centre. This first normal mode dissipates about 2/3 of the total energy. Once the rate of oscillation has exceeded that for the first normal mode, the higher frequency motions will distort the polymer according to the higher odd modes. Hence we can image that for the second normal mode there will be two nodes: the extremes of the chain moving in the same direction with the centre moving in opposite directions. The energy dissipated is proportionately reduced compared with that in the first normal mode. The reduced energy dissipation will lead to a corresponding reduction in the dynamic viscosity and this leads to the well-known phenomenon of shear thinning which is found in all polymer solutions when subjected to high rates of oscillation. In the high frequency limit, the viscosity will drop to a value which corresponds to that of the solvent in which the polymer molecule is dispersed.

Using models in which the motion of the polymer molecule is simulated as a set of normal modes distortions of the polymer chain, it is possible to predict the shape of the shear thinning curves. The factor which needs to be defined is the characteristic length, which is predicted on the basis of the molar mass. Two slightly different approaches have been proposed: the so-called Rouse and Zimm models. These models represent two extreme situations. In the Rouse model it is assumed that there is no interaction between the solvent and the polymer chain. The Zimm model assumes that there is complete interaction. A subsequent model, the Wang–Zimm model, allows for the strength of the interaction to be varied and allows the eigen values to be smoothly varied between the two extreme situations.

9.7.2 Theories of polymer dynamics

At very low shear rates, the polymer coil will move through the solution rather like a colloidal particle. The mean square displacement of the coil can be expressed in terms of the change in its location with time $r\langle 0\rangle - r\langle t\rangle$ and the average distance moved by the particle is proportional to the square root of time:

$$\left\langle \left[r(t) - r(0) \right]^2 \right\rangle = 6Dt \tag{9.27}$$

where D is the diffusion coefficient and t is the time. This is the equation which describes the Brownian motion of a small particle. If a constant force, f, is applied to a small particle, it will be pulled through a liquid and will achieve a constant velocity, v, in the same direction as the applied force. For a given particle the proportionality constant is the friction or viscosity coefficient, ζ, according to:

$$f = \zeta v \tag{9.28}$$

The diffusion coefficient is related to the friction coefficient by the Einstein relation:

$$D = kT / \zeta \tag{9.29}$$

The relationship is an expression of the time, τ, required for the particle to move through its own radius R.

$$R \approx R^2/D \approx R^2\xi/kT \tag{9.30}$$

The time scale for the motion of the particle is proportional to the friction coefficient. The deformation rate of a liquid in shear is defined as $\dot{\gamma} = \partial\dot{\gamma} / \partial t$. In a Newtonian fluid, the stress σ is proportional to this shear rate $\sigma = \eta\dot{\gamma}$ where η is the viscosity coefficient. For a sphere of radius R moving in a Newtonian liquid of viscosity, η, it can be argued dimensionally that $\zeta \approx \eta R$, which leads to the Stokes law:

$$\zeta = 6\pi\eta R \tag{9.31}$$

Combining Equations (9.29) and (9.31) yields the Stokes–Einstein relationship:

$$D = kT / \left(6\pi R\right) \tag{9.32}$$

Rearranging Equation (9.32) gives the *hydrodynamic radius*, R_h:

$$R_h = kT / \left(6\pi\eta D\right) \tag{9.33}$$

9.7.3 Rouse model

The model assumes that the chain is composed of N beads connected by springs of root mean square size b (see Figure 9.7). The beads in the Rouse model only interact with each other through the connecting springs. Each bead is characterised by its own independent friction coefficient, ζ. Solvent is assumed to be freely draining through the chain as it moves. The Rouse

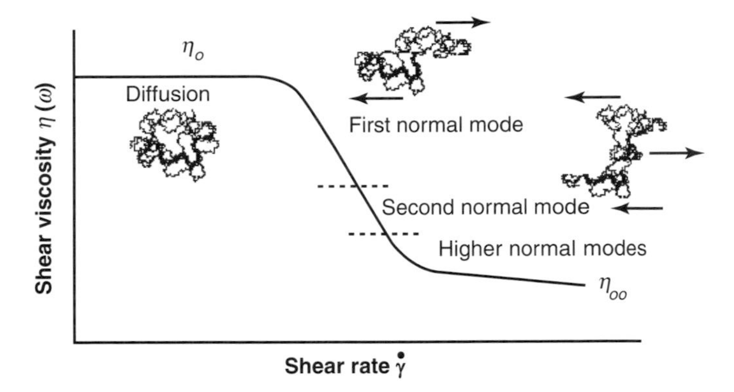

Figure 9.7 Shear dependence of viscosity of an isolated polymer molecule.

model is called the *free draining model*. The total friction coefficient of the whole Rouse chain is the sum of the contributions from each of the N beads:

$$\zeta_R = N\zeta \tag{9.34}$$

The viscous frictional force the chain experiences as it is pulled through the liquid with a velocity v will be $f = -N\zeta v$.

The diffusion coefficient of the Rouse chain is obtained from the Einstein relationship, Equation (9.30):

$$D_R = kT/\xi_R = kT/N\xi \tag{9.35}$$

The polymer diffuses a distance of the order of its size during a characteristic time, called the Rouse time, τ_R, which has the form:

$$\tau_R \approx R^2/D_R \approx R^2\big/\left[(kT)/(N\xi)\right] = (\xi/kT)NR^2 \tag{9.36}$$

On a time scale longer than the Rouse time, the motion of the chain is simply diffusion (see Figure 9.6). In Section 9.7.3 it was found that the size of the sphere which encompasses the polymer coil can be described by:

$$R \approx bN^{\upsilon} \tag{9.37}$$

where υ is the fractal dimensions of the polymer. For an ideal linear chain, $\nu = 1/2$ but it may have a larger value if there are significant interactions between the polymer chain and the solvent. The Rouse time of such a fractal chain can be written as the product of the time scale for the motion of individual beads. The *Kuhn monomer relaxation time is*:

$$\tau_0 \approx \xi b^2/kT \tag{9.38}$$

and a power law in the number of monomers in the chain:

$$\tau_R \approx (\xi / kT)NR^2 = \left((\xi b^2)/kT\right)N^{1+2\nu} \approx \tau_0 N^{1+2\nu} \tag{9.39}$$

For an ideal chain, $\nu = 1/2$ and the Rouse time is proportional to the square of the number of monomers in the chain:

$$\tau_R \approx \tau_0 N^2 \tag{9.40}$$

The full calculation for the relaxation time of an ideal chain was published by Rouse (1953) with a coefficient of $1/6\pi^2$:

$$\tau_R = (\xi b^2)\big/(6\pi^2 kT)N^2 \tag{9.41}$$

This Rouse stress relaxation time is half of the end-to-end vector correlation time because stress relaxation is determined from a quadratic function of the amplitudes of normal modes. The time scale for motion of individual monomers, τ_0, is the time scale at which a monomer would diffuse a distance of the order of its size b, if it were not attached to the chain. In a polymer solution with solvent viscosity η_s, each monomer's friction coefficient is given by Stokes law (Equation (9.31)):

$$\zeta \approx \eta_s b \tag{9.42}$$

The monomer relaxation time, τ_0, and the chain relaxation time of the Rouse model, τ_R, can be written in terms of the solvent viscosity, η_s:

$$\tau_0 \approx \eta_s b^3 / kT \quad \tau_R \approx (\eta_s b^3 / kT) N^2 \tag{9.43}$$

The implications of the above equations are that for times shorter than τ_0 the chain does not diffuse and exhibits elastic response. For times larger than τ_R, the polymer moves diffusively and exhibits the response of a simple liquid. For intermediate times scales $\tau_0 < t < \tau_R$ the chain exhibits a range of motions which are reflected in the profile in Figure 9.6.

9.7.4 Zimm model

The Rouse model assumes that there is no interaction between the polymer and the solvent. However, it is not unreasonable to expect that the polymer coil as it moves will drag some of the solvent with it. The force acting on a solvent molecule at a distance r from the polymer coil becomes smaller as r increases, but only relatively slowly ($\sim 1/r$). This long-range force acting on the solvent is called the *hydrodynamic interaction*. In the case of the bead–spring model (see Figure 9.7), the movement of one bead causes another to move; this is ignored in the Rouse model. If we assume that the motion of an element of the chain drags the solvent; with it then the chain moves as a solid object of size $R \approx bN^{\nu}$. The friction coefficient of the chain of size R being pulled through a solvent of viscosity, η_s, is given by Stokes law:

$$\zeta_Z \approx \eta_s R \tag{9.44}$$

If we assume spherical symmetry for the particle then Equation (9.43) becomes $\zeta_Z = 6\pi\eta_S R$ but in practice the chains may not be exactly spherical and hence the more general form will be used. From the Einstein equation (9.30) the diffusion coefficient becomes:

$$D_Z = (kT)/\xi_z \approx kT/\eta_s R \approx kT/\eta_s bN^{\nu} \tag{9.45}$$

Zimm explored the pre-averaging of the hydrodynamic interaction and obtained an extra coefficient of $8/3\sqrt{6\pi^3}$ for an ideal chain. Thus:

$$D_Z = \left(8/3\sqrt{6\pi^3}\right)(kT/\eta_s R) \cong 0.196(kT/\eta_s R) \tag{9.46}$$

In the Zimm model, the time for the chain to diffuse through its own radius is τ_z, which is defined by:

$$\tau_Z \approx R^2 / D_z \approx (\eta_s / kT) R^3 \approx (\eta_s b^3 / kT) N^{3\nu} \approx \tau_0 N^{3\nu} \tag{9.47}$$

The coefficient relating the relaxation time to a power of the number of monomers in the chain once again is the monomer relaxation time, τ_0. Zimm's full calculation of the chain relaxation time provides an extra coefficient of $1/\sqrt{3\pi}$ for an ideal chain:

$$\tau_Z = \left(1/\sqrt{3\pi}\right)(\eta_s / kT) R^3 \cong 0.163(\eta_s / kT) R^3 \tag{9.48}$$

This Zimm stress relaxation time is half of the Zimm end-to-end vector correlation time. The Zimm time is proportional to the volume occupied by the chain. If we compare Equations (9.47) and (9.39), we see that the Zimm time has a weaker dependence on chain length than the Rouse time: $3\nu < 2\nu + 1$ for $\nu < 1$. Additionally, in a dilute solution, the Zimm time will be shorter than the Rouse time.

9.7.5 Dynamics of polymer molecules in oscillatory shear

Experimentally, it is observed that as the frequency or the shear rate is increased so the effective viscosity falls (see Figure 9.6). At very low frequency or long times the whole polymer molecule is able to undergo diffusion. At a particular time or frequency, the polymer coil is unable to diffuse in the time interval that the shear is reversed, and as a consequence the polymer will no longer diffuse. As the rate/frequency is further increased, a reduction in the effective viscosity is observed, indicative of a reducing interaction between the shear field and the polymer coil. At some high frequency or shear rate, the effective viscosity becomes equal to that of the solvent.

One way to understand the motions of the polymer is to assume that it is like a piece of string. The string is fixed at node points and moves about these fixed points. The motions of the chain can be considered to be analogous to that of a vibrating string (see Figure 9.8).

In all untangled molecular models for polymer dynamics the relaxations are described by N different relaxation modes. These modes are numbered by mode index, $p = 1, 2, 3, \ldots N$. The mode represents coherent motion of a section of the whole chain with N/p monomers, and the corresponding relaxation time of this mode, τ_p, is similar to the longest relaxation time of a chain with N/p monomers. Statistical mechanics tells us that the equipartition principle states that $kT/2$ of free energy is associated with each degree of freedom at equilibrium. Immediately following the unit step strain, the entire chain stores elastic energy of the order of NkT, since there are N independent modes that each store of the order kT. To determine the time-dependent viscosity response, we must calculate the relaxation time of each mode.

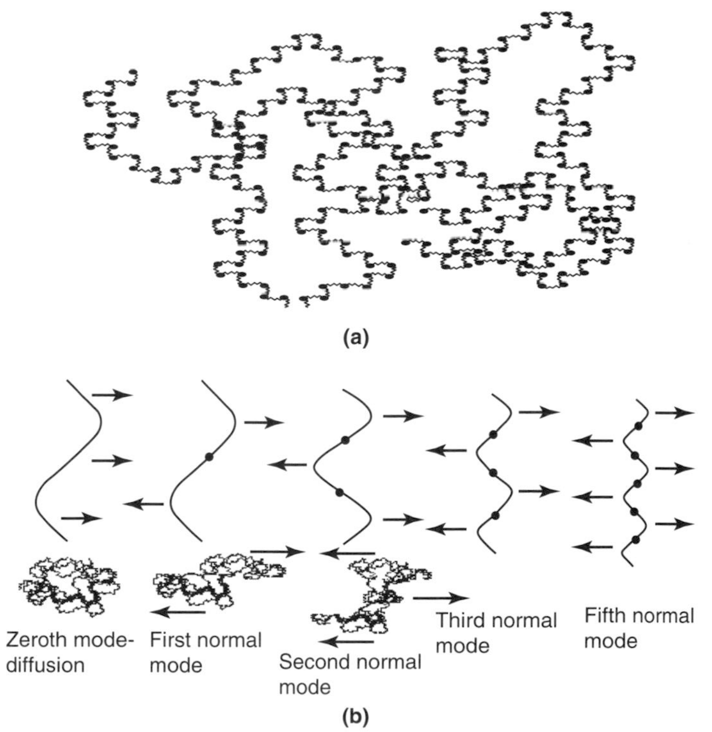

Figure 9.8 Schematic of a bead and spring Rouse chain: (a) chains, note that a random coil model leads to an apparently overlapping coil; (b) normal modes of vibration of a polymer chain.

9.7.6 Rouse model: mode theory

In the Rouse model, the longest relaxation time of the ideal chain is given by:

$$\tau_r \approx \tau_0 N^2 \tag{9.49}$$

The relaxation of a subunit of the chain which contains N/p monomers will have a relaxation time:

$$\tau_p \approx \tau_0 (N/p)^2 \quad \text{for } p = 1,2,3 \ldots N \tag{9.50}$$

where τ_0 is the relaxation time of a monomer with a mode index $p = N$. In terms of Figure 9.7, the longest relaxation mode involves the motion of the whole chain and corresponds to a collective mode with a node at the centre of the chain. The chain does not translate in space but the ends may be undergoing oscillatory motion with the fluctuations of the shear field created by the fluid. At the highest shear rates, the elements of the chain will move about the nearest nodes. In effect, the whole chain acts as if it were composed of independently vibrating elements. Above this limiting condition, $\tau_N = \tau_0$, the polymer chain is unable to move with the fast fluctuating shear field. The mode index $p = 2$ corresponds to the two halves of the chain $N/2$ monomers, each relaxing independently. The mode with index p breaks the chain into p sections of N/p monomers and each of these sections relax as independent chains of N/p monomers on the time scale, τ_p. The number of unrelaxed modes per chain at time $t = \tau_p$ is equal to the mode index p. Each unrelaxed mode contributes energy of the order of kT to the stress relaxation modulus. The stress relaxation modulus at time $t = \tau_p$ is proportional to the thermal energy kT and the number density of sections with N/p monomers, $\varphi / [(b^2 N)/p]$.

Thus:

$$G(\tau_p) \approx (kT/b^3)(\varphi/N)p \tag{9.51}$$

The time dependence of the mode index p for the mode that relaxes at time $t = \tau_p$ has the form:

$$p \approx (\tau_p/\tau_0)^{-1/2} N \tag{9.52}$$

Combining Equations (9.51) and (9.52) gives an expression for the stress relaxation modulus:

$$G(t) \approx (kT/b^3)\varphi(t/\tau_0)^{-1/2} \quad \text{for } \tau_0 < t < \tau_R \tag{9.53}$$

This expression effectively interpolates between a modulus level of the order of kT per monomer at the shortest Rouse mode, $t \approx \tau_R$, to a modulus level of kT per chain at the longest Rouse mode: $t = \tau_R \approx \tau_0 N^2$, using a power law. It is usual to assume simple exponential decay of the chain beyond the longest relaxation time. Combining this constraint with Equation (9.53) yields:

$$G(t) \approx (kT/b^3)\varphi(t/\tau_0)^{-1/2} \exp(-t/\tau_R) \quad \text{for } t > \tau_0 \tag{9.54}$$

where τ_R is the longest stress relaxation time. In practice, most experiments are performed in oscillatory shear rather than by varying the shear rate and are carried out at a frequency ω and can be described by:

$$\gamma(t) = \gamma_0 \sin(\omega t) \tag{9.55}$$

It is possible by increasing the value of ω to step through the relaxation spectrum and probe the response of a specific element of the chain. If the solution is perfectly elastic then the stress is related to the strain through the viscosity, so that:

$$\sigma(t) = \eta(\partial\gamma/\partial t) = \eta_0\gamma_0\omega\cos(\omega t) = \eta_0\gamma_0\omega\sin(\omega t + \pi/2) \tag{9.56}$$

In the case of a polymer solution, the response of the material may be viscoelastic and the stress may lead the strain by a *phase angle*, δ, and Equation (9.54) becomes:

$$\sigma(t) = \sigma_0 \sin(\omega t + \delta) \tag{9.57}$$

where the value of δ will typically be in the range 0–π/2. Since the stress is always a sinusoidal function with the same frequency as the strain it is possible to separate the stress into a component that is in-phase with the strain and another component that is out-of-phase with the strain:

$$\sigma(t) = \gamma_0 \left[G'(w)\sin(\omega t) + G''(\omega)\cos(wt) \right] \tag{9.58}$$

The real $G'(\omega)$ and imaginary $G''(\omega)$ components of the modulus are referred to as the *storage* and *loss* modulus, respectively. The Rouse viscosity for a simple melt has a value:

$$\eta \approx (\xi / b) N \tag{9.59}$$

which indicates that the viscosity is proportional to the number of monomers in the chain. Rouse showed that the exact solution required the addition of a coefficient 1/36 so that:

$$\eta = (\xi / 36b) N \tag{9.60}$$

This shows that there is an exact relationship for the stress relaxation modulus:

$$G(t) = kT(\phi / Nb^3) \sum_{p=1}^{p=N} \exp(-t / \tau_p) \tag{9.61}$$

with:

$$\tau_p = (\xi b^2 N^2)/(6\pi^2 kTp^2) \tag{9.62}$$

The stress relaxation times τ_p are half the correlation times of the normal modes. In oscillatory shear the storage and loss moduli of the Rouse model become:

$$G'(\omega) = kT \frac{\varphi}{Nb^3} \sum_{p=1}^{p=N} \frac{(w\tau_p)^2}{1+(\omega\tau_p)^2} \quad \text{and} \quad G''(\omega) = kT \frac{\varphi}{Nb^3} \sum_{p=1}^{p=N} \frac{(w\tau_p)}{1+(\omega\tau_p)^2} \tag{9.63}$$

where the stress relaxation time of the pth mode is $\tau_p = \tau_R/p^2$.

9.7.7 Zimm model: theory

The scaling analysis applied to the Zimm model indicates that the relaxation time of the pth mode is of the order of the relaxation time of the chain containing N/p monomers, thus:

$$\tau_p \approx \tau_0 (N/p)^{3v} \tag{9.64}$$

The index p of the mode relaxing at time $t = \tau_p$ after a step strain imposed at time $t = 0$ is obtained by solving the above equation for p:

$$p \approx N(\tau_p / \tau_0)^{-1/3v} = N(t / \tau_0)^{-1/3v} \tag{9.65}$$

The number of unrelaxed modes per chain at time $t = \tau_p$ is p. The stress relaxation modulus is proportional to the number density of chain sections with N/p monomers:

$$G(t) \approx (kT / b^3)(\phi / N)p \approx (kT / b^3)\phi(t / \tau_0)^{-1/3v} \tag{9.66}$$

In the case of an ideal polymer chain, $v = 1/2$ and the stress relaxation modulus decays as the 2/3 power of time. Following the approach outlined above for the Rouse model leads to the equation:

$$G(t) \approx (kT / b^3)\varphi(t / \tau_o)^{-1/3v} \exp(-t / \tau_z) \quad \text{for } t > \tau_0 \tag{9.67}$$

If we compare the Rouse and Zimm predictions, we see that the main difference is the power law in time: the Zimm prediction being 2/3 whereas the Rouse one is 1/2, and the Zimm relaxation time is shorter than the Rouse time.

Comparison of experiment and theory for dilute polymer solutions shows that close to ideal conditions, the Zimm model will give a good fit to data whereas the Rouse model tends to be less precise. The difference between the two models is the strength of the interaction between the solvent and the polymer chain. Since the strength of the interaction between polymer and solvent may vary with temperature, etc., it is desirable to have an approach which allows the strength of the interactions to be varied. The Wang–Zimm theory allows for the strength of the interactions to be varied and has as limiting cases the Rouse and Zimm theories. Both the Rouse and Zimm theories are only applicable to polymer chains which have a molar mass below the entanglement value. Surprisingly, it is found that the behaviour of an isolated polymer chain in very dilute solution and a polymer melt are very similar and the shapes of the curves are both well described by the above theories. However, when the molar mass exceeds some critical value, M_c, the simple theory described above ceases to describe the behaviour of the polymer.

9.8 Dynamic rheological behaviour of polymers with molar mass above M_c

Studies of the melt viscosity of a wide range of polymer systems have shown that the viscosity varies with the molar mass in a very similar manner (see Figure 9.9). Below a critical value, M_c, the viscosity exhibits a simple dependence on the molar mass: $\eta \propto M$, whereas above this value the dependence becomes $\eta \propto M^{3.5}$. There is some discussions as to whether the power term is 3.5 or 3.6. Clearly there is some change in the way in which the polymer molecules move once the chain length has exceeded M_c. De Genne and Edwards have considered this problem and deduced that the reasons for the change in behaviour is the presence of chain–chain entanglements for the higher molar mass materials (see Figure 9.10).

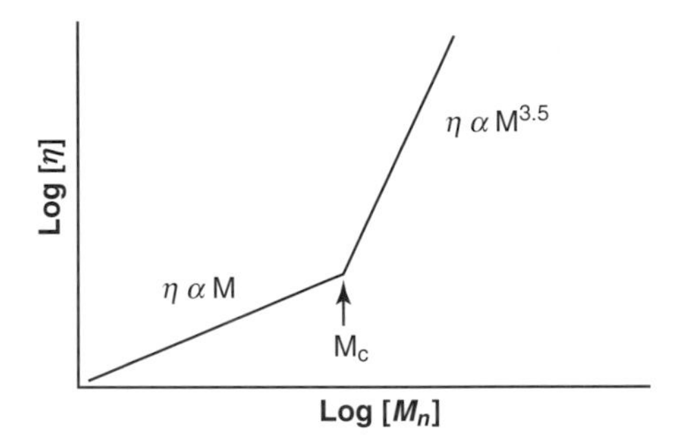

Figure 9.9 Variation of melt viscosity as a function of number average molar mass of a polymer.

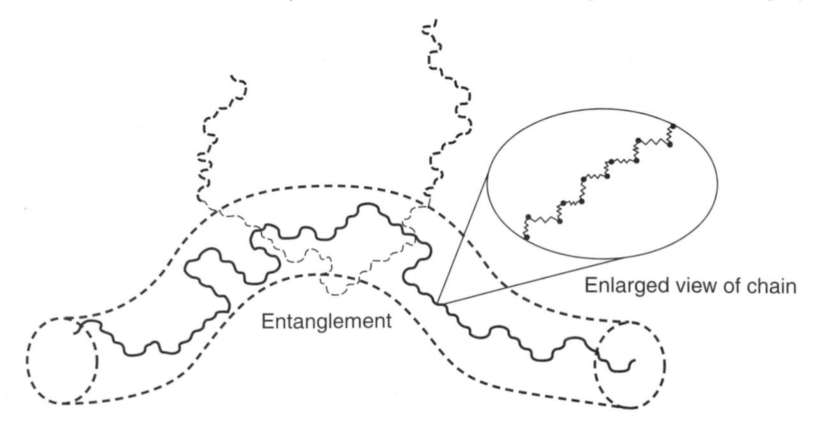

Figure 9.10 Schematic of a polymer chain contained in a theoretical tube formed by constraints of chain entanglement. Inset shows elements of polymer chain in extended and contracted states as polymer reptates along tube.

Table 9.3 Values of plateau modulus and molar mass for entanglement for a series of common polymers

Polymer	Plateau modulus, G_e (MPa)	M_e (g mol^{-1})
Polyethylene	2.60	1000
Polyethyleneoxide	1.80	2000
1,4-Polybutadiene	1.15	1900
Polypropylene	0.47	5800
1,4-Polyisoprene	0.35	6400
Polyisobutylene	0.32	7100
Polydimethylsiloxane	0.20	12,000
Polystyrene	0.20	17,000
Polyvinylcyclohexane	0.068	49,000

The same general plot is observed for many polymer systems; the value of the viscosity will change but the general variations are similar. The value of M_c (the critical molar mass) or M_e (the entanglement molar mass) vary with polymer type. In the literature slightly different values will be observed for M_e and M_c. The value of M_c is usually defined in terms of the change in the viscosity, whereas the value of M_e is defined in terms of the point at which the real and imaginary components of the modulus intersect and reflects the appearance of entanglements in the liquid. Typical values for M_e are listed in Table 9.3.

The reptation of a polymer chain down a tube which is defined by the entanglements has been considered by de Gennes and by Doi and Edwards. It is assumed that the motion of the chain along the contour of the tube is unhindered by topological interactions: other than defining the tube, the entanglements do not affect the motion of the chain.

Displacement of the monomers in the direction perpendicular to the axis of the tube is restricted by surrounding chains to an average distance a called the tube diameter. The number of Kuhn monomers in a strand of size equal to the amplitude of the transverse fluctuations is N_e, the number of monomers in an entangled strand. For a melt, the tube diameter is determined by ideal chain statistics:

$$a \approx b\sqrt{N_c} \tag{9.68}$$

The tube can be thought of as being composed of N/N_e sections of size a, with each section containing N_e monomers. The chain can be considered as either a random walk of entangled strands each of size a or a random walk of N monomers of size b.

$$R \approx a\sqrt{N / N_e} = b\sqrt{N} \tag{9.69}$$

The average contour length, L, of the chain contained in the tube is the product of the entanglement strand length, a, and the average number of entanglement strands per chain, N/N_e:

$$\langle L\rangle \approx a\left(N/N_e\right) \approx \left(b^2N/a\right) \approx \left(bN/\sqrt{N_e}\right) \tag{9.70}$$

The average contour length, L, is shorter than the contour length of the chains, bN, by the factor $a / b \approx \sqrt{N_e}$ because each entanglement strand in a melt is a random walk of N_e Kuhn monomers. One manifestation of entanglement in long chains with $N >> N_e$ is the appearance of a wide spectrum of relaxation times, which is observed experimentally as the modulus changing over a wide shear rate or temperature range. The entanglements act rather like cross-links and a rubbery plateau region is effectively observed in the viscosity time plots. However, the entanglements are physical and the moving chains slip over the entanglement points. The rubbery plateau is assumed to have a modulus, G_e. The volume occupied by an entangled strand with molar mass M_e in a melt with density ρ is the product of the number of Kuhn monomers per strand N_e and the Kuhn monomer volume v_0:

$$\left(M_e/\left(\rho N_{Av}\right)\right) = v_0 N_e \approx v_0\left(a^2/b^2\right) \approx \left(v_0/b^3\right)a^2b \tag{9.71}$$

Since monomers are space filling in the melt, the number density of the entanglement strands is simply the reciprocal of the entanglement strand volume, leading to a simple expression for the plateau modulus of an entangled polymer melt:

$$G_e \approx \left(\rho RT / M_e\right) \approx \left(kT / \left(v_0 N_e\right)\right) \approx \left(b^3 / v_0\right)\sqrt{N_e} \tag{9.72}$$

The number of chains P_e within the confinement volume a^3 is determined from the fact that monomers in the melt are space filling:

$$P_c \approx \left(a^3 / v_0 N_e\right) \approx \left(b^3 / v_0\right)\sqrt{N_c} \tag{9.73}$$

9.8.1 Relaxation times

In the model proposed by de Gennes, the motion of the chain within the tube is likened to that of a snake moving through the grass. This curvilinear diffusion coefficient for the passage of the chain down the tube is simply the Rouse diffusion coefficient of the chain:

$$D_c = kT / N\xi \tag{9.74}$$

The time it takes for the chain to move out of the original tube of average length L is the *reptation time*:

$$\tau_{rep} \approx \langle L\rangle^2 / D_e \approx \left(\zeta b^2 N^3\right) / \left(kTN_e\right) = \left(\zeta b^2 / kT\right) N_e^2 \left(N / N_e\right)^3 \tag{9.75}$$

where Equation (9.65) is used to obtain the averaged contour length of the tube. The reptation time, τ_{rep}, is predicted to be proportional to the cube of the molar mass. The experimentally measured scaling exponent is higher than 3:

$$\tau_{rep} \approx M^{3.4} \tag{9.76}$$

One possible reason for the difference is that in the above model it is assumed that the entanglements defining the tube do not move. In practice we would expect them to move and the tube would be a fluctuating entity rather than having a rigidly defined form. The first part of the relation in Equation (9.70) is the Rouse time of an entanglement strand containing N_e monomers:

$$\tau_e \approx \left(\zeta b^2 / kT\right) N_e^2 \tag{9.77}$$

The ratio of the reptation time τ_{rep} and τ_e is the cube of the number of entanglements along the chain:

$$\left(\tau_{rep} / \tau_e\right) \approx \left(N / N_e\right)^3 \tag{9.78}$$

The chain moves a distance of the order of its own size R in its reptation time τ_{rep}.

Since this is the time scale at which the tube is abandoned:

$$D_{rep} \approx \left(R^2 / \tau_{rep}\right) \approx \left(kTN\right) / \left(\zeta N^2\right) \tag{9.79}$$

The diffusion coefficient of an entangled linear polymer is predicted to be the reciprocal of the square of the molar mass, which disagrees with experimental results which indicate a relationship:

$$D \approx \left(R^2 / \tau\right) \sim M^{-2.3} \tag{9.80}$$

9.8.2 Stress relaxation and viscosity

The stress relaxation modulus, $G(t)$, for an entangled polymer melt is obtained by combining the above time constants with a relaxation process. If we assume that, on a length scale shorter than the tube diameter, the topological interactions are unimportant then it is reasonable to assume that the motion of the chain follows the Rouse model prediction. Experimental observation confirms the validity of this assumption. The entanglement strand of N_e monomers relaxes by Rouse motion with relaxation time τ_e:

$$\tau_e = \tau_0 N_e^2 \quad (9.81)$$

The Rouse model predicts that the stress relaxation modulus on these short time scales decays inversely proportional to the square of time:

$$G(t) \approx G_0 \left(t / t_0\right)^{-1/2} \quad \text{for } \tau_0 < t < \tau_e \quad (9.82)$$

The relaxation time of the Kuhn monomer, τ_0, is the shortest stress relaxation time in the Rouse model:

$$\tau_0 = \left(\zeta b^2\right)/\left(6\pi^2 kT\right) \approx \left(\zeta b^2\right)/\left(kT\right) \quad (9.83)$$

The stress relaxation modulus at τ_0 is the Kuhn modulus (kT per Kuhn monomer):

$$G_0 \approx G\left(\tau_0\right) \approx kT / v_0 \quad (9.84)$$

At the Rouse time of an entanglement strand, τ_e, the chain feels that the entanglements restrict its motion and free Rouse motion is no longer possible. On the time scale we find that $t > \tau_e$. The value of the stress relaxation modulus at τ_e is the plateau modulus, G_e, which is kT per entanglement strand and has the value indicated in Table 9.3.

$$G_e = G\left(\tau_e\right) = G_0 / N_e = \left(kT\right) / \left(v_0 N_e\right) \quad (9.85)$$

In the simple reptation model, there is a delay in relaxation between τ_e and the reptation of the chain, τ_{rep}, restricting the chain's Rouse motions to the imaginary tube. The time that the chain takes to diffuse a distance of the order of its size is longer than its Rouse time by a factor of $6N/N_e$. This slowing of the motion occurs because the chain must move along the confined tube. De Genne envisaged it as sequentially undergoing compression and expansion. Experimentally, τ_{rep} corresponds to the frequency at which $G' = G''$.

Doi and Edwards generated the theory which describes the stress relaxation by solving the so-called first passage problem. As time passes the chain will exit the tube in which it originally resided; this is the relaxation of the stress. The rate is controlled by the chain diffusing in the imaginary tube and so they obtained the expression:

$$G(t) = 8 / \pi^2 G_e \sum_{odd} (1 / p^2) \exp\left(-\left(p^2 t / \tau_{rep}\right)\right) \quad (9.86)$$

The longest relaxation time in this model is the reptation time required for the chain to escape from its tube:

$$\tau_{rep} = 6\tau_0 \left(N^3 / N_e\right) = 6\tau_e \left(N / N_e\right)^3 = 6\tau_R \left(N / N_e\right) \quad (9.87)$$

The reptation model prediction for the viscosity of an entangled polymer melt is determined by integrating Equation (9.81):

$$\eta = \int_0^\infty G(t)\,dt = \frac{8}{\pi^2} G_e \sum_{odd} \frac{1}{p^2} \int_0^\infty \exp\left(-\left(p^2 t / \tau_{rep}\right)\right) dt = \left(\pi^2 / 12\right) G_e \tau_{rep} \quad (9.88)$$

The final result was obtained from the fact that $\sum_{odd} 1/p^4 = \pi^4/96$. Since the stress relaxation is nearly a single exponential, the scaling prediction of the viscosity as the product of the plateau modulus and the reptation time is nearly quantitative:

$$\eta \approx G_e \tau_{rep} \approx G_e \tau_e \left(N / N_e\right)^3 \approx \left(kT / v_o N_e\right)\left(\zeta b^2 N_e^2\right) / \left(kT\right)\left(N / N_e\right)^3 \approx \left(\zeta b^2 N^3\right) / v_0 N_e^2 \quad (9.89)$$

The viscosity of a polymer melt is predicted to be proportional to the molar mass for unentangled melts (the Rouse model), and proportional to the cube of the molar mass for entangled melts (the reptation model).

$$\eta \sim \{M\} \text{ for } M < M_c \qquad M^3 \text{ for } M > M_e \tag{9.90}$$

More refined models have since been produced which attempt to predict the effects of the tube breathing during the diffusion process and obtain a prediction, which matches experimental observation, that, for $M > M_c$, the viscosity varies as the 3.4 power of the molar mass.

The total frequency shear rate response of a polymer with chain length above M_c is composed of two elements: the reptation motion and motion of the chain with length less than M_c. The latter can be accurately described by adding to the predictions of the repatation theory a contribution based on the mode predictions for a polymer of length M_c. Studies of the flexible polymer polydimethylsiloxane over a frequency range 10^{-3}–10^{10} Hz have confirmed the validity of this approach.

9.9 Rubber elasticity

One of the earliest challenges which were addressed by theoreticians was the attempt to model rubber elasticity. In a rubber the chains are unable to reptate as in a viscous fluid as they are chemically cross-linked to form a three-dimensional network. A rubber (elastomer) is a three-dimensional network in which all the elements are connected by chemical bonds (see Figure 9.11). The distance between the chemical cross-links will define the rigidity of the materials. If the length is very short, the chains will be inhibited in their movement and the material will be very stiff. Alternatively, if the distance is long, the chains will behave very much as they would in a viscoelastic fluid and the material will exhibit a high degree of flexibility.

The process of stretching a piece of rubber will cause some of the chains between the cross-links to stretch and others to be compressed. If the material is assumed to be completely elastic, it will shrink in the direction perpendicular to which it is being stretched (see Figure 9.11). The shrinkage is defined by the Poisson's ratio for the material. The rubbery state is unique to polymeric materials. It is characterised by the ability for the material to undergo very large reversible deformation. The stretching and compression of the chains are reflected in the changes in the chain conformations. Stretching will straighten the chain and is associated with a more extended form.

Compression of the chains will produce a more tightly packed structure, which will contain more of the *gauche* conformations. Provided that the chains can switch easily between these structures, then the possibility of storing energy exists. Hence the stretched conformations and *relaxed* structures of the equilibrium state can be seen as states which are in dynamic equilibrium with one another. The energy absorbed by stretching is released when the stress is removed, the chains

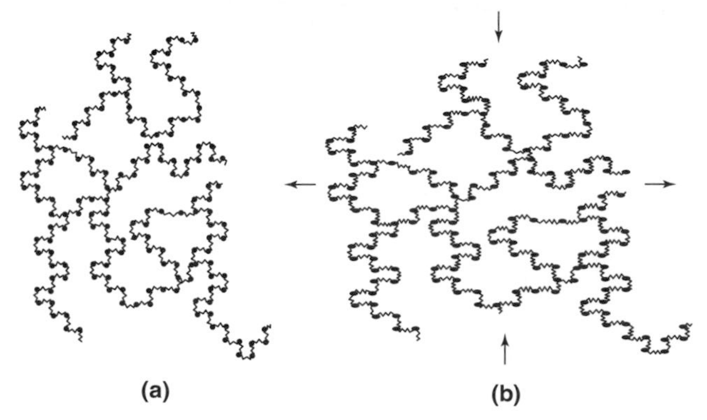

Figure 9.11 Schematic of a cross-linked rubber elastomer network: (a) at equilibrium (unstretched); (b) nonequilibrium (stretched structure).

will adopt their original relaxed structure and there may be some release of thermal energy. The changes which occur during the stretching process are at a molecular level yet they result in macroscopic changes to the sample. The changes that are observed are a consequence of many millions of small conformational changes within the sample. The simple models of a polymer chain assume that it can be pictured as a ball; this is also true in the amorphous solid. The process of stretching the rubber effectively changes the contour of the idealised sphere into an ellipse. For many rubbers, chain extensions of several hundred percent are possible and require large changes in the conformational distribution. The processes associated with stretching can be modelled simply by statistical mechanics. The higher energy *gauche* coiled state of the polymer chain can have a number of different conformations, whereas the extended zig-zag all-*trans* form has essentially only one form. Stretching is marked by a reduction in the number of possible conformations. Boltzmann defined the disorder of a system in terms of the *entropy*, S, which is related to the number of possible arrangements of the system by:

$$S = k\ln(\text{number of arrangements}) \tag{9.91}$$

where S is the entropy and k is Boltzmann's constant. If the number of permitted conformations varies as the number of arrangements then the entropy change on stretching a sample is:

$$\Delta S = k\ln(\text{number of conformations})_s - k\ln(\text{number of conformations})_u \tag{9.92}$$

where subscripts s and u refer to the stretched and unstretched states, respectively. Using the partition function notation Ω and substituting into Equation (9.91), the change in the conformations becomes:

$$\Delta S = k\ln(\Omega_s/\Omega_u) \tag{9.93}$$

Stretching the sample gives rise to a negative entropy change, since Ω_s is less than Ω_u.

The internal energy change for a system is described in terms of the sum of all the contributing energy components: heat added to the system $T\partial S$, work done to change the network volume, $p\partial V$, and work done upon the network deformation, $f\partial L$:

$$\partial H = T\partial S - p\partial V + f\partial L \tag{9.94}$$

Changing to the Helmholtz free energy Equation (9.94) becomes:

$$\partial G = \partial H - \partial(TS) - \partial H - T\partial S - S\partial T = -S\partial T - p\partial V + f\partial L \tag{9.95}$$

The change in the Helmholtz free energy can be written as a complete differential:

$$dG = (dG/dT)_{V,L}\, dT + (dG/dV)_{T,L}\, dV + (dG/dL)_{T,V}\, dL \tag{9.96}$$

Comparing Equations (9.95) and (9.96) we can identify the partial derivatives of the free energy components:

$$(dG/dT)_{V,L} = -S \quad (dG/dV)_{T,L} = -p \quad (dG/dL)_{T,V} = f \tag{9.97}$$

A second derivative of the Helmholtz free energy does not depend on the order of differentiation:

$$\partial^2 G/(\partial T\partial L) = \partial^2 G/(\partial L dT) \tag{9.98}$$

Using the components of the Helmholtz free energy, Equation (9.98) can be written in the form of a Maxwell relation:

$$-(\partial S/\partial L)_{T,V} = (\partial f/\partial T)_{V,L} \tag{9.99}$$

The force applied to the network to produce the extension dL consists of two contributions:

$$f = (\partial F/\partial L)_{T,V} = [\partial(H - TS)/\partial L]_{T,V} = (\partial H/\partial L)_{T,V} - T(\partial S/\partial L)_{T,V} \tag{9.100}$$

The first term describes how the internal energy changes with the change in sample length and the second term describes the entropy change associated with the elongation. It is this latter process which is associated with the change from *gauche* to *trans* conformation and makes the most significant change with change in length. The second term can be written as the Maxwell relation, Equation (9.93):

$$f = (dH / dL)_{T,V} + T(df / dT)_{V,L} = f_E + f_S \tag{9.101}$$

The two contributions to the force are therefore:

$$f_E = (dH / dL)_{T,V} \qquad f_S = T(df / dT)_{V,L} = -T(dS / dL)_{L,V} \tag{9.102}$$

In a typical metal, the energetic terms dominate. In rubbers the entropic terms dominate. For an 'ideal' network there is no energetic contributions to the elasticity and $f_E = 0$.

9.9.1 Separation of energetic and entropic terms

A simple method to separate the terms has been proposed by Flory. If the force to keep a rubber band at a constant extension is measured as a function of temperature, the slope of the line is given by Equation (9.102) (see Figure 9.12).

Experimentally it is found that the value of f_E is very small compared with f_S and it is a valid approximation to neglect the terms associated with f_E. Since the entropy change is negative, so the free energy change on stretching is positive. Thus the direction of spontaneous change is from *stretched* to *unstretched* and is the thermodynamic origin of the restoring force and is an *entropic* force. In the case of metals the energy/enthalpy increases on deformation. Since the entropy in a rubber decreases on stretching, the restoring force increases with the strain. Since the entropy change is multiplied by temperature in Equation (9.102), the restoring force at constant strain increases with increasing temperature. This prediction is in direct contrast to what happens with a metal spring, where the energy-derived restoring force is either independent of or decreases with temperature.

To illustrate the difference between stretching a metal spring and a rubber band let us consider what happens when we apply heat to a stretched rubber band (see Figure 9.13). The length of the rubber band, L_s, is defined by a balance between the mass producing the downward force and the modulus of elasticity which is determined by the network structure of the rubber. Application of heat to the rubber band will cause the polymer chains to adopt more of the higher energy *gauche* conformations and this will cause the rubber to shrink. This is in contrast to what would happen with a metal where heating causes extension of the spring. The shrinkage will produce an upwards force which will in turn raise the mass!

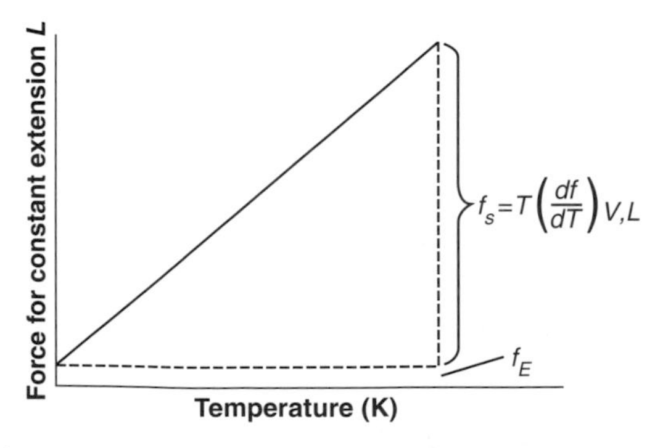

Figure 9.12 Typical force–temperature plot for a rubber.

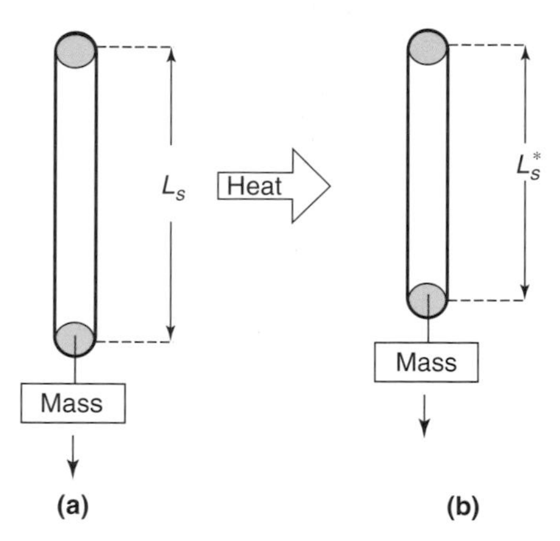

Figure 9.13 Effect of heating on a stretched rubber band: (a) before heat is applied; (b) after heat is applied.

9.9.2 Unentangled rubber elasticity: affine network model

The network depicted in Figure 9.11 is composed of polymer chains, cross-linked at various points but with significant lengths of chain between the cross-links. The affine assumption, originally suggested by Kuhn, implies that the relative deformation of each element of the network is the same as the imposed deformation of the whole network. If the section of the polymer chain between the cross-links before stretching has a length L_{x0} and after stretching this has increased to λL_{x0}, then the dimensions of the deformed network in the x-, y- and z-directions are:

$$L_x = \lambda_x L_{x0} \quad L_y = \lambda_y L_{y0} \quad L_z = \lambda_z L_{z0} \tag{9.103}$$

If we ignore the cross-links, then the extension will have changed the end-to-end distance by a value which is proportional to that of the change in the dimensions of the whole sample.

$$R_x = \lambda_x R_{x0} \quad R_y = \lambda_y R_{y0} \quad R_z = \lambda_z R_{z0} \tag{9.104}$$

The entropy of a chain of N Kuhn monomers of length b with an end-to-end vector $\overline{R}$ is given by:

$$S\left(N,\overline{R}\right) = -3/2k\left(\dot{R}^2/\left(Nb^2\right)\right) + S\left(N,0\right) = -3/2k\left(\left(R_x^2 + R_y^2 + R_z^2\right)/Nb^2\right) + S\left(N,0\right) \tag{9.105}$$

The entropy change is therefore the difference between the initial and final states:

$$S\left(N,\overline{R}\right) - S\left(N,\overline{R_0}\right) = -3/2k\left(\left(R_x^2 + R_y^2 + R_z^2\right)/Nb^2\right) - 3/2k\left(\left(R_{x0}^2 + R_{y0}^2 + R_{z0}^2\right)/Nb^2\right)$$

$$= -3/2k\left(\left(\left(\lambda_x^2 - 1\right)R_{x0}^2 + \left(\lambda_y^2 - 1\right)R_{y0}^2 + \left(\lambda_z^2 - 1\right)R_{z0}^2\right)Nb^2\right) \tag{9.106}$$

The entropy change of the whole network is the sum of the entropy change of all the n strands present:

$$\Delta S_{net} = -3/2k\left[\left(\left(\lambda_x^2 - 1\right)\sum_{i=1}^{i=n}\left(R_{x0}\right)^2 + \left(\lambda_y^2 - 1\right)\sum_{i=1}^{i=n}\left(R_{y0}\right)^2 + \left(\lambda_z^2 - 1\right)\sum_{i=1}^{i=n}\left(R_{z0}\right)^2\right)\Big/ Nb^2\right] \tag{9.107}$$

If the polymer chains are assumed to have dimensions equivalent to those of the chain in the melt phase then:

$$\left\langle R_{x0}^2 \right\rangle = \frac{1}{n}\sum_{i=1}^{n}\left(R_{x0}\right)_0^2 = Nb^2/3 = \left\langle R_{y0}^2 \right\rangle = \left\langle R_{z0}^2 \right\rangle \tag{9.108}$$

Therefore the sum of the squares of the components of the end-to-end vectors of all n strands is equal to:

$$\sum_{i=1}^{n}\left(R_{x0}\right)_i^2=\sum_{i=1}^{n}\left(R_{y0}\right)_i^2=\sum_{i=1}^{n}\left(R_{z0}\right)_i^2=\frac{n}{3}Nb^2 \tag{9.109}$$

The entropy change on deformation of the network can then be written as:

$$\begin{aligned}\Delta S_{net} &= -3/2k\left(\left(\lambda_x^2-1\right)(n/3)Nb^2+\left(\lambda_y^2-1\right)(n/3)Nb^2+\left(\lambda_z^2-1\right)(n/3)Nb^2\right)\Big/Nb^2\\ &= -\frac{nk}{2}\left(\lambda_x^2+\lambda_y^2+\lambda_z^2-3\right)\end{aligned} \tag{9.110}$$

Using this result the Helmholtz free energy change can be calculated as:

$$\Delta G_{net} = -T\Delta S_{net} = nkT/2(\lambda_x^2+\lambda_y^2+\lambda_z^2-3) \tag{9.111}$$

If we assume that the volume of the network does not change during the deformation then the extension in one direction is compensated by a shrinkage in the directions at right angles, then:

$$V = L_{x0}L_{y0}L_{z0} = L_xL_yL_z = \lambda_xL_{x0}\lambda_yL_{y0}\lambda_zL_{z0} = \lambda_x\lambda_y\lambda_zV \tag{9.112}$$

If the volume remains constant then the product of the deformations must be unity:

$$\lambda_x\lambda_y\lambda_z = 1 \tag{9.113}$$

In practice, this is not quite correct as there is a small change but to a first approximation it is a reasonable assumption. For a network stretched in the x-direction then:

$$\lambda_x = \lambda \quad \lambda_y = \lambda_z = 1/\sqrt{\lambda} \tag{9.114}$$

The free energy change for uni-axial deformation becomes:

$$\Delta G_{net} = (nkT/2)\left(\lambda^2+2/\lambda-3\right) \tag{9.115}$$

The force required to deform the network is the rate of change of the free energy with respect to its size along the x-direction:

$$f_x = \partial\Delta G_{net}/\partial L = \partial G_{net}/\partial\left(\lambda L_{x0}\right) = \left(1/L_{x0}\right)\left(\partial\Delta G_{net}/\partial\lambda\right) = \left(nkT/L_{x0}\right)\left(\lambda-\left(1/\lambda^2\right)\right) \tag{9.116}$$

If the cross-sectional area L_y,L_z of the macroscopic network is doubled then twice as large a force is required to obtain the same deformation. The *stress* is the ratio of force and cross-sectional area. Both the force and cross-sectional area have direction and magnitude (the direction of the cross-section area being described by the unit vector normal to its surface), making the stress a tensor. The ij-component of the *stress tensor* is the force applied in the direction in the i-direction per unit cross-section area of a network perpendicular to the j-axis. For example, the xx-component σ_{xx} is the force applied in the x-direction f_x divided by the area L_yL_x perpendicular to the x-axis:

$$\begin{aligned}\sigma_{xx} &= \left(f_x/\left(L_yL_z\right)\right) = \left(nkT/\left(L_{x0}L_yL_z\right)\right)\left(\lambda-\left(1/\lambda^2\right)\right) = (nkT)/\left(L_{x0}L_yL_z\right)\lambda\left(\lambda-\left(1/\lambda^2\right)\right)\\ &= (nkT/V)\left(\lambda^2-(1/\lambda)\right) = \sigma_{true}\end{aligned} \tag{9.117}$$

This is the *true stress* in the network and it is therefore denoted by σ_{true}. Since it is often not easy to measure the cross-sectional area of the deformed network, an engineering stress is also defined. In the *engineering stress* the original cross-sectional area $L_{y0}L_{z0}$ is used instead of the deformed cross-sectional area L_yL_z:

$$\begin{aligned}\sigma_{eng} &= f_x/\left(L_{y0}L_{z0}\right) = (nkT)/\left(L_{x0}L_{y0}L_{z0}\right)\left(\lambda-\left(1/\lambda^2\right)\right)\\ &= (nkT)/V\left(\lambda-\left(1/\lambda^2\right)\right) = \sigma_{true}/\lambda\end{aligned} \tag{9.118}$$

The coefficient relating the stress and the deformation is the shear modulus, G. Using the usual relation between the stress and tension $\sigma_{true} = E\varepsilon$, where $E = 3G$, where G is Young's modulus and $\varepsilon = (L - L_0)/L_0$ is the extensional strain:

$$G = nkT/V = \upsilon kT = \rho RT/M_s \tag{9.119}$$

The number of network strands per unit volume (number density of strands) is $\upsilon = n/V$. In the last equality, ρ is the network density (mass per unit volume), M_s is the number (average molar mass) of a network strand, and R is the gas constant. The network modulus increases with temperature because its origin is entropic, analogous to the pressure of an ideal gas, $p = nkT/V$. The modulus also increases linearly with the number density of network strands: $v = n/V = \rho N_{Av}/M_s$. Equation (9.119) states that the modulus of any network polymer is kT per strand. The affine predictions for both true and engineering stresses in uni-axial deformation at constant network volume are:

$$\sigma_{true} = G\left(\lambda^2 - 1/\lambda\right) \quad \text{and} \quad \sigma_{true} = G\left(\lambda - 1/\lambda^2\right) \tag{9.120}$$

The true stress can be obtained from the free energy density:

$$\begin{aligned}\sigma_{true} &= \left(1/L_y L_x\right)\left(\partial F/\partial L_x\right) = \lambda\left(\partial\left(F/V\right)\right)/\partial V = 2C_1\left(\lambda^2 - 1/\lambda\right) + 2C_2\left(\lambda - 1/\lambda^2\right) + \cdots \\ &= \left(2C_1 + 2C_2/\lambda\right)\left(\lambda^2 - 1/\lambda\right) + \cdots\end{aligned} \tag{9.121}$$

The engineering stress can be calculated from the true stress, given in Equation (9.118):

$$\sigma_{eng} = \sigma_{true}/\lambda = \left(2C_1 + 2C_2/\lambda\right)\left(\lambda - 1/\lambda^2\right) \tag{9.122}$$

This leads to the Mooney–Rivlin equation:

$$\sigma_{true}/\left(\lambda^2 - 1/\lambda\right) = \sigma_{eng}/\left(\lambda - 1/\lambda^2\right) = 2C_1 + 2C_2/\lambda \tag{9.123}$$

For an ideal elastomer the Mooney–Rivlin coefficients C_1 and C_2 are $2C_1 = G$ and $C_2 = 0$. In practice C_2 will be positive and indicates a deviation from ideality. At very high strains the molecules pack together and the stress–strain curves deviate from the predictions of Equation (9.123). The modulus of the crystalline material rises sharply, and this area of the stress–strain curve is known as the stress crystallisation. There is an inverse dependence of M_c, the molecular weight of the chain constrained between cross-links, on C_2. If we 'pull' the chain by the cross-links then the section between the links is the length of chain being deformed. Then, according to the Boltzmann equation, the change in entropy depends on the ratio of the number of conformations, not the simple difference. Then the *relative* decrease in the number of conformations is greater for a short chain than for a long one. Consider 10 conformations being reduced to 5, a relative decrease of 50%. On the other hand 1,000 conformations being reduced by the same amount to 995 is a relative decrease of only 0.5%.

The Mooney–Rivlin theory is a useful description for the behaviour of cross-linked rubber and also phase-separated polyurethane elastomers. In the latter, the hard blocks (see Section 3.12) act as pinning points for the matrix and allow extensions of up to 600% to be achieved in favourable cases. Even with these large extensions the theory is able to usefully describe the observed behaviour.

9.10 Polymer crystal growth

The process of crystal growth is a balance between kinetics and thermodynamics (see Section 3.3). The enthalpy of interaction between neighbouring chains counters the entropy associated with the higher energy *gauche* forms and produces the straight segments which seed the growth of crystals. In practice, the nucleation process, the start of the growth of the crystal, is often seeded by the presence of impurities: residual catalysts fragments, dust, etc. The creation of crystalline regions is associated with an increase in the modulus and reinforcing the polymer matrix. The crystallisation process for polymer systems is influenced by the way in which the polymer chain may have its mobility restricted through the occurrence of chain entanglements and other longer range interactions.

9.10.1 Thermodynamics of polymer molecule in the melt

The energy of a polymer molecule in the molten state will be determined by a combination of inter- and intra-molecular forces. The RISM theory predicts that as the temperature is raised, so the occupancy of the higher energy states is increased, so the size of the polymer coil *shrinks*. In the melt state, the polymer molecule will adopt a minimum coil size. Cooling down the polymer chain will increase the proportion of the chain adopting an extended all-*trans* form and facilitate the creation of the *stem* structures which are required for crystal growth. The polymer chains are not static entities and changes in the location of the *gauche* conformations can disrupt the formation of favourable interactions for crystal growth, suppress nucleation or alternatively assist the creation of *stems* of the correct length and aid nucleation. If two or more *gauche* sequences become located close together on the chain a *hair pin* or *folded* structure is generated, which aids the growth process. The number of such folds will have an equilibrium value at a particular temperature. Although intuitively one might expect polyethylene chains to adopt an extended regular structure, both solution-grown and melt-grown crystals are chain *folded*. The attachment process may involve either the end of chains or a low energy all-*trans* section interacting with a preformed surface. The ends of chains will have a higher energy than the central sections, hence ends tend to be located at surfaces. Clearly, the process of attaching the polymer chain to the surface involves not just a change in enthalpy but also subtle changes in the total entropy of the system. The process of nucleation should ideally only involve interactions between polymer molecules, but in practice residual catalysts and other solids present in the melt help the process.

The concept of chain folding is central to understanding the nature of the crystalline polymeric state. As the polymer chain is being laid down on a substrate, folding will occur and energetically the process would drive the folds to be located at a surface. Thus, the polymer tends to form stacks of extended polymer chains which form the *lamellae* and these have a thickness which is indicative of the temperature at which they were grown. This observation is consistent with the polymer chain having a distribution of conformational states and entropy defined by that population. Disordered *amorphous* material, in a nonequilibrium state, will be trapped within the crystalline material.

9.10.2 Nucleation

The nucleation and growth processes depend on the degree of supercooling of the melt or solution phase. The crystal thickness, or alternatively the thickness of each new crystalline layer in a growing crystal, is determined by the one that grows fastest rather than the one that is at equilibrium. Once the chain has attached itself to the substrate, it is possible that a neighbouring chain will start to build a subsequent layer of the crystal and the energy balance will change. As a consequence, the original structure may change to a new more stable structure and the phenomenon of *polymorphism* will be observed. The process of change in the structure is known as *Oswald ripening*.

In the melt phase, thermodynamics would indicate that chains with different chain lengths should have different energies. As a consequence, it is possible that chains with a similar energy will tend to segregate together in space, to *phase separate*. In the case of a broad molecular mass distribution polymer system, the individual chains will fractionate according to their molar mass and the morphology of the crystals can reflect this effect. However, the viscosity of the melt during the crystallisation process will usually stop an equilibrium state being achieved. Hence phase diagrams often give little insight into the state of the material during the crystallisation process.

Whilst it is relatively straightforward to describe the processes involved in polymer crystallisation, it is far more complex to depict it in terms of a model which allows the quantitative prediction of the crystallisation process.

9.10.3 Minimum energy conditions and simple theory of growth

Simplistically, we can consider the relative energy of attachment to a cube with faces 1,2,3 (see Figure 9.14). The dimensions of the crystal will be determined by the relative energy of each surface and will be a minimum with respect to the melt (ΔG) at a given volume:

$$\Delta G = V\Delta G_m^0 + 2L_1L_2\sigma_3 + 2L_1L_3\sigma_2 + 2L_2L_3\sigma_1 \tag{9.124}$$

where ΔG_m^0 is the specific free energy of melting and σ_i are the specific surface free energies. This equation is true for any crystal growth process and is not specific for a polymer system. One of the L_i terms can be eliminated from the equation by considering that $V = L_1L_2L_3$ is constant, thus:

$$L_3 = V / (L_1L_2) \tag{9.125}$$

Inserting Equation (9.120) into Equation (9.119) yields:

$$\Delta G = V\Delta G_m^{\ 0} + (2V / L_3)\sigma_3 + 2L_1L_2\sigma_2 + (2V / L_1)\sigma_1 \tag{9.126}$$

By taking the derivatives of ΔG with respect to L_1 and L_3 and setting them equal to zero, the following expression is obtained:

$$\partial(\Delta G) / \partial L_1 = L_3 - (V / L_1^2)\sigma_1 = 0 \quad \Rightarrow \quad L_1 / \sigma_1 = L_2 / \sigma_2 \tag{9.127}$$

and

$$\partial(\Delta G) / \partial L_3 = L_1\sigma_2 - (V / L_3^2)\sigma_3 = 0 \quad \Rightarrow \quad L_2 / \sigma_2 = L_3 / \sigma_3 \tag{9.128}$$

Combining Equations (9.122) and (9.123) gives:

$$L_1 / \sigma_1 = L_2 / \sigma_2 = L_3 / \sigma_3 \tag{9.129}$$

Equation (9.129) indicates that the dimensions of the equilibrium crystal in different directions are proportional to the surface free energies (σ_i) of the perpendicular surfaces. In the case of polymer crystals, two of the surfaces will contain chain folds having significantly different energies from the others elements of the chain. For polyethylene the specific surface energy (σ_e) of the folded surfaces is about 60–70 mJ m^{-2}, which is about five times greater than the value for the surface containing the aligned chains (σ), which has a value of about 15 mJ m^{-2}. The ratio of the equilibrium thickness (along the chain axis direction) to the width perpendicular to the chain direction is consequently close to >5, The prediction of this ratio being >5 is three to four

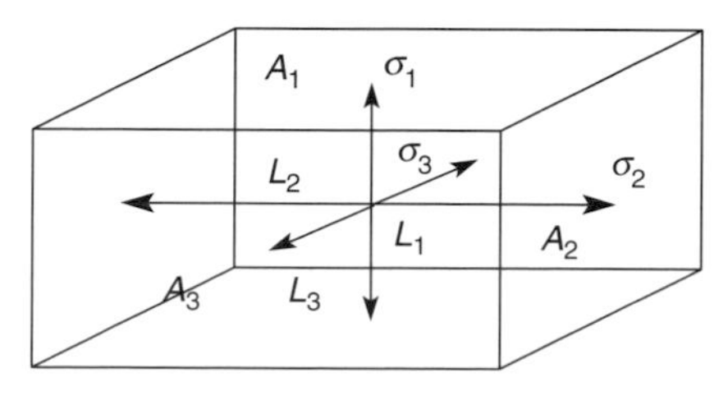

Figure 9.14 Equilibrium shape of crystal with three different surfaces (i =1, 2, 3 with different surface free energies σ_i).

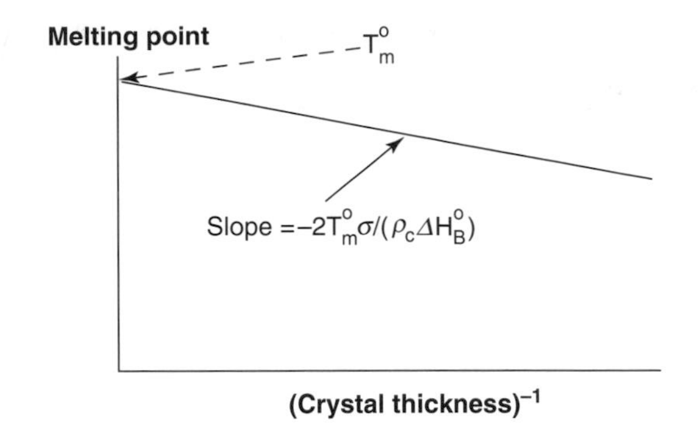

Figure 9.15 Plot of variation of melting point with reciprocal of crystal thickness.

orders of magnitude greater than that observed experimentally. The difference between experiment and prediction could be a consequence of crystals not being in equilibrium with the melt and that crystals will rearrange when given enough 'thermal stimulation'. The thickness of solution growth lamellae (L_c^*) depends on the degree of supercooling $\left(\Delta T = T_m^0 - T_c\right)$, where T_m^0 is the equilibrium melting point and T_c is the crystallisation temperature.

$$L_c = C_1/\Delta T + \delta L \tag{9.130}$$

where C_1 and δL are constants for a particular polymer system. The effects of supercooling on the lamellae thickness for the case of linear polyethylene are shown in Figure 9.15.

The Thompson–Gibbs equation allows the melting point to be related to the lamellae thickness. The change in free energy on melting (ΔG_m) is given by:

$$\Delta G_m = \Delta G^* + \sum_{i=1}^{n} A_i \sigma_i \tag{9.131}$$

where ΔG^* is the surface independent change in free energy and σ_i is the specific surface energy of surface i with area A_i. At equilibrium:

$$\Delta G_m = 0 \Rightarrow \Delta G^* = \sum_{i=1}^{n} A_i \sigma_i \tag{9.132}$$

For a simple polymer crystal, the lamella growth dominates and the two surfaces which contain the folded chains dominate the total surface energy term. It can be assumed that there is a free energy term that is independent of this fold surface and represents the interaction between the all-*trans* linear chain *stems* and has the form:

$$\Delta G^* = \Delta G_B^* A L_c \rho_c \tag{9.133}$$

where ρ_c is the density of the crystal plane. Since both ΔH_B and ΔS_B can be regarded as temperature-independent bulk parameters, and the specific bulk free energy change (ΔG_B^*) is given by:

$$\Delta G_B^* = \Delta H_B^0 - T_m \Delta S_B^0 = \Delta H_B\left(1 - \left(T_m / T_m^0\right)\right)$$
$$= \Delta H_B^0\left(\left(T_m^0 - T_m\right)/T_m^0\right) \tag{9.134}$$

By inserting Equation (9.132) into Equation (9.133) we obtain:

$$\Delta G^* = \Delta H_B^0\left(T_m^0 - T_m\right)\left(A L_c \rho_c\right)/T_m^0 \tag{9.135}$$

Inspection of the lamellae would indicate that the surfaces containing the chain *stems* are usually the short sides and have a small area compared with the sides that contain the folded chains.

As indicated previously, the surfaces contain the folded chains which make the major contribution to the surface free energy. It is therefore appropriate to assume that the total area of the four lateral surfaces is small compared to the area of the fold surfaces and their contribution to the total surface free energy of the crystal can be neglected:

$$\sum_{i=1}^{n} A_i\sigma_i \approx 2\sigma A \tag{9.136}$$

Combination of Equations (9.132), (9.135) and (9.136) gives:

$$\Delta H_B^0\left(T_m^0 - T_m\right)\left(AL_c\rho_c\right)/T_m^0 = 2\sigma A$$

$$T_m^0 - T_m = \left(2\sigma A T_m^0\right)/\left(AL_c\rho_c\Delta H_B^0\right) = \left(2\sigma T_m^0\right)/\left(L_c\rho_c\Delta H_B^0\right) \tag{9.137}$$

which may be simplified to give the Thompson–Gibbs equation:

$$T_m = T_m^0\left(1 - \left(2\sigma/\left(L_c\rho_c\Delta H_B^0\right)\right)\right) = -2T_m^0\sigma/\left(\rho_c\Delta H_B^0\right) \tag{9.138}$$

The Thompson–Gibbs equation predicts a linear relationship between the melting point and the reciprocal of the crystal thickness (see Figure 9.15). In practice, it is difficult to obtain the melting point and crystal thickness data as the experiments involved in determining the melting point allow crystal thickening to occur.

9.10.4 Nature of chain folding

If the chain folding were regular and the chain *stems* were to lie adjacent to one another then the density would be close to that predicted from the crystal lattice structure. Measurement of any crystalline polymer will indicate that the density is less that that from the predictions for the perfect crystalline material, thus indicating the presence of amorphous material. It is proposed that part of the amorphous content arises from disorder, irregular folding of the polymer chains (see Figure 9.16). The irregular folds in which the chains do not re-enter the lamellae at the next layer lead to the concept of the so-called *switch board model*. Early telephone switchboards were formed from a matrix of connections into which a connecting cable could be plugged. It was therefore possible to connect adjacent points (tight re-entry) or in a random fashion. Flory argued that *random* re-entry was normal and regular, tight folding was rare. His argument was based on the melt behaviour where the polymer chains follow closely random coil statistics and

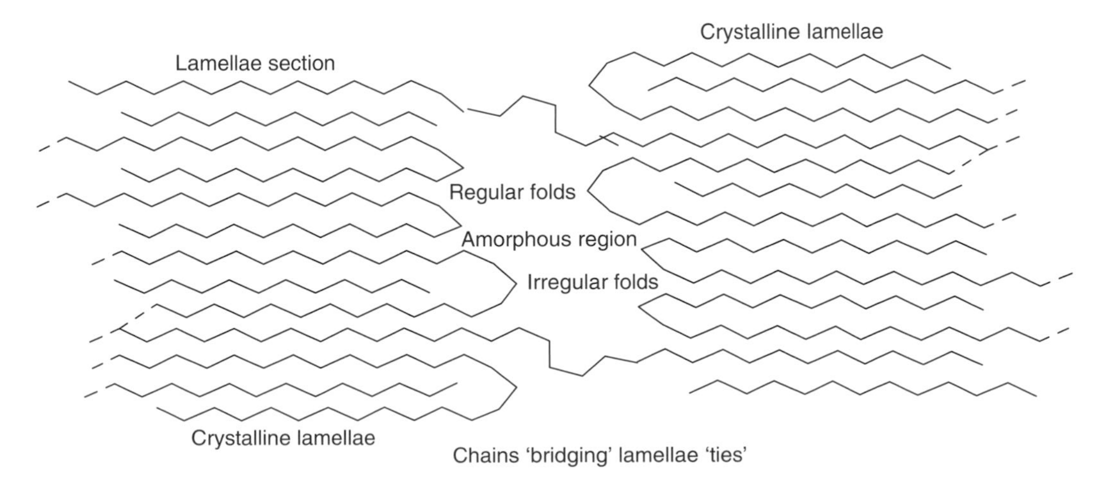

Figure 9.16 Schematic of folded surface showing regular folds, irregular folds and bridging 'tie' molecules.

the RISM model has been every successful. It is only with the advent of small angle neutron scattering (SANS) of blends of deuterated and protonated polymers, e.g. $(\text{-}CD_2\text{-})_n$ and $(\text{-}CH_2\text{-})_n$, that the validity of this model can be tested. The tight loops re-enter the lamellae from which they originate.

For solution-grown single crystals of linear polyethylene, the average radius of gyration ($\langle s \rangle$) of the molecules is proportional to $M^{0.1}$. A much higher molar mass dependence of $\langle s \rangle$ is observed for the polymer coil in solution $\langle s \rangle \propto M^{0.5}$ and indicates that the dimensions of the chains decrease markedly during crystallisation. For the polymer chain to have a smaller value of $\langle s \rangle$ it has to fold like an accordion and the *gauche* sequences have to be closely coupled in order to achieve close packing.

Flory had based his ideas of re-entry on the tried and tested assumption that the coil follows random coil statistics in the melt. It is clear from the SANS data that the chains fold much more tightly than would be predicted by a random re-entry model. Infrared spectroscopy showed that 75% of the folds in solution-grown single crystals of polyethylene led to adjacent re-entry (tight folds) and that single molecules were diluted by 50% along the {110} fold plane. Both observations are consistent with the *superfold* model where the chains behave more like the regular folds than the random re-entry model.

9.10.5 Crystals grown from the melt and lamellae stacks

The *superfolding* model indicates that in dilute solution the polymer molecules add to the growing face of a single crystal without significant competition from other polymer molecules. Crystal growth in concentration solutions and the melt will be very different and chain entanglement will influence the growth process. When polymers are entangled, then several chains may be added to the surface at the same time, the process becomes more chaotic and the folding behaviour may tend more towards the switch board model. In general, low molecular weight polyethylene, because it is not entangled, can readily crystallise, exhibiting a well-defined lamellar morphology with a thickness–width ratio of 0.01–0.001. The lamellae often align parallel to one another to form regular stacks. This behaviour is particularly apparent with metallocene polyethylene, where there are very few branched chains to inhibit chain folding. The lamellae formed with low molecular weight materials can exhibit thickness–width ratios which are smaller than 0.001. Once entanglement can occur then growth of the large lamellae is inhibited and smaller, less regular lamellae are usually observed. Branched and/or high molar mass polymers exhibit lamellae that tend not to be able to form regular stacks and exhibit much higher ratios. The regular tightly folded lamellae can form roof-shaped lamellae (pyramidal structures), whereas intermediate molar mass materials tend to form C-shaped lamellae and high molar mass polymers form S-shaped lamellae (see Section 3.3.1). Branching will introduce defects. Depending on its location and the length of the branches, so different effects on the lamellae growth process will occur. Branched polyethylene mostly shows C- and S-shaped lamellae.

Polymers are always polydisperse with a distribution in molar mass and often contain chain branches, either introduced specifically during the synthesis or as a consequence of synthetic defects. Both these effects will influence the observed morphology. The introduction of low levels of comonomers can lead to behaviour which is rather like that of random side chains. Different molecular species crystallise in different stages indicating the thermodynamic control on the overall process, i.e. they are incorporated into the crystal structure at different temperatures and times. The intermediate and high molar mass components crystallise early in the stacks of thick, dominant crystals. Small pockets of rejected molten low molar mass material remain after crystallisation of the dominant lamellae. The low molar mass species crystallise in separate crystal lamellae, favouring stacks of so-called subsidiary crystal lamellae. The process leading to the separation of high and low molar mass material that accompanies crystallisation is referred to as a *molar mass segregation* or *fractionation*. Some polymers show a preference for segregation of low molar mass species to the spherulitic boundaries.

For melt-grown polymers, the amorphous density is 10–20% lower that the crystal density. Flory would argue that all the chains entering the amorphous phase would take a random walk before re-entering the crystal lamellae. However, a significant fraction of the chains must be folded directly back into the crystal in order to account for the observed low amorphous density.

An expression can be derived for the fraction of tight folds. Usually it is assumed that all of the chains can be divided into two groups: those that form tight folds and those that do not. The latter are assumed to exhibit random re-entry behaviour. In Figure 9.16, the tight regular folds have a density that is essentially that of the lamellae and the irregular folds are much larger, enter further into the amorphous region and exhibit random re-entry behaviour. If the length of chain in between the lamellae is designated as L_a, and there are n bonds in a typical Gaussian amorphous chain sequence, then:

$$n = L_a^2/Cl^2 \tag{9.139}$$

where C is a characteristic constant for a given polymer–temperature combination and l is the monomer bond length. Regular chain folding constitutes a fraction (f_{fold}) of the entries and these will *not* contribute to the fraction of the amorphous chains. The number of chain segments in an average amorphous entry is given by:

$$n = \left(L_a^2/Cl^2\right)\left(1 - f_{fold}\right) \tag{9.140}$$

The number of chain segments (n^0) in a straight chain stem is given by:

$$n^0 = L_a/l \tag{9.141}$$

The ratio of segments in the two phases will be in the ratio of the amorphous (ρ_a) to crystalline density (ρ_c), given by:

$$\left(\rho_a/\rho_c\right) = \left(n/n^0\right) \tag{9.142}$$

The amorphous density is then obtained by a combination of Equations (9.140)–(9.142):

$$\rho_a = \rho_c\left(L_a/Cl\right)\left(1 - f_{fold}\right) \tag{9.143}$$

Careful X-ray diffraction studies have shown that the chains do not match exactly and as a consequence they are tilting by an angle, θ. To allow for this effect, Equation (9.143), can be modified to give:

$$\rho_a = \rho_c\left(L_a/Cl\right)\left(1 - f_{fold}\right)\cos\theta \tag{9.144}$$

For a typical linear polyethylene, the ratio for the densities ρ_a/ρ_c is 0.85 and the amorphous distance is approximately 5 nm and l has a value of 0.127 nm. The value of C = 6.85 and θ = 30°: using these data the value of the f_{fold} is 0.83; i.e. 83% of all chain stems are expected to be tightly folded and 17% are expected to be statistically distributed chains in the amorphous layer. The amorphous phase can have a dominating influence on the physical properties of many semicrystalline polymers, determining the ultimate strength, ductility and diffusion coefficients for small molecules. The strength and ductility of semicrystalline polymers of high molar mass arise from the presence of interlamellar tie chains connecting adjacent crystal lamellae and are critical in determining the resistance to environmental stress cracking (see Section 7.2.6.3).

9.10.6 Location of chain ends

Any discussion of the lamellae structure raises the question: where are the chain ends?

In Figure 9.16 some of the chains ends are shown to be close to the surface of the crystal. It has been estimated that about 90% of the chain ends are located in the amorphous phase, the surface of the lamellae. The concentration of interlamellar *tie* chains in a given sample depends on the molar mass which, in turn, determines the spatial distribution of the chains and long period, i.e. the sum of crystal and amorphous layer thickness. If the polymer chains are shorter than the critical length for entanglement then the chains can act independently and become separately incorporated in the lamellae. For low molar mass chains, whose length is twice as long as the long period, the chains will fold into the same lamellae and there will be fewer bridging *tie* molecules and a brittle crystalline material will develop. For high molecular mass materials, an increasing proportion of the chains will leave one lamella and enter the adjoining lamella. The bridging section will often be entangled and the *tie* molecules will strengthen the material.

9.10.7 Crystallisation kinetics

A core assumption when discussing the crystallisation kinetics of polymers is that the theories must consider the effects of chain folding. The principle consideration for crystallisation is that the interaction between neighbouring chains leads to a lower energy state. The crystallisation process involves nucleation and diffusion of the relevant entity to the surface site. The formation of the critical nucleus will be controlled by thermodynamics. The total change in free energy, ΔG, is the sum of contributions from the bulk and surface energies. Using the convention which we introduced in Figure 9.14, where σ_i is the specific surface energy of surface i and A_i is the area, then the free energy can be described by:

$$\Delta G = \Delta G_B V_{crystal} + \sum_i A_i \sigma_i \tag{9.145}$$

where ΔG_B is the change in the specific free energy in transformation of stems from the solid to the melt and $V_{crystal}$ is the volume of the crystal. For simplicity it will be assumed that the crystal being considered has a spherical form. The free energy on crystallisation, ΔG, is then given by:

$$\Delta G = 4/3\pi r^3 \Delta G_B + 4\pi^2 \sigma \tag{9.146}$$

where r is the radius of the spherical crystal and σ is the average specific free energy of the surface. The radius of the sphere, r^*, associated with the free energy barrier is obtained by setting the derivative of ΔG with respect to r equal to zero:

$$\partial \Delta G / \partial r = 4\pi^{*2} \Delta G_B + 8\pi r^* \sigma = 0 \tag{9.147}$$

$$r^* = 2\sigma / \Delta G_B \tag{9.148}$$

The temperature dependence of this equation lies in ΔG_B:

$$\Delta G_B = \left(\Delta H_B^0 \Delta T\right) / T_m^0 \tag{9.149}$$

where ΔH_B^0 is the heat of fusion per unit volume, T_m^0 is the equilibrium melting point, $\Delta T = T_m^0 - T_c$ is the degree of *supercooling* and T_c is the crystallisation temperature. Equation (9.144) is valid provided that ΔH_B^0 and the entropy of fusion, ΔS_B^0, are temperature independent, which is a good approximation over a limited temperature range near the equilibrium melting temperature. Insertion of Equation (9.149) into Equation (9.148) yields:

$$r^* = -\left(2\sigma T_m^0 / \Delta H_B^0 \Delta T\right) \tag{9.150}$$

Since ΔH_B^0 is negative then the radius of the critical nucleus increases with decreasing degree of *supercooling*. Inserting Equation (9.145) into Equation (9.143) gives an expression for the free energy barrier (ΔG^*):

$$\Delta G^* = \frac{4\pi\left(-2\sigma T_m^0\right)^3}{3\left(\Delta H_B^0 \Delta T\right)^3} \frac{\Delta H_B{}^0 \Delta T}{T_m^0} + \frac{4\pi\left(-2\sigma T_m^0\right)^2 \sigma}{\left(\Delta H_B^0 \Delta T\right)^2} \tag{9.151}$$

which simplifies to:

$$\Delta G^* = \left(16\pi\sigma^3 \left(T_m^0\right)^2\right) / \left(3\left(\Delta H_B^0\right)^2 \Delta T^2\right) \tag{9.152}$$

Equation (9.152) predicts that nucleation occurs more readily at lower crystallisation temperatures because of the lower critical nucleus size and the lower free energy barrier associated with the crystallisation process. As with the case for simple crystal growth, nucleation can take a number of forms:

- Primary nucleation will involve the formation of the first nuclei and involves six new surfaces being formed.
- Secondary nucleation will usually occur on a surface and typically involves four surfaces being formed.
- Tertiary nucleation will typically involve two surfaces being formed and represents the stem attaching to an edge face.

The various forms of nucleation are presented schematically in Figure 9.17.

The free energy barrier is highest for primary nucleation and this seldom occurs in practice, heterogeneous nucleation being the normal mechanism. It has been proposed that the following equation can be used to describe the temperature dependence of both diffusive transport and nucleation. The overall crystallisation rate, w_c, at a general temperature, T_c, is a combination of several factors. The crystallisation process involves the diffusion of the stem to the surface and then its nucleation or attachment:

$$\left(\dot{w}_c\right) = C\exp\left(-\left(U^*/\left(R\left(T_c - T_\infty\right)\right)\right)\right)\exp\left(-\left(K_g/\left(T_c\left(T_m^0 - T_c\right)\right)\right)\right) \tag{9.153}$$

diffusion nucleation

where C is a rate constant, U^* is an energy constant, R is the gas constant, T_∞ is the temperature at which all segmental mobility is frozen in, and K_g is a kinetic constant for the secondary nucleation.

The pre-exponential factor C depends on the regularity and flexibility of the polymer segments and will be a large value for a flexible all-*trans* polymer. $C = 0$ for an atactic polymer and is low for an inflexible polymer such as isotactic polystyrene. The second term in Equation (9.148) describes the temperature dependence of the short-range motion of the stems to the surface and is expressed by the Williams–Landau–Ferry (WLF) equation. The WLF equation describes the slowing down of the segmental motion of the chain backbone at the glass transition temperature and reflects the restriction of the volume available for motion to occur. At $T_c = T_\infty$ this term becomes zero. The third term describes the temperature dependence of the nucleation rate and has a zero value above the equilibrium melting temperature, $T_c = T_m^0$. The bell-shaped curve of the temperature dependence of the rate of crystallisation is shown in Figure 9.18. The form of the crystallisation rate cure is common to many systems and similar expressions have been found for metals, inorganic compounds, sulfur, selenium, antimony, proteins and carbohydrates, graphite silicates and also polymers.

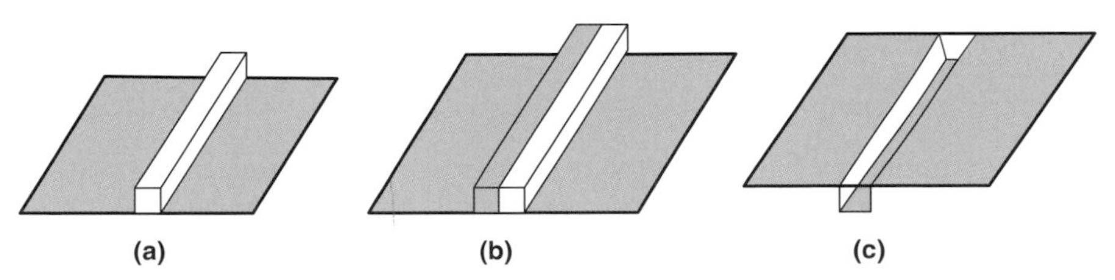

Figure 9.17 Schematic of types of nucleation for stems adding to crystal surface: (a) primary nucleation, $N = 6$; (b) secondary nucleation, $N = 4$: (c) tertiary nucleation, $N = 2$. *Stem* adding to surface is shown in white except in tertiary case where addition is to trough.

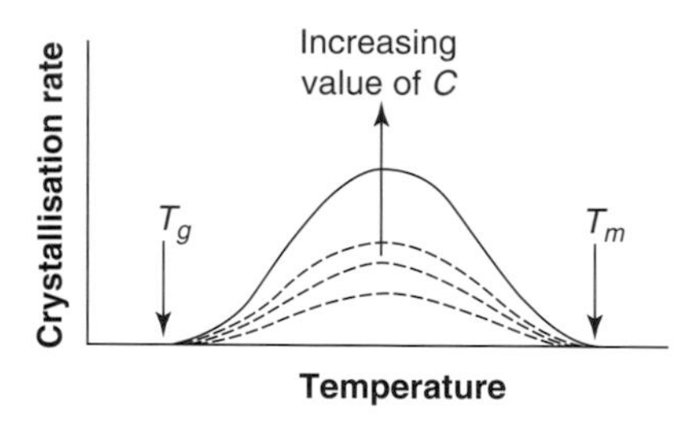

Figure 9.18 Variation of crystallisation rate with temperature. T_g and T_m are, respectively, glass and melt temperatures of material.

9.10.8 Equilibrium melting temperature

Central to most crystallisation is the idea of an equilibrium melting temperature (T_m^0), above which crystallisation does not occur. The rate of crystal growth is therefore related to the extent to which supercooling (ΔT) occurs and is defined by:

$$\Delta T = T_m^0 - T_c \tag{9.154}$$

where T_c is the crystallisation temperature. The equilibrium melting temperature for a polymer system normally refers to the growth of a theoretical crystal of infinite thickness in which the chains are fully extended. Homopolymers of intermediate or high molar mass can grow crystals of practically infinite thickness. The fully extended chain length of a polyethylene of $M = 100{,}000\,\mathrm{g\,mol^{-1}}$ is $100{,}000/14 \times 0.127\,\mathrm{nm}$ (~900 nm) and is several orders of magnitude larger than the thickness of the lamellae. The usual dimensions are ~10 nm and for these dimensions to be achieved chain folding must occur. The melting point depression arising from the finite crystal thickness (900 nm) is predicted by the Thompson–Gibbs equation. However, low molar mass homopolymers can obviously only grow crystals of a limited thickness, equal to their fully extended length. The melting point of these materials corresponds to the predictions for the fully extended chain. Copolymers are more complicated, the change in the chemical sequence structure representing a disruptive element inhibiting chain packing. Depending on the nature of the comonomer, the 'foreign' moieties with a 'statistical' placement in the polymer chains may not be able to crystallise or they may be included to some extent in the crystals. As a consequence the equilibrium thickness will reflect the disruptive effects of these moieties.

The equilibrium melting temperature can be determined in a number of different ways. First, the melting point, T_m, of samples with a well-defined crystal thickness, L_c, can be measured and the data extrapolated to $L_c^{-1} = 0$ using the Thompson–Gibbs equation:

$$T_m = T_m^0\left[1-\left(2\sigma/\left(\Delta H_B^0 \rho_c L_c\right)\right)\right] \tag{9.155}$$

where ρ_c is the crystal density, ΔH_B^0 is the heat of fusion per unit mass and σ is the specific fold surface energy. Certain polymers, such as polyethylene, can be crystallised at elevated pressures to form extended chain crystals and these micrometre-thick crystals have melting points which are close to the theoretical value of T_m^0.

Secondly, the value of T_m^0 can be derived from a study of low molar mass analogues and then extrapolated to the infinitely thick crystal value. Using data for the enthalpy (ΔH_B) and entropy (ΔS_B) of fusion for oligomers, T_m^0 is obtained by extrapolation to infinite molar mass and for linear polyethylene has the form:

$$T_m = 414.3\left[(x-1.5)/(x+5.0)\right] \tag{9.156}$$

Table 9.4 Equilibrium, melting point and selected thermodynamic data for selected polymers

Polymer	T_m^0 (K)	ΔS_B^0 (J K mol^{-1})	ΔH_B^0 (kJ mol^{-1})
Polyethylene	414	9.6	4.01
Polytetrafluoroethylene	600	5.7	3.42
Isotactic polypropylene	463	7.5	2.31
Polyoxymethylene	457	10.7	4.98
Polyethyleneoxide	342	8.4	2.89
Nylon 6,6	553	10.2	4.85
Polyisoprene 1,4-*cis*	301	14.5	4.39
Polyisoprene 1,4-*trans*	347	36.6	12.70
Polychloroprene 1,4-*trans*	353	23.7	8.36
Isotactic polystyrene	516	16.3	8.38
Polydecamethylene adipate	352.5	121.2	42.64
Polydecamethylene sebacate	353	142.1	50.16
Polyteramethylene terephthalate	411	63.1	31.76
Cellulose tributyrate	480	33.8	12.54

where x is the degree of polymerisation of the polymer. For high molecular mass materials this value melting point has a limiting value of 414.3K. Similar relationships have been observed with other polymer systems (see Section 3.7).

Equilibrium melting points of a few selected polymers are presented in Table 9.4. It is clear from these examples that there is no correlation between the melting temperature and the enthalpy of fusion. Clearly, the values of ΔH_B do not dictate the values of the melt temperature. The balance between the enthalpy and entropy is the controlling factor. Nylon 6,6 has a high enthalpy of fusion due to the strong hydrogen bonds between the amide groups. In contrast, the higher melting point of polytetrafluororethylene (PTFE) is due to its low entropy of fusion. At high temperature, PTFE crystals show considerable segmental mobility and a relatively small increase in entropy on melting. PTFE exhibits a lattice change in the solid state just below its melting point. This lattice change is consistent with a high mobility in PTFE below its melting point. The high melting point of polyoxymethylene is due to the high enthalpy of fusion arising from intermolecular interactions involving the ether groups. The addition of the extra methylene group in polyethyleneoxide leads to a lowering of the effects of the ether groups and a consequent reduction in the enthalpy of fusion.

The introduction of a ring structure into a linear chain substantially increases the melting temperature relative to the aliphatic chain as would be expected from the decreased conformational entropy of the melt. Striking examples of this phenomenon are found in comparison of the melting temperature of aliphatic and aromatic polyester and polyamides. Cellulose derivatives usually have a highly extended structure with low entropy and consequently have high melting points. The chain structure influences the melting temperature through its conformational properties.

9.10.9 General Avrami equation

A generalised approach to the description of the crystallisation process was proposed by Avrami. Without prior knowledge of the molecular mechanism involved in the crystallisation process, the Avrami equation gives a convenient means of empirically describing the process. The model assumes that crystallisation starts randomly throughout the sample, nucleation often being a heterogeneous process. The theory attempts to accommodate the effects of the growth rate on the shapes of the crystals formed. It assumes that the crystals that are seeded will grow smoothly in all three dimensions. All nuclei are formed and start to grow at time $t = 0$ and with a rate that is the same in all directions (spherical growth) and is equal to $\dot{\omega}$. The theory considers the growth front, $E(t)$, emanating from the central nucleus:

$$E(t) = 4/3\pi(\dot{\omega}t)^3 q \tag{9.157}$$

where q is the volume concentration of nuclei. Statistical analysis of the crystal growth problem considers the process in terms of an expanding circular wavefront which mimics spherulite growth. The number of waves which pass a particular point at time t can be described in terms of a Poisson distribution which has the form:

$$p(c) = (\exp(-E)E^c)/c! \tag{9.158}$$

where E is the average value of the number of waves passing a point, P, after some time, t, and the number of waves are constrained to be an exact number, c. The probability that no fronts pass through the point, P, is given by:

$$p(0) = \exp(-E) \tag{9.159}$$

In the context of the crystallisation process, $p(0)$ is equivalent to the volume fraction $(1 - v_c)$ of the polymer which is still in the molten state:

$$p(0) = 1 - v_c \tag{9.160}$$

where v_c is the volume fraction of crystalline material. Combining Equations (9.160) and (9.157) yields:

$$1 - v_c = \exp\left(-4/3\pi\dot{\omega}^3 q t^3\right) \tag{9.161}$$

This equation indicates that in the case of spherical growth, the growth rate depends on the cube of the time. If the growth rate is assumed to be constant and linear in space and time then the number of waves, dE, which pass the arbitrary point, P, for nuclei within the spherical shell confined amid the radii r and $r + dr$ is given by:

$$dE = 4\pi r^2\left(t - (r/\dot{\omega})\right)I^* dr \tag{9.162}$$

where I^* is the density, the number of nuclei per cubic metre per second. The total number of waves passing (E) is obtained by integration of dE between 0 and $\dot{\omega}t$:

$$E = \int_0^{r_t} 4\pi r^2 I^*\left(t - (r/\dot{\omega})\right)dr = \left(\pi I^*\dot{\omega}^3\right)t^4 \tag{9.163}$$

where r_t is the confined radius of the sphere and after combining with Equations (9.157) and (9.158) gives:

$$1 - v_c = \exp\left(-\left(\left(\pi I^*\dot{\omega}^3\right)/3\right)t^4\right) \tag{9.164}$$

It should be noted that the time is raised to the fourth power. In general it is found that crystallisation based on different nucleation and growth mechanisms can be described by the same general formula, the general Avarmi equation:

$$1 - v_c = \exp\left(-Kt^n\right) \tag{9.165}$$

where K and n are constants typical of the nucleation and growth mechanisms. The growth geometry describes the characteristics of the dominant growth process.

Equation (9.165) can be expanded according to $\exp(-Kt^n) \approx 1 - Kt^n + \cdots$ and for the early stages of crystallisation where there is little restriction of crystallisation due to impingement:

$$v_c = Kt^n \tag{9.166}$$

Table 9.5 Avrami exponent, n, for different nucleation and growth mechanisms

Growth geometry	Athermal	Thermal*	Thermal**
Linear growth	1	2	1
Two-dimensional (circular)	2	3	2
Three-dimensional			
Spherical	3	4	5/2
Fibrillar	≤1	≤2	
Circular lamellar	≤2	≤3	
Solid sheaf	≤5	≤6	

Thermal* indicates without diffusion control and
Thermal** indicates diffusion controlled.

The Avrami exponent (n) increases with increasing 'dimensionality' of the crystal growth (see Table 9.5) Diffusion controlled growth reduces the value of the exponent by a factor of 1/2 compared with the corresponding 'free' growth case. There are certain limitations and special considerations for polymers with regard to the Avrami analysis:

- The solidified polymer is always only semicrystalline because as discussed previously the crystals are never 100% perfect or completely volume filling. This effect can be taken into account by a modification of Equation (9.161) to:

$$1-\left(v_c/v_\infty\right)=\exp\left(-Kt^n\right) \tag{9.167}$$

  where v_∞ is the volume crystallinity finally reached.
- The volume of the system studied changes during crystallisation as a consequence of the difference in density between the melt and solid:

$$1-v_c=\exp\left(-K\left[1-v_c\left(\left(\rho_c-\rho_l\right)/\rho_l\right)\right]t^n\right) \tag{9.168}$$

  where ρ_c is the density of the crystal phase and ρ_l is the density of the melt.
- The nucleation is seldom either athermal or simple thermal. A mixture of the two is common and reflects thermal diffusion in the melt as crystallisation occurs.
- Crystallisation always follows two stages: first, there is the primary crystallisation, characterised by radial growth of spherulites or axialites. Secondary crystallisation, i.e. the slow crystallisation behind the crystal front caused by crystal thickening, then follows. The formation of subsidiary crystal lamellae from secondary crystallisation is associated with the fractionated low molar mass material and results in crystal imperfections.

The constants in the Avrami equation are obtained by taking the double logarithm of Equation (9.167):

$$\ln\left[-\ln\left(1-\left(v_c/v_\infty\right)\right)\right]=\ln K+n\ln t \tag{9.169}$$

Differential scanning calorimetry can be used to study the crystal growth kinetics (see Section 3.7). The nature of the experiment leads to data which are defined in terms of mass rather than volume. In order to convert the data the following expression is used to relate the mass crystallinity, w_c, to a volume fraction, v_c:

$$v_c=\left(\left(w_c/\rho_c\right)/\left(\left(w_c/\rho_c\right)+\left(\left(1-w_c/\rho_a\right)\right)\right)\right)=\left(w_c\right)/\left(w_c+\left(\rho_c/\rho_a\right)\left(1-w_c\right)\right) \tag{9.170}$$

where ρ_a is the amorphous density. Typically, the theory is found to fit well with the initial crystallisation data but deviations are observed once the growth effects of neighbouring entities start to impinge. The crystal impingement effects are observed as a decrease in the rate of crystallisation.

9.10.10 Comparison of experiment with theory

Studies of low molar mass polyethylene, $\overline{M_c} \leq 10{,}000$, show an exponent of 4 consistent with the theory for spherulitic growth. The low molar mass samples display sheaf-like (axialitic) morphology and as expected a high value of the exponent is observed. Polymers with intermediate molar mass $10{,}000 < M_w < 1{,}200{,}000$ have an Avrami exponent near 3 and for higher molar mass samples $M_w \geq 3{,}000{,}000$ the exponent is further reduced to a value of ~2. These observations are consistent with growth occurring predominately in one direction. The crystallisation of high molar mass polymers is strongly influenced by chain entanglements and the slow and incomplete crystallisation leads to small, uncorrelated crystals, i.e. to so-called random lamellar structures. A low value of n is expected for a 'low dimensional' platelet or fibrillar-like growth. It is common to observe that the Avrami exponent decreases with increasing molar mass, reflecting the influence of the differences in morphology and crystal growth mechanisms on the observed behaviour.

9.11 Determination of molar mass and size

One of the important differences between polymers that are produced via a vinyl synthetic route and a condensation or step growth route is the molar weight distribution of the polymer molecules produced (see Section 1.10). The molar mass will be reflected in the size of the polymer molecule. Molar mass can be determined by either absolute or relative methods:

$$M_n = \sum_{ii} M_i N_i \Big/ \sum_i N_{ii}$$

where N_i is the number of polymer chains with molar mass M_i.

$$M_w = \sum_i M_i W_i \Big/ \sum_i W_i$$

where W_i is the weight of polymer with molar mass M_i. However, if $W_i = M_i N_i$, then:

$$M_w = \sum_i M_i^2 N_i \Big/ \sum_i N_i M_i$$

The molar mass distribution is defined as M_w/M_n and is sensitive to the method of synthesis. An ionic initiation vinyl polymerisation can give values of M_w/M_n in the range 1.05–1.2. A typical radical polymerisation process would produce polymers with values of M_w/M_n equal to 1.3–1.5. Condensation or step growth polymerisation methods will generally produce broad distributions of molar mass and values of M_w/M_n in the range 2.0–24.0 are not unusual.

9.11.1 Absolute methods of determination

These involve the counting of either the number of polymer chains with a particular length or the weight of those chains.

9.11.2 Number average molar mass

The number average molar mass is usually determined by counting the number of chain ends, using either spectroscopic or titration methods. The primary method for the determination of the number average is by osmotic pressure measurements. Osmotic pressure is a colligative property and is proportional to the number of chains present in the polymer solution. The osmotic pressure is the pressure difference created when a polymer solution is placed in contact with its solvent, but separated by a semipermeable membrane. The semipermeable membrane stops the polymers diffusing into the solvent but allows the solvent to attempt to dilute the solution. The dependence of the osmotic pressure on the concentration of the polymer in solution is described by the equation:

$$(\pi / c) = (RT / M) + RTA_2c^2 + \cdots \qquad (9.171)$$

As c_2 is reduced to zero, then:

$$limit_{c\to 0}(\pi / c) = (RT/M) \tag{9.172}$$

For a polydispersed polymer in solution then:

$$c = \sum_i c_i \quad \pi = \sum_i \pi_i \quad \pi_i = RTc_i / M_i \tag{9.173}$$

Using the above equation one obtains:

$$limit_{c\to 0}(\pi / c) = \left(RT\sum_i (c_i / M_i)\right) \Big/ \sum_i c_i \tag{9.174}$$

but $c_i = N_iM_i$, so:

$$\sum_i c_i/M_i \Big/ \sum_i c_i \quad \sum_i N_i \Big/ \sum_i N_iM_i = 1/\overline{M_n} \quad \text{thus} \quad (\pi / c) = (RT)M_n \tag{9.175}$$

Osmotic pressure yields the *number average* molar mass. The slope from the plot is indicative of the quality of the solvent polymer interaction: $A_2 > 0$ is indicative of a good solvent. An ideal solvent will have $A_2 = 0$ and a poor solvent will have $A_2 < 0$.

9.11.3 Scattering methods for molar mass determination

The intensity of radiation scattered by a polymer molecule in solution can be used to determine the molar mass of the polymer and also the mean square radius of gyration, $\langle s^2 \rangle$. Studies can be carried out using various forms of radiation: X-rays, light and neutrons. For simplicity we will only consider light.

9.11.4 Light scattering by small particles (size compared to the wavelength of light)

The scattering from a liquid or solution is described by the Rayleigh ratio, R_θ, where θ is the angle of observation of the scattered intensity, i_θ, at a distance r from the specimen upon which a beam of light of intensity I_0 is incident.

$$R_\theta = \left(i_\theta r^2 / I_0\right) \tag{9.176}$$

For a polymer solution of concentration c (mass/unit volume) the quantity of interest is the difference between the Rayleigh ratio of the solution and that of the solvent, ΔR_θ. For small molecules, i.e. those with dimensions $<\lambda/20$, where λ is the wavelength of light *in vacuo*, then the scattering angle θ used is generally 90° and vertically polarised light with scattered light detection in the horizontal plane then:

$$\Delta R_{90} = \frac{4\pi^2 \bar{n}^2 (d\bar{n}/dc)^2 c}{\lambda^4\left[-(1/\bar{V}_1 k_B T)(d\mu/dc)_{T,P}\right]} \tag{9.177}$$

where $\bar{n}$ is the refractive index of the solvent, $(d\bar{n} / dc)$ is the specific refractive index increment of the polymer in the solvent and depends on temperature, k_B is the Boltzmann constant and $\bar{V}_1$ is the partial molar volume of the solvent. Using the Flory–Huggins theory for the thermodynamics of polymer solutions:

$$\mu_1 - \mu_1^0 = RT \ln a_1 = -\pi\overline{V_1} = -\left((RTc / M) + RTAc^2 + \cdots\right)\overline{V_1} \tag{9.178}$$

$$(\partial\mu_1 / \partial c)_{T,P} = -\bar{V}_1\left((RT / M) + 2RTAc + \cdots\right)$$
$$-\left(1/(\bar{V}_1 k_B T)\right)(\partial\mu_1 / \partial c)_{T,P} = N_A\left((1/M) + 2Ac + \cdots\right) \tag{9.179}$$

hence

$$\Delta R_\theta = \left(4\pi^2\bar{n}^2 \left(d\bar{n} / dc\right)^2 c\right) \Big/ \lambda^4 N_A \left(\left(1 / M\right) + 2Ac\right) \tag{9.180}$$

Collecting the constants into one term, K^*, sometimes known as the optical constant, we obtain:

$$K^* = \left(4\pi^2\bar{n}^2 \left(d\bar{n} / dc\right)^2 c\right) \Big/ \lambda^4 N_A \tag{9.181}$$

Rearranging the above equations:

$$\left(K^* c / \Delta R_{90}\right) = \left(1 / M\right) + 2A / c + \cdots \tag{9.182}$$

If the left-hand side is plotted against concentration then the slope is $2A$ and the intercept is:

$$limit_{c\to 0}\left(K^* c / \Delta R_{90}\right) = 1/M \tag{9.183}$$

For a polydispersed polymer dissolved in the solvent:

$$limit_{c\to 0}\Delta R_{90} = K^* \sum_i c_i M_i \quad \text{with } c = \sum_i c_i \tag{9.184}$$

$$limit_{c\to 0}\left(K^* c / \Delta R_{90}\right) = \sum_i c_1 \Big/ \sum_i c_i M_i = \sum_i N_i M_i \Big/ \sum_i N_i M_i^2 = 1 / \overline{M_w} \tag{9.185}$$

The light scattering experiment gives the weight average molar mass.

9.11.5 Light scattering from molecules larger than $\lambda/20$

In large molecules, light is scattered from different points on the same molecule leading to a phase difference between the scattering from those points. This leads to constructive and destructive interference and will have an effect on the intensity measured at different angles. The equations required to describe this situation are more complex than those presented above but can be simplified to the following:

$$\left(K^* c / \Delta R_{90}\right) = \left(1/\bar{M}_w P(\theta)\right) + 2Ac + \cdots \tag{9.186}$$

where $P(\theta)$ is a function which contains information on the shape of the polymer molecule producing the scattering. For a random coil (i.e. a spherical polymer molecule) Debye calculated that the function should have the form:

$$P(\theta) = 2 / u^2 \left(\exp(-u) - 1 + u\right) \tag{9.187}$$

where

$$u = \left(4\pi\bar{n}\sin(\theta / 2) / \lambda\right)^2 \left\langle s^2 \right\rangle \tag{9.188}$$

When $u \leq 1$, the exponential term in Equation (9.182) can be expanded and the whole expression simplified to:

$$P(\theta) = 1 - u / 3 \tag{9.189}$$

Inverting Equation (9.184) we obtain:

$$P(\theta)^{-1} = 1 + u / 3$$

$$K^* c / \Delta R_\theta = \left(1 / \bar{M}_w\right)\left(1 + u / 3\right) + 2Ac \tag{9.190}$$

This equation depends on the scattering angle and the concentration of the polymer molecules in solution. Zimm proposed that the data could be plotted as the right-hand side $K^* c/\Delta R_\theta$ against a function $\sin^2(\theta / 2) + kc$, where K is a constant that contains the various constants such as: the wavelength of light, the refractive index of the solvent, and the partial volume (see Figure 9.19).

The intercept where both the angle and the concentration tend to zero is the inverse of the molar mass weight average ($1/M_w$). The slope of the line for $\theta = 0$ is equal to $2A/k$ and hence

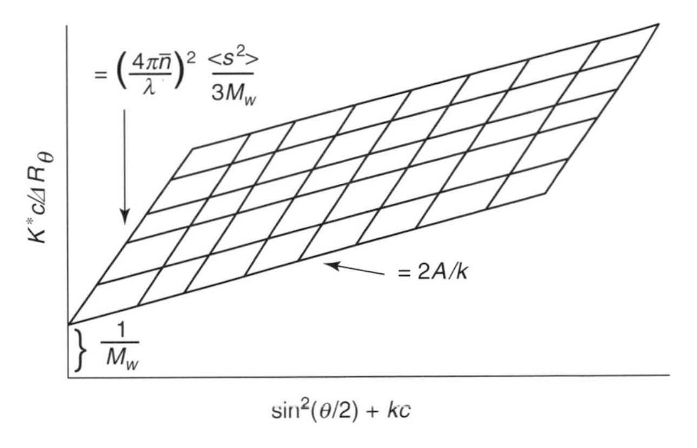

Figure 9.19 Zimm plot for light scattering data.

reflects the nature of the interaction of the polymer with the solvent. The slope of the line for the concentration equal to zero is:

$$\text{slope as } c \to 0 \quad = \left(4\pi\bar{n} / \lambda\right)^2 \left(\left\langle s^2 \right\rangle / 3\bar{M}_w\right) \tag{9.191}$$

The intercept is equal to $1 / \bar{M}_w$, hence:

$$\left\langle s^2 \right\rangle = 3\lambda^2 / \left(16\pi^2\bar{n}^2\right) slope\left(c = 0\right) / intercept \tag{9.192}$$

The experimental determination of the size of the polymer molecule in solution provides information against which the ideas of the models presented above can be checked.

9.11.6 Viscosity measurements

The first of the relative methods of measurement which we will consider is the viscosity method. The viscosity of a polymer solution can be expressed in terms of a series expansion in the concentration:

$$\eta_{ps} - \eta_s\left(1 + ac + bc^2 + cc^3\right) \tag{9.193}$$

Rearranging the above equation:

$$\left(\left(\eta_{ps} - \eta_s\right) / \eta_s c\right) = a + bc + \cdots \tag{9.194}$$

Einstein proposed that the viscosity of a spherical particle should be $\Delta\eta = 2.5nV_m$. A plot of $(\eta_{ps} - \eta_s) / \eta_s c$ against c gives the intrinsic viscosity $[\eta]$ as the intercept (see Figure 9.20). The volume occupied by the polymer coil is defined by:

$$v_0 = 4 / 3\pi\left\langle \bar{r} \right\rangle^3 \tag{9.195}$$

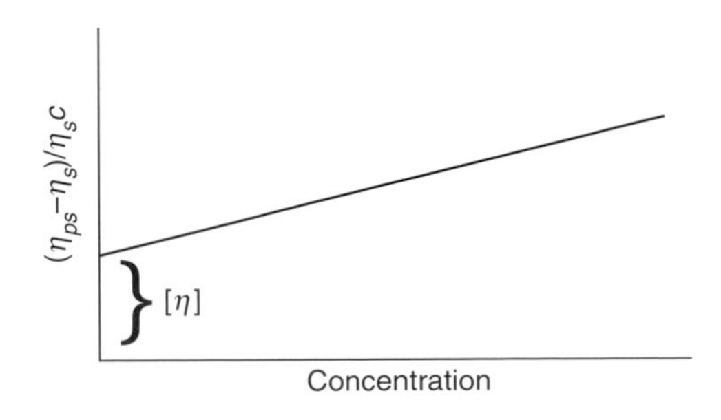

Figure 9.20 Plot of variation of $\left(\eta_{ps} - \eta_s\right) / \eta_s c$ with concentration.

Therefore:

$$[\eta] = 2.5\frac{4}{3}\pi\langle\bar{r}\rangle^3\frac{c}{M} \propto K\left(\left\langle\sqrt{M}\right\rangle^3\right)\Big/M \propto KM^{1/2} \tag{9.196}$$

The relation $[\eta] = KM^{\alpha}$ is the so-called Mark–Houwink relationship. Values for the constants K and α are tabulated for most common polymers in Brandrup *et al.* (1999).

The constant α has a value of 0.5 for an ideal polymer chain and approaches a value of 1 for a rigid rod. Values for K include various scaling factors and reflect the extent to which the polymer in dilute solution approximates to a random coil structure.

9.11.7 Relative methods of molar mass determination: gel permeation chromatography and size exclusion chromatography

Gel permeation chromatography (GPC) is also known as size exclusion chromatography (SEC). The method involves fractionation of the polymer molecules according to molar mass or more strictly speaking their size (hydrodynamic volume) in solution. The column used is packed with porous beads of polystyrene or silica and the solution is added to the top of the column and eluted (see Figure 9.21). The column is packed with porous beads which will retain small molecules but allow larger molecules to pass through the column quickly because they are not able to enter the porous breads. The ability of the polymer to enter the porous bead is determined by the effective radius in solution of the polymers, which is defined by the *hydrodynamic radius*.

The *hydrodynamic radius* is defined by the RISM theory and can be calculated from the mean radius of gyration of the polymer. It is usual to detect the amount of polymer leaving the column by observing the refractive index of the media. The change in the refractive index is proportional to the concentration of the polymer in solution and hence can be used to create elution profiles for the polymer sample. The smaller molecules will be able to permeate more extensively into the porous structure of the bead and will be retained on the column. In order to determine the molar mass of a polymer sample it is necessary to calibrate the column that is being used. This process involves determining the elution times for a number of narrow molar mass distribution

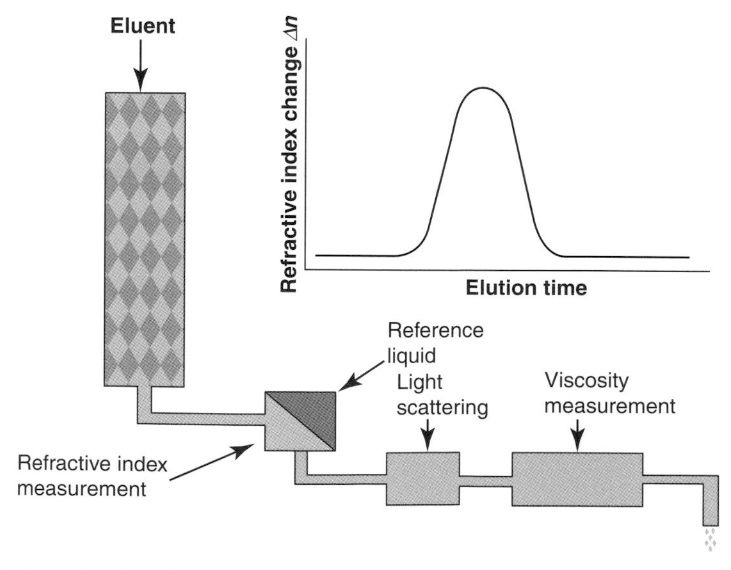

Figure 9.21 Schematic of a GPC/SEC apparatus and illustration of refractive index plotted against elution time output.

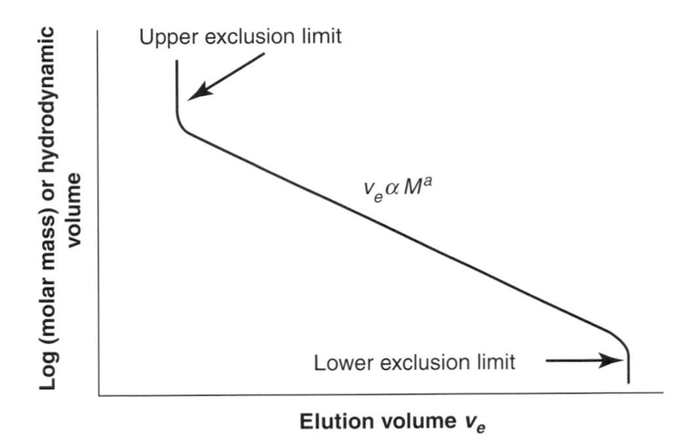

Figure 9.22 Typical molar mass–elution plot for a polymer system.

calibration samples. Using these data it is possible to construct a molar mass–elution time plot which will typically have the form shown in Figure 9.22.

Larger molecules will have restricted permeation and will be washed through the column more quickly than smaller ones. Molecules larger than some critical value, *the upper exclusion limit*, will not enter the beads and will be washed through the column at a rate which reflects the movement of the solvent front through the column. There is also a lower exclusion limit. It reflects a point beyond which impurities in the solvent used to elute the column may influence the data. The elution volume can be described by an equation of the form:

$$V_e = A - B\log[M] \tag{9.197}$$

The elution curves can be calibrated for known samples of polymers. The intensity of the detected signal is directly proportional to the number of polymer molecules and hence the output can be converted into a profile of the number of polymer molecules against molar mass. Instead of using the molar mass it is also possible to use the *hydrodynamic volume* which can be calculated from the Mark–Houwink parameters obtained from viscosity measurements.

Triple detector instruments have appeared on the market recently and incorporate refractive index detection, viscosity measurements and light scattering. The light scattering allows the absolute determination of the molar mass, the viscosity the Mark–Houwink coefficients, and the refractive index the overall distribution of molar mass. The determination of the molar mass of a polymer is very important in understanding the effects of molar mass on physical properties.

Worked examples

Question 1

The concentration dependence of the osmotic pressure of solutions of a macromolecule at 20°C was found to be:

c/g dm^{-3}	1.21	2.72	5.08	6.60
π/Pa	134	321	655	898

Find the molar mass of the macromolecule and the osmotic virial coefficient.

Answer

$$\pi / c = \left(RT / M_n\right)\left(1 + A\left(c / \bar{M}_n\right) + \cdots\right)$$

Therefore to determine $\bar{M}_n$ and A we need to plot π / c against c. The data can be tabulated as follows:

$c/(\mathrm{g\ L^{-1}})$	1.21	2.72	5.08	6.60
$(\pi/c)(\mathrm{Pa/g\ L^{-1}})$	111	118	129	136

The graph will give an intercept of 105.4 and a slope of 4.64. It follows that: $RT/M_n = 105.4\ \mathrm{Pa\,g^{-1}}$ and hence the value of M_n is:

$$8.3141\ \mathrm{J\ K^{-1}\ mol^{-1}}\ 293\mathrm{K} / 105.4\ \mathrm{Pa\,kg^{-1}\ m^3} = 23.1\ \mathrm{kg\,mol^{-1}}.$$

The slope of the graph is equal to $RTA / \bar{M}_n^2$ so $RTa / \bar{M}_n^2 = 4.64\ \mathrm{Pa\,g^{-2}\,L^2} = 4.64\ \mathrm{Pa\,kg^{-2}\,m^6}$, therefore $A = (23.1\ \mathrm{kg\,mol^{-1}})^2 \times 4.64\ \mathrm{Pa\,kg^{-2} m^6} / (8.314\ \mathrm{J\ K^{-1}\ mol^{-1}} \times 293\mathrm{K}) = 1.02\ \mathrm{m^3\ mol^{-1}}$.

Question 2

Polyethyleneoxide was studied in xylene at 90°C using osmotic pressure measurements. The following data were obtained:

$c/\mathrm{kg\ m^{-3}}$	2×10^3	4×10^3	6×10^3	8×10^3
h/m	2.58×10^{-2}	5.965×10^{-2}	9.66×10^{-2}	13.935×10^{-2}

The density of xylene is $801.4\ \mathrm{kg\,m^{-3}}$. Calculate the number average molar mass and the second virial coefficient.

Answer

In this case the pressure is given in the height of the solvent. The osmotic pressure is then calculated using $\pi = \rho hg$, where g is the acceleration due to gravity and h is the height of the solvent column.

$c/\mathrm{kg\ m^{-3}}$	2×10^3	4×10^3	6×10^3	8×10^3
$\pi/c\ \mathrm{m\ kg\ m^{-3}}$	1.292×10^{-2}	5.965×10^{-2}	9.66×10^{-2}	13.935×10^{-2}

Intercept = 1.1665×10^{-5}; slope = 7.345×10^{-10};

$$M_n = 8.314 \times 363 / 1.1665 \times 10^{-5} \times 801.4 \times 9.81 = 3.29 \times 10^4$$

$$A = 7.345 \times 10^{-10} \times 801.4 \times 9.81 / 8.314 \times 363 = 1.91 \times 10^{-9}\mathrm{m^3 mol\,kg^{-2}}$$

Question 3

The data below were obtained by light scattering from a polymer in dilute solution; the wavelength of light used was 441.0 nm, the refractive index of the solvent was 1.340 and $dn/dc = 0.1912\ \mathrm{mL\,g^{-1}}$. Draw a Zimm plot and calculate the weight average molar mass, the second virial coefficient and the value of $\langle s^2 \rangle$.

	$\Delta R_\theta/10^{-4}$			
$\theta/c/\mathrm{g\ mL^{-1}}$	0.315×10^{-3}	0.69×10^{-3}	1.115×10^{-3}	1.62×10^{-3}
30	0.88	1.32	2.72	3.66
45	0.69	1.44	2.18	2.95
60	0.52	1.09	1.66	2.27
75	0.40	0.84	1.29	1.76
90	0.34	0.72	1.11	1.52
105	0.33	0.70	1.09	1.52
120	0.36	0.76	1.18	1.65
135	0.40	0.81	1..35	1.88

$$K^* = \frac{4\pi^2 n_o^2 (dn/dc)^2}{\lambda^4} = \frac{2.613}{3.78\times10^{-18}} = \frac{6.908,10^{17}}{N_A} = 1.14\times10^{-6}\,\text{cm}^{-1}\,\text{gm}^{-3}$$

	$C_2/\Delta R_\theta$			
$\theta/c/10^{-3}$ g mL^{-1}	7.94	6.34	4.76	3.17
30 [0.067]	328.5	320.9	311.1	299.85
50 [0.179]	342.1	328.2	316.0	307.5
70 [0.329]	352.5	339.0	326.2	314.9
90 [0.5]	363.4	349.2	336	324.9
110 [0.67]	372.8	359.5	348.4	335.8
130 [0.92]	383.6	368.9	357.2	339.8

[...] are values of $\sin^2(\theta/2)$ at which the values of θ/c have been calculated.

Extrapolate the data to values of $\theta = 0$ and $c = 0$.

$c/10^{-3}$ g mL^{-1}	0.1962	0.3924	0.589	0.785	0.98
$c/\Delta R_\theta$	399.9	503.8	573.9	711.9	788.7
θ	30	50	70	90	110
$c/\Delta R_\theta$	378	493.9	656.9	834.6	1009

Intercept is 274.

$\bar{M}_w = 1/274\times1.1\times10^{-7} = 33\times10^4$; $A = 10/2\times1.1\times10^{-7}\times(343.5-274)/0.1 = 3.8\times10^{-4}\times$ $\langle s^2\rangle = (3\lambda^2/(16\pi^2 n_0^2))(slope/intercept) = 5.5\times10^{-12}\,\text{cm}^2 = 5.5\times10^4\,\text{Å}^2$

The size of a polymer is approximately 250 Å across.

The viscosities of solutions of polyisobutylene in benzene were measured at 24°C (the θ temperature for the system) with the following results.

c (g/100 cm^3)	0	0.2	0.4	0.6	0.8	1.0
$\eta/10^{-3}$ kg m^{-1}s^{-1}	0.647	0.690	0.733	0.777	0.821	0.865

Given that the Mark–Houwink coefficients are 8.3×10^{-4} (K 100 cm^3 g^{-1}) and $\alpha = 0.50$, calculate the molar mass of the polymer. Comment on the concentration dependence of the viscosity.

Draw up a table.

c (g /dm^{-3})	0	0.2	0.4	0.6	0.8	1.0
$(\eta-\eta_s)/\eta_s$		0.332	0.332	0.334	0.336	0.336

The intercept (intrinsic viscosity) is 0.332 dm^{-3}g^{-1}; $[\eta] = KM^\alpha$. Therefore:

$$M = \{[\eta]/K\}^{1/\alpha} = (0.332/8.3\times10^{-4}\}^{1/0.5} = 160{,}000$$

The slope of the curve is zero and this is consistent with the θ condition.

Brief summary of chapter

The models discussed in this chapter contain different constraints: some include the detail of the conformational distribution, some do not. If the polymer is ideally flexible then it is possible to describe much of its dynamic behaviour using a simple string-type analogy. However, such models do not account for some of the particular facets of polymers, such as the negative coefficient of expansion.

Additional reading

Brandrup J, Immergut E.H. and Grulke E.A. (Eds.) *Polymer Handbook*, 4th edn., Wiley, Hoboken, NJ, USA, 1999.

Flory P.J. *Statistical Mechanics of Chain Molecules*, Wiley Interscience, New York, NY, USA, 1969.

Gedde U.W. *Polymer Physics*, Chapman and Hall, London, UK, 1995.

Mattice W.L. and Suter U.W. *Conformational Theory of Large Molecules*, Wiley, New York, NY, USA, 1994.

Pethrick R.A. *Polymer Structure Characterization: From Nano to Macro Organisation*, Royal Society of Chemistry Publishing, Cambridge, UK, 2007.

Rouse P.E. A theory of the linear viscoelastic properties of dilute solutions of coiling polymers. *Journal of Chemical Physics*, 1953, **21**, 1272–1280.

Rubinstein M. and Colby R.H. *Polymer Physics*, Oxford University Press, Oxford, UK, 2003.

10

Polymers for the electronics industry

10.1 Introduction

Polymers are used extensively in the electronics industry for a variety of applications. They are used to create the cases for mobile phones, telephones, laptop computers and personal computers. They are also critical for the creation of electronic circuits, the supporting structures for components, and are integral elements in the development of display technology. Many computers are found in cars and white goods (fridges, washing machines, dishwashers and vacuum cleaners). Polymers play an important role in determining the durability of electronic components when used in harsh environments. They can be used in a variety of different ways:

- *Structural materials*: For various reasons household, commercial and industrial wiring may need to be contained in conduits. Traditionally, such structures were made from metal pipe but are now more commonly formed from PVC pipes and moulded PVC junction boxes. Zinc die cast boxes have been replaced with mouldings created from sheet moulding compound (SMC). SMC is created from polyester which is reinforced with glass fibre and other fillers. If the material is to be white, the fillers will be silica or talc. However, if electrical conductivity is required, as is the case when the box is used to shield electrically delicate circuits, carbon black will be used as a filler and imparts conductivity to the moulding, thus avoiding static build-up. The moulded box can act as a Faraday screen for the circuits. Similarly carbon black filled PVC will impart a level of electrical screening. SMC is manufactured by dispersing long stands (>2–3 cm) of chopped glass fibres in a bath of polyester resin. The longer glass fibres in SMC result in better strength properties. Typical applications include demanding electrical applications, such as in automobiles, where corrosion resistance or the ability to withstand a harsh environment is needed. The SMC is heat cured using a transfer moulding process (see Section 5.8.4). The SMC product has the advantages over the metal equivalent of being lighter and more easily produced. However, one of the main problems is the ability to use fixtures with such mouldings. This problem is usually solved by inserting the metal fixtures into the moulding, thus avoiding the need to tap into the plastic housing itself.
- *Wiring and related structures*: Polymers are extensively used as the sheathing for wire used in electrical connections and as such it is usually the insulation characteristics of the polymer which are important. However, in the complex wire bundles which are used offshore, it is desirable to be able to discharge the electrical charges created by operating the cable. Carbon black loaded 'conducting' coatings are often used to prevent the cables attracting fish and marine creatures that are sensitive to electrical fields. In normal electrical cabling, it is desirable to achieve a high level of fire retardancy and this can be achieved by the use of clay nanoparticles which assist the char formation, suppress the tendency for the molten plastic to flow and improve mechanical characteristics.
- *Circuit boards*: Most electronic components are mounted on printed circuit boards (PCBs). PCBs are complex structures which are made up of multilayers of patterned metal circuits which connect individual components and integrated circuits (ICs). A typical structure is shown in Figure 10.1. The IC is constructed from a base board which is usually constructed from a glass-filled epoxy resin central layer onto which is bonded the patterned layers which contains the various metal interconnects.

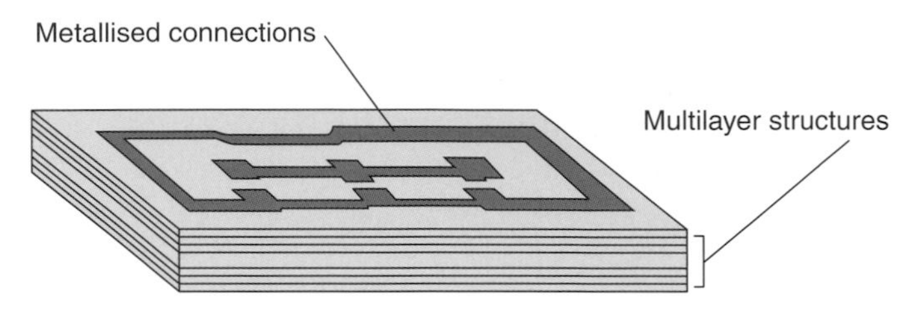

Figure 10.1 Schematic of a printed circuit board.

The layers are made up of curable epoxy laminates which are highly filled with silica so as to produce a material which has a low coefficient of thermal expansion (CTE). The low CTE is necessary to ensure that the metal elements do not 'move' significantly as the temperature is changed. The dimensions of the metal interconnections are typically of the order of micrometres and the spacing between the interconnections is usually of the same order as the metal strips. It is therefore very important that the PCBs have a very low CTE in order to avoid stressing the fine wires and causing problems with reliability. The wiring is often distributed in four or five layers located either side of the central board and each layer is about 10–20 μm in thickness. To avoid overheating, the dielectric constant and resistance of the matrix have to be kept low. The resins typically used are epoxy-based materials.

- *Encapsulation*: A major use of polymers in electronics is as encapsulants. All semiconductor devices are surrounded by a resin casting which is usually an epoxy resin which has been cured by a resin transfer process. The resin will often be loaded with carbon black to impart some electrical conductivity, silica to increase the toughness of the resin and other fillers which may act as scavengers for moisture and increase the barrier characteristics of the resin. Epoxy resins are also used as 'potting' compounds to encapsulate high voltage transformers, etc. The epoxy resin used in these applications is often an anhydride cured material and the initiator used is often dicyanodiamide (DICY). The cured resin contains predominantly ether linkages, can have a glass transition point, T_g, which is of the order of 150°C and has good mechanical properties. Similar resins systems are used in the resin transfer process for semiconductor encapsulation.
- *Polymeric semiconductor photoresist and electron beam resist materials*: One of the most important uses of polymeric materials is to generate the pattern of the features fabricated during semiconductor manufacture. The processes involved are summarised later in this chapter. Without these materials semiconductor manufacture as we know it would not be possible. Polymers have played a critical role in the development of semiconductor technology.
- *Intrinsically conducting polymer systems*: Although traditionally we consider polymers to be insulators, in the last 20 years there has been increasing interest in the study of polymers which are capable of carrying electrical charges. These so-called intrinsically conducting polymers were originally considered as possible replacements for silicon in semiconductor applications. However, the processes associated with electrical conductors involve the creation of a localised electron which can very effectively react with oxygen leading to problems with the stability of these materials. The initial promise of these materials as electrical conductors has never been realised but they are used to suppress corrosion passively and are the basis of the organic light-emitting diode (OLED) display systems, which are currently becoming popular.

Polymers are used in a variety of different ways and it is appropriate to consider two specific areas of application in detail: lithography and OLED materials.

10.2 Lithographic materials

Semiconductor devices consist of a series of layers of materials of different conductivities which are put together to create the active electronic device. The typical device will involve two different types of materials called p-type and n-type semiconductors. A p-type semiconductor contains holes, mobile vacancies in the electronic structure that simulate positively charged particles, whereas n-type semiconductors contain free electrons. Most devices are created by doping silicon with various ions. Application of an external bias voltage can stimulate electronic charge to flow more easily across such a junction in one direction than in the other (see Figure 10.2). The typical semiconductor device, an integrated circuit or 'chip', will contain many thousand such devices. The device will require the creation of a gate area which will usually by amorphous silicon oxide (see Figure 10.2).

The device fabrication steps involve a multiple-step sequence of photographic and chemical processing steps during which electronic circuits are gradually created on a silicon wafer. A typical silicon wafer is produced from extremely pure silicon and is produced by slicing a monocrystalline cylinder (boule) of diameter ~300 mm, into slices which are about 0.75 mm thick. The wafer is carefully polished to obtain a very flat surface. In general, semiconductor processing can be divided into:

- *Front-end processing*: referring to the creation of structures prior to the deposition of metal interconnects and the creation of gate areas.
- *Back-end processing*: referring to processing after the metal interconnects have been deposited.

The reason for the separation of processing in this way is that front-end processing can involve the use of aggressive gases for implantation, etching and depositing areas of film, whereas back-end processing involves metals which can easily be stripped or oxidised by these gases. Back-end processing usually involves the use of less aggressive liquid phase chemistry.

10.2.1 Semiconductor processing

In semiconductor device fabrication, the various processing steps fall into four general categories: deposition, removal, patterning, and modification of electrical properties.

- Deposition is any process that grows, coats, or otherwise transfers a material onto the wafer. Available technologies consist of physical vapour deposition (PVD), chemical vapour deposition (CVD), electrochemical deposition (ECD), and molecular beam epitaxy (MDE).
- Removal processes which create the required structures and involve wet or dry etching and chemical–mechanical planarisation (CMP).
- Patterning is involved at virtually every step and is generally referred to as lithography. Conventionally, lithography involves coating the wafer with a layer of polymer which is known as a 'resist'. This material is sensitive to either light, electrons or X-ray radiation. The resist layer will mask an area during the subsequent processing and allow the creation of the desired structure or stop deposition of material or injection of impurities in particular areas. The photoresist is deposited on the wafer using a spinner and the viscosity of the solution

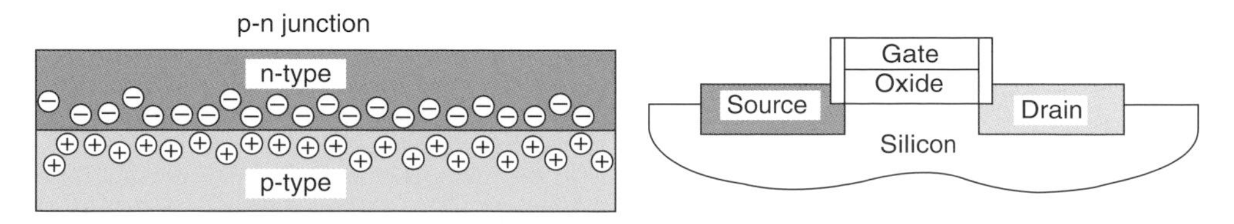

Figure 10.2 Schematic of a p-n junction and indication of an equivalent semiconductor structure.

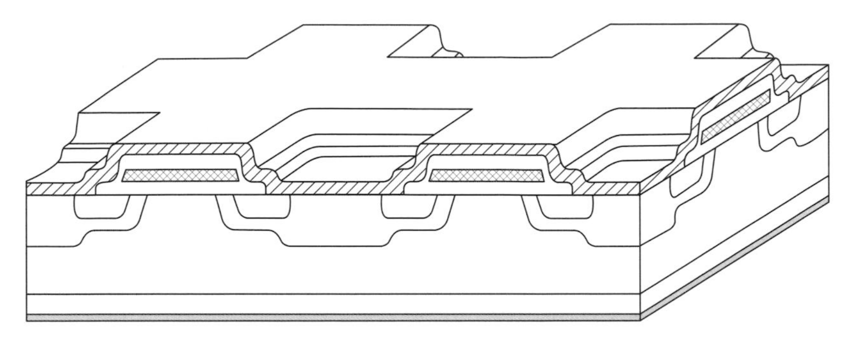

Figure 10.3 Schematic of section through a typical CMOS structure.

has to be correct to give a very uniform film when dried. By changing the molar mass/concentration of the polymer it is possible to change the viscosity and hence the thickness of the spun film that is created. The photoresist is exposed by a 'stepper'. A stepper is a machine that focuses, aligns, and moves the mask, exposing select portions of the wafer to short wavelength light. The mask is like a photographic negative and contains the information for the pattern required for that particular process. Each of the negatives that are used have to be very precisely positioned, so that the patterns are aligned at each stage in the manufacturing process. Misalignment would create a jumble of patterns which would not create the desired device structures. In the typical lithographic process, unexposed regions are washed away by a developer solution. After etching or other processing, the remaining photoresist is removed by plasma ashing in front-end processing and using resist strippers in back-end processing.

- Modification of electrical properties has historically consisted of doping transistor sources and drains, originally by diffusion furnaces and later by ion implantation. These doping processes are followed by furnace annealing or, in advanced devices, by rapid thermal annealing (RTA), which activates the implanted dopants.
- Many modern chips have eight or more levels produced in over 300 sequenced processing steps. Each level may involve a number of steps, for instance a complementary metal–oxide semiconductor (CMOS) device may involve a wafer going through up to 50 photolithographic cycles, as shown in Figure 10.3.

10.2.2 Front-end processing

'Front-end processing' involves directly forming the transistor on the silicon wafer. The process essentially involves doping to create areas with p- and n-characteristics. The front-end process is followed by: growth of the gate dielectric, which is traditionally silicon dioxide (SiO_2), patterning the gate, source and drain regions, and subsequent implantation or diffusion of dopants to obtain the desired electrical properties. In memory devices, the storage cells, which are usually capacitors, are either created in the silicon surface or are stacked above the transistor.

10.2.3 Metal layers

Once the various semiconductor devices have been created they must be interconnected to form the desired electrical circuits. This 'back-end' processing involves creating metal interconnecting wires that are isolated by insulating dielectrics. Interconnect wires were traditionally created from deposited aluminium, titanium, and, more recently, copper. A blanket of aluminium was deposited and then a resist layer placed on top. The resist was patterned and the exposed aluminium layer etched away to create interconnecting wires. Dielectric material is then deposited over the exposed wires. The various metal layers are interconnected by etching holes, known as 'vias', in the insulating material and depositing tungsten in them with a CVD technique. This approach is still used in the fabrication of many memory chips such as dynamic random access memory (DRAM) as the number of interconnect levels is small, usually no more than four. More recently, as the number of interconnect levels for logic has substantially increased due to the large number of transistors that are now interconnected in a modern microprocessor, the timing delay in the interconnect wiring has become a significant issue, prompting a change of material from aluminium to copper and the dielectric material from silicon dioxides to newer low-dielectric constant materials. Recently the so-called 'Damascene' process has been introduced which eliminates a number of the conventional processing steps. In Damascene processing, the dielectric material is first deposited as a blanket film and is then patterned and etched leaving holes or trenches. In 'single Damascene' processing, copper is deposited in the holes or trenches surrounded by a thin barrier film resulting in filled vias or wire 'lines', respectively. In the 'dual Damascene' technology, both the trench and via are fabricated before the deposition of copper resulting in formation of both via and line simultaneously, further reducing the number of processing steps. A thin barrier film, called a copper barrier seed (CBS), is necessary to prevent copper diffusion into the dielectric. As the number of interconnect levels increases, planarisation of the previous layers is required to ensure a flat surface prior to subsequent lithography. Without planarisation, the levels would become increasingly crooked and extend outside the depth of focus of available lithography, interfering with the ability to pattern. Chemical metal polishing (CMP) is the primary processing method to achieve such planarisation although dry 'etch back' is still sometimes employed if the number of interconnect levels is no more than three. Even in this more recent process the importance of photolithography cannot be underestimated.

Semiconductor fabrication involves the use of a range of hazardous materials which include:

- poisonous elemental dopants such as arsenic, antimony and phosphorous
- poisonous compounds like arsine, phosphine and silane
- highly reactive liquids, such as hydrogen peroxide, fuming nitric acid, sulfuric acid and hydrofluoric acid

It is vital that workers are not directly exposed to these dangerous substances. Most fabrication facilities employ exhaust management systems, such as wet scrubbers, combustors, heated absorber cartridges, etc. It is very important that the engineers involved fully appreciate that they are dealing with hazardous materials.

10.2.4 Photolithography

Photolithography is a process used in microfabrication to selectively remove parts of a thin film. Stereolithography, a variant of photolithography, is now used to create three-dimensional structures by the polymerisation of photoactivated monomer liquids. Photolithography uses light to transfer a geometric pattern from a photomask to a light-sensitive chemical photoresist on the substrate. The term 'resist' indicates that the layer has to be able to resist the action of the chemicals used in the subsequent processing steps. However, it is also important that the resist can be cleanly removed after the processing has been completed. The process of photolithography is summarised in Figure 10.4. The light source used to produce the exposure of the photoresist depends on the size of the feature which is being developed. The light source is filtered to produce a single spectral line; the 'g-line' (436 nm) or 'i-line' (365 nm). Other sources use excimer lasers: krypton fluoride (248 nm) and argon fluoride (193 nm). For the finest structures a UV source that is ~0.1 μm deep is required.

Changing the wavelength is not a trivial matter. Air begins to absorb significantly around the 193 nm wavelength. Thus, moving to shorter wavelengths would require installing vacuum pumps and purge equipment on the lithography tools. Insulating materials, such as silicon dioxide, when

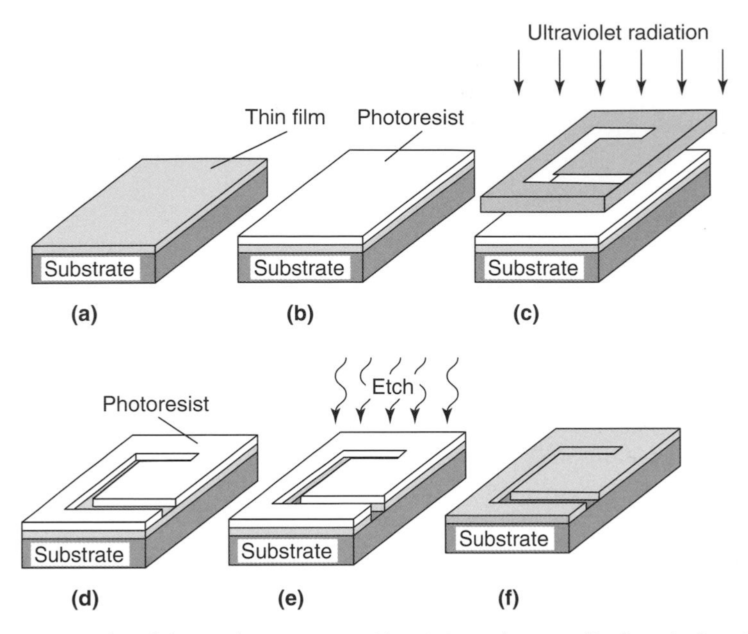

Figure 10.4 Steps in photolithographic process: (a) pristine substrate; (b) deposited resist; (c) exposed photoresist; (d) developed photoresist; (e) etch of substrate; (f) strip resist and clean.

exposed to photons with energy greater than the bandgap, release free electrons and holes which subsequently cause adverse charging of the substrate. The optics systems used are designed to overcome the problems associated with the diffraction limiting condition encountered with simple printing systems and achieve patterning below the diffraction limits of the light source.

The photolithographic process involves a number of steps which are carried out in sequence: spinning of the resist to produce a thin uniform film, backing of the film to remove solvent and aid exposure, exposure to the mask, development of the pattern using some form of solvent treatment, and cleaning the structure generated prior to using the wafer in the processing step (see Figure 10.3). The cleaning may involve a second solvent stripping process, light gas phase etching to sharpen the profile and remove residual material and baking to harden the resist to make it more resistant to the subsequent processing.

The processing usually involves:

- *Cleaning*: For a pristine wafer, if organic or inorganic contaminations are present on the wafer surface, they are usually removed by wet chemical treatment, e.g. washing with a solution containing sulfuric acid and hydrogen peroxide.
- *Preparation of wafer*: After cleaning the wafer is heated to a temperature sufficient to drive off any moisture on the wafer surface. An adhesion promoter may be applied: bistrimethylsilylamine (hexamethyldisilazane (HMDS)), is commonly used to promote adhesion of the photoresist to the wafer. A reaction occurs between the hydroxyl groups of the silicon dioxide, forming a methyl-coated surface. This water-repellent layer prevents the aqueous developer from penetrating between the photoresist layer and the wafer's surface, thus preventing so-called lifting of small photoresist structures in the developing pattern.
- *Spin coating*: The resist in the form of a viscous solution of polymer in solvent is deposited on the wafer by spinning the wafer at high speed (see Figure 10.5). The spinning process is usually carried out at speeds in the range 1200–4800 rpm for 30–60 s, and produces a layer 0.5–2.5 μm thick. A correctly deposited film will have a uniform thickness within 5–10 nm.

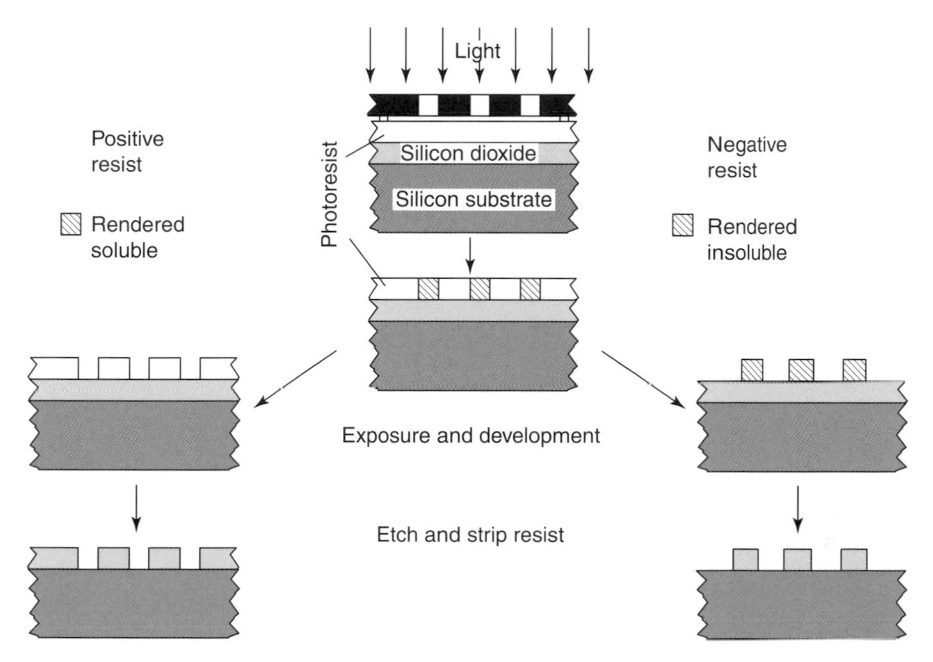

Figure 10.5 Schematic of photolithography and pattern development using positive and negative resists.

- *Baking*: The resist-coated wafer is then baked in an oven to drive off excess solvent, typically at 90–100°C for 5–30 min in an oven.
- *Exposure and developing*: The wafer is next placed in a stepper mask aligner where it is exposed to light through an appropriate mask (see Figure 10.4). Positive photoresist becomes soluble in the basic developer when exposed; negative photoresist becomes insoluble in the organic developer. This chemical change allows some of the photoresist to be removed by a special solution, known as 'developer' by analogy with photographic developer (see Figure 10.5).

The chemistry of this process is reviewed in Section 10.2.6. A post-exposure bake is performed before developing, typically to help reduce standing wave phenomena caused by the destructive and constructive interference patterns of the incident light. Developers originally used sodium hydroxide. However, sodium residues can create problems with electrolytic metal migration and it is desirable to eliminate possible contamination for the process. Metal-ion-free developers such as tetramethylammonium hydroxide (TMAH) are now used.

- *Postbake*: The resulting wafer is then baked, typically at 120–180°C for 20–30 min. The process solidifies the remaining photoresist to increase its ability to protect the underlying structures during ion implantation, wet chemical etching, or plasma etching.

10.2.5 Wafer processing

The processing of a wafer will usually involve the following processes:

- *Etching*: A reactive liquid ('wet') or plasma ('dry') process is used to remove the uppermost layer of the substrate in the areas that are not protected by resist. Dry etching techniques allow anisotropic removal of material and avoid significant undercutting of the pattern. Wet etch processes are generally isotropic in nature.
- *Photoresist removal*: The next step is the removal of the resist, which is usually carried out using a liquid 'resist stripper'. Alternatively, the photoresist may be removed by a plasma containing

oxygen, which oxidises and degrades the polymer. This process is called ashing, and resembles dry etching. After metal has been deposited plasma stripping cannot be used as it dramatically affects the metal lines which have been carefully deposited and liquid strippers are usually used.

- *Photomasks*: The image for the mask originates from a computerised data file. The mask is often a piece of fused quartz onto which a layer of chrome has been deposited. The pattern is created by exposing a resist layer using an electron beam in a machine which resembles an electron microscope. The electron beam is steered by the computer and either degrades or cross-links the polymer film producing either a positive or negative image.

10.2.6 Chemistry of photoresists

Photoresist materials can be divided into those which act negatively and those which act positively. Whether it is positive or negative, the first step is exposure to light, which induces a chemical change which renders the area illuminated more or less susceptible to attack by the solvent which is used to develop the image of the mask (see Figure 10.5). Negative resists are the most widely used, being used for both semiconductor manufacture and PCB fabrication. The exposure of polymers with UV light in the range 200–300 nm (~4–6 eV), leads to radical formation, which can be used to produce cross-linking, increases in molecular mass and the creation of insoluble and brittle films.

10.2.6.1 *Two-component negative resist*

This resist system was at one time the most popular resist system, but has been replaced by other systems. The resist is based on a synthetic rubber which is produced using Zeigler–Natta polymerisation of isoprene (see Figure 10.6), which forms poly-*cis*-isoprene, an elastomeric material with a low T_g. The elastomer is usually treated to induce partial cyclisation of the polymer to produce material with a higher glass transition temperature and greater structural integrity than its precursor. The cyclised rubber matrix materials are extremely soluble in nonpolar, organic solvents such as toluene, xylene or halogenated aliphatic hydrocarbons. The photo sensitivity is achieved by the use of bisarylazides (see Figure 10.6). Changes in the structure of the bisazide allow the sensitiser to be tuned to the exposure wavelength.

The irradiation of the bisarylazide or some suitably modified version of this molecule undergoes a series of chemical reactions, which leads to cross-linking and hardening of the exposed

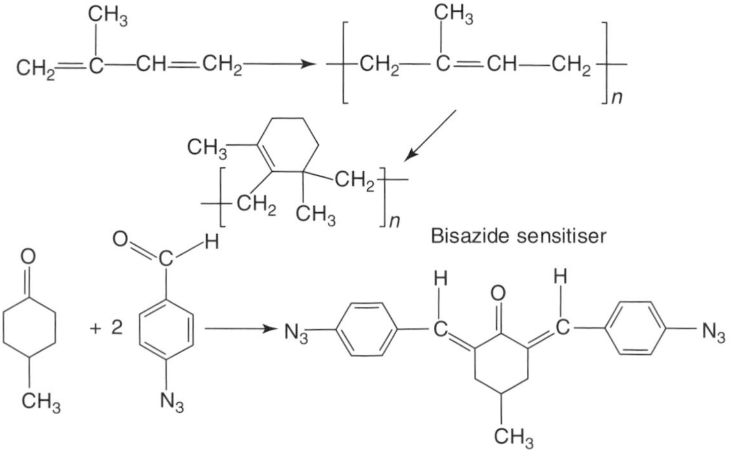

$CH_2{=}C(CH_3){-}CH{=}CH_2 \longrightarrow -[CH_2{-}C(CH_3){=}CH{-}CH_2]_n-$

Bisazide sensitiser

Figure 10.6 Bisarylazide-rubber resist. Matrix resin is cyclised poly-*cis*-isoprene; photo-active sensitiser is bisarylazide.

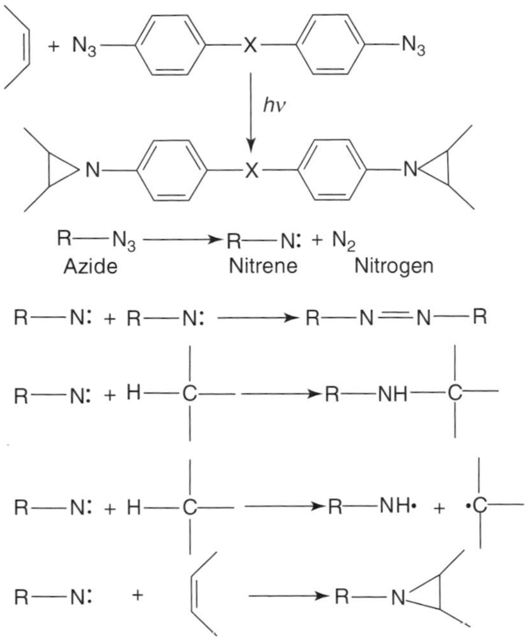

+ N_3 — X — N_3

$h\nu$

N — X — N

$R{-}N_3 \longrightarrow R{-}N\colon + N_2$

Azide Nitrene Nitrogen

$R{-}N\colon + R{-}N\colon \longrightarrow R{-}N{=}N{-}R$

$R{-}N\colon + H{-}C \longrightarrow R{-}NH{-}C$

$R{-}N\colon + H{-}C \longrightarrow R{-}NH\cdot + \cdot C$

$R{-}N\colon$ + $\longrightarrow$ R—N

Figure 10.7 Cross-linking reactions induced by photodecomposition of bisarylazide.

areas which are rendered less soluble to solvent exposure. The process can be summarised by the reaction scheme shown in Figure 10.7.

The primary photoevent is the creation of a nitrene, which then undergoes a variety of reactions that result in covalent, polymer–polymer linkages. A schematic representation of the cross-linking reactions via the nitrene insertion involves the formation of aziridine linkages and is shown together with several other reaction modes available to the nitrene. This resist has several disadvantages:

- the presence of oxygen acts as a free-radical scavenger, inhibiting cross-linking, and the resist must be exposed under vacuum or nitrogen
- the solvent developer swells the cross-linked negative image, causing degradation of the pattern and limiting resolution to 2 μm in a coating that is 1 μm thick
- the aromatic solvent developer may pose environmental, health and safety concerns

10.2.6.2 *Two-component positive resist*

Positive toning resists provide an alternative approach to the fabrication of microstructures. The diazoquinone/novolac materials (see Figure 10.8) decompose when irradiated and change the solubility of the resist.

The basic resist is a two-component system where a low-molecular-weight phenolic-based resin is mixed with a diazoketone derivative. The phenolic resin provides excellent film-forming properties and is highly soluble in basic solutions. The addition of a diazonapthoquinone photosensitiser acts as a dissolution inhibitor, and dramatically reduces the solubility of the unexposed film in basic solutions. Upon exposure, the diazo derivative undergoes molecular rearrangement to form a carboxylic acid, and the resist becomes base soluble, forming a positive image. The structure of the photoactive diazoketone compounds can be optimised to match the excitation line.

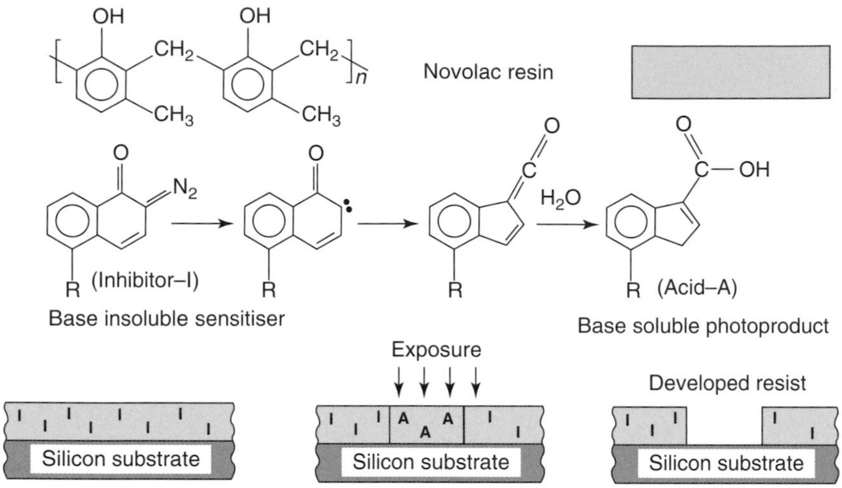

Figure 10.8 Schematic representation of positive resist action of diazonaphthoquinone-novolac resists.

An interesting variation on the chemistry can be achieved by the incorporation of a small amount of a basic additive such as monoazoline (1-hydroxyethyl-2-alkylimidazoline), imidazole or triethanolamine into the diazoquinone–phenolic resin photoresist. The change in the process is shown in Figure 10.9. After the initial exposure using the mask, the resist is baked and in the presence of the base decarboxylation of the diazoquinone photoproduct (the indenecarboxylic acid) occurs, producing a base insoluble product. The unexposed resist is now flood exposed, making it base soluble. Development of the resist will now produce a pattern which is a negative image of the mask.

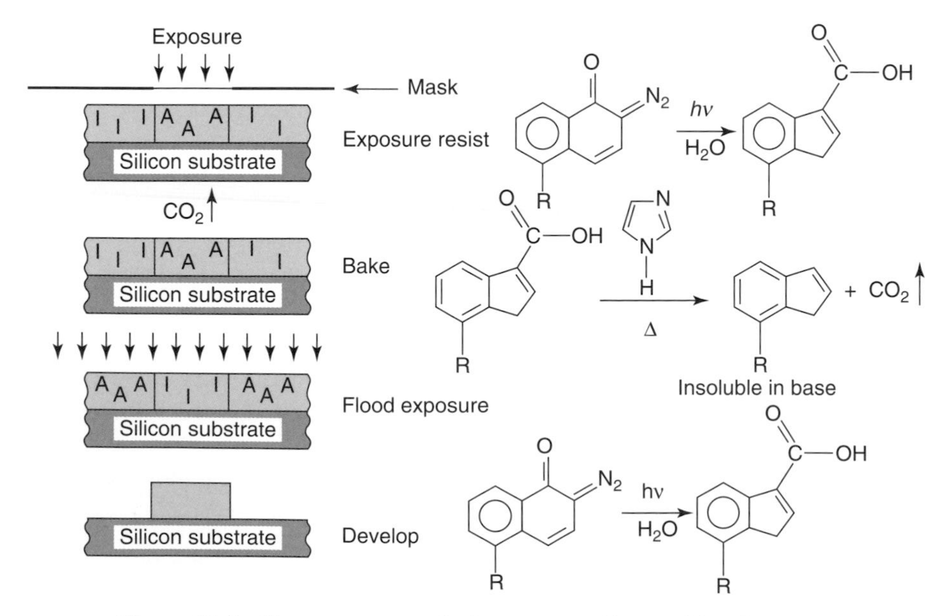

Figure 10.9 Process sequences for image reversal in a positive photoresist.

10.2.6.3 *Photocross-linking of reactive monomer*

A variation on the above scheme is the photocross-linking of a solution of a reactive monomer enhanced with a styrene–methacrylic acid copolymer via a photosensitive initiator (see Figure 10.10). The system is usually adjusted so that the solution plus initiator is a solid. This photo resist is capable of <75 μm lithography and is often used as a film resist, but is not capable of being used to produce very high resolution lithography.

The process involves the initiator absorbing radiation and forming an imidazole radical, which abstracts hydrogen from a tertiary amine to form a radical which initiates the polymerisation process. Polyfunctional photopolymerisable monomers react to form a highly cross-linked, chemically resistant structure and a binder to provide toughness and film-forming properties, typically a styrene–maleic acid polymer. This photo resist system is capable of <75 μm lithography.

10.2.6.4 *Chemically amplified negative resists*

The requirement to produce resists capable of achieving smaller pattern sizes led to the development in the late 1970s of chemically amplified negative resists. The photolysis of certain thermally stable onium salts produces an unstable intermediate which ultimately generates strong acids. The chemically amplified resist uses this acid to transform the resist and render it capable of development. The acid acts as a catalyst to initiate a sequence of chemical reactions. Compared to the conventional free-radical initiators, the onium salts have excellent thermal stability, are not sensitive to oxygen, and exhibit no 'dark' or side reactions in solution. The resist film is changed either by an increase in the molecular mass or the solubility of the polymer. The increase in molecular weight could be achieved by *cationically polymerising* monomers such as epoxies and vinyl compounds, or by enabling *condensation* reactions between phenol formaldehyde resins and amino-based cross-linkers. Changes in polarity could be achieved through the *acid-catalysed deprotection* of a variety of esters.

Acid-catalysed deprotection

The acid created by illumination changes the solubility of the resin, as shown in Figure 10.11. The photoacid generator (PAG) is usually triphenylsulfonium hexafluoroantimonate onium salt which

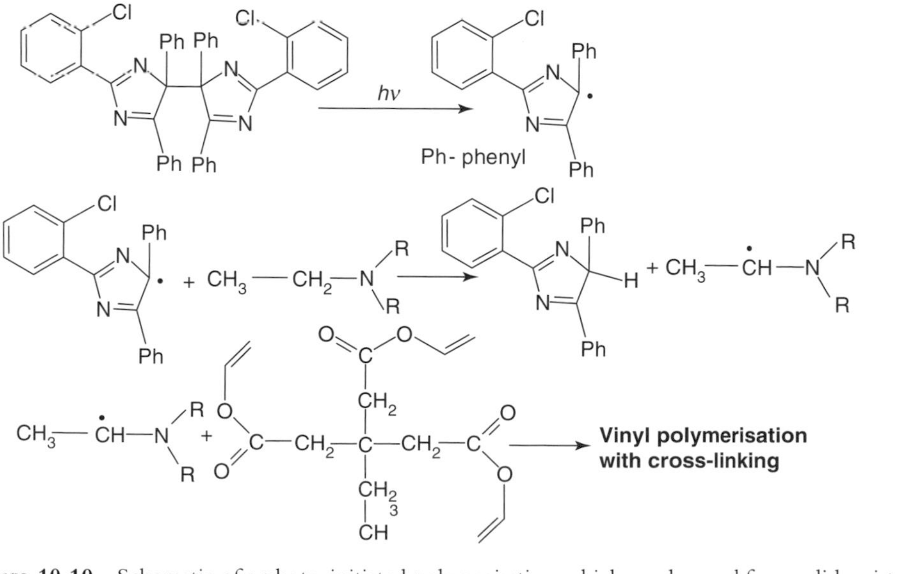

Figure 10.10 Schematic of a photo-initiated polymerisation which can be used for a solid resist.

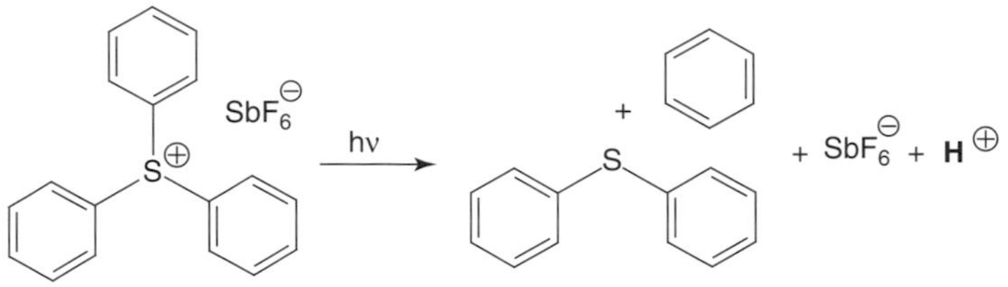

Figure 10.11 Photoacid generation from triphenylsulfonium hexafluoroantimonate onium salt.

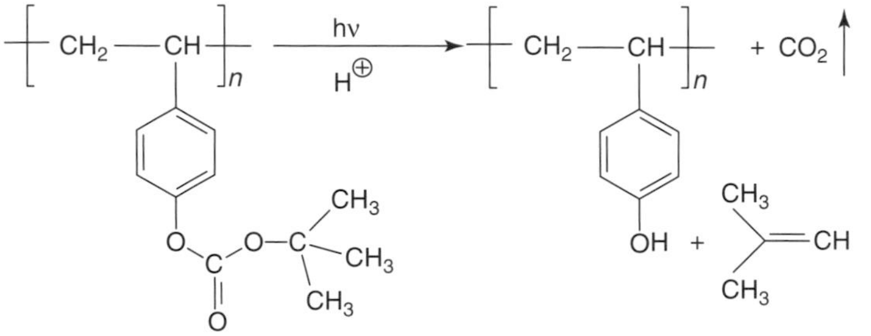

Figure 10.12 Schematic of acid catalysed decarboxylation of tBOC.

forms $HSbF_6$, which is a very strong acid. The liberated proton is (H^+). The first deep UV (DUV) negative resist system used in semiconductor manufacturing was based upon poly-4-*t*-butoxycarbonyloxystyrene (tBOC) and a triphenylsulfonium hexafluoroantimonate onium salt (see Figure 10.12).

Upon exposure in the DUV, and subsequent baking to diffuse the photogenerated acid and complete the reaction, the acid cleaves the labile tBOC-protecting groups to form a polar polyvinyl phenolic polymer (see Figure 10.12). The unexposed resist is removed by using a nonpolar solvent, forming a negative image. This system, known as tBOC resist, provided the exposure sensitivity required to produce 1Mb DRAMs in the late 1980s. However, it was difficult to control the width of the line of this new chemically amplified system in a manufacturing environment. The sensitivity of the photoresist was affected by airborne chemical contaminants. Special carbon filters had to be installed, and coatings had to be used to protect the resist film from the diffusion of contaminants, in particular amines. The diffusion of the acid, sensitivity and resolution of the system were found to depend upon precise control of the prebake temperature, time between exposure and postexposure bake (PEB), and the time and temperature of the PEB process itself.

Polymerisation catalysts

The PAG initiator system also opens up the possibility of using light to initiate polymerisation. A series of resist systems has been developed using this approach. The incorporation of an epoxy group into an acrylic polymer creates a system which can readily be transformed from a thermoplastic to a thermoset. The transformation changes the solubility in the areas illuminated and allows pattern generation. An electron-beam resist based on a high-molecular-weight epoxy resin, a glycidylacrylate–ethylacrylate copolymer has been used for a number of years (see Figure 10.13).

When exposed to high-energy e-beam radiation, anionic species are created which can react with the epoxy functions forming a cross-linked matrix. Although this system has been used for electron beam lithography, it is not adequate for fabrication of high-resolution semiconductor devices. The developer induces swelling in the pattern because of the high molecular weight of the resin and the thermal stability and plasma resistance of the system are not adequate.

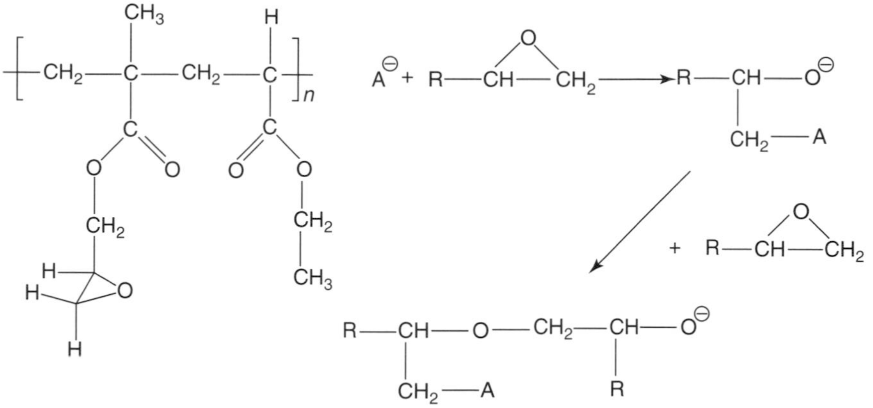

Figure 10.13 Acrylic epoxy glycidylacrylate–ethyl acrylate copolymer resist.

Adaptation of the acrylic resist for DUV application using a PAG initiator has led to the development of systems based on SU-8 (novalac epoxy) which involve the cross-linking of a highly epoxy functionalised aromatic prepolymer (see Figure 10.14). The aromatic backbone provides the required thermal stability and plasma etch resistance. Upon exposure to either UV, e-beam, or X-ray radiation, it forms a ladder-like structure with a high cross-linking density and a T_g of ~200°C. This resist has been used to fabricate structures with feature sizes in the range 0.25–0.1 μm using e-beam lithography. The low molecular weight of the resin produces high contrast and excellent solubility, and the high epoxy functionality provides sensitivity.

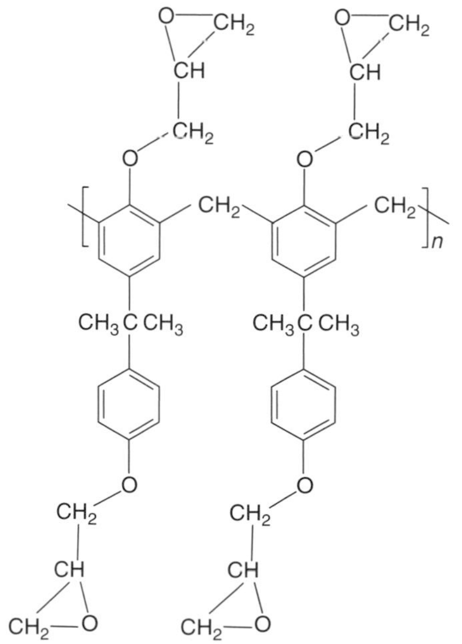

Figure 10.14 Glycidyl ether of bisphenol A, SU-8 resist.

Acid-catalysed condensation reactions

Following on the general theme of requiring a resist to operate to smaller feature sizes a system was developed based on an *acid-catalysed condensation reaction* between an aminoplast cross-linker such as a urea or melamine formaldehyde with a base-developable resin such as a novolac or polyhydroxystyrene (see Figure 10.15).

However, this resist was prone to self-reaction and was not widely adopted. But, the concept of acid-catalysed reactions has been developed to provide other resist systems. Many of the problems associated with the stability of this initial resist are associated with the melamine. An alternative resist, known as CGR, has been developed and is used for complimentary metal-oxide semiconductor (CMOS) logic device fabrication. The cross-linking which is used in this system is shown in Figure 10.16. The condensation reaction shown in Figure 10.17 involves the reaction of the methoxy leading to a condensation reaction with the aromatic resin, which is still operative.

The cross-linker (see Figure 10.16) can, however, easily be purified and has a well-defined melting point. The performance enhancements using this compound allow good lithographic performance to be achieved. This negative-resist chemistry has a number of advantages. The resist

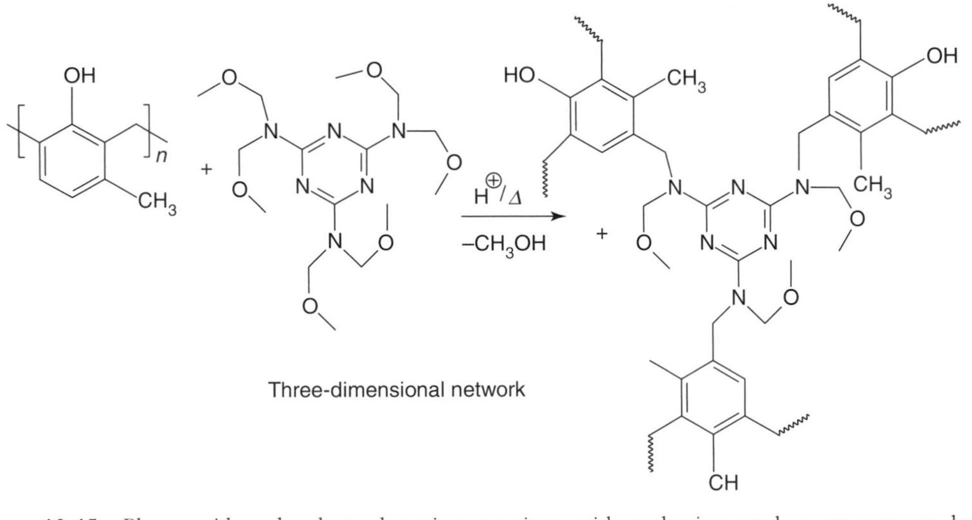

Figure 10.15 Photo acid-catalysed condensation reactions with melamine used as an aqueous base developed negative resist.

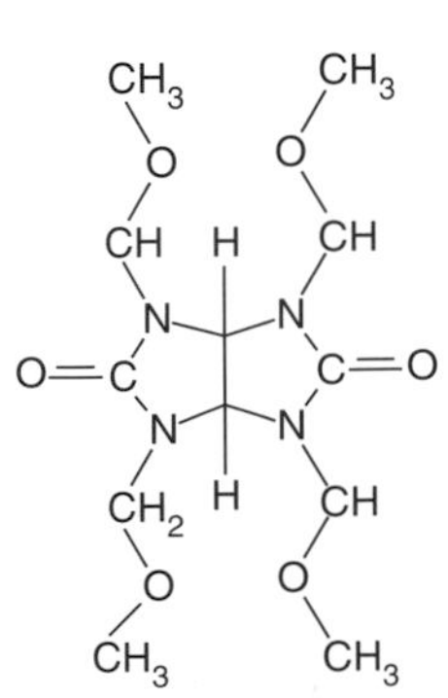

Figure 10.16 Alternative to melamine for promotion of condensation.

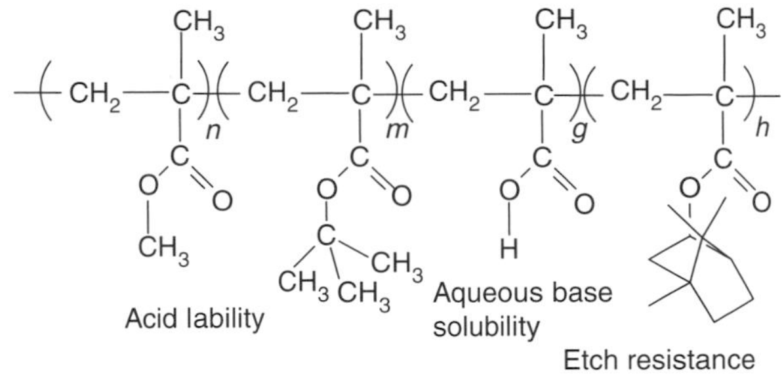

Figure 10.17 Schematic of a 193 nm resist indicating design features.

does not require a topcoat to protect it from chemical vapours, which can affect other resist systems. It is less sensitive to basic contaminants because aminoplast resins already contain a tertiary amine functionality, and very few reactive sites are required to turn the exposed area into an infinitely cross-linked network, which dramatically increases differential solubility. Because the exposed image is cross-linked, there is less dependence on developer time and temperature conditions, and the exposed image can be overdeveloped without affecting the width of the line. This negative resist has been used to fabricate 0.35 μm CMOS devices. A fundamental problem with these resists is the absorption of energy which occurs below 300 nm due to the presence of the aromatic ring structures. As a consequence, a new approach has to be adopted for 245 and 193 nm lithography.

10.2.6.5 *Deep UV resist systems*

The APEX positive resist system and related resists attempt to address the problem of achieving transparency below 300 nm. The optical properties of a resist which can operate at 248 nm or 193 nm are very different from those systems discussed above. At 193 nm a high level of transparency can only be achieved if the aromatic groups in the resists are omitted. At 248 nm and above, incorporation of aromatics is possible whilst retaining an acceptable transmission level and equates to the conventional DUV resists which contain hydroxystyrene polymers (see Figure 10.12). To achieve the required level of transparency acrylic-based resists are used as they have the required level of transparency for use at 193 nm (see Figure 10.17). Each monomer serves a separate function in the terpolymer; *t*-butylmethacrylate (TBMA) provides an acid-cleavable side group which is responsible for creating a radiation-induced solubility change. Methylmethacrylate (MMA) promotes hydrophilicity for photoinitiator solubility and positive-tone development characteristics, while also improving adhesion and mechanical properties and minimising shrinkage after exposure/bake processing steps. Methacrylic acid (MAA) controls aqueous development kinetics. This polymer is prepared in a single step from readily available, inexpensive components. By selecting the terpolymer composition and molecular weight, the imaging properties (such as dissolution properties, photospeed and contrast) can be altered to a significant extent.

The etch resistance of acrylic resists can be substantially improved by increasing the carbon content of the polymer via the incorporation of cyclic aliphatic functionality. For example, the etch rate of polyisobornylmethacrylate is less than half that of polymethylmethacrylate in a hydrogen bromide plasma. To provide greater flexibility in resist design, a three-component resist has been developed and consists of a *methacrylate polymer*, an *alicyclic dissolution inhibitor* compound, and a *photoacid generator*. Studies indicate that 5-*B* steroids from natural sources possess the desired properties (see Figure 10.18).

To achieve the highest resolution sophisticated imaging and alignment systems have to be used. The total process (spinning, prebake, exposure, development and postbake) has to be carefully optimised. A 0.1 μm lithographic structure is shown in Figure 10.19.

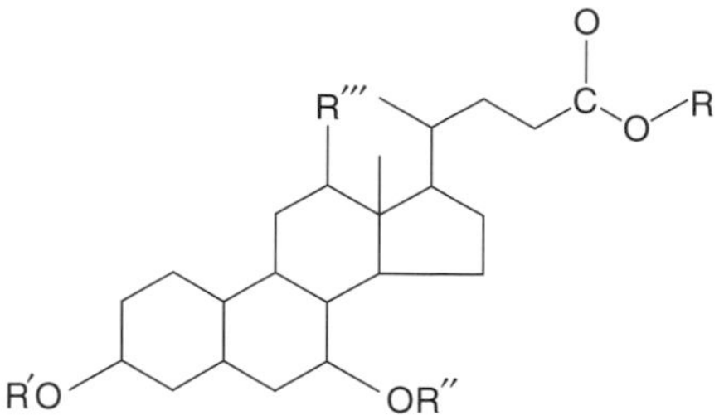

Figure 10.18 Structure of a typical 5-*B* steroid used as a dissolution inhibitor.

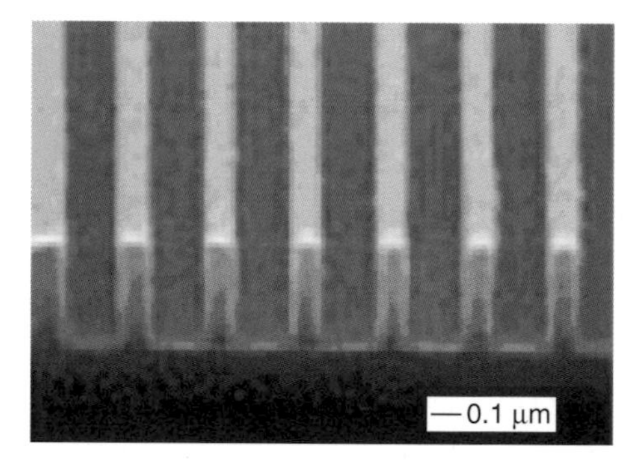

Figure 10.19 Example of a 0.1 μm structure.

The plasma etch resistance can be increased by silation, using a treatment which involves reaction with tetraelhoxysilane (TEOS) to produce a silated layer close to the surface. It is possible to improve the etch resistance by incorporation in the base polymer of an α-trifluoromethylacrylic moiety as a repeat unit. Four platforms are currently available:

- all-acrylic
- all-alicyclic
- acrylic–alicyclic
- acrylic–aromatic systems

While the all-alicyclic-norbornene polymers are synthesised by transition-metal-initiated addition polymerisation, all other polymers involving α-trifluoromethylacrylic monomers are prepared by conventional radical copolymerisation.

10.2.7 Applications of lithography

Lithography is used for a variety of applications:

- *PCB fabrication*: Common applications include PCB fabrication and involve the application of photoresist, exposure to the mask, development, followed by an etching step in which ferric chloride is used to remove the copper-clad substrate.
- *Sand carving*: Materials are sand blasted after a photolithographically printed pattern has been applied as a mask.
- *Microelectronics*: This application is used chiefly for silicon wafers/silicon ICs. It is the most developed and the most specialised of the available technologies.
- *Glass, or other substrate, patterning and etching*: This includes specialty photonics materials, microelectromechanical systems, glass PCBs, and other micropatterning tasks.

10.3 Intrinsically conducting polymers

In the early 1980s, it was suggested that silicon technology would be replaced by organic-based electronics. This claim was based on the discovery of a means whereby plastics could be made to conduct electricity. The incorporation of carbon black into a plastic can introduce paths for electron conduction and this approach has been used for a considerable length of time to create materials which can be used as electrostatic screens and shields. The work in the 1980s was oriented around the concept that it was possible to create polymers which would act rather like wires. These materials need to have a backbone structure which had delocalised π-electrons connected so as to allow a high degree of mobility within the polymer (see Figure 10.20).

Acetylene on polymerisation produces a long chain linear polymer backbone in which the π-electron clouds overlap. Whilst the electrons are bound to individual carbon atoms, it is easy to envisage that either the absorption of light might provide the electrons with enough energy to hop to a neighbouring site or the injection of an additional electron would create a carrier. Both mechanisms are possible. Polyacetylene is termed an intrinsically conducting polymer: it has the ability to allow an electron to move 'freely' along the polymer backbone. The realisation of a conducting polymer system was achieved by doping polyacetylene, which is a black powder, with an appropriate dopant. Varieties of different systems have been used and include perchlorate and BF_3. The main problem with these polymer materials is their sensitivity to oxygen. Oxygen can react relatively easily with the π-electron structure, destroying the extended conjugation of the backbone and leading to a dramatic drop in the conductivity of the materials. Subsequent studies have indicated that it is possible to achieve appreciable levels of conductivity by doping other polymer systems which possess long-range conjugation (see Figure 10.21).

All these polymers have the required degree of conjugation, π-electron cloud overlap, but are significantly less susceptible to oxidation than polyacetylene. Polypyrrole and polyaniline contain nitrogen which can be quaternised, by the addition of a proton, and in this form can conduct charge. As with polyacetylene, the polymers have to be doped to allow charge injection to the conjugated backbone (see Figure 10.22).

The anion (A^-) is able to move between polymer chains and provides the mobile species which aid charge mobility within the polymer system. Although the objective of replacing silicon or electrical wiring by a polymer is very unlikely, these intrinsically conducting polymers have found a niche market as anticorrosion coatings, the conductivity reducing the overpotential which aids corrosion when insulating polymer coatings are applied to metallic substrates. The conductivity observed in these polymers is associated with the creation of free electrons or holes associated with the delocalised π-electron structure. These states are not easily quenched and provide the opportunity for creating applications which use these interesting electroluminescent effects.

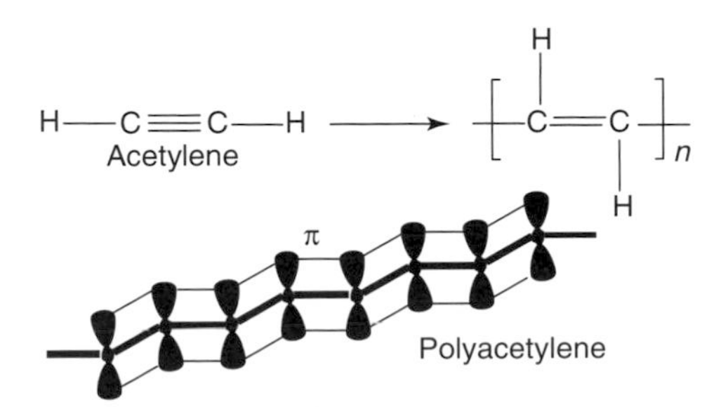

Figure 10.20 Polymerisation of acetylene to produce polyacetylene.

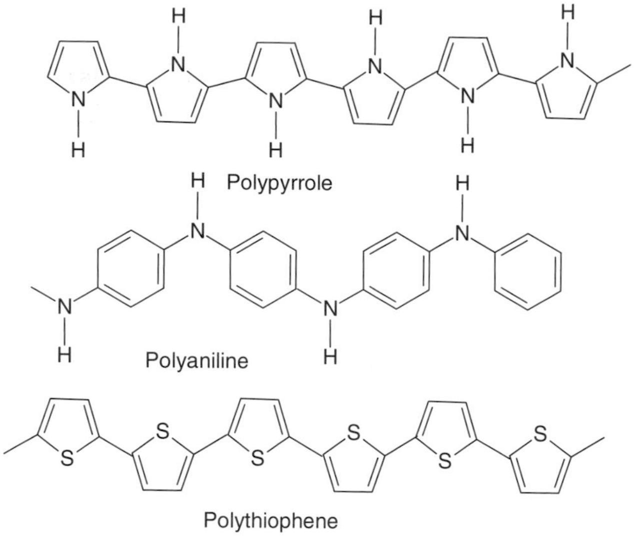

Figure 10.21 Structures of polypyrrole, polyaniline and polythiophene.

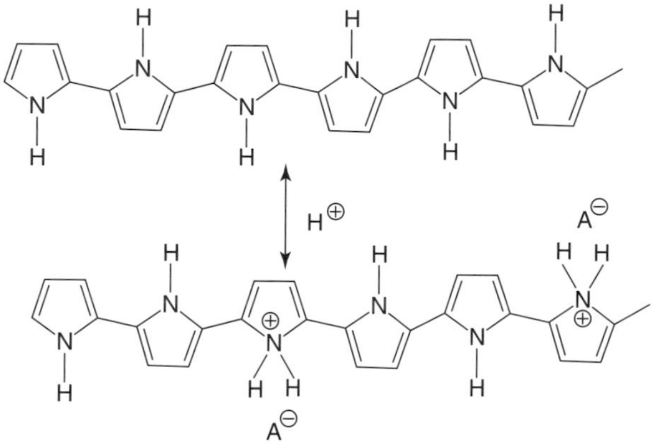

Figure 10.22 Quaternisation of polypyrrole.

10.4 Organic light-emitting polymers

Leading on from the studies of the intrinsically conducting polymers was the observation that the application of an electrical field to certain of these materials allowed light to be emitted. Similarly, it is possible to expose some of these materials to light and to obtain spontaneous emission at a different wavelength. In a polymer, such as polyacetylene or the polyacene homologous series of compounds, the π-overlapping conjugated electrons can be excited into a localised state and become conducting electrons by the application of an external electrical field or the absorption of light (see Figure 10.23).

The energy required to promote the electron from the bound (bonding) to the conduction band (antibonding) state depends on the number of conjugation bonds. The larger the number of bonds conjugated, the smaller the energy gap. The orange colour of carrots is an indication of the effect of joining a number of double bonds together. Light absorbed by this conjugated system excites the electrons to a higher energy state, whereupon it is emitted at a different wavelength and a coloured emission is observed. The colour depends on the number of bonds connected together.

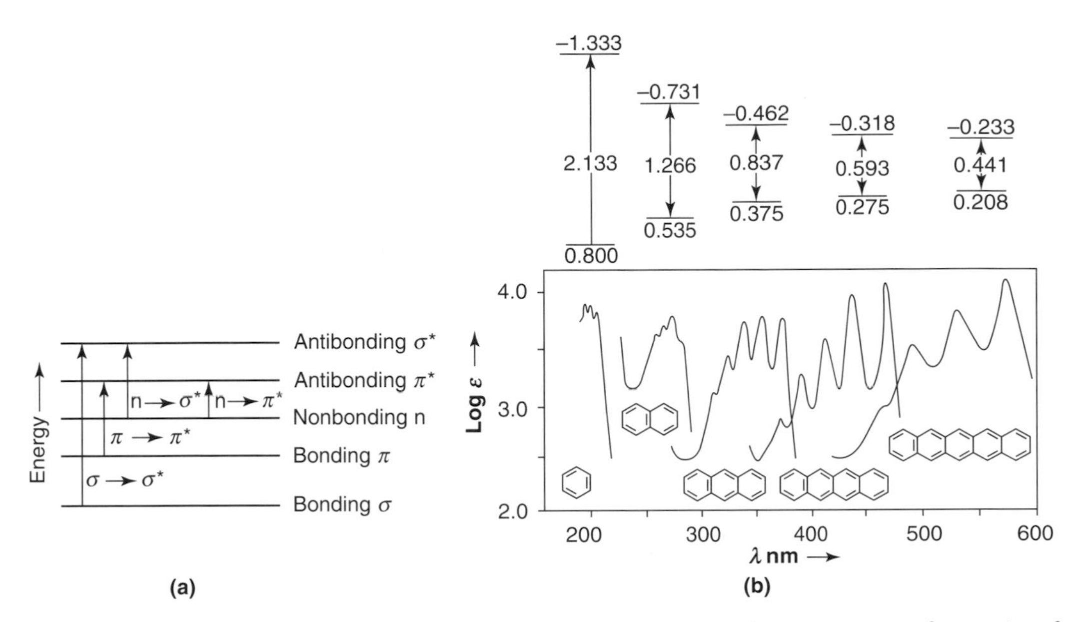

Figure 10.23 Types of electronic transition: (a) energy levels; (b) associated emission spectra for a series of polyacene compounds.

The peaks in the emission spectra (see Figure 10.23) reflect the various π-states involved in the electronic excitation. In the case of polyacetylene, the number of bonds is large and the energy gap is low and the material looks black. In the case of carrots, the number of bonds is smaller and the gap larger and hence the orange colour. In the case of benzene (C_6H_6), the emission is in the UV range, naphthalene ($C_{10}H_{10}$) is in the near-UV range, phenanthrene ($C_{14}H_{14}$) is in the blue range, etc. Provided the delocalised state extends over a sufficient range of bonds then quenching does not occur and high levels of emission are achieved.

A variety of molecules have been identified which possess the required degree of conjugation and hence can be excited either by application of external electrical fields or pumped with an external light source (see Figure 10.24). These materials are currently being used to construct

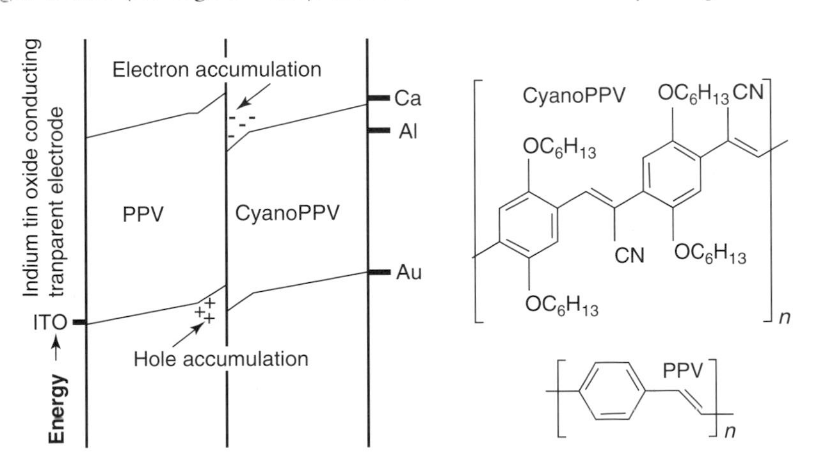

Figure 10.24 Schematic of structure of a typical OLED device constructed from PPV and cyanoPPV showing relative energy levels involved.

OLEDs and have the potential of being used to harvest light (for use in fabricating solar cells). A detailed discussion of this topic is beyond the scope of this textbook. Samsung SDI is the world's largest OLED manufacturer, producing nearly 50% of the total of OLED displays made worldwide. In October 2008, it unveiled the world's largest OLED TV at 40-inch with a full high definition resolution of 1,920 × 1,080 pixel.

The light is emitted at the junction between one polymer which acts as a donor and the other which acts as an acceptor. Changes in the chemical structure alter the bandgap and the wavelength of the light emitted. A wide range of molecules have now been identified which have the potential to be used in this electronic application.

Brief summary of chapter

- Polymers are used extensively in the electronics industry, with applications ranging from simple insulation to electronic conduction.
- Polymer materials play a critical role in the fabrication of semiconductor devices and are now making an impact on display technology.

Additional reading

Müllen K. (Ed.) *Organic Light Emitting Devices: Synthesis, Properties and Applications*, Wiley-VCH, Weinheim, Germany, 2006.

Shinar J. (Ed.) *Organic Light-Emitting Devices: A Survey*, Springer-Verlag, New York, NY, USA, 2004.

Nalwa, H.S. (Ed.) *Handbook of Organic Electronics and Photonics, Vols. 1–3*, American Scientific Publishers, Los Angeles, CA, USA 2008.

11

Medical applications of polymers

11.1 Applications in medical devices

A major use of polymers is in the area of medical devices. Engineers may be involved in the design of medical devices and therefore it is appropriate to include some introductory comments on this very important area of the application of polymeric materials in this textbook. Polymers used in medical applications can be broadly divided into two areas.

Polymers used in devices and therapy

In these applications the polymers are selected on the basis of their physical properties and their compatibility with the application area. The main additional criterion in selecting a material is the effect that the polymer might have on the media it encounters. For instance, if the polymer is to be in contact with blood then it must not promote clotting. Tubes and related surfaces used in blood dialysis equipment must therefore not promote blood clotting in any way, and are classed as being haemocompatibile. The other important factor is that the polymer must not contaminate the fluids with which it is in contact. Contamination can arise as a consequence of the loss of plasticiser or other additives, and, most importantly, this can include processing aids added to the polymer material.

Polymers used in the creation of implants

There has been a long and unfortunate history of polymers used as medical implants. Polymer containing devices are used in breast implants, heart valve replacements, hip joint replacements and a number of other surgical operations. The materials used in this context have to be carefully screened to ensure that they cannot cause the body to react against them and produce disadvantageous reactions. In a number of cases the polymer or implanted material may be coated with a substance such as heparin which suppresses the rejection process. The selection of these materials is therefore not only in terms of their mechanical properties but also needs to consider the way in which the body reacts to their presence.

11.1.1 Polymers used in devices and therapy

It is impossible to cover the topic of medicinal polymers in any detail, but it is appropriate to consider a few selected topics which will illustrate the issues which the engineer may need to consider.

11.1.2 Silicone breast implants

Silicone-gel breast implants are made of an outer shell of silicone elastomer filled with silicone gel and implanted in the breast (see Figure 11.1). The silicone elastomer and gel are essentially cross-linked PDMS (see Section 4.3.2). There is a long trail of litigation associated with these implants. The first generation of implants had a smooth outer surface and was constructed from a thick silicone shell which was filled with a gel. The composition and structure of the shell are important. The main problems were associated with significant loss of the gel from the implant and its influence on the shrinkage of the scar tissue associated with the operation. The second

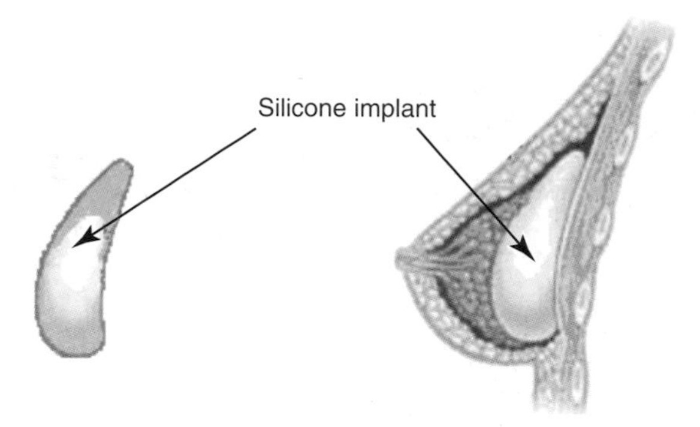

Figure 11.1 Location of silicone implant within breast.

generation used a thinner silicone shell to reduce the amount of capsular contraction; the devices proved to be more susceptible to rupture. Subsequent devices have had textured surfaces to reduce capsular contraction and thicker silicone shells to reduce the gel bleeding.

Silicone fluids at low level of exposure are not perceived to be harmful and are used in personal products, including hand creams, hair and skin products, and antiperspirants, and will be absorbed through the skin. Silicone oil is commonly used as a lubricant in syringes. People with insulin-dependent diabetes are therefore exposed to small but regular doses of silicone oil, resulting in a large, cumulative exposure to silicone over a period of time. Liquid silicone is injected into the eye during surgery to treat retinal detachment. The problem with implants has been mainly from secondary issues. This example highlights the difficulty of engineering materials which are being placed in the body.

The nature of the inflammatory response to silicones is the same as would be expected with any other foreign material. If a prolonged inflammatory reaction occurs, fibrous tissue forms to isolate the site of the injury. Immune reactions lead to the formation of antibodies and to the generation of reactive T-lymphocytes (a type of white blood cell). It has also been suggested that women with silicone implants may develop immune responses against their own tissues (autoimmune disorders) for example, rheumatoid arthritis and other connective tissue diseases. These disorders are characterised by the presence of antibodies to particular components of the patient's own tissues (autoantibodies), specifically by autoreactive T-lymphocytes, and will be encountered when the body attempts to reject a foreign body. There is little evidence that the chemistry of the polymer causes problems; it appears that either the implant or the release of the gel promotes the body to react.

11.1.3 Hip joints

Joints have a tendency to wear and it is now quite common for the head of the bone which forms the hip to be replaced. This process usually involves the use of a metal insert (see Figure 11.2). The implanted device will usually be manufactured from a titanium hip prosthesis (the section which is inserted into the bone), with a ceramic head and polyethylene acetabular cup into which the ball fits. There have been a number of variations on this device. The issues which are involved in the design are:

- How does one ensure that the titanium rod is securely bonded to the bone? A number of adhesives have been used to achieve this bond. The problem usually encountered is having an adhesive which will develop mechanical properties quickly enough yet still leave time to the surgeon to correctly manipulate the joint. A fast curing PMMA-based material is usually used as the adhesive.

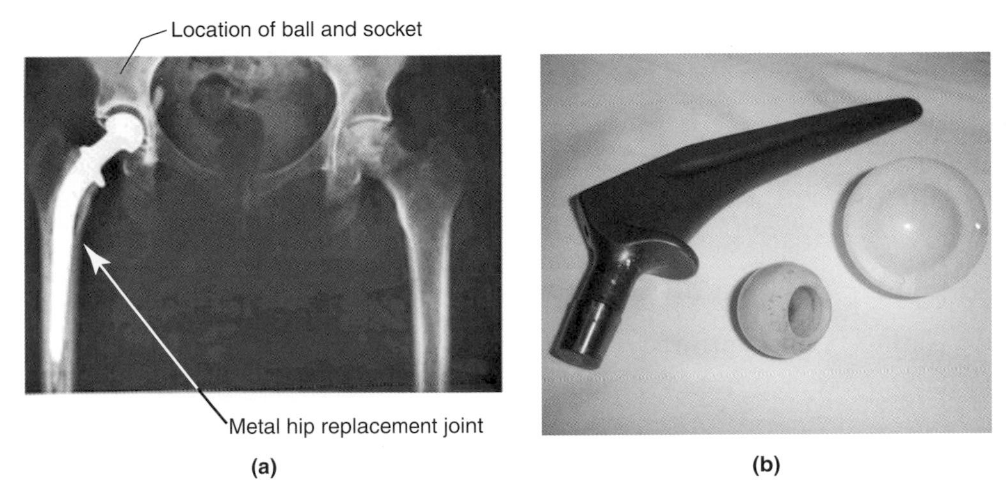

Figure 11.2 Hips: (a) X-ray of a hip replacement; (b) typical implanted device.

- How can the joint be lubricated so as to reduce wear? The polyethylene may be reinforced to increase its wear resistance and the surface of the ball is carefully crafted to help reduce the wear. Initially, PTFE was used in the replacements joints because of its low friction characteristics but it produced small particles as it wore leading to osteolysis. Use of very high molar mass polyethylene has proved to be very successful and illustrates how the long chain polymers are able to resist fatigue.
- To reduce wear and increase the compatibility of the device, hydroxyapatite is used as the ceramic.

The development of longer lasting and more effective hip joint replacements is a continuing area of medical materials development.

11.1.4 Heart valve replacement

The replacement of heart valves has become a relatively routine operation. The earliest heart valves were of a cage-like structure (see Figure 11.3). The ball will usually be titanium coated, allowing it to be seen by X-rays. The first artificial heart valve was the caged ball, which utilises a metal cage to house a silicone elastomer ball. When blood pressure in the chamber of the

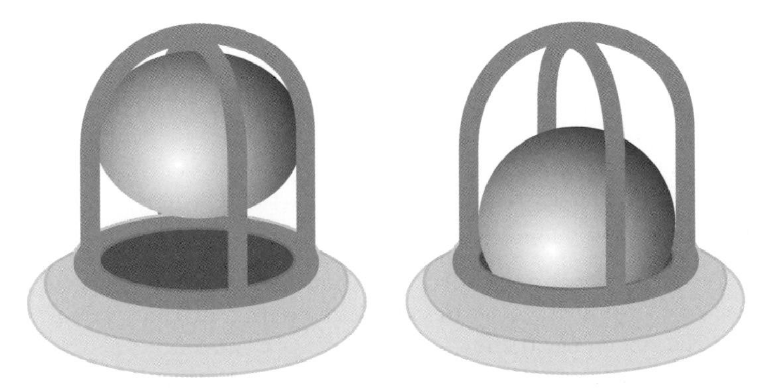

Figure 11.3 Schematic of a simple heart valve.

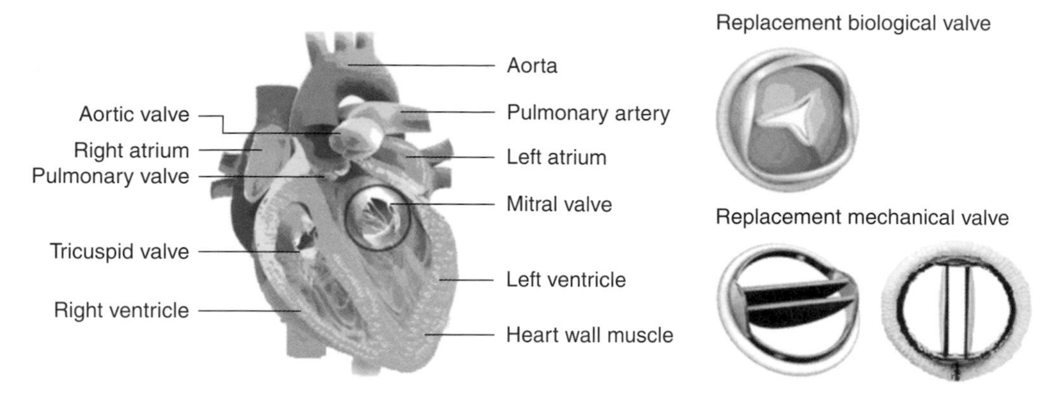

Figure 11.4 Schematic of heart and two types of artificial valves.

heart exceeds that of the pressure on the outside of the chamber, the ball is pushed against the cage and allows blood to flow. At the completion of the heart's contraction, the pressure inside the chamber drops and is lower than beyond the valve, so the ball moves back against the base of the valve forming a seal. Caged ball valves have a high tendency to form blood clots, so the patient must have a high degree of anticoagulation medication (usually warfarin). Developments in heart valve technology have produced two variants: the mechanical valve and the biological valve (see Figure 11.4).

The valves shown in Figure 11.4 attempt to mimic the natural heart valve. Mechanical valves use pivoted discs to control the blood flow. Modern mechanical valves can last indefinitely (the equivalent of over 50,000 years in an accelerated valve wear tester). Tilting disk valves have a single circular occluder controlled by a metal strut. They are made of a metal ring covered by a PTFE fabric into which the suture threads are stitched in order to hold the valve in place. The metal ring holds, by means of two metal supports, a disc which opens and closes as the heart pumps blood through the valve. The disc is usually made of an extremely hard carbon material (pyrolytic carbon), in order to allow the valve to function for years without wearing out. The alternative is the bileaflet valve (see Figure 11.4), which consists of two semicircular leaflets that rotate about struts attached to the valve housing. The bileaflets are vulnerable to backflow. However, they provide much more natural blood flow than caged ball or tilting disc implants. The main advantage of the bileaflet valve is that it is easily tolerated by the body. Mechanical heart valves are today very reliable and allow the patient to live a normal life. Most mechanical valves will last for at least 20–30 years.

Mechanical heart valves are very durable; the struts and occluders are made out of either pyrolytic carbon or titanium coated with pyrolytic carbon, and the sewing ring cuff is made of Teflon, polyester or dacron. The major load arises from transvalvular pressure generated at and after valve closure, and in cases where structural failure does happen, it is usually as a result of occluder impact on the components. Impact wear usually occurs in the hinge regions of bileaflets, between the occluder and ring in the tilting discs, and between the ball and cage in caged ball valves. Friction wear occurs between the occluder and strut in tilting discs, and between the leaflet pivots and hinge cavities in bileaflets.

Biological or tissue valves

Biological valves created form the valves of animals, like pigs, undergo several chemical procedures in order to make them suitable for implanting in the human heart (see Figure 11.3). The porcine (or pig) heart is most similar to the human heart, and represents the best anatomical fit for replacement. Implantation of a porcine valve is a type of xenotransplantation, or xenograft (a transplant from one species (in this case a pig) to another). There are some risks associated with a xenograft, such as the human body's tendency to reject foreign material. Medication can be used to retard this effect, but is not always successful. Another type of biological valve utilises biological tissue to make leaflets that are sewn into a metal frame. This tissue is typically harvested from the *pericardial sac* of either bovine (cow) or equine (horse) species. The pericardial sac is useful for valve construction due to its extremely durable physical properties. The leaflets are flexible and durable and do not require the patient to take blood thinners for the rest of their life.

11.1.5 Contact lenses

Contact lenses have to be permeable to oxygen but must not absorb potential irritant materials, such as smoke and chemicals. In the 1970s PMMA lenses, which were rigid, were introduced but have been now replaced by oxygen permeable equivalents. Rigid lenses offered a number of unique properties. The lens was able to replace the natural shape of the cornea with a new refracting surface, allowing correction of vision for persons who have astigmatism or distorted corneal shapes. The current soft lenses are permeable to oxygen and avoid the problems of the earlier lenses. The soft lenses are based on hydrogels, polymers which can swell in water but have a cross-linked structure. The modern lens is a complex material which is often made from silicone hydrogels. Silicone hydrogels have both the extremely high oxygen permeability of silicone and the clinical performance of the conventional acrylic-based hydrogels. The silicone provides the oxygen permeability but makes the lens surface highly hydrophobic and less 'wettable'. This frequently results in discomfort and dryness during lens wear. In order to compensate for the hydrophobicity, hydrogels are added to make the lenses more hydrophilic. However, the lens surface may still remain hydrophobic. Hence some of the lenses undergo surface modification processes which cover the hydrophobic sites of silicone.

11.1.6 Polymers used in devices and therapy

Polymers are used extensively in applications where the polymer may not be implanted in the body but nevertheless have contact with body fluids. Examples are membranes used in dialysis, sutures which may be inserted into veins, etc. These materials will all require to be screened both for haemocompatibility and for the mechanical properties which make them fit for purpose. Sutures can be made from PVC or polyurethanes. Studies have shown that by changing the molar mass of the polyol used in the PU very significant differences can be observed in the haemocompatibility. If the molar mass is low, the surface of the polymer will contain a number of areas of 'bare' potential hydrogen bonding urethane sites and these can interact with the blood protein. As the molar mass of the polyol is increased, so the hydrophilic polymer blooms from the surface and screens the urethane sites. Above a particular molar mass the polyol has the ability to form crystalline phases in the soft block and the screening effect is lost.

The selection of polymers for medical applications requires not only an understanding of the way in which the characteristics of the material are influenced by the chemical structure of the polymer but also how the materials interact with the surrounding biological media

Brief summary of chapter

- From an engineering point of view plastics can be used either as implants or in reconstructive surgery. The polymers not only need to have the desired physical properties but they must also be compatible with the task for which they are intended. Whilst, from an engineering point of view, a polymer may have appropriate mechanical characteristics for being used as a heart valve replacement material, if it is not blood compatible then it cannot be used.

Additional reading

Mahapatro A. (Ed.) *Polymers for Biomedical Applications*, ACS Symposium Series, vol. 977, ACS, Washington, DC, USA, 2008.

Lambert B.J., Tang F.-W. and Rogers W.J. *Polymers in Medical Applications*, RAPRA Technology Limited, Shawbury, UK, 2001.

Ottenbrite R.M. *Frontiers in Biomedical Applications*, Technomic Publishing Co. Inc., Lancaster, PA, USA, 1990.

12

Recycling of plastics and environmental issues

12.1 Introduction

The growth in the use of plastics has created the problem of disposal after they have exceeded their useful life. The world's annual consumption of plastic materials has increased from around 5 million tonnes in the 1950s to above 100 million tonnes today. In the UK, a total of approximately 4.7 million tonnes of plastic products are used in various economic sectors. In the USA, the use of HDPE (high-density polyethylene) bottles increased by 20 million lb (9.1 million kg), a rise of 27.1%, in 2004 and continues to increase. Polypropylene recycling, in the same year, increased by 4.1 million lb (1.85 million kg) to reach 10.1 million lb (4.85 million kg). One tonne of plastics is equivalent to 20,000 two-litre drinks bottles or 120,000 carrier bags. Packaging represents the largest single sector of plastics use in the UK. The sector accounts for 35% of UK plastics consumption and is the material of choice in nearly half of all packaged goods. Plastics are used across a broad range of industrial sectors (see Table 12.1).

The considerable growth in plastic use is due to the beneficial properties of plastics, which include: being lightweight, durable, resistant to chemicals, water and impact, and having electrical and thermal insulation properties. The approach which is taken to recycling plastics depends on which class of material is being dealt with. In general, thermosets cannot be effectively reused and are either used as a source of fuel or to produce particulate fillers for use in road building, floors or other low-level structural applications.

In principle, thermoplastics can be reused, however, in practice, this is only effective when the material is obtainable as a single polymer type. Unfortunately, a significant amount of the plastics used in food packaging is in the form of multilayered laminates, which makes recycling problematic. Simple food trays, cutlery, etc. are often a single plastic and could potentially be recycled. However, hygiene regulations require that packaging should not be reused, as effective sterilisation of plastic materials is problematic. In general, the preferred approach is to recover the plastic and reprocess it into similar or lower specification applications.

12.2 Recycling plastics

The problem of plastics waste can be addressed in several ways:

- First, does an economic route exist for the reuse of material? In the case of many thermoset materials and also complex mixtures of materials the answer is no and the most effective route for the disposal of the material is their reuse as fillers or other lower specification applications.
- Secondly, the material can be used for its basic value as a fuel. This approach is not encouraged as the byproduct of this process is a carbon dioxide emission, but as sources of fuel become more scarce this approach to waste may be reviewed.
- Thirdly, if the material is a single thermoplastic then the possibility of reuse becomes a potentially viable route. Plastic bottle recycling is developing into a common practice and highlights what is possible. Many industries, including the automobile, electronic and construction industries, are developing methods of reclaiming and appropriate disposal of plastics, primarily with the aim of reducing landfill.

Table 12.1 Use of plastics in various industrial sectors

Product area	Use (%)	Product area	Use (%)
Packaging	35	Footwear	1
Medical	2	Toys/sport	3
Building and construction	23	Electronics	8
Transport	8	Agriculture	7
Engineering	2	Other	3

- Fourthly, new plastics are being developed which are essentially biodegradable and hence do not present a long-term problem for landfill. These materials are produced form natural product sources and hence are not consuming the precious hydrocarbon resources.

It will be clear from the above comments that any recycling process requires that the nature of the plastic in any article is readily identifiable. In recent years a number of rapid scanning spectroscopic methods have been developed to allow sorting of plastics. Because there are differences in density between materials, in principle plastics which have been ground down into a fine powder can be sorted out. Whether the grinding process is economically viable depends on the value of the polymer and/or the tax which is placed on the recycling process. In many countries, recycling of electrical goods is being promoted by a tax being placed on the disposal of computers and similar devices.

12.3 Issues of plastic identification

To aid the recycling of plastics, manufacturers are encouraged to label the type of plastic from which a component has been fabricated. There are about 50 different groups of plastics, with hundreds of different varieties. All types of plastic are recyclable. To make sorting and thus recycling easier, the American Society of Plastics Industry developed a standard marking code to help consumers identify and sort the main types of plastic. These types, their most common uses, the US labels and European codes are listed in Table 12.2.

12.4 Why do we need to recycle plastics?

The overall environmental impact which plastics can have depends on the type of plastic and the production method employed. Plastics may contain additives which could be potentially harmful

Table 12.2 US and European labels for types of polymers and their most common uses

US Label	Polymer	European number code	US Label	Polymer	European number code
♳ 1	*Polyethyleneterephthalate*: fizzy drink bottles, oven-ready meal trays	1	♴ 2	*HDPE*: milk and washing-up liquid bottles	2
♵ 3	*PVC*: food trays, cling film, squash, mineral water and shampoo bottles	3	♶ 4	*LDPE*: carrier bags and bin liners	4
♷ 5	*Polypropylene*: margarine tubs, microwaveable meal trays	5	♸ 6	*Polystyrene*: yoghurt pots, meat/fish trays, hamburger cartons, egg cartons, cups, plastic cutlery, packaging for electronic goods and toys	6
♹ 7	*Any other plastics*: e.g. melamine, used in plastic plates and cups	7			

chemicals: dyes, stabilisers, plasticisers and processing aids. For example, phthalates, which are added as plasticisers to PVC, can be released into the environment and can be potentially harmful. Because most plastics are nondegradable, they take a long time to break down, possibly up to hundreds of years. With more and more plastic products, particularly plastic packaging, being disposed of soon after their purchase, the landfill space required by plastics waste is a growing concern. Plastics waste, such as plastic bags, often becomes litter. Nearly 57% of litter found on beaches in 2003 was plastic. According to a 2008 UK Environment Agency report, 64% of post-consumer plastics waste was sent to landfill, 12% was incinerated and only 17% was recycled. Reusing plastic is preferable to recycling as it uses less energy and fewer resources. Long life, multitrip plastic packaging has become more widespread since then, replacing less durable and single-trip alternatives, so reducing waste. For example, the major supermarkets have increased their use of returnable plastic crates for transport and display purposes four-fold from 8.5 million in 1992 to an estimated 35.8 million in 2002. They usually last for up to 20 years and can be recycled at the end of their useful life.

12.5 Methods for recycling plastics

The ability to usefully recycle plastics depends on the 'source' of the material.

Recycling of plastic process scrap

Many of the processes discussed in Chapter 5 will produce 'process scrap' and it is relatively simple to collect this material and return it to a compounder, allowing it to be blended with original material and used again. The material is relatively uncontaminated and about 250,000 tonnes of the plastic waste is recycled or reprocessed annually. The use of recycled plastic is restricted to about 10% due to the deterioration in mechanical properties when compared to the virgin material.

Post-use plastics recycling

Plastics arising predominantly from households is the biggest source of plastics waste. In the UK over 20 million households regularly dispose of mixed plastics waste, which includes food packaging and bottles. In 2003, 24,000 tonnes of plastic bottles were collected but this represented only 5.5% of all plastic bottles sold. In 2006, 2,700 million plastic bottles were recycled, which represents 108,000 tonne of plastic and a major increase in the use of plastic bottles, however the recycling rate has been increased to 20%. There is still a significant gap between the production and recycling of plastics and clearly an opportunity for future activity.

However, recycling of plastic bottles represents one of the best illustrations of recycling of plastics. The plastic bottles can be sorted according to type and melted to form an acceptable feed stock for future bottle production. In most cases the physical properties of the recovered polymer are degraded relative to those of the original material and the compounder will adjust the blend to recover the original specification.

Mechanical recycling

If the type of polymer is unknown, mechanical recycling may be used. This usually involves melting, shredding or granulating the waste plastics. The product of this process has ill-defined properties and is only used for the lowest grade applications such as chips for floor coverings.

Chemical or feedstock recyling

Whether or not a polymer is capable of being degraded back to its constituent monomers depends on the chemistry of the polymer. The degradation paths for a number of common polymers were considered in Section 8.9. Some materials, such as polystyrene, will return to their constituent monomers. Other materials, such as polyethylene, possess more complex degradation paths and the fragments created on heating, whilst being useful chemicals, are not directly suitable for conversion into usable polymeric materials. A range of feedstock recycling technologies are currently being explored and include pyrolysis, hydrogenation, gasification and thermal cracking. Feedstock recycling has a greater flexibility over composition and is more tolerant to impurities than mechanical recycling, although it is capital intensive and requires very large quantities of used plastic for reprocessing to be economically viable (e.g. 50,000 tonnes per year).

In 1991, LINPAC Plastics Recycling opened a unique recycling plant with the ability to handle postconsumer polystyrene products. Based in Allerton Bywater, West Yorkshire (UK), it has a capacity of over 14,000 tonnes per year. It can accept fast food boxes, meat trays, egg cartons, yoghurt pots, vending cups, and a range of other polystyrene products. In addition, the plant processes a range of polyethylene and polypropylene goods, such as bottles, crates, sheets, caps, pipes and fibres.

12.6 Degradation and bioplastics

Since landfill and disposal are the main concerns with regards to plastics, the concept of a plastic which eventually will disappear as a consequence of biodegradation is very attractive. Most hydrocarbon-based polymers, polyethylene, polystyrene, etc., are not subject to biological attack. Cellulose and other naturally occurring polymers are subject to biological attack and will, in time, disappear from the face of the earth. Wood is a good example of a structural material which can have a very acceptable working life but we know that ultimately it will biodegrade.

For a number of years, there has been interest in producing a range of biodegradable thermoplastics. One approach has been to incorporate cornstarch and other photodegradable plastics into oil-based materials, thus aiding breakdown by sunlight and the creation of sites on the hydrocarbon polymers for subsequent biological attack. Biodegradable cutlery and carriers for beer cans are now being manufactured in a plastic which photodegrades in six weeks.

However, there are concerns about the use of degradable plastics. First, the plastics will only degrade if exposed to light. Hence burial in landfill sites, where there is no light, does not represent an acceptable method of disposal. Secondly, the anaerobic biodegradation will lead to methane production, a greenhouse and potentially explosive gas. Thirdly, the mixture of degradable and nondegradable plastics may complicate systems for sorting the plastics.

12.7 Bioplastics

Historically, some of the earliest commercially used plastics were created by chemical modification of natural products. Cellulose nitrate and cellulose acetate are prime examples of plastics which come from a natural source and are also biodegradable. However, the base material, cellulose, tends to have characteristics which vary depending on the way it is produced and hence the lack of consistency in comparison to oil-based materials has led to the virtual disappearance of these materials from plastics. It is highly probable that in future they will be 'rediscovered', as, with the increase in the price of oil, they may become more commercially attractive. Alternative sources of bioplastics are also being developed in which the polymer is made from plant sugars and plastics are grown inside genetically modified plants or microorganisms. For almost 20 years, research has been carried out on the potential creation and use of bioplastics. These materials all contain oxygen atoms in their polymer backbones and are thus capable of being attacked by bacteria.

12.8 Polyhydroxyalkonates

Polyhydroxyalkonate (PHA) (also known as biopol) was first described by Lemoigne in 1926. It is commonly used as a storage material in many microorganisms. Polybetahydroxybutyrate (PHB) accumulates in many microorganisms, such as *Alcaligenes*, *Azotobacter*, *Bacillus*, *Nocardia*, *Pseudomonas* and *Rhizobium*. PHB consists of repeated units of $-CH(CH_3)-CH_2-CO-O-$ and has properties comparable to polypropylene but is biodegradable. *Alcaligenes eutrophus* and *Azotobacter beijerinckii* can accumulate up to 70% of their dry weight of PHB. Extraction of PHB is carried out using halogenated hydrocarbons. Direct moulding and extrusion of dried cells is possible when the PHB contents are high. PHB is suitable for specialised areas like biomedical use and speciality coatings. For industrial applications it is desirable to be able to change the basic physical properties of the polymer, and copolymers incorporating betahydrovalerate (PHA) have been developed. The incorporation of the comonomer creates polymers with enhanced physical properties relative to those of the homopolymer. PHA copolymers comprising medium-chain monomers (six or more carbon lengths) are more elastomeric than PHB and may contain unsaturated carbon bonds. These materials are better for coating and production of film materials, and offer possibilities for chemical modifications. The short-chain and medium-chain PHA arise from different biosynthetic routes and therefore are made by different microorganisms. The composition of culture medium (particularly carbon substrate) influences the microbial polymer (e.g. range of polymers formed, molecular weight and crystallinity), which in turn determines the physical properties (e.g. mechanical and tensile strength). Polyhydroxybutyratecohydroxyvalerate (PHBV) is a copolymer comprised of short-chain monomers of 4–5 carbon lengths. These materials have been extensively studied over the last 20 years and their potential has been extensively explored. However, their commercial acceptance has been limited as they are unable to demonstrate significant technical, performance and economic advantages to make them competitive with oil-based materials.

12.9 Polylactides and polyglycolides

Recently there has been significant interest in the use of bioplastics based on polylactides (PLA) and polyglycolides (PGA). Vicryl, developed by Dupont, is a 92:8 glycolic acid:lactic acid copolymer and is a thermoplastic biodegradable polyester. The feedstock can be created from maize and related corn-based products. Low molar mass polylactic acid and polyglycolic acid are made by direct polymerisation of the respective acids. The high molar mass PLA and PGA are made by ring opening polymerisation of lactide and glycolide, which are cyclic diesters of the respective acids. Polyglycolic acid and polylactic acid have degradation times of a few days and a few weeks, respectively, while PLA and PGA will degrade in a few months to years. The potential application areas for these materials are in the disposable cutlery area of food distribution. Initial indications are that these materials may become commercially more successful than biopol.

12.10 Issues with recycling

Whilst in principle the reuse of thermoplastics is possible, the main problem is the loss of physical properties during the melting and reforming. A plastic will typically lose its physical properties. Figure 12.1 shows the ultimate tensile strength which was achieved by blending various amounts of a reclaimed HDPE with pristine material.

It can clearly be seen that the ability to achieve the original performance of the material is compromised by degradation, which reduces the chain length and lowers the molar mass average. Unfortunately all vinyl addition polymers are subject to the same issues in recycling via a melt process. It is typical for there to be an ~30% reduction in tensile properties on melt recycling. Usually recycled material can only be used for a lower grade application where the physical property requirements are less demanding. However, PTFE, which is the most commonly recycled plastic and the main component of soft drink bottles, can be effectively recycled. As it is a condensation polymer, the reclaimed bottle material can be put through a process which allows the molar mass to be increased and the original physical properties recovered.

In principle, all of the condensation polymers can have their properties recovered but whether this is economically practical depends on the cost of plastic recovery and the volume of material to be treated.

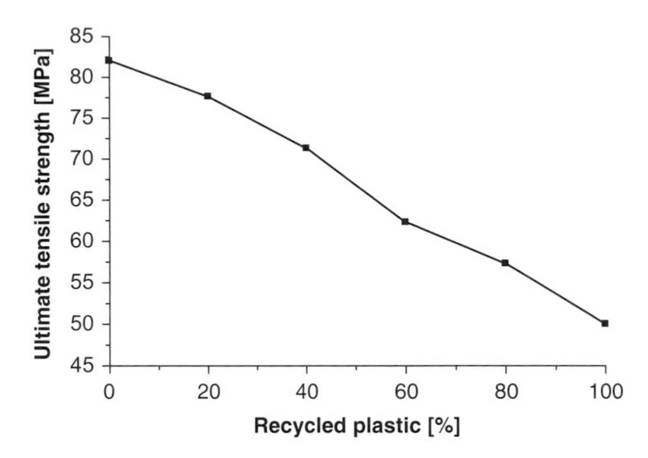

Figure 12.1 Ultimate tensile strength versus recycled plastic (%) for HDPE blends.

12.11 Feedstock recycling

Feedstock recycling in which the polymer is degraded to a lower molar mass material is seen as a possible alternative use for plastic waste. However, since the mixture is often complex, this may not be economically viable. The presence of halogenated polymers such as PVC can render the products of pyrolysis toxic and hazardous to handle. A significant amount of effort is currently being put into the development of procedures which would allow economic treatment of plastic waste via the feedstock recycling route.

12.12 Conclusions

Recycling of thermoplastics is a viable route for a limited range of plastics provided that they are available as a well-defined single-component waste material. This is the case for bottle recovery, where sorting of PTFE and PE is relatively easy. When the base material is a complex laminate or a thermoset then recycling is not an effective option. The alternatives are to attempt to use the plastic waste as a feedstock or alternatively to granulate and use the material as a filler.

Brief summary of chapter

The increasing use of plastics raises questions about recycling to conserve a scarce resource. Twenty years ago recycling was not considered an issue for plastics, which were considered to be cheap and disposable. Today recycling is now an important issue which engineers need to consider when they include the materials in a particular design application. In this context, the durability of the product is important and needs to be carefully considered.

Additional reading

Colom X., Canavate J., Carrillo F. and Sunol J.J. Effect of the particle size and acid pretreatments on compatibility and properties of recycled HDPE plastic bottles filled with ground tyre powder. *Journal of Applied Polymer Science* 2009, **112**(3), 1882–1890.

Stromberg E. and Sigbritt K. The design of a test protocol to model the degradation of polyolefins during recycling and service life. *Journal of Applied Polymer Science* 2009, **112**(3), 1835–1844.

Vilaplana F. and Sigbritt K. Quality concepts for the improved use of recycled polymeric materials: a review. *Macromolecular Materials and Engineering* 2008, **293**(4), 274–297.

Index

acid-catalysed condensation 336
acrylic polymers 2
addition polymerisation 230
adhesion 198
adhesive bond design 209–10
adhesive bond testing 210
adhesive joints 211
adhesive properties 159, 160–1, 198
adipic acid 8
adsorption 201
affine network 299
air entrapment 209
aligned composites 155–6
alternating copolymers 15
aluminium oxide 154, 219
amine 9–10
amorphous 4, 65
anisotropy 22, 81, 162
anticlastic surface 27
antioxidants 99
Argand diagram 29
atactic 12, 66
atom transfer polymerisation 253
autoclave moulding 137
Avrami equation 311–12
azimuthal angle 11
Azo initiators 231

backbone motion 41–2
Baekeland, Leo 1
Bayer 2
beta abstraction 258, 267
bismaleimides 95
blends 15
block copolymers 80
blow moulding 123, 187
bonding 208
branched 15
break point 33
brittle failure 37, 61, 160, 191
brucite 101
bubble removal 107
Buckminster fullerene 100
bulk polymerisation 246
butt joints 211

cantilever 26
Carruthers, Wallace 1, 7
carbon black 91, 99–100, 323
carbon fibres 147, 161, 174
carboxy terminated butadiene 220
  crylonitrile (CTBN) 220
cationic polymerisation 19, 333
ceiling temperature 48–9, 256
cellulose 1–2
  acetate 1
  nitrate 1
chain growth 225
chain motion 41
chain scission 92, 98, 256
characteristic ratio 281, 284
char formation 96–7
char yield 269
Charpy test 35
chemical amplified resists 333
chemical metal polishing 327
chemical modification 200
chiral 11
clay 101
cobalt-catalysed polymerisation 254
cohesive failure 202
collagen 1, 8
compliance 162
composites 2, 145, 147
composite fabrication 136
composite repair 142
complex modulus 29–30
compliance constants 185–6
compression 166
  moulding 89, 128–9
  testing 38
  zone 112
compressive strength 39
condensation polymerisation 8, 225
cone calorimeter 268
configuration 17
contact lenses 347
continuous composites 155
coordination catalyst 248
copolymers 6, 14, 78
corrosion protection 220

coupling coefficient 174
crack-growth rate 214
creep 37
creep modulus 51
critical molecular mass 49, 292
cross-linking 197
crystal growth 301, 303
crystalline 4, 11
crystallinity 65, 191
crystallisation kinetics 308
cure 97–8
cure monitoring 141
cyanate ester 95, 137

damage assessment 144
damascene process 327
deep UV resists 337
deformation 39
de Gennes 292
degradation 93, 110, 255, 352
degree of polymerisation 227, 229, 236
dehydrochlorination 260
dendrimer 15
depolymerisation 259
development of plastics 1
diblock copolymers 14, 82
Dickson 2
differential scanning calorimetry (DSC) 76
diffusion 201, 222
diffusion control 225
discontinuous composites 155–6
dispersed phase 147
disproportionation 7
DNA 1, 8
Doi and Edwards 292, 295
drag flow 113
drawing 125
draw ratio 74
drop test 36
ductile failure 37, 190
dynamic contact angle 208
dynamic mechanical thermal analysis (DMTA) 40, 60
dynamic modulus 27

Edwards *see* Doi and Edwards 292, 295
effect of water 61
E-glass 154, 161
elastic constant 50, 162, 167
elastic limit 33, 38–9
elasticity 92, 165, 172
electron beam resist 324
electron microscopy 72, 74, 80, 150, 190
electrostatic interactions 199, 203
elongation 33, 50
emulsion polymerisation 247
end-to-end distance 278
encapsulation 324
entanglement 49–50, 292–5
energy 24
energy balance 213
energy absorption 194
energy dissipation 29
engineering constant 174
engineering stress 300
environmental issues 349
environmental stress cracking (ESC) 62, 187, 191
ethanediol 8
epoxy resin 2, 8, 61, 93–4, 96, 99, 160
expansion 22
extensometers 33
extruder 111, 120, 131
extruder volume efficiency 118

fabrication 158
fractionation 307
fracture mechanics 212
failure 158
failure criteria 181
fatigue resistance 160
fatigue testing 38
Farben 2
Fawcett 2
fibre fracture 176
fibre orientation 155
fibre performance 154
fibre-reinforced composites 147, 153
fillers 99
fire 267
fire retardancy 270
flame chemistry 271
flexural modulus 34
Flory–Huggins theory 315
filament winding 138
film formation 124
flow into a mould 117
folding 71, 302, 305
fracture mechanisms 176, 178–9
free draining model 287
free flow 117
freely jointed random coil model 277, 284
freely rotating chain 280
free volume 41–2
free radical polymerisation 7, 233
free radical copolymerisation 242
front-end processing 326

gas diffusion 222
gauche conformation 11, 65, 282, 296, 301
gauge length 32
Gauss formula 279

gelation 95, 97
gel permeation chromatography (GPC) 318
gibbsite 101
Gibson 2
GLARE 150
glass fibre 99
glass-filled coatings 211
glass transition 41, 43, 48, 60–1, 82–3, 85, 97
Goodyear 1
graphite 100, 153
Griffiths 213
grit blasting 200
group transfer polymerisation (GTP) 253
gutta percha 89

habit 70
Halpin–Tsai equation 171
hand lay-up 134
heart valves 343, 345
heterogeneous catalysis 249
high density polyethylene (HDPE) 57, 125, 187, 192, 350
Hill's criterion 183
hip joint implants 343–4
homogeneous catalysis 251
homogeneous strain 22
homopolymers 6
honeycomb structures 144, 151
Hooke's law 21, 33, 162
Hookian solid 214
high-temperature vulcanisation (HTV) 91
hydraulic grips 32
hydrodyamic radius 318
hydrogen abstraction 260
hydrophilic glass 158
hydrophilic polymers 61, 125
hydrophobic glass 158
hydrophobic polymers 61, 125
hydroysis 7, 267
hyperbranched 15

impact energy 37
impact velocity 37
implants 343
inhibition 237
initiation 230–1
injection moulding 89, 110, 130
Instron wedge clamps 32
intrinsic fracture energy 214
intrinsically conducting polymers 324, 339
intumescent systems 270
isocyanate 9, 84
isotactic 12, 66
isotropic 22
  material 165
IUPAC 4
Izod test 35

jaw breaks 32

kaolinite 101
Kelvin–Voigt model 50–1, 5–7
Kevlar 2, 99, 154
kinetics of copolymerisation 243
kinetics of polymerisation 225–6, 230, 235
knitted composite structures 181
Kuhn length 285
Kuhn segment 281, 295

lamellae 69, 71–2, 191, 302, 306
layered laminates 152
leakage flow 113, 115
Lewis acid 8
light scattering 277, 315–16
limiting oxygen index (LOI) 269, 272
linear solid 53
lithographic materials 325
long fibres 155
longitudinal Poisson ratio 170
longitudinal shear modulus 170
long range interactions 283
longitudinal tension 177
low density polyethylene (LDPE) 76, 125, 192, 350

macromolecules 3
Mark–Houwink relationship 318
matrix materials 157
maximum stress criterion 181
Maxwell model 50, 55–7
mean sequence length 245
mechanical interlocking 200
mechanical properties 21, 30, 159
medical applications 343
melt casting 89
melting 108, 310
melting point 18
metal layers 326
metallocene catalysis 251
metering zone 113
microcracking 160
mixing zone 112
modulus 21, 166
molar mass 17, 191, 314
  averages 18
  control 229
  distribution 17–18, 227
monoclinic material 163
montmorillonite 102
Mooney–Rivlin coefficients 301
morphology 65

moulding 104
moulding bottles 125

nanocomposites 147, 272
nanofillers 149, 220, 272
nanotubes 150
Natta, Giulio 1, 248–9, 330
negative resist 330
nonbonding interactions 282–3
nondestructive testing (NDT) 145
normal mode 285, 289–90
nucleation 302
number average molecular weight 18, 314
nylon 1, 9, 61, 67–8, 78–9, 225, 311

off-axis failure criterion 182
oligomers 17, 227
organic light-emitting polymers 324, 340
orthotropic composites 171
orthotropic material 163
oscillatory shear 289
osmosis 161
osmotic pressure 314
Oswald ripening 302
oxygen permeability 222

parison 125–6
Parks 1
particulate composites 147, 149
peel ply 203
peel test 217
pendulum impact test 35–6
permanent set 52
permeability 222
peroxide initiators 231
Perspex® 6
phase lag 29
phase separation 78, 97, 302
phase shift 41, 290
phenolic 95
photo-acid generator (PAG) 334, 337
photolithography 327
pi-bond (π-bond) 6
plasma etching 200
plastic identification 350
plasticisation 47, 102
plastisol 89, 103, 132–3
plateau modulus 293
Plexiglass 1
pneumatic grips 32
Poisson's ratio 23, 27, 175, 216
polyacetylene 339
polyacrylonitrile 45
polyamide 9, 70
polybutadiene 13, 68, 196, 285
polybutene 68, 70
polybutylacrylate 44, 47
polycarbonate 2, 43, 61, 86–7, 187, 189
polychloroprene 90, 311
polydimethylsiloxane (PDMS) 43, 47, 91, 285, 343
polyethylene 1, 5–6, 18, 45, 47, 62, 65, 68, 71–2, 76, 99, 108, 192, 257, 285, 311
polyether ether ketone (PEEK) 86–7
polyethersulfone (PES) 86–7
poly(ethyleneadipate) 8
polyethyleneoxide 9, 67–8, 285, 311
polyester 9, 61, 86, 93–4, 99, 160, 225, 323
polyethyleneterephthalate (PET) 2, 68, 70, 125, 127, 265, 350
polyethylacrylate 44
polyglycolides (PGA) 353
polyheptylacrylate 44
polyhydroxyalkonates 352
polyhexylacrylate 44
polyimide 9, 87, 95, 99
polyisoprene 68, 89, 196, 198, 285, 311
polylactides 353
polymer classification 17
polymer coating 123
polymer dynamics 286
polymer identification 273
polymer processing 89
polymethacrylate 5, 44
polymethylmethacrylate 1, 5–6, 43–6, 61, 187, 202, 259, 263, 285
poly(4-methyl-1-pentene) 68, 70
polyα-methylstyrene 46
poly2-methylstyrene 46
poly3-methylstyrene 46
poly4-methylstyrene 46
polymorphism 66, 302,
polynonylacrylate 45
polyoctylacrylate 44
polyoxymethylene 67–70, 311
polypentylmethacrylate 44
polypropylene 1, 5, 11–12, 45, 62, 66, 68, 70, 125, 285, 311, 350
polypropylmethacrylate 44
polypyrrole 340
polystyrene 1, 5, 12, 43, 45, 62, 68, 70, 125, 187, 266, 285, 311, 350
polysulfone 86–7, 220, 259
polytetrafluoroethylene 67, 70, 77, 311, 346
polythiophene 430
polyurea 9, 84–5
polyurethane (PU) 2, 9, 84
polyvinyl acetate 263
polyvinyl alcohol 68, 125
polyvinylchloride (PVC) 1, 5, 45, 68, 125, 260, 323, 350
polyvinylidenechloride 45, 67

popper beads 4
positive resists 331
power requirements 118
preform 127
prepreg 96, 136
pressure flow 113–14
pressure forming 129
printed circuit boards 323–4
propagation 230–1
pultrusion 139

quartz 101

radius of gyration 278, 280
random coil 278
random copolymers 15
RAPRA 187
Rayleigh ratio 315
reactivity ratio 244
recombination 7
recycling (of plastics) 349, 351
reinforcing fillers 149
relaxation time 29, 294
resilience 195, 197
resin infusion processing 136
resin transfer moulding (RTM) 135
retardation 237
rigidity modulus 23
ring opening polymerisation 8
RNA 1
room temperature vulcanisation (RTV) 91
rotational isometric states model (RISM) 282, 284, 302
rotational moulding 89, 103
rotating sector method 238
Rouse model 286, 290
rubber 1–2, 89, 197
rubber balls 195
rubber elasticity 92, 296, 299
rule of mixtures 152
rupture point 33

sample clamping 31
sandwich structures 151
sectorisation 71
selection of materials 223
semiconductor processing 325
semiconductors 324
service life 218
shear 22
shear modulus 23, 34, 166
shear stress analysis 215
shish-kebab structure 75
short fibres 155
sight glasses 187
sigma-bond (σ-bond) 6
silane modification 159
silica 101
silicone breast implants 343
silicon carbide 154
silicon nitride 154
siloxanes 91
single lap joint 213
single site catalyst 252
size exclusion chromatography 318
slippage 31
smectic 67
smoke density 269
solidification 110
solubility 222
solution casting 89
solution polymerisation 246
specific modulus 153
specific strength 153
spherulitic structure 67, 71
spray lay-up 135
spring-dash pot models 50
stabilisers 272
statistical models 277
Staudinger 1
steady state assumption 234
step-growth polymerisation 7
stereochemistry 17
storage modulus 291
stiffness 164–5, 173
stiffness constants 185
strain 21, 34, 50
stress 21, 50
stress relaxation 38, 290, 294
stress–strain measurements 21, 60, 300
stretching of a wire 24
styrene–butadiene–styrene 14, 80–3, 147
surface energy 206
surface modification 203
surface roughness 208
surface tension 110, 204
suspension polymerisation 246
switch board model 305
syndiotactic 11, 47

telomers 17
temperature rate effects 241
tensar process 128
tensile creep 57
tensile modulus 23
tensile testing 30
termination 7, 230, 233
thermal expansion coefficient 42, 324
thermogravimetric analysis (TGA) 256
thermoplastic 1, 4, 17, 89, 276
thermoplastic processing 103
thermoset 1, 17, 89, 93, 134, 274

thin beam 25
time-dependant modulus 55
time–temperature transformation (TTT) 95–6
toughening 220
tortuous path 221
trans conformation 11, 65, 282, 297
transfer 232, 259
transfer constants 240
transverse isotropic material 164
transverse Young's modulus 169
triblock copolymers 14, 82
twin-screw extruder 119
tyre technology 195, 197

ultimate tensile strength (UTS) 34
uniaxial tension 166
unidirectional material 164

vacuum bagging 135
vacuum forming 129
valence constrained random coil model 280, 284
Van der Waals interactions 199
vented barrel 120
vermiculite 101
vibration damping 194
vinyl ester 95, 99, 160
vinyl polymerisation 230
viscoelastic behaviour 50
viscosity 49, 294
volume change 22
vulcanisation 89

wafer processing 329
warp direction 172
water ingress 161
water permeability 222
wedge clamps 32
weft direction 171–2
weight average molecular weight 19
wetting 207
Whinfield 2
Williams–Landel–Ferry (WLF) equation 61, 309
working temperature 47
woven fabric 171

X-ray scattering 65

yield point 38–9
yield strength 33, 38
yield stress 33
Young's equation 205
Young's modulus 21, 23, 33, 50, 148, 175, 215

Zeigler 1, 248, 249, 330
Zener model 53
Zimm model 288, 291
Zimm plots 317
Zwick test machine 30